Advances in Aerospace Guidance, Navigation and Control

Florian Holzapfel and Stephan Theil (Eds.)

Advances in Aerospace Guidance, Navigation and Control

Selected Papers of the 1st CEAS Specialist Conference on Guidance, Navigation and Control

Springer

Editors

Florian Holzapfel
TU München
LS für Flugsystemdynamik
Boltzmannstr. 15
85748 Garching
Germany
E-mail: florian.holzapfel@tum.de

Stephan Theil
DLR Institute of Space Systems
GNC Systems Department
Robert-Hooke-Str. 7
28359 Bremen
Germany
E-mail: Stephan.Theil@dlr.de

ISBN 978-3-642-19816-8 e-ISBN 978-3-642-19817-5

DOI 10.1007/978-3-642-19817-5

Library of Congress Control Number: 2011923796

Typesetting: Data supplied by the authors

Cover Design: Scientific Publishing Services Pvt. Ltd., Chennai, India

Printed on acid-free paper

9 8 7 6 5 4 3 2 1

springer.com

Preface

Over the last few decades, both aeronautics and astronautics have been strong motors for the advancement of control systems theory and application, as well as the fields of sensors, data fusion and navigation. Many of those achievements that earned aerospace the reputation as a synonym for high tech and progress have been enabled by innovations in guidance, navigation and control.

Today, aerospace is still one of the driving applications in those fields, which stems from the special characteristics and needs of that segment. In an airplane seating hundreds of people, you cannot use the latest control algorithm to find out if it works by trial and error. In a deep space probe sailing for new shores, you need to do things right on the first and only attempt; (unlike a car or a microwave oven) you cannot test the integrated system in real operation before its actual mission. You cannot simply put a bulky machined box with standard components which works in "living room" environmental conditions in the slender body of an agile missile.

Things are also quite different on the algorithm side. Aerospace systems can have highly nonlinear and strongly coupled dynamics. The ranges of altitude, Mach number, center of gravity and weight are enormous and the dynamics can significantly change with those parameters. Huge uncertainties can still remain in spite of costly modeling efforts. The range of time scales contributing to the system dynamics is large, speeds are higher, the environment is harsher and changing more rapidly, the distances travelled are much larger and the operation times for some systems are much longer than in other fields. To summarize, the challenges in the aerospace disciplines are unique and more demanding than in other domains.

If these challenges were not enough, appropriate solutions must also be reliable, highly accurate, highly available, safe and must guarantee a well-defined performance level, even under a large variety of circumstances like system failure. An airplane cannot turn right and stop at the next cloud if things go wrong. All these challenges must be accomplished under mass, volume, power and cost constraints.

This may sound like praise for aerospace, its scientists and engineers, but however you wish to see it, it is a viable explanation as to why, in contrast to other fields, there have always been dedicated conferences on "flight control", "space navigation" and "missile guidance", as specialized sessions at general conferences are not enough. The American AIAA Guidance, Navigation and Control conference serves as a brilliant example where the community of "rocket scientists" gathers to present on and to discuss these specific topics.

Europe has seen a strong trans-national consolidation process in aerospace over the previous decades. Most of the visible products, be it commercial aircraft, fighters, helicopters, satellites, launchers or missiles, are not made by a single country – they are the fruits of cooperation. No European country by itself hosts a specialized GNC community large enough to cover the whole spectrum of disciplines. However, on a European scale, mutual exchange of ideas, concepts and solutions is enriching for all. Thus, the interest in having a truly European GNC conference collecting the ideas of the scientific community within the continent and inviting the whole world to join has been articulated quite frequently. The first CEAS Specialist Conference on Aerospace Guidance, Navigation and Control (EuroGNC) is an attempt to follow this request and turn the vision into reality. It is the hope of the organizers and the Technical Committee members from all CEAS Member states that this conference establishes itself as a high class biannual recurring event at changing locations, which brings together the researchers, scientists, developers and engineers who have dedicated their professional life and their passion to the arts of guidance, navigation and control.

We are very thankful to AIAA for acknowledging our effort as a worthy complement to their GNC conference and for co-sponsoring the event.

Maybe – to draw a vision – an Asian aerospace GNC conference could join the European event, taking turns year by year. This would allow the global community – beyond the yearly "jour fixe" at the AIAA GNC conference – to come together for a second time and to intensify information exchange and cooperation in a truly global field of science.

At this point, it is important to thank all of those who took the burden to make a beginning and to organize the first EuroGNC conference. First of all DGLR, the German member society of CEAS which dared to take the risk of being the first organizer and then all the members of the organizing and technical committees who bravely accepted the challenge of tampering with systems and processes for paper review and others which are still in their prototype stages, smoothing the path for those to follow. Without their time, patience, dedication and willingness the event would never have been possible.

But now enjoy the book which summarizes selected contributions of scientists to the first CEAS EuroGNC – the European Specialist Conference on Aerospace Guidance, Navigation and Control, giving proof to the fact that Europe and its friends from around the world do make a valuable contribution to the progress of the field.

Munich and Bremen
February 2011

Florian Holzapfel
Stephan Theil

About the Book

For the 1st CEAS Specialist Conference on Guidance, Navigation and Control the International Technical Committee established a formal review process. Each paper was reviewed in compliance with good journal practices by at least two independent and anonymous reviewers. At the end of the review process papers were selected for publication in this book.

All submitted papers were assigned to one of the four areas:

- atmospheric applications,
- guidance and control,
- sensors, data fusion and navigation,
- space applications.

The selected papers from each area appear in this book in alphabetical order.

The members of the International Technical Committee are:

Mark Balas	University of Wyoming, U.S.A.
Samir Bennani	ESA/ESTEC, The Netherlands
Daniel Choukroun	Ben Gurion University of the Negev, Israel
Jussi Collin	Tampere University of Technology, Finland
John Crassidis	University of Buffalo, U.S.A.
Jörg Dittrich	DLR, Germany
Alexej Efremov	Moscow Aviation Institute, Russia
Patrick Fabiani	ONERA, France
Walter Fichter	Universität Stuttgart, Germany
Chris Fielding	BAE Systems, United Kingdom
Benoit Frapard	Astrium Satellites, France
Luisella Giulicchi	ESA/ESTEC, The Netherlands
Martin Hagström	Swedish Defense Research Agency, Sweden
Florian Holzapfel	Technische Universität München, Germany
Eric Johnson	Georgia Institute of Technology, U.S.A.
Philipp Krämer	Eurocopter Deutschland, Germany
Hugh Liu	University of Toronto, Canada
Marco Lovera	Politecnico di Milano, Italy
Robert Luckner	Berlin Technical University, Germany
Bob Mulder	TU Delft, The Netherlands
Stephan Theil	DLR, Germany
Andrzej Tomczyk	Rzeszów University of Technology, Poland

Contents

Part I: Atmospheric Applications

Part I
Atmospheric Applications

A Decoupled Approach for Trajectory Generation for an Unmanned Rotorcraft

Sven Lorenz and Florian M. Adolf

Abstract. A decoupled approach to trajectory generation based on a cubic spline geometry formulation is introduced. The distinct consideration of boundary conditions yields a continuously differentiable trajectory definition such that path tracking errors are minimized during flight. A curvature-based, dimensionless space-filling curve allows to determine a suitable velocity profile along the path for hover-capable vehicles. Tracking of the trajectory is enabled by a conversion between the spline parameters and the arc length of the spline. In the past years, this approach in combination with a suitable trajectory tracking control has been successfully flight tested with an unmanned helicopter.

1 Introduction

Autonomous flight in densely populated environment like urban terrain generally requires excellent maneuverability. Hence, helicopter-based UAV platforms are preferred. Especially for dynamic high speed flight, changes like moving obstacles or mission updates can require such complex control platforms to be equipped with an onboard motion planning system.

As the survey by Goerzen [1] points out, a sole precision tracking of a precomputed trajectory is not a feasible overall solution. Dynamic constraints, atmospheric conditions, uncertainty in the vehicle state estimates, and limited knowledge about the environment may leave no chance to follow a precomputed plan precisely. Approaches for path smoothing[1] exist [2] that allow to generate continuously differentiable, timely annotated paths. Moreover, given a feasible flight control system, a number of path following solutions exist [3, 4] even for commercial, black box autopilots [5].

Sven Lorenz · Florian M. Adolf
Research engineers at the German Aerospace Center (DLR), Institute of Flight Systems, Department of Unmanned Vehicles, Braunschweig, Germany
e-mail: sven.lorenz@dlr.de, florian.adolf@dlr.de

[1] Functions with derivatives of all orders fit a smooth curve to a set of points.

However, the integrated optimization of a motion planning task can comprise a whole set of hard realtime computations when aspects like task sequence ordering, route planning, reactive obstacle avoidance, and vehicle control are considered. Thus, this work proposes a decoupled approach to motion planning. Previous work by Andert and Adolf [6] indicates that a thorough problem decomposition is one key to enable realtime sensor fusion, obstacle modeling, and 3D motion planning during flight. This work develops this idea further by decomposing the trajectory generation problem in decoupled subsequent layers. Although path and trajectory planning may have the same meaning, there is a fundamental difference between them. In this context a path is defined as the interpolation of position coordinates. Trajectories refer to timely annotated paths, e.g. represented by a velocity profile over a discretized path geometry.

Even a hover-capable vehicle cannot perform arbitrary yaw turns during fast forward flight, such that an adequate yaw attitude command must be determined. A common approach is to use an instantaneous tangent along a path. A continuously differentiable path geometry should be preferred to enable smooth transitions. The trajectory will be vehicle specific. The configuration space[2] may be altered if the vehicle properties or the characteristics of the environment change.

Generally, path search uses only simplified dynamic constraints and concentrates on collision-free path segments, often with various geometries (e.g. linear, circle, spline). Then a time dimension is added to the path yielding a trajectory by defining the velocity profile over the path geometry. Instantaneous vehicle state estimates are used together with the path slope in order to provide a feasible input to a trajectory following controller. The nonlinear plant will be controlled by a baseline control system that maintains desired velocities. More details on the control system can be found in [7, 8].

The following section starts with the path definition. Based on the geometry of the path the trajectory definition will be presented in section 3. The trajectory definition includes the determination of the velocity profile as well as an approach to account for acceleration limitations. Section 4 presents the trajectory following control system. A summary of the overall approach and future research directions are discussed in section 5.

2 Path Definition

In general, a smooth path should be continuously differentiable. Due to the resulting complexity of generating a collision-free path through narrow passages a function of reduced order is selected. Most path planners avoid obstacles by generating coordinates that divide the path from start to end position into consecutive segments. These path segments use cubic splines that are continuously differentiable up to the third derivative. A large safety margin accounting for path following errors can therefore be avoided. Furthermore, the path planner utilized in this work assures a

[2] Algorithmic search space of possible configurations that a physical system may attain.

collision-free sphere volume (e. g. safety radius) of a selectable size. Note that it may insert other path geometries into spline segments (e. g. through narrow corridors).

A univariate, polynomial spline is defined as a piecewise polynomial function. In its most general form a polynomial spline S: $[a,b] \rightarrow \Re$ consists of polynomial pieces $P_i : [\tau_i, \tau_{i+1}] \rightarrow \Re$, where a strictly increasing sequence of real numbers is used between the boundaries a and b:

$$a = \tau_0 < \tau_1 < \cdots < \tau_{k-2} < \tau_{k-1} = b. \tag{1}$$

That is,

$$S(\tau) = P_0(\tau)\ ,\ \tau_0 \leq \tau < \tau_1, \tag{2}$$
$$S(\tau) = P_1(\tau)\ ,\ \tau_1 \leq \tau < \tau_2, \tag{3}$$
$$\cdots$$
$$S(\tau) = P_{k-2}(\tau)\ ,\ \tau_{k-2} \leq \tau \leq \tau_{k-1}. \tag{4}$$

The given k points τ_i are called knots. The vector $\underline{\tau} = [\tau_0, \ldots, \tau_{k-1}]$ is called a knot vector for the spline. The knots are not equidistantly distributed in the interval $[a,b]$, the spline is therefore called to be non-uniform.

If the above described interpolation method is applied to a large number of waypoints, comparatively high-order splines have to be selected for a feasible interpolation. Consequently, oscillations between the support points may occur. Therefore, a cubic spline is applied piecewise for each segment. This way, a transition condition at each segment boundary ensures smoothness up to the second derivative. A third-order spline is selected for each segment k and each degree of freedom $i = [x,\ y,\ z]$:

$$S_{i,k}(\tau) = a_{i,k} + b_{i,k} \cdot (\tau - \tau_k) + c_{i,k} \cdot (\tau - \tau_k)^2 + d_{i,k} \cdot (\tau - \tau_k)^3 \tag{5}$$

$S_{i,k}(\tau)$ represents the spline for the dimension i in segment k for a specific τ. Its derivatives w. r. t. τ are given by:

$$S'_{i,k}(\tau) = b_{i,k} + 2 \cdot c_{i,k} \cdot (\tau - \tau_k) + 3 \cdot d_{i,k} \cdot (\tau - \tau_k)^2 \tag{6}$$
$$S''_{i,k}(\tau) = 2 \cdot c_{i,k} + 6 \cdot d_{i,k} \cdot (\tau - \tau_k) \tag{7}$$
$$S'''_{i,k}(\tau) = 6 \cdot d_{i,k}. \tag{8}$$

The n waypoints $p_j(x,\ y,\ z)\ ,\ j = 1 \ldots n$ are the support points of the spline interpolation. A number of n points causes $k = n - 1$ segments. Each segment contains three spline functions including four spline parameters. To determine four parameters four equations are required. The following requirements for each segment boundary are defined:

- Consecutive spline segments must be connected with each other.
- The first and second derivatives must match the previous segment's derivatives.

Note that path segment boundaries at a and b may have an enforced slope as well. Otherwise, the second derivatives at the boundaries a and b are set to zero. Figure 1

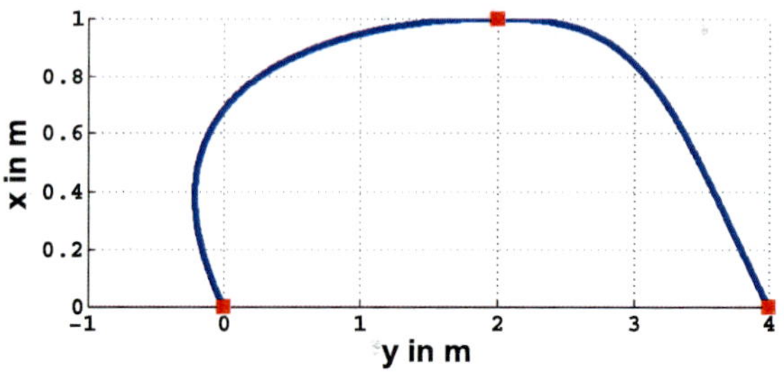

Fig. 1 Example: Path based on splines supported by three waypoints, resulting in two segments, with unspecified boundaries at the coordinates (0,0) and (0,4).

shows a path result without enforced slopes at a and b. The approach presented enables a smooth transition between the segments up to the second derivative. The parameter sets defined for each spline yield the coordinates between the knots. In addition to the desired overall path shape, vehicle configurations on that path must be determined. The following section focuses on this task.

3 Trajectory Definition

The three-dimensional path defined in the section above has to account for obstacles and to ensure the attainment of a set of global mission goals. Given a sufficiently low velocity, helicopters are able to fly along paths with arbitrarily sharp turns. Thus, solely the velocity is selected as the timely annotation of a path.

The curvature of the path is used to determine the velocity. However, curvature changes may yield larger velocity changes than the vehicle can perform. Hence, a search for velocity minima along the path must be performed before the vehicle starts to fly the trajectory. From points, where the deceleration to the velocity minimum has to begin, the velocity will be reduced while the vehicle still follows the given path.

In this section first the determination of the velocity based on the curvature will follow. To adopt the velocity profile for the limited acceleration, a search algorithm for velocity minima will be presented afterwards.

3.1 Quantification of the Velocity

Based on the curvature κ the velocity will be calculated for certain points $\tau_{k+1} = \tau_k + \Delta\tau$. The curvature κ for a space curve $\underline{S}(\tau)$ is defined at a point τ as:

$$\kappa(\tau) = \frac{1}{r_{circle}(\tau)} = \frac{\left\| \underline{S}'(\tau) \times \underline{S}''(\tau) \right\|}{\left\| \underline{S}'(\tau) \right\|^3}. \tag{9}$$

Based on Tietze [9], the following relations will lead to the largest velocity command based on the thrust available for a helicopter. The sum of the forces acting on the center of gravity is used to determine the ability to compensate for gravity and centripetal force, and to change the velocity if desired. With the symbols $\underline{F}_G$ for the gravity force vector, $\underline{F}_A$ for the aerodynamic force vector, $\underline{F}_S$ for the thrust vector of the helicopter and $\underline{F}_K$ for the inertial force vector the sum of the forces results in

$$0 = \underline{F}_G + \underline{F}_A + \underline{F}_S + \underline{F}_K. \tag{10}$$

The main thrust is produced by the main rotor of the helicopter. Due to the small magnitude of the aerodynamic forces it is assumed that the magnitude of the thrust of the main rotor must equal the sum of gravity and inertial forces:

$$|\underline{F}_S| = |\underline{F}_K| + |\underline{F}_G| \tag{11}$$

$$|\underline{F}_S| = \sqrt{(m\dot{V}_K)^2 + (m\dot{\chi}V_K\cos\gamma)^2 + (m\dot{\gamma}V_K)^2} + mg. \tag{12}$$

In eq. (12), m represents the vehicle mass, V_K the inertial velocity value, γ and χ the orientation of the path w. r. t. the geodetic frame, and g the magnitude of the acceleration due to the gravity force.

Replacement of $\dot{\chi}$ by $\frac{V_K\cos\gamma}{r_{circle}}$ and $\dot{\gamma}$ by $\frac{V_K}{r_{circle}}$ in eq. (12) results in

$$\left(|\underline{F}_S| - mg\right)^2 = \left(m\dot{V}_K\right)^2 + \frac{1}{r^2_{circle}}\left(mV_K^2\right)^2\left(\cos^4\gamma + 1\right). \tag{13}$$

The term $\frac{1}{r^2}$ can be replaced with κ^2, cf. eq. (9). A definition of the largest acceleration $a_{max} = \frac{|\underline{F}_S|_{max}}{m_{min}}$ which can be produced by thrust enables the elimination of the dependency on mass m. The velocity-curvature relation results in:

$$V_{K,max} = \sqrt{\frac{(a_{max} - g)^2 - \dot{V}_K^2}{\kappa^2 \cdot (cos^4\gamma + 1)}}. \tag{14}$$

Therefore, eq. (14) defines the largest velocity depending on the current acceleration in path direction and the curvature of the spline.

Figure 2 shows a typical velocity profile for a spline curve over τ. The solid (green) line represents the largest velocity limited to 20 m/s and is based on eq. (14). As explained in more details in the following subsection, the dotted (blue) line defines the velocity command for a deceleration with a specified deceleration limitation.

Due to the neglected aerodynamic forces and the issue that the vehicle is not able to change the flight path arbitrarily fast a scaling factor will be introduced. Beside the flight velocity the path deviation must be provided to follow the path e. g. in the presence of disturbances. A trajectory following control system to guide the vehicle based on the desired path and the path deviation will be presented subsequently to the following subsection.

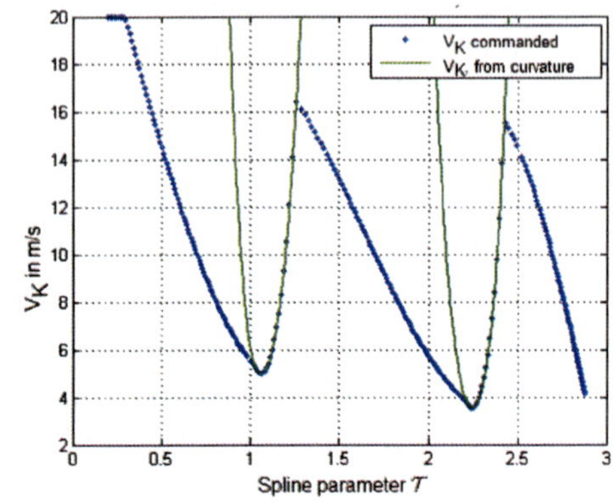

Fig. 2 Velocity profile for a spline curve. Based on the curvature the maximum velocity is determined (solid). Due to the deceleration limitation the dotted velocity profile is commanded to slow down to the velocity minima or stop at the end.

3.2 Velocity Minima Search Algorithm

In general, vehicles have acceleration and deceleration limitations. A sole path geometry cannot account for such limitations. Thus, based on the largest velocity defined in eq. (14) the occurrence of minima, as illustrated in fig. 2, must be determined. The first challenge is the search for a minimum since the curvature is only known at discrete points. Moreover, the velocity profile may have multiple local minima which self-denies the use of gradient based search algorithms only.

Hence, locally bounded gradient searches are performed on a sampled spline. Changes in the slope of the velocity are used to detect minima. For each detected change a gradient based minimum search is performed. Similarly to the approach in [10], the minima detection can be ensured by the property of resolution completeness that defines a sufficient distance between each sample and a sufficient number of samples for each path section.

The search through the spline results in a table of velocity minima with respect to their corresponding position on the spline. Based on a model for the deceleration, a point for the initiation of a slow-down maneuver can be determined. This method gives a velocity command depending on the distance to a minimum. In this implementation a constant velocity slope of 0.3 s^{-1} is used. By applying a method to convert distances into changes of τ, cf. section 4.3, the table is supplemented with the specific slow-down points. Preliminary to the selection of the first entry a sort by the slow-down points results in a list, which has the crucial points in the right order.

In this work this global search is enhanced to become a quasi incremental search. Similar to the approach in [4], a velocity minimum nearest to the helicopter position defines the initial search horizon w. r. t. τ. This decreases the initial search time. The vehicle can start or continue flying while searching for minima that are further away. Of course, saving initial time comes at the price of additional computational effort during flight. Nevertheless, this becomes particularly useful when unforeseen obstacles require path re-planning: Low level flight paths through urban terrain may have a large set of velocity minima to account for.

4 Trajectory Following Control

After the definition of a path a feasible trajectory was defined in the previous section. The aim of the trajectory following control system is to guide the vehicle to stay on the given trajectory. Due to unconsidered disturbances and simplifications made in the model used for trajectory generation, the vehicle will not follow the trajectory without error feedback. Therefore, a control system for trajectory error reduction must be designed.

Based on the velocity profile a combination of feedforward and error feedback signals is used to constitute the desired velocity vector. In addition to the control gain determination for error compensation, the determination of the feedforward velocity vector has major influence on the trajectory following performance. Moreover, as the spline-based path is nonlinear, the calculation of path errors is nontrivial. By expressing the path error in a path fixed frame and transforming the resulting velocity vector afterwards into the geodetic frame the control problem can be simplified. The compensation of path errors will be done by proportional feedback.

4.1 Definition of Frames and Transformations

For the application for an UAV, used to demonstrate this method, some simplifications are made. The operation in a local urban terrain allows the assumption of a flat Earth. Therefore, the position is described in a Cartesian coordinate system with respect to an initial point $\underline{p}_0 = [x_0,\ y_0,\ z_0]^T$. From this origin to the current position in the geodetic frame (Index "g") the vector $\underline{p}$ represents the position of the UAV:

$$\underline{p} = \begin{bmatrix} x - x_0 \\ y - y_0 \\ z - z_0 \end{bmatrix}_g . \tag{15}$$

A path-fixed frame (Index "k") is defined as a coordinate system where the x_k axis points into the direction of the flight velocity as depicted in fig. 3. The angles γ and χ specify the orientation of the y_k and z_k axes with respect to the geodetic frame.

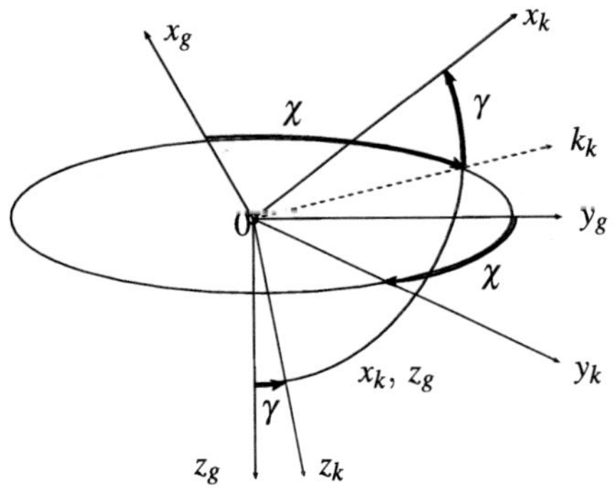

Fig. 3 Definition of frames and angles for a path fixed frame with respect to the geodetic frame, cf. Brockhaus [11]

The transformation of a velocity vector represented in the Cartesian coordinate system into polar coordinates is defined as:

$$\begin{bmatrix} V_K \\ \chi \\ \gamma \end{bmatrix} = \begin{bmatrix} \|\underline{V}\| \\ atan2(v_{Kg}, u_{Kg}) \\ -atan\left(\frac{w_{Kg}}{\|u_{Kg}+v_{Kg}\|}\right) \end{bmatrix}, \quad with \quad \|u_{Kg}+v_{Kg}\| \neq 0, \tag{16}$$

where the two-argument function $atan2$ returns the angle in radians between the positive u_{Kg}-axis of a plane and the point given by the coordinates (u_{Kg}, v_{Kg}) on it.

A deviation from the path, e. g. due to a sudden change of wind or a gust, will be compensated by the feedback of the path error. These errors with respect to the path fixed frame Δx_k, Δy_k and Δz_k are defined to be positive when the vehicle is left, behind and above the desired position. An error vector between a commanded position (Index "c") and the current position (Index "s") is defined as:

$$\Delta \underline{p}_g = \begin{bmatrix} x_c - x_s \\ y_c - y_s \\ z_c - z_s \end{bmatrix}_g . \tag{17}$$

The transformation of the error $\Delta \underline{p}_g$ into the path-fixed frame is achieved by:

$$\begin{bmatrix} \Delta x \\ \Delta y \\ \Delta z \end{bmatrix}_k = T_{gk}^T \cdot \Delta \underline{p}_g, \quad T_{gk} = \begin{bmatrix} cos\gamma cos\chi & -sin\chi & sin\gamma cos\chi \\ cos\gamma sin\chi & cos\chi & sin\gamma sin\chi \\ -sin\gamma & 0 & cos\gamma \end{bmatrix} . \tag{18}$$

4.2 Determination of the Velocity Command

The first derivative of a spline, as defined in eq. (6), leads to a vector which represents a tangent on the path for a given parameter τ. The direction of this vector is used to generate a velocity vector, whose orientation can be expressed in terms of the angles χ and γ by using eq. (16). The absolute value of the velocity V_c is determined based on the velocity profile described in section 3 and fig. 2.

The resulting velocity vector represents the command for a vehicle which is exactly on the desired path. To compensate path errors this vector must be modified to guide the vehicle back to the path. When expressing the path error in the path-fixed frame, as defined in eq. (18), the modified velocity command in the geodetic frame will be calculated with:

$$\begin{bmatrix} u \\ v \\ w \end{bmatrix}_g = T_{gk}(\chi_c, \gamma_c) \left(\begin{bmatrix} V_c \\ 0 \\ 0 \end{bmatrix}_k + \mathbf{K} \begin{bmatrix} \Delta x_k \\ \Delta y_k \\ \Delta z_k \end{bmatrix}_k \right), \quad \mathbf{K} = \begin{bmatrix} k_x & 0 & 0 \\ 0 & k_y & 0 \\ 0 & 0 & k_z \end{bmatrix} . \tag{19}$$

A path error Δx_k will therefore result in a larger velocity in path direction. Path errors Δy_k and Δz_k will result in a rotated and in its absolute value modified velocity command. With regard to the velocity limitation based on the formerly defined criteria the velocity command V_c should be less than $V_{K,max}$ from eq. (14). The limited velocity is defined as:

$$V_{K,lim} = SF \cdot V_{K,max}, \quad SF \leq 1. \tag{20}$$

The velocity command obtained from the velocity profile and the tangent on the path is scaled to $V_{K,lim}$ with respect to its orientation. The values of the gains k_x, k_y and k_z will define the response to path errors. Therefore their determination was done by adjusting the gains for a step in a (simulated) path error to result in an asymptotic error reduction with overshoot less than 3%.

4.3 Determination of the Desired Position on a Spline

The position of the vehicle can be determined based on sensors suitable for the application[3]. However, to calculate the path error at a certain time the corresponding commanded position must be determined. For a spline, which is a piecewise parametric space curve, the position is expressed in terms of the formerly defined parameter τ. Compared to a path length between two points a change in this parameter has a nonlinear relation to a distance in the 3D space. In order to determine the desired position for an increasing τ a method to convert between a distance in 3D and a change in the parameter τ must be found.

The length of a space curve can be obtained by its arc length. From a partition $a = 0 < \tau_1 < ... < \tau_{n-1} < \tau_n = b$ of the interval $[a,b]$ we obtain a finite collection of points $\underline{p}(\tau_0), \underline{p}(\tau_1), \dots, \underline{p}(\tau_{n-1}), \underline{p}(\tau_n)$ on the curve $\underline{S}$. Denote the distance from $\underline{p}(\tau_i)$ to $\underline{p}(\tau_{i+1})$ by $d(\underline{p}(\tau_i), \underline{p}(\tau_{i+1}))$, which is the length of the line segment connecting the two points. The arc length l of $\underline{S}$ is then defined to be

$$l(\underline{S}) = \sup_{a=\tau_0<\tau_1<\cdots<\tau_n=b} \sum_{i=0}^{n-1} d(\underline{p}(\tau_i), \underline{p}(\tau_{i+1})). \tag{21}$$

By taking two points of a spline in terms of τ_a and τ_b the length can be approximated by using a step size controlled ordinary differential equation (ode) solver. The solver is supported by the first derivatives as well as an accuracy ε which must be defined. For this application a Runge-Kutta solver with adaptive step size control was chosen. An initial step size of 10^{-3} and an accuracy of $\varepsilon = 10^{-6}$ turned out to be suitable.

Moreover, the conversion of a distance into an appropriate change of the spline parameter τ is done by using a combination of bisection and Newton-Raphson search algorithm. With the support of the length approximation method discussed before the search algorithm returns the spline parameter $\tau = \tau_0 + \Delta s$ corresponding to an initial τ_0 and a distance Δs. While Newton-Raphson's global convergence properties are poor, fail-safe routines exist that utilize a combination of bisection

[3] Usually GPS-aided sensor fusion algorithms can achieve position estimates of submeter accuracy.

and Newton-Raphson [12]. The hybrid algorithm takes a bisection step whenever Newton-Raphson would take the solution out of bounds, or whenever Newton-Raphson is not reducing the size of the brackets rapidly enough.

Based on the value of the commanded velocity V_c and the time Δt, elapsed since the last update, the passed distance Δs is approximated by

$$\Delta s = V_c \cdot \Delta t. \tag{22}$$

Any error that occurs due to this approximation results in a change in the desired position on the path. Therefore, by feeding back the error between the position of the vehicle and the desired position, this error will not have significant influence as long as the update time will be sufficiently short.

The velocity profile shown in fig. 2 does not account for limitations in the acceleration (only for deceleration). Hence the path error in path direction Δx_k would increase reasonably fast due to the limitations of the vehicle. Instead of accounting for this property in the trajectory definition, the modification of the reference position on the path by extending eq. (22) was selected. Depending on the distance error Δx_k the change of the reference point τ is reduced. By an additional design parameter K_r the progress of the reference point is controlled:

$$\Delta s = V_c \cdot \Delta t - K_r \cdot \Delta x_k. \tag{23}$$

Prior to the conversion of Δs to τ, Δs is limited to 10 m. The design parameter K_r is adjusted to 3 % overshoot. Finally, by using τ in eq. (5), the desired position command $\underline{p}_c$ can be provided.

This path control system was validated in real-time simulations and flight tests. The algorithms turned out to be a feasible solution. However, in case of a large curvature the velocity provided by eq. (14) will be small. Note that the velocity command can become infeasible as soon as it is in the order of the velocity state estimation noise. Moreover, this effect can be reinforced by the limitation as

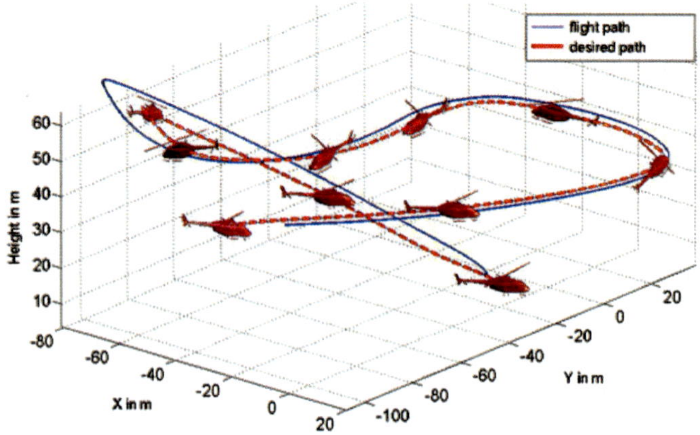

Fig. 4 Flight test result for a spline based trajectory

presented in eq. (20). Thus, a minimal velocity is enforced until the end of the path is reached.

For the application to a helicopter UAV, the path tracking showed an acceptable path following performance. However, a systematic path error is observable (fig 4) and thus needs further investigation. One reason for this effect seems to be the time the helicopter needs to adjust to the new velocity command. Due to the constantly changing command for the velocity, the flight state lags behind. Moreover, the path error compensator appears to be not fast enough to compensate this error. Thus, research is underway to prevent this lagging.

5 Summary

This work presents a decoupled trajectory generation approach for motion planning. Cubic spline segments are used to generate a collision-free path through narrow passages. Based on the curvature of the path the velocity profile is defined. This way, the complex nonlinear dynamics of a VTOL vehicle can be separated from geometric path planning and trajectory generation. The system has been successfully flight tested on an unmanned helicopter.

Motivated by the physical model of the vehicle, the trajectory curvature and the vehicle acceleration limits are used to compute the maximum velocity along the trajectory. To account for path changes during flight (e. g. obstacle avoidance) the computational efficiency for replanning is addressed by incrementally determining the velocity profile. It is updated incrementally towards a virtual horizon moving along the trajectory.

Furthermore, a trajectory following system is presented which utilizes instantaneous trajectory tangents to generate a feasible velocity command vector for the control system. It compensates potential path errors, e. g. due to wind disturbance. Finding the desired position on the path is a nontrivial task and is caused by the parameterization of the space curve by a knot vector. Thus, a conversion method for the arc length of a curve and a change of the spline parameters is proposed. Moreover, a linear control law is used for path error compensation. To prevent exceeding the maximum velocity limit, a scale factor reduces the velocity command.

Further research will have to focus on a reduction of the path deviation. This reduces worst case safety margins needed for path collision checking.

References

1. Goerzen, C., Kong, Z., Mettler, B.: A Survey of Motion Planning Algorithms from the Perspective of Autonomous UAV Guidance. Intell. Robot Syst. (2009)
2. Mattei, M.: Smooth Flight Trajectory Planning in the Presence of No-Fly Zones and Obstacles. Journal of Guidance, Control, and Dynamics 33 (March 2010)
3. Gates, D.J.: Nonlinear Path Following Method. Journal of Guidance, Control and Dynamics 33 (March 2010)
4. Ogren, P.: Receding Horizon Control of UAVs using Gradual Dense-Sparse Discretizations. In: AIAA Guidance, Navigation and Control Conference (August 2010)

5. Kaminer, I.: Path Following for Unmanned Aerial Vehicles Using L1 Adaptive Augmentation of Commercial Autopilots. Journal of Guidance, Control and Dynamics 33 (March 2010)
6. Andert, F., Adolf, F.: Online world modeling and path planning for an unmanned helicopter. Auton. Robots 27, 147–164 (2009)
7. Lorenz, S.: Open-loop reference models for nonlinear control with applications to unmanned helicopter flight. In: AIAA Guidance, Navigation and Control Conference and Exhibit, No. AIAA-2010-7860, Toronto, Canada (2010)
8. Lorenz, S.: Adaptive Regelung zur Flugbereichserweiterung des Technologiedemonstrators ARTIS. Ph.D. thesis, TU-Braunschweig (2010)
9. Tietze, S.: Autonome Bahnführung eines Hubschrauber UAV's. Master's thesis, TU-Berlin, Institut für Luft- und Raumfahrttechnik (2005)
10. Bhatia, A., Frazzoli, E.: Resolution-complete safety falsification of continuous time systems. In: Proceedings of the 45th IEEE Conference on Decision and Control, December 2006, pp. 3297–3302 (2006)
11. Brockhaus, R.: Flugregelung, 2nd edn. Springer, Berlin (2001)
12. Press, W.H., Flannery, B.P., Teukolsky, S.A., Vetterling, W.T.: Numerical Recipes in C: The Art of Scientific Computing. Cambridge University Press, Cambridge (1992)

A Linear Parameter Varying Controller for a Re-entry Vehicle Benchmark*

Andrés Marcos and Samir Bennani

Abstract. In this article the design of a linear parameter varying controller for an atmospheric re-entry vehicle benchmark is presented. The control design approach used is based on the Single Quadratic Lyapunov Function approach. The re-entry vehicle used is a high-fidelity benchmark that includes full nonlinear motion, detailed aerodynamic database, nonlinear actuators, colored sensor models, realistic uncertainties and a control-surface mix logic. The latter logic fully couples the longitudinal and lateral/directional motions and together with the noise and uncertainties used result in a challenging and representative atmospheric re-entry benchmark. The results indicate that the LPV controller satisfies all the performance and robustness objectives and alleviates the designer task due to the automated gain-scheduled nature of the approach.

1 Introduction

Gain scheduling [1, 16, 17] is probably the most widespread approach in the Space industry to design controllers for systems undergoing large dynamical changes or requiring complex mode switching implementations. It can be considered an incremental step to local synthesis that uses linear time invariant (LTI) point-designs together with standard gain scheduling concepts, such as interpolation, to obtain a global controller based on the local designs. Despite its wide spread use there are some drawbacks due to its ad hoc validation character, mainly related to the effect that the selection of the design points and of the scheduling rule (in terms of parameters and complexity) has on the performance of the global gain-scheduled controller.

Linear parameter varying (LPV) synthesis techniques have positioned themselves as an alternative, or more advanced, step to gain scheduling and can be considered

---

Andrés Marcos
Deimos Space S.L.U., Madrid, 28760, Spain

Samir Bennani
ESA ESTEC, Noordwijk, 2200, The Netherlands

* The LPVMAD study is funded by the European Space Agency under ESA-ESTEC contract 20565/07/NL/GLC.

as a type of automated gain scheduling approach [1, 18]. These techniques are at the center of a European Space Agency (ESA) study established to address an envisioned need for advanced gain scheduling techniques in Space. The "LPV Modeling, Analysis and Design (LPVMAD)" consortium led by Deimos Space (Spain) and composed by research teams from the Computer and Automation Research Institute (SZATKI, Hungary), Delft University of Technology (The Netherlands) and Leicester University (United Kingdom) was formed to address the development of an industrial LPV control design framework supported by reliable LPV software tools. Specifically, the objectives of the study were:

1. To assess the possibility, needs and impact of LPV techniques in the control design process for space systems
2. To propose a control design LPV framework
3. To develop reliable LPV tools for modeling, analysis and design in support of such a framework
4. To demonstrate the developed framework and tools for a relevant space system

Results from Phase I of the study [10, 11, 12, 13, 14, 15] detailed the successful accomplishment of the first three points above. Phase II was tasked with addressing the 4th point above using a more challenging benchmark that included practical issues such as saturation, motion coupling, controller scheduling and time-varying behaviour. The selected benchmark was an enhanced version of the well-known NASA HL-20 atmospheric re-entry vehicle.

This article presents results from the fourth objective of the project. In particular, it details the design of an LPV controller for the low supersonic to subsonic phase of the selected re-entry benchmark. The layout of the article is as follows. Section 2 presents the LPVMAD re-entry benchmark. Section 3 details the decoupled longitudinal and lateral/directional LPV controller designs. Section 4 presents the analyses performed during the design cycle as well as the final validation campaign, based on a Monte Carlo campaign using the full, i.e. coupled, nonlinear vehicle under parametric (center of gravity, moments of inertia, mass and aerodynamic coefficients) and time-varying (Mach number) changes. Section 5 concludes with a summary of the results.

2 LPVMAD Re-entry Benchmark

The NASA HL-20 lifting-body vehicle was proposed as a substitute of the Space Shuttle Orbiter and although it was finally de-commissioned, many research efforts were performed to develop a fairly detailed baseline aerodynamic database for the complete velocity range of the vehicle [2, 3, 4, 5].

For the present study, a 3 Degrees-of-Freedom optimized guidance trajectory adapted from reference [7] is considered. The study focuses in the pure aerodynamic surfaces phase corresponding, approximately from low supersonic to sub-sonic speeds, i.e. Mach$\in$[3,0.8]. This selected phase is characterized by the end of a bank reversal and a similar maneuver for the angle-of-attack, see Figure 1. Notice that these almost simultaneous maneuvers results in strong coupling of the longitudinal and lateral/directional motions (especially severe due to the control surface mix logic contained in the benchmark –more on this below).

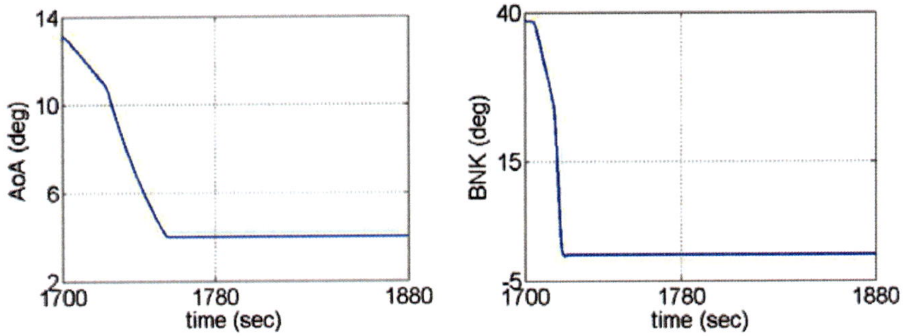

Fig. 1 Reference trajectory: angle-of-attack and bank angle versus time

The full nonlinear equations of motion, see reference [2] for details, are used together with a representative aerodynamic database [2, 3, 4, 5]. This HL-20 implementation is a quite more advanced, and representative, model than the publicly available Matlab model [6], which uses the polynomial simplifications given in NASA report [3]. The complete aerodynamic database is formed by nonlinear look-up tables (LUT) dependent on Mach number, angle of attack, sideslip and control surface deflections.

The available physical control surfaces are upper left and right flaps (DUL and DUR), lower left and right flaps (DLL and DLR), wing left and right flaps (DEL and DER) and rudder (DR). The deflection is positive downward for the upper/lower flaps (+TED and -TEU), away from the vehicle for the wing flaps (+TED) and to the left for the rudder (+TEL). The system implements a nonlinear control surface mix logic that transforms the controller-commanded elevator δ_{ele}, aileron δ_{ail}, speed-brake δ_{sbk} and rudder δ_{rud} deflections into the previous physical surfaces based on Mach number and complex nonlinear relationships. The output of the control-mixer is passed to the actuation system. Each actuation system is composed of a series interconnection of a second order filter, a magnitude limiter, a rate limiter and a time delay (of 0.005 seconds). The sensors are implemented as first order coloring filters and include measurement bias.

A multiplicative uncertainty model is considered around nominal values of the parameters u, based on given percentage uncertainty ranges $\Delta_{\%}$ and normalized random but bounded gain δ_U, i.e. $u_\Delta = u(1+ \delta_U \Delta_{\%})$. For those magnitudes with nominal value equal to zero, an additive model is employed (the multiplicative model would not introduce uncertainty in the parameter for these cases). Reference [19] details the nominal values and percentage uncertainty $\Delta_{\%}$ used in the re-entry vehicle benchmark for the main vehicle geometry and aerodynamic parameters: center of gravity (e.g. $\Delta_{\%}$ of ±0.5% for x_{CG}), moments of inertia (±5%), mass (±5%) and aerodynamic coefficients –including a $C_{L/D}$ uncertainty profile

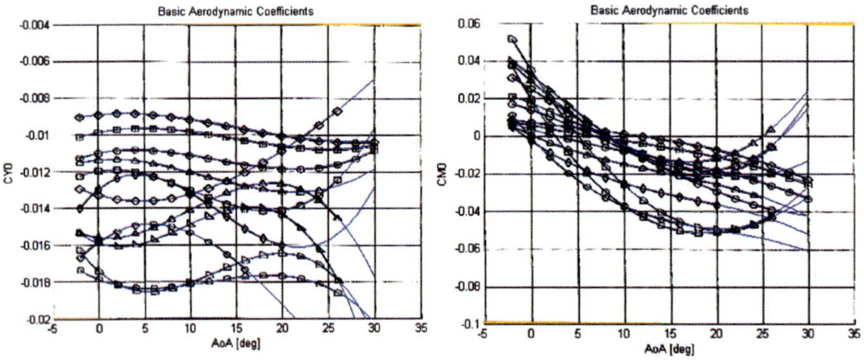

Fig. 2 Basic aerodynamic coefficients in term of angle-of-attack and Mach

that serves to physically relate the uncertainties from C_L and C_D. Further, the aerodynamic database is quite nonlinear and alternates stable with unstable areas – notice the positive and negative slopes in C_{MO} from Figure 2.

A detailed analysis of the dynamic characteristics of the benchmark is given in [9, 19], and shows that the vehicle is quite challenging with shifts in stability for the lateral/directional and longitudinal motions and strong natural frequency and damping changes as the vehicles flies the trajectory, see for example Figure 3.

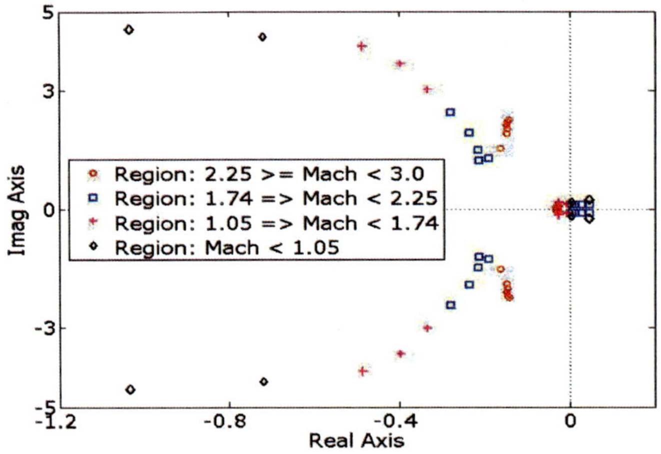

Fig. 3 Lateral/Directional motion: Pole/Zero map based on Mach regions

3 LPV Control Design

This section presents the application of advanced gain-scheduling control design methods to the presented high-fidelity benchmark. The LPV technique used is the so-called Single Quadratic Lyapunov Function (SQLF) approach [18]. Essentially, this technique synthesizes a LPV controller by means of finding a single quadratic Lyapunov function, not depending on the varying parameter vector, that satisfies an LPV version of the well-known bounded real lemma (BRL). In practice, it searches for a Lyapunov function valid for a set of linear matrix inequalities (LMIs) corresponding to the satisfaction of the BRL at a grid of values of the varying parameter vector that defines the LPV plant.

The principal advantage of this method is that it builds up from LTI H_∞ synthesis. And actually, the current software tools used to synthesize the LPV controller directly takes the weights and interconnection used in the standard LTI H_∞ design set-up and facilitates its tuning to arrive at the desired global performance. Indeed, the synthesis of design-point LTI H_∞ controllers can be seen as a first step in the application of this LPV technique. LTI H_∞ optimization is a design technique where specification of performance and robustness objectives is the main driver in correctly posing the mathematical optimization problem –as opposed to other control synthesis techniques where satisfaction of the objectives is evaluated after the design. LTI H_∞ control synthesis is one of the cornerstones of modern control and is widely used across industry [16, 17, 18]. Thus, in order to assess the advantages and shortcomings of the advanced LPV control solution, a standard gain-scheduling control design based on LTI H_∞ design-point controller can be used. Such a baseline controller was developed and presented in reference [19]. Since this baseline controller follows a decoupled motion design approach, the LPV design process also uses a motion decoupling approach during synthesis –but both are always validated using the full, i.e. coupled, nonlinear vehicle.

The control design objectives are to track the reference angles of attack, sideslip and bank in the face of the established noise and uncertain levels with desired deviations of less than 2 degrees, and with acceptable short-term deviations of less than 4 degrees.

A. Longitudinal Motion Controller Design

The design rationale for the longitudinal motion is that of angle-of-attack tracking through an ideal-model formulation. The same design rationale and plant set-up as for the baseline control design are used [19].

Since only inner-loop control is considered, the open-loop plant can be reduced to the short-period motion. In this manner, the state dimension of the controller is largely reduced (recall that for LTI H_∞-based synthesis the obtained controller has the same number of states as the design interconnection). The open-loop plant used for design has 2 states (angle-of-attack and pitch rate), two outputs (same as the states) and two inputs (elevator and speed-brake deflections). The defined H_∞ design interconnection is given in Figure 4. *Wact* penalizes the actuators magnitude; *Wrob* is used to add uncertainty at the plant input; *Wperf* and *Wid* serve to capture the angle-of-attack tracking formulation and to penalize its error; *Wcmd* shapes the angle-of-attack input; and finally, *Wnoise* is used for robustness by shaping the noise effect on the system. The *I*-block is introduced to indicate that no actuator model is used for the longitudinal motion.

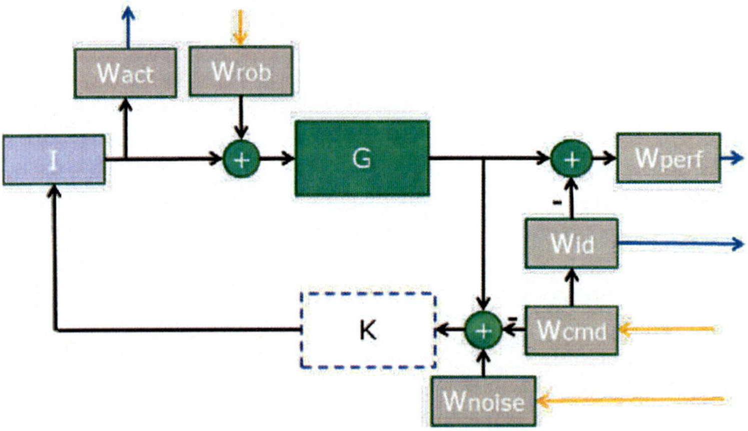

Fig. 4 LPV controller design interconnection

The only modification with respect to the baseline control design [19] is the use of a more refined grid of LTI plants around which to synthesize the LPV controller and also a slight change in the design weights. There is an apparent increase of complexity, four different sets of weights are used for the LPV controller design versus three for the baseline, but the resulting LPV controller is indeed a single dynamic system and no effort on ad hoc manual gain-scheduling is required – which is after all the greatest advantage of LPV synthesis techniques. Table 1 shows the weights for the four longitudinal design-point controllers in terms of the generic weight transfer function *Weight = K(as+1) / (bs+1)*.

Table 1 Longitudinal LPV controller: H_∞ weights per Mach region
[– := same weight as on the Mach region to the left; d2r:=180/π]

Type	Weight	M =[3.959, 2.913]			M =[1.457]			M =[1.052, 0.922]			M =[0.821]		
		K	a	b	K	a	b	K	a	b	K	a	b
Noise	$W_{\alpha\text{-noise}}$	0.01*d2r	0	0	--	--	--	--	--	--	--	--	--
	$W_{q\text{-noise}}$	0.001*d2r	1000	100	0.01*d2r	100	10	--	--	--	--	--	--
Ideal/cmd	W_{ideal}	1.5² / (s² + 2*0.9*1.5*s +1.5²)			--	--	--	--	--	--	--	--	--
	W_{cmd}	3*d2r	0	0	5*d2r	0	0	4*d2r	0	0			
Performance	W_{perf}	150	0	20	--	--	--	150	0	10	150	0	20
Robustness	$W_{rob\text{-}ele}$	0.05*d2r	0	100	--	--	--	--	--	--	--	--	--
	$W_{rob\text{-}sbk}$	0.05*d2r	0	20	--	--	--	--	--	--	--	--	--
Actuation	$W_{act\text{-}ele}$	1/40/d2r	0	0	--	--	--	--	--	--	--	--	--
	$W_{act\text{-}sbk}$	1/20/d2r	0	0	--	--	--	--	--	--	--	--	--

B. Lateral/Directional Motion Controller Design

The lateral/directional LPV controller is more complex than the previous LPV longitudinal controller, which is a single controller obtained using 4 sets of weights. Nevertheless, the use of the LPV synthesis approach simplifies the global lateral/directional controller complexity as for the LPV case is composed by two different LPV controllers straightforwardly "blended" (i.e. linearly interpolated) in contrasts to three different controllers for the baseline control design [19].

The lateral/directional open-loop plants used for the design consist of four states (yaw rate r, roll rate p, sideslip angle β and bank angle σ), two input channels (aileron δ_{ail} and rudder δ_{rud} deflections) and four outputs (yaw rate, roll rate, lateral acceleration n_y and bank angle). The interconnection used is as in Figure 4 but with the weights based on a model-matching approach for bank angle commands (*Wcmd* and *Wid*) and including two additional objectives in the performance weight *Wperf*: lateral acceleration minimization and turn-coordination [8].

Table 2 Lateral/Directional LPV controller: H_∞weight per Mach region
[– := same weight as on the Mach region to the left; d2r:=180/π]

Type	Weight	Klat-LPV-1			Klat-LPV-2					
		M=[2.91→2.09] at M every 5seconds			M=[2.25, 2.09, 2.03, 1.93, 1.74]			M =[1.05, 0.92, 0.80]		
		K	a	b	K	a	b	K	a	b
Noise	$W_{p=} W_r$	0.01*d2r	0	0	--	--	--	--	--	--
	W_σ	0.01*d2r	0	0	--	--	--	--	--	--
	W_{ny}	0.01/9.8	0	0	0.5/9.8	0	0	--	--	--
Ideal/cmd	W_{ideal}	1.03	0	0.5886	1.16	0	0.166	1.03	0	0.588
	W_{cmd}	15*d2r	1	10	25*d2r	0.05	0.5	25*d2r	1	1.6667
Performance	W_{ny}	0.1	0	0	3.5	0	0	1.5	0	0
	W_ϕ	103	0.5	10.3	120	0.5	24	50	0	20
	W_{TC}	2	0	0	--	--	--	--	--	--
Robustness	W_{ail}	0.5*d2r	0	2	0.01*d2r	0	0.1	--	--	--
	W_{rud}	0.5*d2r	0	2	0.02*d2r	0	0.2	--	--	--
Actuation	W_{ail}	0.05/d2r	0	0.1	--	--	--	--	--	--
	W_{rud}	0.05/d2r	0	0.1	--	--	--	--	--	--
	$W_{ail\text{-}dot}$	0.025/d2r	0	0	0.06/d2r	0	0	--	--	--
	$W_{rud\text{-}dot}$	0.025/d2r	0	0	0.06/d2r	0	0	--	--	--

Additionally, and critical in this case, it is the inclusion of magnitude and rate first-order models for the actuator weights (in the position of the I-block)-Table 2.

From a qualitative aspect, and based on a comparison with the experience of designing the baseline controller [19], it is remarked that the tuning of the weights can be performed much faster for the LPV controller than for the baseline due to the global automated scheduling nature of the approach. That is, the LPV synthesis adds an additional design degree of freedom through the optimized search of a common Lyapunov function valid at the selected grid points. This optimized search compensates for any 'rough' weight definition at the local points; alternatively, this can be seen as allowing the engineer to focus on performance when defining the weights at the local points while robustness is achieved by the automated scheduling of the LPV approach. It is worth mentioning that the use of the final LPV weights for the redesign of the ad-hoc gain-scheduling does not result in similar performance and robustness properties. This is the result of the non-global nature of the ad hoc gain-scheduling and the manual selection of the interpolation scheme. In summary, the systematic character of the LPV synthesis techniques can save a lot of time in defining the weights and in achieving the required control design objectives when compared to traditional ad-hoc gain-scheduling.

4 Control Analysis

A. *Linear (During-Design) Analysis*

During the design phase a set of linear analytical analysis and pseudo-linear time-simulations (linear but with some selected nonlinearities) are performed to converge to a controller that fulfills the desired objectives. Among these analyses:

- Weighted performance versus inverse of the performance weight.
- Weighted actuation versus noise and inverse of the actuation weight.
- Controller GK transfer functions for each of the tested plants.
- Sensitivity transfer functions and flying qualities.
- Time-domain simulation using the LTI plants and corresponding extracted LTI controllers together nonlinear actuators (dynamics, rate and magnitude saturations) and sensors (noise and dynamics).

A summary of results is given in this section. For example, Figure 5 shows the longitudinal weighted and controller transfer functions (TFs). As it is seen, they are below the inverse of the weights which means that the controller is unaffected by the corresponding noise channels and have the appropriate shape. Figure 6 on the other hand shows longitudinal controller linear step response simulations, which serve to evaluate its flying qualities (this is tested for a set of LTI plants including those used for design as well as other points in the flight envelope).

This is then validated through pseudo-linear (i.e. linear plant with selected nonlinearities) time domain simulations, Figure 7. A doublet-plus-ramp profile for the angle of attack reference command is used to determine the controller tracking of step and ramp profiles. This is especially important in the benchmark as it is noticed that good tracking for steady-state commands (e.g. step responses) is easily achievable in the full nonlinear simulation but that ramp-type profiles, which are very characteristics for unpowered vehicles in approach configuration, is more difficult to achieve while satisfying actuation limits. The results in Figure 7 shows that for the linear plus nonlinear actuation/sensor, the controller is valid –still, note the increase in elevator actuation for the ramp tracking.

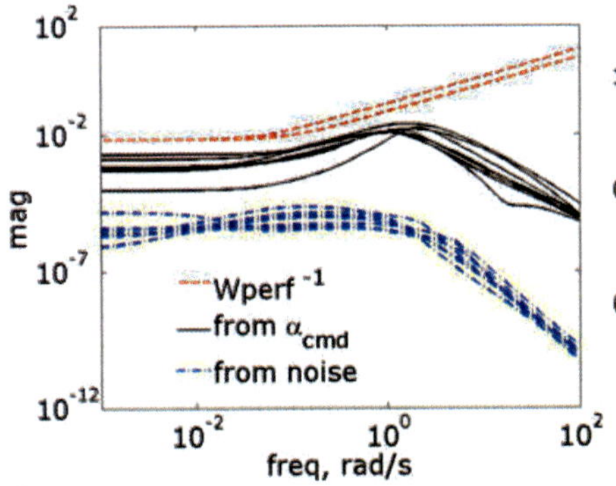

Fig. 5 Longitudinal LPV controller: weighted TF to α_{error}

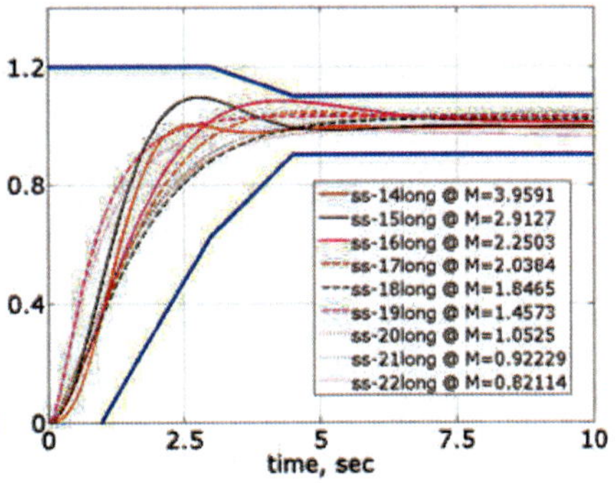

Fig. 6 Longitudinal LPV controller: Flying Qualities

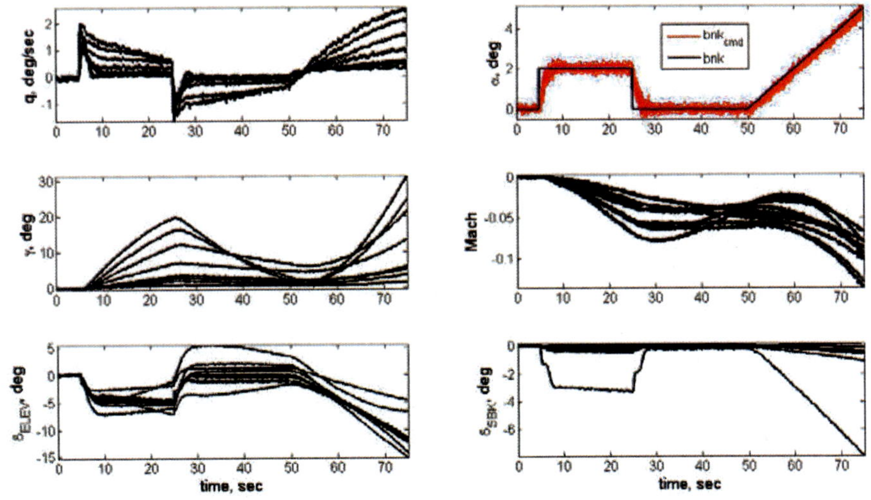

Fig. 7 Longitudinal LPV controller: pseudo-linear responses (LTI plant with full actuators & sensors)

With respect to the lateral/directional controller, the analyses performed during the design process also show the satisfaction of the objectives. Nevertheless, in this case it is observed some noise influence at high frequencies –see Figure 8 for

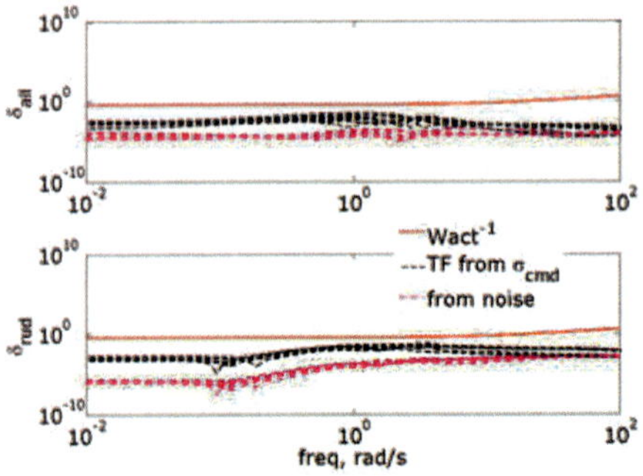

Fig. 8 Lateral/Direction LPV controller: weighted actuation

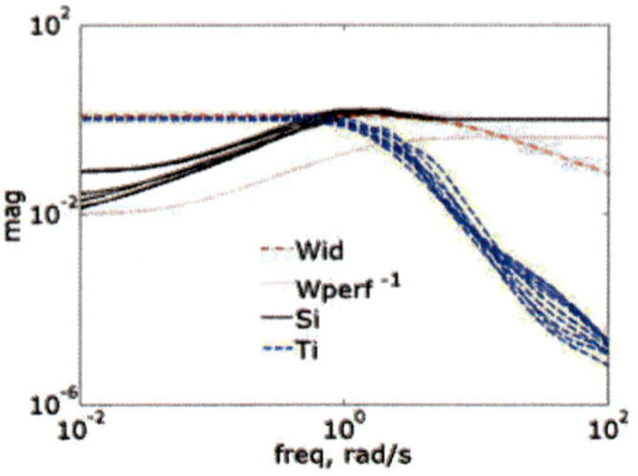

Fig. 9 Lateral/Direction LPV controller: S&T

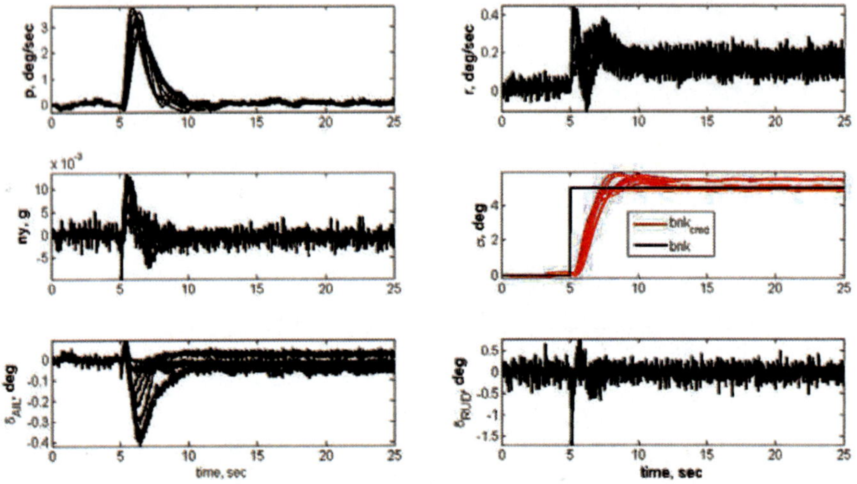

Fig. 10 Lateral/Direction LPV controller: pseudo-linear responses (LTI plant with full actuators & sensors)

an example using the weighted actuation linear analysis. Although this coupling is not critical it is reflected in the sensitivity analysis (S and T) and in the time-domain simulations, see Figure 9 and Figure 10. Thus, the conclusion from the analyses is that the lateral/direction LPV controller has the desired frequency and flying quality characteristics and should fulfill, in as much as linear analyses can guarantee, the objectives in the nonlinear simulation.

In summary, for both LPV controllers it can be seen from the linear analytical analysis perspective, that all objectives are satisfied. In the following section, it is shown that the conclusion from these during-design analyses is validated in the full nonlinear time-domain simulations through a Monte Carlo campaign.

B. Nonlinear Full-Motion Monte Carlo Analysis

A Monte Carlo campaign is performed using the full-motion nonlinear simulator of the high-fidelity benchmark and the uncertainty parameters and ranges defined in Section 2, see also [9, 19]. No failed cases, out of >1400 runs, are found indicating that the LPV controller achieves the desired robust objectives while all errors are within the limits (see Figure 11) indicating that the performance objectives are also achieved. When compared with the results from the baseline controller – not shown due to space limitations, see [19]— the LPV controller compares very well and improves both performance and robustness.

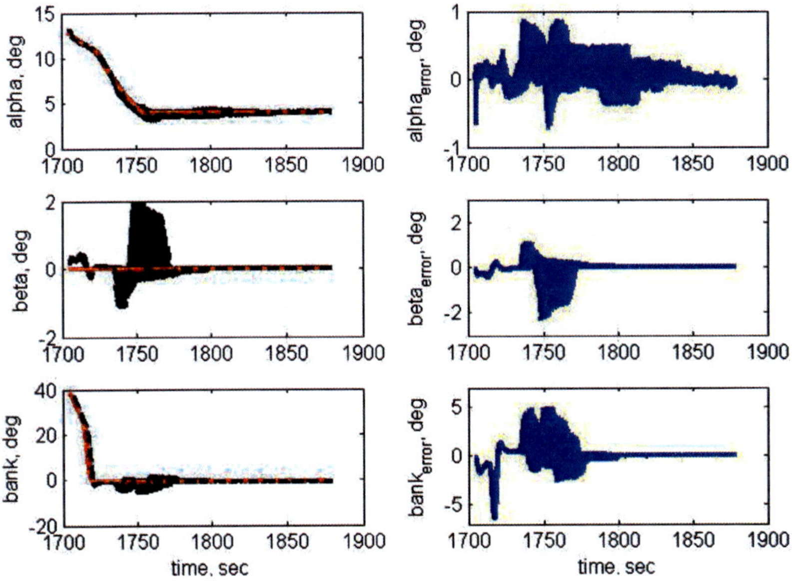

Fig. 11 Monte Carlo full-motion: tracking signals and errors

5 Conclusion

In this article, longitudinal and lateral/directional LPV controllers have been designed and analyzed for a high-fidelity re-entry benchmark. The LPV design uses the so-called Single Quadratic Lyapunov Function (SQLF) approach, which builds up on LTI H_∞ point-designs. The assessment of the full-motion LPV controller has been performed using a wide array of linear analytical techniques (from sensitivity transfer functions to flying quality responses) and full nonlinear time-domain Monte Carlo campaigns. The results indicated that that LPV controller designs fulfill all the performance and robustness objectives. Further, in comparison to a traditional gain-scheduling controller [19], the systematic character of the LPV synthesis techniques can save a lot of time in defining the weights and in achieving the required control design objectives.

Acknowledgements

The authors would like to thank Dr. Luis F. Peñín, head of the GNC/AOCSS division at Deimos Space for their help and comments, and Professor Gary Balas from the University of Minnesota for providing the LPV SQLF software tools used for the LPV controller design.

References

1. Balas, G.J.: Flight control law design: An industry perspective. In: European Control Conference, Cambridge, UK (September 2003)
2. Jackson, E.B., Cruz, C.I., Ragsdale, W.A.: Real-Time Simulation Model of the HL-20 Lifting Body. NASA Technical Memorandum 107580 (July 1992)
3. Jackson, E.B., Cruz, C.I.: Preliminary Subsonic Aerodynamic Model for Simulation Studies of the HL-20 Lifting Body. NASA Technical Memorandum 4302 (1992)
4. Cruz, C.I., Ware, G.M.: Control Effectiveness and Tip-Fin Dihedral Effects for the HL-20 Lifting-Body Configuration at Mach Numbers from 1.6 to 4.5. NASA Technical Memorandum 4697 (1995)
5. Scallion, W.I.: Aerodynamic Characteristics and Control Effectiveness of the HL-20 Lifting Body Configuration at Mach 10 in Air. NASA Technical Memorandum 209357 (September 1999)
6. Gage, S.: HL20 aeroblockset MATLAB. The Mathworks
7. Powell, R.W.: Six-Degree-of-Freedom Guidance and Control-Entry Analysis of the HL-20. Journal of Spacecraft and Rockets 30(5) (1993)
8. Doyle, J.C., Lenz, K., Packard, A.: Design examples using mu-synthesis: Space shuttle lateral axis FCS during re-entry. In: IEEE Control Decision Conference, vol. 25, pp. 2218–2223 (1986)
9. Marcos, A., Veenman, J., Scherer, C., De Zaiacomo, G., Mostaza, D., Kerr, M., Bennani, S.: Application of LPV modeling, design and analysis methods to a re-entry vehicle. In: AIAA GNC 2010 (2010)
10. Szabó, Z., Marcos, A., Mostaza, D., Kerr, M., Rödöny, G., Bokor, J., Bennani, S.: Development of an Integrated LPV/LFT Framework: Modeling and Data-based Validation Tool. In: IEEE Control System Technologies (2011) (to be published)
11. Marcos, A., Bennani, S.: LPV Modelling, Analysis and Design in Space Systems: Rationale, Objectives and Limitations. In: AIAA GNC 2009 (2009)
12. Menon, P.P., Prempain, E., Postlethwaite, I., Bates, D., Bennani, S.: An LPV loop shaping controller design for the NASA HL-20 re-entry vehicle. In: AIAA GNC 2009 (2009)
13. Menon, P.P., Prempain, E., Postlethwaite, I., Bates, D., Bennani, S.: Nonlinear Worst-Case Analysis of an LPV Controller for Approach-Phase of a Re-entry Vehicle. In: AIAA GNC 2009 (2009)
14. Veenman, J., Köro lu, H., Scherer, C.W.: Analysis of the Controlled NASA HL20 Atmospheric Re-entry Vehicle based on Dynamic IQCs. In: AIAA GNC 2009 (2009)
15. Veenman, J., Scherer, C.W., Köro lu, H.: IQC-Based LPV Controller Synthesis for the NASA HL20 Atmospheric Re-entry Vehicle. In: AIAA GNC 2009 (2009)

16. Rugh, W.J., Shamma, F.: Research on Gain Scheduling. Automatica 36, 1401–1425 (2000)
17. Leith, D.J., Leithead, W.E.: Survey of Gain-Scheduling Analysis and Design. International Journal of Control 73(11), 1001–1025 (2000)
18. Balas, G.J.: Linear, parameter-varying control and its application to a turbofan engine. International Journal of Robust and Nonlinear Control 12, 763–796 (2002)
19. Marcos, A.: A gain scheduled H-infinity controller for a re-entry benchmark. In: AIAA GNC Conference 2010 (2010)

A Low Cost Small UAV Flight Research Facility

Austin M. Murch, Yew Chai Paw, Rohit Pandita, Zhefeng Li, and Gary J. Balas

Abstract. This paper presents an overview of the low-cost, open source small Unmanned Aerial Vehicle (UAV) flight research facility at the University of Minnesota. A detailed description of the facility, its components, and its capabilities is presented, as well as applications of the UAV platforms to research, education, and collaboration. Flight results from controls research is presented, followed by a brief summary of current research and future directions.

1 Introduction

Unmanned Aerial Vehicles (UAVs) are used worldwide today for a broad range of military, civil, and research applications. There continues to be a growing demand for reliable and low cost UAV systems. This is especially true for small to miniature-sized UAV systems (less than 2 meter wing span) where the majority of systems are still deployed as prototypes due to demand and lack of reliability. Improvements in the modeling, testing and flight control for these vehicles would help to increase their reliability and performance of small UAVs during operations. The traditional approach used in the development cycle [1, 2] for manned aircraft is time consuming and resource intensive. Applying the same techniques to the small UAVs is not realistic.

The UAV Research Group in the Department of Aerospace Engineering and Mechanics (AEM) at the Univeristy of Minnesota is focused on development and implementation of a low-cost, open source small Unmanned Aerial Vehicle (UAV) flight research facility. The goal of this facility is to support research activities within the department including control, navigation and guidance algorithms, embedded fault detection methods, and system identification tools. The system is built mostly out of commercial-off-the-shelf (COTS) components to minimize the overall material and development costs. In addition, the entire architecture is open and available

Austin M. Murch · Yew Chai Paw · Rohit Pandita · Zhefeng Li · Gary J. Balas
University of Minnesota, Minneapolis, MN 55455
e-mail: {murch,paw,pandita,zhefeng,balas}@aem.umn.edu

to any researchers or organizations who wish to collaborate on the development or application of the UAV capabilities. (http://www.aem.umn.edu/~uav/) In addition to the researchers from AEM department, the UAV Research Group is collaborating with researchers at the Budapest University of Technology and Economics in Hungary and the University of Sannio at Benevento, Italy.

2 UAV Testbeds

The UAV Research Group uses COTS R/C fixed wing aircraft modified to carry the necessary avionics and instrumentation payloads. Several different aircraft models have been used during the development process, but the primary test aircraft is the Ultra Stick 25e [3] (shown in Figure 1(a)). Two additional aircraft in use are an Ultra Stick 120 (shown in Figure 1(b)) which can handle significantly more payload than the Ultra Stick 25e, and the Mini Ultra Stick, the smallest version of the Ultra Stick family.

(a) Ultra Stick 25e (b) Ultra Stick 120 (FASER)

Fig. 1 Ultra Stick UAV Testbeds

The Ultra Stick 120 (aka FASER; Free-flying Aircraft for Subscale Experimental Research) has a conventional horizontal and vertical tail with rudder and elevator control surfaces, respectively. The aircraft has a symmetric airfoil wing with aileron and flap control surfaces. All six control surfaces are actuated by Hitec HS-5625MG servos. The plane is propelled by a 1900W Actro 40-4 brushless electric motor with a Graupner 14 x 9.5 folding propeller. Power for the motor comes from two 5000mAh 5-cell lithium polymer batteries connected in series. The servos are powered by a separate 1350 mAh 3-cell lithium polymer battery. The main internal payload bay is located directly under the wing in the fuselage and measures approximately 35cm L x 10cm H x 10cm W; additional payloads may be accomodated in the aft fuselage or externally. The 120 class can carry approximately 2.5kg of payload.

The Ultra Stick 25e is an approximately 65% scale model of the Ultra Stick 120, with the same basic configuration. The UAV Research Group maintains three Ultra Stick 25e aircraft, named 'Odin', 'Loki', and 'Thor'. All six control surfaces are actuated by Hitec HS-225BB servos. The plane is propelled by a 600W E-Flite Power

Table 1 Summary of Aircraft Parameters

Parameter	Mini Ultra Stick	Ultra Stick 25e	FASER
Wing span	$0.985\,m$	$1.27\,m$	$1.92\,m$
Wing chord	$0.21\,m$	$0.3\,m$	$0.43\,m$
Length	$0.865\,m$	$1.05\,m$	$1.32\,m$
Wing reference area	$0.21\,m^2$	$0.32\,m^2$	$0.769\,m^2$
MTOW (Tested)	N/A	$2.04\,kg$	$9.07\,kg$
Empty weight	$0.62\,kg$	$1.50\,kg$	$6.35\,kg$
Endurance	$10-15\,min$	$15-20\,min$	$15-20\,min$
Cruise speed	$10-15\,m/sec$	$15-20\,m/sec$	$20-30\,m/sec$

25 brushless electric motor with an APC 12 x 6 propeller. Power for the motor and servos comes from a 4200mAh 3-cell lithium polymer battery. The main internal payload bay is located directly under the wing in the fuselage and measures approximately 22cm L x 6cm H x 7.5cm W; additional payloads may be accomodated in the aft fuselage or externally. The 25e class can carry approximately 0.55kg of payload.

The Mini Ultra Stick is an approximately 50% scale model of the Ultra Stick 120, with the same basic configuration. The UAV Research Group currently uses this aircraft as a wind tunnel model, as the smaller size enable it to fit into the AEM department's low speed wind tunnel.

The specifications of the three Ultra Stick aircraft are given in Table 1. The Ultra Stick aircraft family is marketed as a trainer-level aerobatic aircraft, so it is relatively easy for a moderately skilled R/C pilot to fly, but is still capable of highly dynamic manuevering flight.

3 Onboard Avionics

The current architecture of the onboard avionics is shown in Figure 2 and Table 2 gives a listing of the individual components in the system. The flight computer uses a real-time operating system and flight software written in C. The flight computer handles data collection from each sensor, performs attitude and position estimation, executes flight control algorithms, stores relevant data, outputs PWM servo commands, and sends information to the ground control station via the data modem. A failsafe switching board is used to switch control of the aircraft between manual mode (human R/C pilot) and the flight computer. The hardware interface to the flight computer is handled via a custom-designed interface board.

3.1 Sensors

The sensor suite for the UAVs is focused on measuring the aircraft state data needed for normal flight guidance, navigation, and control algorithms. Position, velocity,

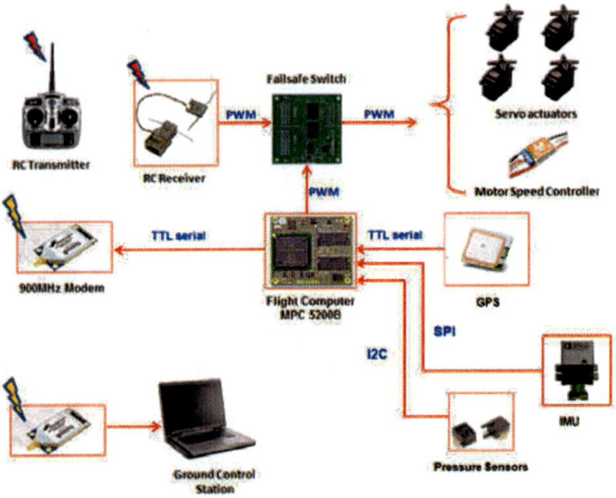

Fig. 2 Onboard avionics architecture

Table 2 Onboard avionics system components

Component	Module	Cost
Flight computer	Phytec MPC5200B microcontroller	\$300
Interface board	AEM custom design	\$250
IMU	Analog Devices iSensor ADIS16405	\$800
GPS	USGlobalSat EM-406A	\$50
Pressures	Honeywell ASDX	\$60
Data Telemetry	Free Wave MM2 900 Mhz modem	\$375
Failsafe Switch	AcroName RxMux	\$300
Manual Control	Spektrum DX-7 2.4 Ghz R/C system	\$300

and accelerations in the aircraft body frame and navigation frame are the normal quantities of interest. In general, the minimum sensor suite to achieve these data with sufficient accuracy is the combination of an IMU, GPS, and pitot-static system. Prior work has utilized an integrated sensor suite (Crossbow Micronav) that combined all of these functions, but this product is no longer available or supported. An affordable replacement for the Micronav has been difficult to find, and as a result, individual sensor components have been selected and integrated into the system. While this process does incur development and testing overhead, it has the advantage of allowing the researcher to select optimal sensors for the given application, and also offers a simple path for upgrading sensor capabilities. The replacement sensor components are described as follows:

IMU sensor: The Analog Devices iSensor ADIS16405 is a small, low-cost, temperature compensated, tri-axial accelerometer, rate gyro, and magnetometer. Data is provided to the flight computer via SPI bus at 50Hz.

GPS module: The USGlobalSat EM-406A is a small 25mm square unit with an integrated antenna. The GPS circuitry is located directly underneath the patch antenna, allowing a very compact operation. The EM-406A uses the SiRF StarIII GPS engine and data is provided to the flight computer via TTL serial at 1Hz.

Pressure sensors: The Honeywell ASDX is a small IC-based, digital output, temperature-compensated pressure sensor available in differential and absolute versions. These units are used to measure total and static pressure to determine the airspeed and altitude of the aircraft, and are connected to a Pitot-static probe that protrudes forward of the right wing of the aircraft. Pressure data is provided to the flight computer via I2C bus at 50Hz.

3.2 Flight Computer

The current flight computer is a phyCore MPC5200B-tiny 32-bit PowerPC microcontroller. Prior work has utilized an phyCore MPC55, but limitations in processing power motivated an upgrade. The MPC5200B has a clock frequency of 400 Mhz, 760MIPS of processing power, and performs floating point computation. Current flight software utilizes about 2% of the CPU capacity. It has a wide range of I/O capabilities to support communication with external devices in addition to onboard data storage capacity. Details on the specifications of the MPC5200B can be found on the Phytec website [4]. Data is stored onboard the MPC5200B in the 64MB SRAM and downloaded after the flight via an Ethernet connection to the ground station.

3.2.1 Flight Software

The onboard flight computer utilizes a real-time operating system (eCos) and flight software written in C. In addition to being open source and freely available, the eCos operating system provides a real-time kernel, can be configured to minimize the computing overhead required for the operating system, supports multi-threading, and is POSIX C compatible [5].

The flight software uses multiple prioritized threads running at different rates to perform the necessary tasks for flight. Prioritizing multiple threads helps to ensure the most important time critical tasks are performed first (e.g. data acquisition).

Table 3 Flight software thread description

Priority	Thread	Description	Frequency (Hz)
1	DAQ	data acquisition for each sensor component	50
2	AHRS	attitude determination using EKF	50
3	CLAW	flight control law and actuator commands	25
4	INS-GPS	INS/GPS navigation filtering algorithms	10
5	DATA	onboard data storage	50
6	TELE	packing and sending of telemetry data	20

Currently, there are 6 threads, listed in Table 3, which describes their function, priority, and update rate.

The data acquisition, data storage, and telemetry threads are relatively stable and few changes are made to these software functions from flight to flight. Most research activity occurs in the AHRS, flight control law, and INS/GPS software functions.

4 Ground Control Station

The Ground Control Station (GCS) is used during flight testing to monitor the aircraft state and health status. It consists of a laptop computer running the GCS software connected via serial to a data modem. The GCS software is a Java-based program inspired by the Open Source Glass Cockpit Project. It is designed to give vital flight information in real time to observers in order to assess the flight performance and maintain situational awareness of the aircraft during the flight test. The GCS software includes a Heads-Up Display, a moving map showing the location of the aircraft, commanded actuator positions, and indicators to display flight control mode information. Figure 4 shows a screenshot of the GCS software.

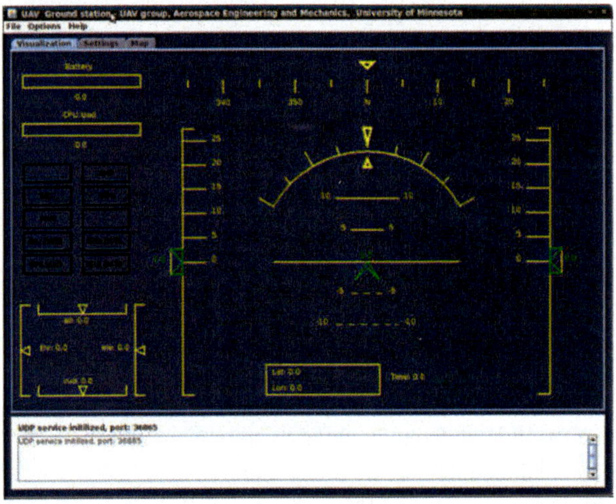

Fig. 3 Ground control station

5 Simulation Testing

The use of simulation-based development and testing prior to actual flight testing reduces the total development time and helps ensure the algorithm under development is validated and bug-free, reducing costly debug time in the field and minimizing risk to the UAV platforms. The UAV Research Group maintains an integrated framework of three simulation environments used during the development process (Figure 4). The UAV simulation model is constructed in the Matlab/Simulink

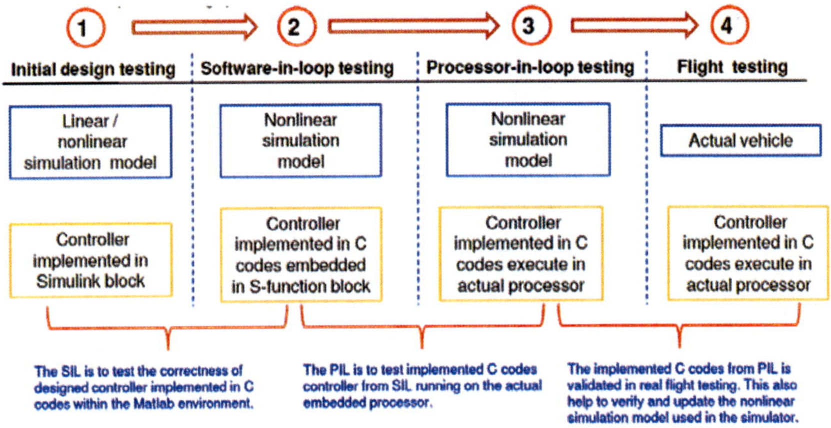

Fig. 4 Application of simulation testing for flight control development

environment using the Aerospace Blockset [6]. All three simulations share the same nonlinear dynamics model.

5.1 Nonlinear Simulation

The 6-DOF nonlinear simulation model uses the full nonlinear equations of motion, linear derivative aerodynamics, and table lookup propulsion models. Models of relevant aircraft subsystems such as actuators, motor, propeller, sensor dynamics, and noise are included. The environmental model includes a detailed model of Earth's atmosphere, gravity, magnetic field, wind, and turbulence. The aerodynamic derivatives were derived from first principles and empirical methods, and then updated using flight test data [7]. Bifilar swing tests were used to determine the moments of inertia, and wind tunnel testing was used to characterize the motor and propeller thrust, torque, and power.

5.2 Software-in-Loop Simulation

The Software-in-the-Loop (SIL) simulation contains the nonlinear simulation with the C-code implementation of the control law in a Simulink S-function block. Only the flight control law is used at this stage; the remainder of the flight software is not included. This step is primarily used for debugging and to verify the C implementation of the algorithm under development matches the designer's expectations.

5.3 Processor-in-Loop Simulation

The PIL simulation is an extension of SIL simulation that includes the MPC5200B flight computer into the simulation setup. Figure 5 shows the differences between

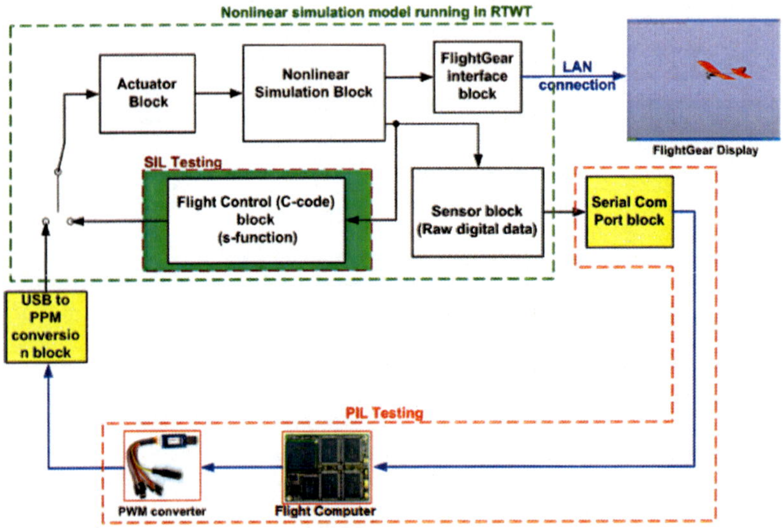

Fig. 5 Block structure of SIL and PIL simulation

the SIL and PIL simulations. The Mathwork's Real-Time Windows Target toolbox [8] is used to ensure the simulation runs in real time on a Windows PC which is crucial for meaningful results when the real-time flight computer is included in the simulation loop. The PIL simulation also has an interface for a R/C pilot via a USB R/C-transmitter-style interface. Aircraft state data can be visualized on the GCS software or via FlightGear, a free open-source flight simulator [9]. The PIL simulation offers the following additional benefits to SIL simulation:

- Ability to test and identify controller implementation issues before the flight testing. This helps to determine the limitations on actual hardware and provides important information for controller redesign.
- Provides a real-time testing environment for synthesized controllers.
- Provides a testbed for integration and testing of hardware and software subcomponents at the system level.
- Provides an environment for the pilot and flight test engineers to prepare and understand the scope of the flight test and gain confidence in the overall system.

Beside testing, debugging and validating the control design and implementation, the PIL simulation is used for post-flight analysis in validating the simulation model that is recursively updated from flight test data. Once the simulation model has been sufficiently validated, it can be used to augment and substitute much of the flight testing which helps to reduce the risk and developmental costs of the system.

6 Flight Testing

Flight testing is the final stage of testing in the development cycle. Under the current concept of operations, a R/C pilot is the pilot in command and has authority over what has control of the aircraft (human vs. flight computer). The R/C pilot performs the takeoff and landing of the aircraft and transfers control to the flight computer once at an appropriate flight condition. A toggle switch on the R/C transmitter is used to transfer control of the aircraft to the flight computer via the RxMux fail-safe switch. At anytime during testing, the R/C pilot can take over control from the flight computer by toggling this switch on the R/C transmitter. Future plans include development of autonomous takeoff and landing capabilities, but the R/C pilot will remain as a safety backup.

Flight operations are contained to be within visual range of the R/C pilot and under 400 feet AGL altitude. Typical flight durations are 15-20 minutes, depending on the purpose of the flight. A normal deployment will last 3-4 hours and have 6-9 flights.

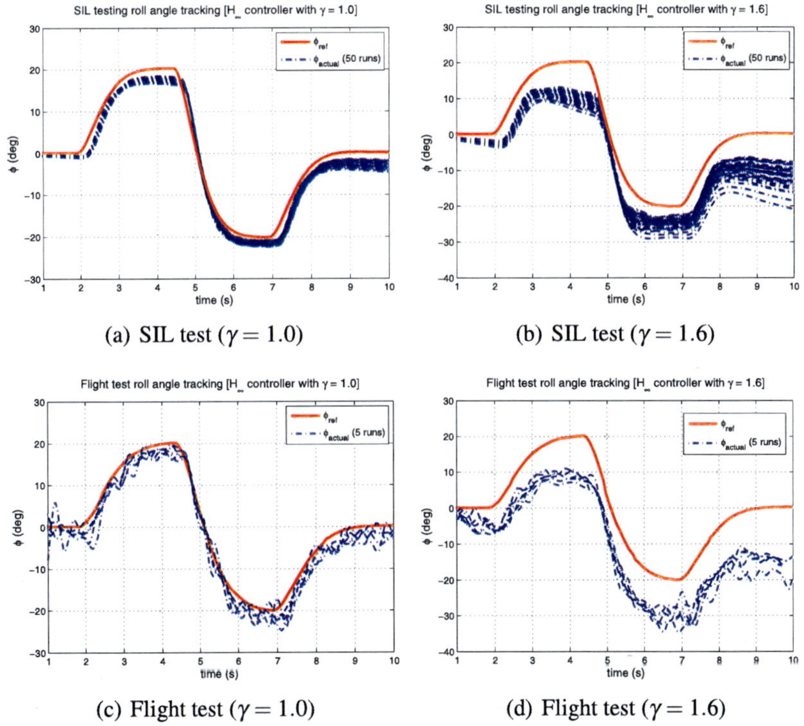

(a) SIL test ($\gamma = 1.0$)

(b) SIL test ($\gamma = 1.6$)

(c) Flight test ($\gamma = 1.0$)

(d) Flight test ($\gamma = 1.6$)

Fig. 6 SIL and flight test result of H_∞ controllers

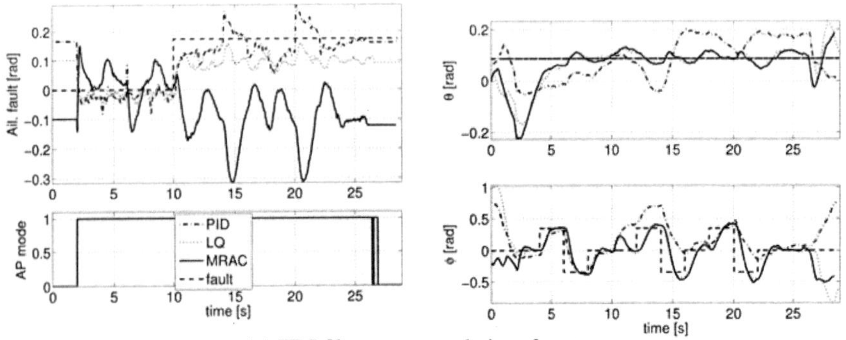

(a) FDI filter output and aircraft states

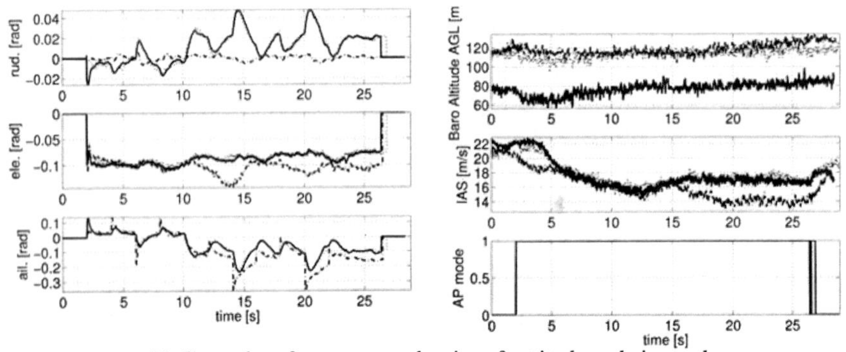

(b) Control surface commands, aircraft attitude and airspeed

Fig. 7 Experimental flight data analysis

Flight testing to date has primarily been focused on robust control law and fault detection filter research. Figure 6 shows a comparison of SIL simulation data and flight data for an H_∞ control design [7]. This research effort was successful in using the entire UAV framework to develop, test, and fly a flight control law experiment. This effort also validated the integrated framework design approach and the concept of using a low-cost UAV platform for flight controls research. An acceptable match between the simulation and flight results was achieved despite the low fidelity simulation model and the relatively low quality sensors on the UAV platform.

The model-based fault detection and isolation (FDI) algorithms evaluated using the UMN flight test facility were based on robust H_∞ filter designs. The effect of various closed-loop controllers on the FDI filter performance for an aileron fault were investigated. Three lateral-directional axis controllers were considered: a classical PID design, a LQ optimal multivariable design and a direct model-reference adaptive controller (MRAC). The experimental objective were to compare the robustness of the three controllers and the performance of the FDI filter to detect the aileron fault in the presences of the three controllers. Figures 7a and 7b show the flight test tracking performance of the three controllers. Recall that the controllers being investigated are associated with the lateral-directional axes of the aircraft. An identical pitch controller is used in all experiments to track 5 deg pitch command reference. The parameter *'AP mode'* indicates the status of FCC, i.e. whether the pilot or onboard flight program (OFP) is in control. The experimental results show the varied behavior attained with the robust filter for the different controllers. The benefit of having a flight test platform is the ability to validate theory and simulations on the test aircraft. In 2010, a total of 69 research flights were conducted.

7 Current Activities and Future Directions

Currently, the UAV Research Group is pursuing several research appliations for the UAV platforms and development framework.

- The National Science Foundation (NSF) has recently started a program focused on Cyber-Physical Systems (CPS) research [10]. As part of this research, the UAV platforms will play a key role in testing embedded fault detection algorithms including model based methods, software methods, and data-driven anomaly detection methods. These fault detection methods will also be applied to real-world industry problems in a collaborative effort with local industry to a production UAV platform and air data sensors.
- Precision landing of small UAVs is a significant challenge given the low quality sensors typically used. Many operational losses of UAVs result from the inability of the UAV to return reliably to a small protected location. The UAV simulation and platform will be used to develop precision landing algorithms, sensor requirements, and test results in an effort to improve the operational efficiency and loss rate of small UAVs.
- Extensive wind tunnel data is expensive to collect and is rarely openly available. Researchers at the NASA Langley Research Center performed significant static wind tunnel tests of an Ultra Stick 120 as part of the FASER development effort [11, 12]. This data is being implemented into the UAV simulation and will be openly available for research, education, and collaboration. This data set includes control surface and thrust effects over an angle of attack range of -5 to 40 degrees and an angle of sideslip range of +/- 30 degrees. As mentioned previously, the UAV Research Group operates an Ultra Stick 120 that was donated by NASA LaRC. This aircraft is instrumented with two angle of attack/sideslip vane sensors on wingtip booms, which expands the possible research applications of this platform to include aerodynamic model identification, gust alleviation, and high angle of attack/post-stall flight.
- The AEM department has developed a "Design, Build, Simulate, Test and Fly a UAV" course focused on the use of rapid prototyping software tools for vehicle modeling, guidance, navigation, and flight control, real-time implementation, hardware-in-the-loop simulation and flight tests. The UAV platforms and simulation framework was used for the class. Students repeated the design process of the UAV, including component selection, simulation, flight controller design, testing and implementation. At the end of the class, student-designed controllers were flight tested on the UAV platform.

8 Conclusion

A low-cost, open source small UAV flight research facility has been developed at the University of Minnesota. The UAV Research Group is actively applying the UAV platform to flight controls, guidance, navigation, and fault detection research. The AEM department is integrating the UAV platforms into the educational curriculum, giving students a unique opportunity to work with real flight data and have access to a flight test capability. The total cost required to field the UAV aircraft, avionics, and sensors described in this paper is under 3,000USD, which is less than the cost of most COTS autopilot systems marketed at small UAVs!

References

1. Brian, S., Lewis, F.L.: Aircraft Control and Simulation. Wiley Interscience, Hoboken (2003)
2. Pratt, R.W.: Flight Control Systems: Practical Issues in Design and Implementation. Institute of Engineering and Technology (2000)
3. E-Flite, Ultra stick 25e arf, http://www.e-fliterc.com/Products/Default.aspx?ProdID=EFL4025
4. Phytec. MPC5200B microcontroller datasheet, http://www.phytec.com/products/sbc/PowerPC/phyCORE-MPC5200B-tiny.html
5. eCos, Homepage, http://ecos.sourceware.org/
6. I. The MathWorks, Matlab and simulink, http://www.mathworks.com
7. Paw, Y.C.: Synthesis and validation of flight control for UAV. Ph.D. thesis, University of Minnesota (2009)
8. Matlab, Simulink, Real-Time Windows Target 3 User's Guide. The MathWorks Inc. (2008)
9. FlightGear, Homepage, http://www.flightgear.org/
10. National Science Foundation, Cyber-physical systems (CPS) program, http://www.nsf.gov/pubs/2010/nsf10515/nsf10515.htm
11. Owens, D.B., Cox, D.E., Morelli, E.A.: Development of a low-cost sub-scale aircraft for flight research: The FASER project. In: 25th AIAA Aerodynamic Measurement Technology and Ground Testing Conference (2006-3306)
12. Morelli, E.A., De Loach, R.: Wind tunnel database development using modern experiment design and multivariate orthogonal functions. In: 41st AIAA Aerospace Sciences Meeting and Exhibit (2003-0653)

Adaptive Nonlinear Flight Control and Control Allocation for Failure Resilience

Thomas Lombaerts, Michiel van Schravendijk, Ping Chu, and Jan Albert Mulder

Abstract. In this publication, reconfiguring control is implemented by making use of Adaptive Nonlinear Dynamic Inversion (ANDI) for autopilot control. The adaptivity of the control setup is achieved by making use of a real time identified physical model of the damaged aircraft. In failure situations, the damaged aircraft model is identified by the so-called two step method in real time and this model is then provided to the model-based adaptive NDI routine in a modular structure, which allows flight control reconfiguration on-line. Three important modules of this control setup are discussed in this publication, namely aerodynamic model identification, adaptive nonlinear control, and control allocation. Control allocation is especially important when some dynamic distribution of the control commands is needed towards the different input channels. After discussing this modular adaptive controller setup, reconfiguration test results are shown for damaged aircraft models which indicate satisfactory failure handling capabilities of this fault tolerant control setup.

1 Introduction

Recent airliner accident statistics ([1]) show that a significant share of these accidents, namely 17%, is caused by loss of control in flight, where control of an aircraft is lost in flight due to a technical malfunction, a piloting mistake or an external upset. This observation motivates flight control research in the field of so-called fault tolerant flight control (FTFC). Fault Tolerant Flight Control aims to increase the survivability of a failed aircraft by reconfiguring the control laws rather than by means

Thomas Lombaerts
German Aerospace Center DLR, Institute of Robotics and Mechatronics,
Münchner Straße 20, 82234 Weßling, Germany
e-mail: thomas.lombaerts@dlr.de

Thomas Lombaerts · Michiel van Schravendijk · Ping Chu · Jan Albert Mulder
Delft University of Technology, Faculty of Aerospace Engineering,
Kluyverweg 1, 2629 HS Delft, The Netherlands
e-mail: {t.j.j.lombaerts,q.p.chu,j.a.mulder}@tudelft.nl

of hardware redundancy only, which is current practice. There are many control approaches possible in order to achieve fault tolerant flight control. An important aspect of these algorithms is that they should not only be robust, but even adaptive in some way, in order to adapt to the faulty situation, see ref. [7, 15]. In the category of adaptive control algorithms, a distinction is made between indirect adaptive control and direct adaptive control. Indirect adaptive control involves two stages. First, an estimate of the plant model is generated online. Once the model is available, it is used to generate the controller parameters. Instead of estimating a plant model, a direct adaptive control algorithm estimates the controller parameters directly in the controller. This can be done via two main approaches: output error and input error. Of both main categories mentioned here, indirect adaptive control is preferable due to its flexibility and its property of being model based. In both categories, there are also two sub-versions, namely model reference adaptive control (MRAC) and self-tuning control (STC). In the former, one relies on a reference model and works on minimizing the tracking error between plant output and reference output (such as the concept of sliding mode control). With model reference indirect adaptive control it is feasible to achieve three important goals, namely trim value adjustment for the inputs, decoupling of inputs and outputs and closed loop tracking of pilot commands, see ref. [7]. Self-tuning control focuses on adapting the (PID) control gains of the controller by making use of the estimated parameter values and is known to be more flexible, see ref. [39]. Currently, much research is performed in the field of indirect adaptive control, where the adaptation is more extensive than only tuning the PID control gains. One of these new indirect control possibilities is adaptive model predictive control (AMPC), which is an interesting algorithm thanks to its nature to deal with (input) inequality constraints. These constraints are a good representation for actuator faults. It should be noted that there have already been some successful applications of MPC in the field of fault tolerant flight control, see ref. [2, 23, 34]. An alternative indirect adaptive nonlinear control approach is discussed in this publication, which allows a reconfigurable control routine placing emphasis on the use of physical models to be developed, and thus producing internal parameters which are physically interpretable at any time. This technique can deal with control surface failures as well as structural damage resulting in aerodynamic changes.

This publication discusses the combination of the two step method as an identification procedure, and nonlinear dynamic inversion as a control method in order to obtain a model based fault tolerant flight controller. Another important module of this control setup is control allocation. Control allocation is especially important when some dynamic distribution of the control commands is needed towards the different input channels. The structure of this publication is as follows. Section 2 provides information on the high fidelity RECOVER simulation model which has been used in this research project. The identification procedure is described in section 3, while section 4 discusses the nonlinear adaptive control method. Further explanations about control allocation can be found in section 5. Simulation results are discussed in section 4.1 for the autopilot. Finally, section 6 presents some conclusions and recommendations for future research.

2 RECOVER Simulation Model

The presented work is part of a research project by the Group for Aeronautical Research and Technology in Europe (GARTEUR). This group has established flight mechanics action group FM-AG(16) with the specific goal to investigate the possibilities of fault tolerant control in aeronautics and to compare the results of different reconfiguring control strategies applied to a reference benchmark flight trajectory. That benchmark scenario is inspired by the so-called Bijlmermeer disaster of EL AL flight 1862, where a Boeing 747-200 Cargo aircraft of Israel's national airline EL AL lost two engines immediately after take-off from Amsterdam airport Schiphol in the Netherlands and crashed into an apartment building in the neighbourhood while trying to return to the airport. A detailed simulation model of this damaged aircraft is available from the National Aerospace Laboratory NLR. This RECOVER (REconfigurable COntrol for Vehicle Emergency Relief) benchmark model is discussed in detail in ref. [40, 41] and has been used (also in earlier versions) by a number of investigators and organizations ([34, 35, 42]). More information about the reference benchmark scenario can be found in ref. [27, 29]. Other control strategies and results as part of the framework of FM-AG(16) applied to the same benchmark model can be found in ref. [3, 10, 16, 21, 22]. Related Fault Detection and Isolation (FDI) work can be found in ref. [44].

The simulation benchmark for evaluating fault tolerant flight controllers as discussed in ref. [41] contains six benchmark fault scenarios of varying criticality. Fig. 1 shows the failure modes and structural damage configuration of the Flight 1862 accident aircraft, which is the most important fault scenario in the simulation benchmark. This publication will focus primarily on the El Al engine separation scenario, which is an accurate simulation of flight 1862, as explained above, which

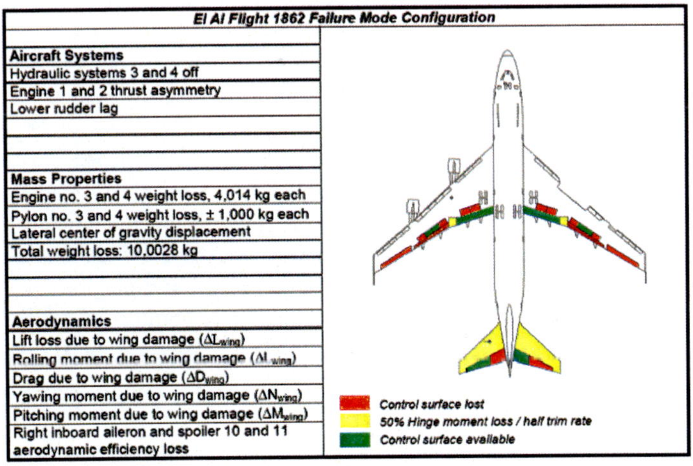

Fig. 1 Failure modes and structural damage configuration of the Flight 1862 accident aircraft, suffering right wing engine separation, partial loss of hydraulics and change in aerodynamics, source: [40]

has been validated with data from the digital flight data recorder DFDR and where the loss of hydraulics is taken into account. Another scenario is the loss of the vertical tail, where the vertical tail plane and rudders are detached, resulting in a loss of lateral static stability.

3 Aerodynamic Model Identification

The identification method considered in this study is the so-called two step method, which has been continuously under development at Delft University of Technology over the last 20 years, see ref. [8, 24, 36]. There are many other identification algorithms mentioned in the literature like maximum likelihood identification (MLI) and other one step identification routines, but not all of them are applicable on line. One of the few procedures which can be implemented in real time is the so-called filtering method developed at DLR, see ref. [20]. This is a joint state and parameter estimation algorithm, but very complex. The advantage of the two step method is that it is easier to implement on-line. The key concept of the two step method, is that the identification procedure has been split into two consecutive steps, as substantiated in ref. [9]. More precisely, the global non-linear one step identification method is decomposed in two separate steps, where the nonlinear part is isolated in the aircraft state estimation step. Consequently, the aerodynamic model parameter identification procedure in the second step can be simplified to a procedure which is linear in the parameters, but involves nonlinear regressors. The aim is to update an a priori aerodynamic model (obtained by means of wind tunnel tests and CFD calculations) by means of on-line flight data. The first step is called the Aircraft State Estimation (ASE) phase, where the second one is the Aerodynamic Model Identification (AMI) step. In the Aircraft State Estimation procedure, an Iterated Extended Kalman Filter is used to determine the aircraft states based upon redundant but perturbed information from all sensors (air data, inertial, magnetic and GPS measurements). By means of this state information, it is possible to construct the combined aerodynamic and thrust forces and moments acting on the aircraft. By means of a recursive least squares operation, the aerodynamic derivatives can be deduced. Especially in case of structural damage, the aerodynamic properties can change drastically, resulting in the need for joint aerodynamic model structure selection and parameter estimation. This can be done on-line by means of Adaptive Recursive Orthogonal Least Squares (AROLS), see ref. [31]. Validation tests have shown that these methods are very accurate. More details about the identification procedure can be found in ref. [13]. The identification method is triggered by a monitoring algorithm of the innovation (residual).

4 Adaptive Nonlinear Control

A major advantage of dynamic inversion is its ability to naturally handle changes of operating conditions, which removes the need for gain scheduling, e.g. as is the case for classical control methods. This is especially advantageous for control of space re-entry vehicles, due to their extreme and wide operating conditions which vary between hypersonic speed during re-entry and subsonic regions during the terminal glide approach phase to the runway. Another advantage is its natural property of decoupling the control axes, i.e. no coupling effects remain between steering channels and the different degrees of freedom. NDI control has been implemented in the Lockheed F-35 Lightning II, ([45, 5]). Dynamic inversion is a popular control method for flight control and aircraft guidance, ([4, 11, 43]) as well as reconfiguring control, ([14, 37]).

The main assumption in NDI is that the plant dynamics are assumed to be perfectly known and therefore can be cancelled exactly. However, in practice this assumption is not realistic, not only with respect to system uncertainties but especially to unanticipated failures for the purpose of fault tolerant flight control. In order to deal with this issue, one can make use of robust control methods as outer loop control or neural networks to augment the control signal. However, another solution is the use of a real time identification algorithm, which provides updated model information to the dynamic inversion controller. These augmented structures are called adaptive nonlinear dynamic inversion (ANDI). The latter procedure is the method of choice for this research.

Figure 2 gives a high-level logical structure overview of the control strategy, involving the two step method algorithm and adaptive nonlinear dynamic inversion.

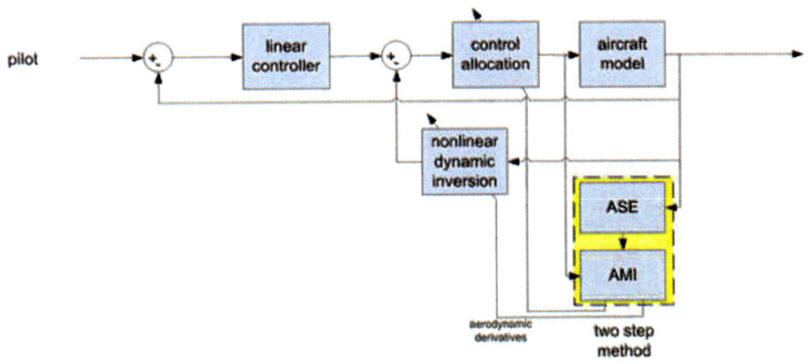

Fig. 2 Layout of Adaptive Nonlinear Dynamic Inversion FTFC

Three consecutive inversion loops have been implemented, namely a body angular rate loop, an aerodynamic angle loop and a navigation loop, which can be placed in a cascaded order based upon the time scale separation principle. The body angular rate loop tracks roll rate p, pitch rate q and yaw rate r by commanding the control surfaces aileron δ_a, elevator δ_e and rudder δ_r. This control law is based upon the dynamic equation of an aircraft: $\dot{\omega} = \mathbf{I}^{-1}\mathbf{M}_a - \mathbf{I}^{-1}\omega \times \mathbf{I}\omega$, where ω is the angular rate vector, $\mathbf{I}$ is the inertia matrix and $\mathbf{M}_a$ are the aerodynamic moments and relies on the identified stability and control derivatives for the three angular moments. The second loop is the aerodynamic angle loop which tracks roll angle ϕ, angle of attack α and sideslip angle β by commanding the body angular rates in the inner loop. The navigation loop tracks course χ, flight path angle γ and speed V by commanding the roll angle ϕ, angle of attack α and throttle setting T_c. The control law consists of a kinematics based inversion and a subsequent aerodynamic forces based inversion, as elaborated in [19, 18]. Linear controllers act on each separate NDI loop. They involve proportional and proportional-integral control, and gains have been selected to ensure favourable flying qualities by means of damping ratio ζ and natural frequency ω_n while complying with the time scale separation principle. Optimization of these gain values has been achieved by means of multi-objective parameter synthesis (MOPS) optimization, see ref. [26, 32, 33]. The detailed description of the control laws for this project can be found in [30].

4.1 Evaluations of Autopilot Experiments on RECOVER: Engine Separation Scenario

The engine separation scenario is a very sensitive situation to combine commands in heading, altitude and speed simultaneously. Crucial in this context is to avoid engine throttle saturation. Therefore, in this experiment only a heading change has been considered, as shown in fig. 3(a). Moreover, a limited maximum roll angle of 20° has been imposed, due to the restricted safe flight envelope. It has been found that altitude and speed changes are also feasible separately, but these are not discussed in this publication.

The time histories of the states in fig. 3(b) reveal that the aircraft in post failure conditions flies with a small nonzero roll angle and sideslip angle, due to the asymmetric damage, despite a zero commanded sideslip angle. The control surface deflections in figure 4 confirm the cessation of functioning of the control surfaces which are powered by the hydraulic circuits connected to engines number 3 and 4, as illustrated in fig. 1. The remaining operative surfaces are successful in keeping the aircraft in equilibrium and under control, although with restricted authority.

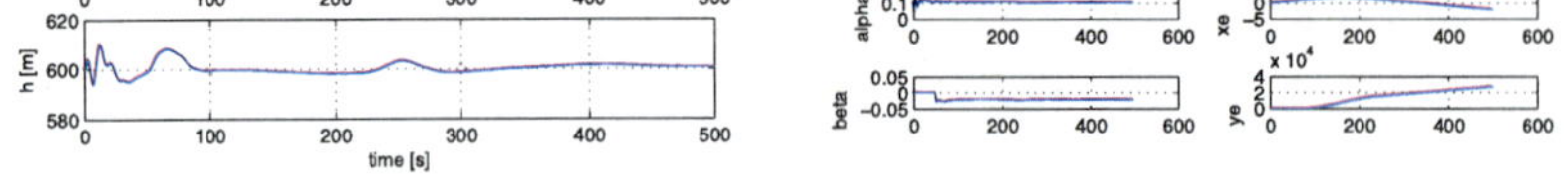

(a) tracking quantities

(b) states (angular rates in (rad/s), velocities in (m/s), angles in (rad), distances in (m))

Fig. 3 Tracking quantities and states for the engine separation scenario

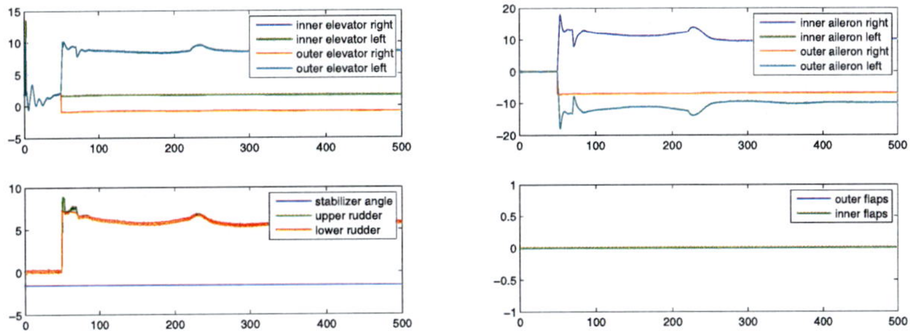

(a) deflections of elevators, stabilizer and rudders (deg)

(b) deflections of ailerons and flaps (deg)

Fig. 4 Deflections of elevators, stabilizer, rudders, ailerons and flaps for the engine separation scenario

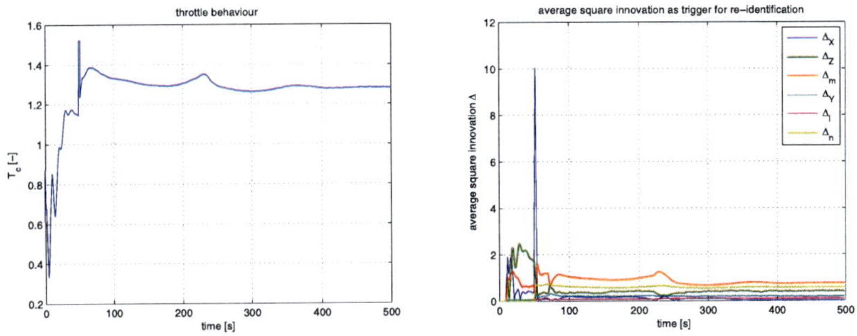

(a) throttle behaviour

(b) average square innovation as trigger for re-identification

Fig. 5 Spoilers and specific forces for the engine separation scenario

Two additional interesting quantities to investigate are the throttle setting and the average square innovation $\overline{\Delta} = \frac{1}{m}\sum_{i=1}^{n} \Delta(i)$, with $\Delta = \mathbf{b} - \mathbf{A}\mathbf{x}$ the innovation (fitting error), which triggers the re-identification routine as explained in ref. [25, 28]. Figure 5(a) confirms that the throttle setting does not saturate[1], however the remaining control margins in order to remain inside the safe flight envelope are severely restricted. This is due to the asymmetric thrust which needs to be compensated by the control surfaces. The spike at $t = 50s$ is caused by the feedforward path in the controller, which is needed to compensate for the instantaneous speed loss of the two lost engines. Figure 5(b) depicts the values for the average square innovation for each force and moment channel separately. At $t = 50s$, it can be seen that the threshold $\Delta_{th} = 10$ for $\overline{\Delta}_X$ is exceeded, and a re-identification procedure is triggered for C_X. It has become necessary to include the sideslip angle β, which has become

[1] More information about throttle saturation in the RECOVER simulation model can be found in [13].

significant due to the sideslipping flight, as an additional regressor in the identification procedure. This leads to a successful new identification procedure which is performed extremely quickly as can be seen in this figure. This result confirms the beneficial contribution from the identification routine in this fault tolerant flight control setup. The used control allocation strategy is static and equal distribution of the control commands over all relevant control surfaces. More advanced techniques are considered in section 5.

5 Control Allocation

This section describes ways to enhance the performance of Fault-Tolerant Flight Control (FTFC) Systems by augmenting them with different Control Allocation (CA) methods, based upon a study published in [38]. These methods can be used to distribute the desired control forces and moments over the different control effectors available, i.e. control surfaces and engines, which makes the control system more versatile when dealing with in-flight failures. In a first stage, a number of different control allocation methods were compared using a simplified aircraft model. This yielded two promising methods: Weighted Least Squares (WLS) and Direct Control Allocation (DCA).

In the WLS method, [17], the cost function is written as follows:

$$\|\mathbf{W}_u(\mathbf{u}-\mathbf{u}_d)\|^2+\gamma\|\mathbf{W}_v(\mathbf{B}\mathbf{u}-\mathbf{v})\|^2=\left\|\underbrace{\begin{pmatrix}\gamma^{\frac{1}{2}}\mathbf{W}_v\mathbf{B}\\ \mathbf{W}_u\end{pmatrix}}_{\mathbf{A}}\mathbf{u}-\underbrace{\begin{pmatrix}\gamma^{\frac{1}{2}}\mathbf{W}_v\mathbf{v}\\ \mathbf{W}_u\mathbf{u}_d\end{pmatrix}}_{\mathbf{b}}\right\|^2 \tag{1}$$

Then the optimization problem can be written as:

$$\mathbf{u}_\Omega = \arg\min_{\mathbf{u}} \|\mathbf{A}(\mathbf{u}^i+\mathbf{p})-\mathbf{b}\| \tag{2}$$

$$\mathbf{u}=\mathbf{u}^i+\mathbf{p} \tag{3}$$

$$\underline{\mathbf{u}}\leq\mathbf{u}\leq\bar{\mathbf{u}} \tag{4}$$

and the residual is computed as: $\mathbf{d}=\mathbf{b}-\mathbf{A}\mathbf{u}^i$

This method is in theory able to find a feasible solution $\mathbf{u}$ for all attainable virtual controls $\mathbf{v}$. This is due to the fact that it is able to free up previously saturated actuators. This gives this method much more flexibility compared to other methods.

Direct Control Allocation (DCA) is an alternative approach to Optimization Based Control Allocation. Instead of optimizing some criterion, such as the ℓ_2-norm, DCA produces a solution based on geometric reasoning. Durham was the first one to describe Direct Control Allocation [12]. Ref. [6] described the Direct Control Allocation problem as:

$$\max_{a,\mathbf{u}^*} a \tag{5}$$

$$\text{subject to: } \mathbf{B}\mathbf{u}^* = a\mathbf{v} \tag{6}$$

$$\underline{\mathbf{u}}\leq\mathbf{u}^*\leq\bar{\mathbf{u}} \tag{7}$$

Then determine the control $\mathbf{u}$ as:

$$\mathbf{u} = \begin{cases} \frac{1}{a}\mathbf{u}^* \text{ if } a > 1 \\ \mathbf{u}^* \text{ if } a < 1 \end{cases} \tag{8}$$

The DCA method looks for the largest virtual control $a\mathbf{v}$ that can be produced in the direction of the desired virtual control $\mathbf{v}$. When the maximum virtual control is larger than the desired virtual control $(a > 1)$, it scales back the true control input to match the two. If the maximum virtual control is smaller than the desired virtual control $(a < 1)$, the true control input is not scaled. The benefit of DCA is that it will always produce a control effort in the same direction in the control dimension space as the desired control effort $\mathbf{v}$.

These two methods were subsequently implemented in the aforementioned Adaptive Nonlinear Dynamic Inversion (ANDI) controller, discussed in section 4. Using the RECOVER model described in section 2, both methods were subjected to the engine separation scenario and a few other failures. Both methods were scored on a number of metrics and compared with some other methods. For each method, the contribution of an Actuator Health Monitoring System (AHMS) was evaluated. This system updates the control efficiencies in the mathematical model. The results of these tests in fig. 6 reveal that DCA is the best method to be implemented with an NDI based FTFC System. Regarding maneuverability, DCA is consistently the top performer, especially outperforming the other methods when it comes to turn radius, see fig. 6(a), and Dutch Roll damping, as shown in fig. 6(b).

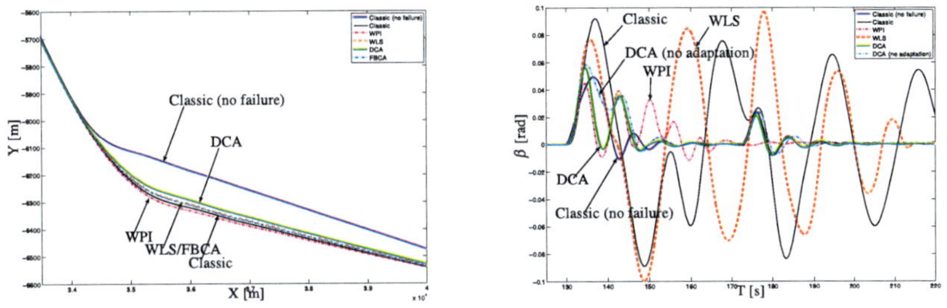

(a) Turn radius in the engine separation scenario

(b) Dutch roll damping in the vertical tail loss scenario

Fig. 6 Performance comparisons between control allocation variants for the engine separation scenario and the vertical tail loss

For DCA, the addition of AHMS had little impact, although for some other control allocation methods there were clear benefits when using this system. The lack of effect the AHMS has on the performance of DCA can be explained by the fact that this method takes the current actuator position and the rate limits into account when computing the solution to the control effort distribution problem, which serves the same purpose as AHMS, although in a different way.

This study has shown that a significant improvement in the performance of a FTFC System can be achieved by including a control allocation method, specifically DCA has shown to be very promising in this context as motivated above. The ability to use the full potential of the actuator suite helps to keep control over the aircraft when a large part of the control power for one of the aircraft axes is lost. More details about this study can be found in ref. [38].

6 Conclusions and Future Work

Summarizing, it can be stated that, following experiments in a high fidelity simulation model including validated structural failure modes based upon past accidents, the fault tolerant flight control approach based upon the real time physical model identification integrated with adaptive nonlinear dynamic inversion is successful in recovering damaged aircraft. The designed methods are capable of accommodating the damage scenarios which have been investigated in this project. Another important result is that model identification using the two step method has proven to be real time implementable in practice. It has been found that adaptive control allocation is another crucial module for a successful fault tolerant flight controller. An extensive comparative study has shown that the Direct Control Allocation method is the best method to be combined with an ANDI based FTFC system. The DCA method looks for the largest virtual control that can be produced in the direction of the desired virtual control.

It has been found that damage induced flight restrictions are very important during post failure flight. Therefore, future efforts should also be put into the estimation of the post-failure safe flight envelope and subsequent optimization of the reference trajectory taking into account the new envelope.

References

1. Civil aviation safety data 1993-2007. Technical report, Civil Aviation Authority of the Netherlands (CAA-NL) (2008)
2. de Almeida, F.A., Leißling, D.: Fault-tolerant model predictive control with flight test results on ATTAS. In: AIAA Guidance, Navigation and Control Conference, number AIAA 2009-5621 (2009)
3. Alwi, H.: Fault Tolerant Sliding Mode Control Schemes with Aerospace Applications. PhD thesis, University of Leicester (February 2008)
4. Balas, G., Garrard, W., Reiner, J.: Robust dynamic inversion control laws for aircraft control. In: Proceedings of the AIAA Guidance, Navigation and Control Conference, Washington, DC, pp. 192–205. AIAA (1992)
5. Balas, G.J.: Flight control law design: An industry perspective. European Journal of Control, special issue 9(2-3), 207–226 (2003)
6. Bodson, M.: Evaluation of optimization method for control allocation. In: Proceedings of AIAA Guidance, Navigation and Control Conference, number AIAA-2001-4223 (August 2001)
7. Bodson, M., Groszkiewicz, J.E.: Multivariable adaptive algorithms for reconfigurable flight control. IEEE Transactions on Control Systems Technology 5(2), 217–229 (1997)

8. Chu, Q.P.: Lecture Notes AE4-394, Modern Flight Test Technologies and System Identification. Delft University of Technology, Faculty of Aerospace Engineering (2007)
9. Chu, Q.P., Mulder, J.A., Sridhar, J.K.: Decomposition of aircraft state and parameter estimation problems. In: Proceedings of the 10th IFAC Sympium on System Identifiation, vol. 3, pp. 61–66 (1994)
10. Cieslak, J., Henry, D., Zolghadri, A., Goupil, P.: Development of an active fault-tolerant flight control strategy. AIAA Journal of Guidance, Control and Dynamics 31, 135–147 (2008)
11. da Costa, R.R., Chu, Q.P., Mulder, J.A.: Re-entry flight controller design using nonlinear dynamic inversion. Journal of Spacecraft and Rockets 40(1), 64–71 (2003)
12. Durham, W.C.: Computationally efficient control allocation. Journal of Guidance, Control and Dynamics 24(3), 519–524 (2001)
13. Edwards, C., Lombaerts, T.J.J., Smaili, M.H.:Fault tolerant control - a Benchmark Challenge. Lecture Notes in Control and Information Sciences, vol. 399. Springer, Heidelberg (2010)
14. Ganguli, S., Papageorgiou, G., Glavaski, S., Elgersma M.: Aircraft fault detection, isolation and reconfiguration in the presence of measurement errors. In: AIAA Guidance, Navigation and Control Conference and Exhibit, number AIAA-2006-6551, Keystone, Co. (August 2006)
15. Groszkiewicz, J.E., Bodson, M.: Flight control reconfiguration using adaptive methods. In: Proceedings of the 34th Conference on Decision and Control, pp. 1159–1164 (1995)
16. Hallouzi, R., Verhaegen, M.: Fault-tolerant subspace predictive control applied to a boeing 747 model. AIAA Journal of Guidance, Control and Dynamics 31, 873–883 (2008)
17. Harkegard, O.: Backstepping and Control Allocation with Applications to Flight Control. PhD thesis, Linkoping University, Sweden (2003)
18. Holzapfel, F.: Dynamic inversion based control concept with application to an unmanned aerial vehicle. In: AIAA Guidance, Navigation and Control Conference and Exhibit, number AIAA-2004-4907 (2004)
19. Holzapfel, F.: Nichtlineare adaptive Regelung eines unbemannten Fluggerätes. PhD thesis, Lehrstuhl für Flugmechanik und Flugregelung, Technische Universität München (2004)
20. Jategaonkar, R.: Flight Vehicle System Identification: A Time Domain Methodology, 1st edn. Progress in Astronautics and Aeronautics Series, vol. 216. AIAA (2006)
21. Joosten, D.A., van den Boom, T.J.J., Lombaerts, T.J.J.: Effective control allocation in fault-tolerant flight control with MPC and feedback linearization. In: Proceedings of the European Conference on Systems and Control, Kos, Greece, pp. 3552–3559 (July 2007)
22. Joosten, D.A., van den Boom, T.J.J., Lombaerts, T.J.J.: Fault-tolerant control using dynamic inversion and model-predictive control applied to an aerospace benchmark. In: The Proceedings of the 17th IFAC World Congress, vol. 17, pp. 12030–12035 (2008)
23. Kale, M.M., Chipperfield, A.J.: Stabilized MPC formulations for robust reconfigurable flight control. Control Engineering Practice 13(6), 771–788 (2005)
24. Laban, M.: Online aircraft aerodynamic model identification. PhD thesis, Delft University of Technology (1994)
25. Lombaerts, T., Chu, Q.-P., Mulder, J.A., Joosten, D.: Flight control reconfiguration based on a modular approach. In: Proceedings of the 7th IFAC SAFEPROCESS Symposium on Fault Detection, Supervision and Safety of Technical Processes, pp. 259–264 (July 2009)

26. Lombaerts, T.J.J.: Fault Tolerant Flight Control. A Physical Model Approach. PhD thesis, Delft University of Technology (2010)
27. Lombaerts, T.J.J., Breeman, J., Joosten, D.A., van den Boom, T.J.J., Chu, Q.P., Mulder, J.A., Verhaegen, M.: Specifications modelling document for Garteur AG16 fault tolerant control. Technical report, Delft University of Technology (2005)
28. Lombaerts, T.J.J., Chu, Q.P., Mulder, J.A., Joosten, D.A.: Modular flight control reconfiguration design and simulation. Control Engineering Practice (2011); invited for a special SAFEPROCESS 2009 section, (accepted for publication since December 18, 2010)
29. Lombaerts, T.J.J., Joosten, D.A., Breemand, J.H., Smaili, H.M., Chu, Q.P., van den Boom, T.J.J., Mulder, J.A., Verhaegen, M.: Assessment criteria as specifications for reconfiguring control. In: Proceedings of the AIAA Guidance, Navigation, and Control Conference and Exhibit. Keystone, CO (August 2006)
30. Lombaerts, T.J.J., Looye, G.H.N., Chu, Q.P., Mulder, J.A.: Fault tolerant flight control using a physical modular approach. Aerospace Science and Technology (2010) (under review)
31. Lombaerts, T.J.J., Van Oort, E.R., Chu, Q.P., Mulder, J.A., Joosten, D.A.: Online aerodynamic model structure selection and parameter estimation for fault-tolerant control. Journal of Guidance, Control and Dynamics 33(3), 707–723 (2010)
32. Looye, G., Joos, H.-D.: Design of autoland controller functions with multiobjective optimization. AIAA Journal of Guidance, Control and Dynamics 29(2), 475–484 (2006)
33. Looye, G.H.N.: An Integrated Approach to Aircraft Modelling and Flight Control Law Design. PhD thesis, Delft University of Technology (2007)
34. Maciejowski, J.M., Jones, C.N.: MPC fault-tolerant flight control case study: Flight 1862. In: Proceedings of the 5th IFAC Symposium on Fault Detection, Supervision and Safety of Technical Processes SAFEPROCESS, Washington DC, USA, pp. 121–126 (June 2003)
35. Marcos, A., Balas, G.: A Boeing 747–100/200 aircraft fault tolerant and diagnostic benchmark. Technical Report AEM–UoM–2003–1, Department of Aerospace and Engineering Mechanics, University of Minnesota (2003)
36. Mulder, J.A.: Design and evaluation of dynamic flight test manoeuvers. PhD thesis, TU Delft, Faculty of Aerospace Engineering (1986)
37. Oppenheimer, M.W., Doman, D.B.: Efficient reconfiguration and recovery from damage for air vehicles. In: Proceedings of the AIAA Guidance, Navigation and Control Conference and Exhibit, number AIAA-2006-6552 (2006)
38. Van Schravendijk, M.: Fault-tolerant flight control with control allocation. Master's thesis, Delft University of Technology (April 2010)
39. Slotine, J.-J.E., Li, W.: Applied Nonlinear Control. Prentice Hall, Englewood Cliffs (1991)
40. Smaili, H.M., Breeman, J., Lombaerts, T.J.J.: A simulation benchmark for aircraft survivability assessment. In: Proceedings of the International Congress of Aeronautical Sciences, number 2008-9.3.2 (2008)
41. Smaili, M.H., Breeman, J., Lombaerts, T.J.J., Joosten, D.A.: A simulation benchmark for integrated fault tolerant flight control evaluation. In: Proceedings of the AIAA Modelling and Simulation Technologies Conference and Exhibit, number AIAA-2006-6471 (2006)
42. Szaszi, I., Ganguli, S., Marcos, A., Balas, G.J., Bokor, J.: Application of FDI to a nonlinear Boeing 747 aircraft. In: 10th Mediterranean Conference on Control and Automation, Lisbon, Portugal (July 2002)

43. van Soest, W.R., Chu, Q.P., Mulder, J.A.: Combined feedback linearization and model predictive control for re-entry flight. AIAA Journal of Guidance, Control and Dynamics 29(2), 427–434 (2006)
44. Varga, A.: Detection and Isolation of Actuator/Surface Faults for a Large Transport Aircraft. In: Edwards, C., Lombaerts, T., Smaili, H. (eds.) Fault Tolerant Flight Control - a Benchmark Challenge. LNCIS, vol. 399, pp. 423–448. Springer, Heidelberg (2010)
45. Walker, G.P., Allen, D.A.: X-35B STOVL flight control law design and flying qualities. In: Proceedings of the Biennial International Powered Lift Conference and Exhibit, number AIAA-2002-6018 (2002)

Coordinated Road Network Search for Multiple UAVs Using Dubins Path

Hyondong Oh, H.S. Shin, A. Tsourdos, B.A. White, and P. Silson

Abstract. This paper proposes a coordinated road network search algorithm for multiple heterogeneous unmanned aerial vehicles (UAVs). The road network search problem can be interpreted as the problem to seek minimum-weight postman tours with a graphic representation of the road network. Therefore, the conventional Chinese Postman Problem (CPP) is first presented. We, then, consider physical constraints of UAVs into the search problem, since they cannot be addressed in the typical CPP. This modified search problem is formulated as Multi-choice Multi-dimensional Knapsack Problem (MMKP), which is to find an optimal solution minimising flight time. The Dubins path planning produces the shortest and flyable paths in consideration of physical constraints, so that the Dubins path is used to calculate the cost function of the modified search problem. The performance of the proposed multiple UAVs road network search algorithm is verified via numerical simulation for a given map.

1 Introduction

Recently, there have been great efforts to develop cooperative systems of multiple UAVs and investigate their benefits. For instance, in military operation, the cooperative system is able to not only enhance the survivability, but also increases chance to

Hyondong Oh
Autonomous Systems Group, Department of Informatics and Systems Engineering, Cranfield University, Swindon SN6 8LA, United Kingdom
e-mail: h.oh@cranfield.ac.uk

H.S. Shin
e-mail: h.shin@cranfield.ac.uk

A. Tsourdos
e-mail: a.tsourdos@cranfield.ac.uk

B.A. White
e-mail: b.a.white@cranfield.ac.uk

P. Silson
e-mail: p.m.g.silson@cranfield.ac.uk

succeed its missions. In terms of practical value, one of the greatest benefits of the cooperative system is that it can provide users with better information superiority. However, developing such an autonomous cooperative system is quite challenging because of technical and operational issues such as decision making and information fusion to be solved.

For reconnaissance, inspection, and intelligence mission, UAVs need to patrol some region and gather information. These missions often constraint UAVs path and trajectory. For instance, if UAVs are utilized to get some information of enemy's activities on the specific roads and military bases, or to observe the traffic of ports or roads, they should fly over only those roads or region rather than patrol whole terrain. This problem is defined as the road search problem, and also known as the vehicle routing problem. The cooperative systems of UAVs can considerably improve information superiority in this problem.

Road search problem has mainly been handled in the operational research area [1, 2, 3, 4], and this can be generally classified by two categories: one is Traveling Salesman Problem (TSP) which finds a shortest circular trip though a given number of cities, and the other Chinese Postman Problem (CPP) finding again the shortest path but with path constraints on an entire network of road. The TSP using multiple UAVs can be considered as task assignment problem to minimise the cost (time or energy) by assigning each target to the UAV and a various methods have been developed such as binary linear programming based optimisation [5, 6], heuristic method including iterative network flow [7] and tabu search algorithm [8]. On the other hand, the CPP is normally used for ground vehicle application such as road maintenance, snow disposal [9] and boundary coverage [10].

Vehicle routing algorithms usually approximate their path to a line for reducing the computational load, so the physical constraints imposed on the vehicle are not addressed. Although some algorithms for the TSP consider the physical constraints, they are mostly developed for a single vehicle [11]. For multiple vehicles, only heuristic method [12] is implemented due to the complexity of the problem. Moreover, these constraints have been rarely considered in the CPP. This paper proposes a road network search synthesis which considers the physical constraint, as well as provides a sub-optimal minimum time solution. We first define the road network problem using a graph representation, implement the conventional CPP which seeks minimum-weight postman tours of the road network, and then modify the search problem in consideration of the constraints of the multiple UAVs. This modified search problem is formulated as Multi-choice Multi-dimensional Knapsack Problem (MMKP) [5, 13], which is to find an optimal solution minimising flight time. MMKP formulation allows to consider fuel or energy capacity of each UAV and different types of roads and vehicles. The Dubins path is used to evaluate the performance index of MMKP since it generates the shortest and flyable paths between road junctions by using simple geometry [14].

An overview of road network search problem is introduced in Section 2. Section 3 presents two heuristic algorithms to solve the conventional road search problem for multiple vehicles. In Section 4, this paper addresses the constraints of UAVs and proposes a road search synthesis using the cooperating UAVs. At the last section, the performance of the proposed synthesis is verified via the numerical examples.

2 Road Network Search Problem

To search the roads identified in the map, a road network should be established, which consists of a set of straight line joining waypoints. These waypoints are located either on road junctions or are located along the roads at sufficient separation to allow fairly accurate representation of the curved road by a set of straight lines. The road network is chosen for this study as shown in Fig. 1(a). It shows the Google map [15] of some village in the UK. The road network can be converted to a graph as shown in Fig. 1(b). In order to search all roads within the map, there are two typical routing problems [2].

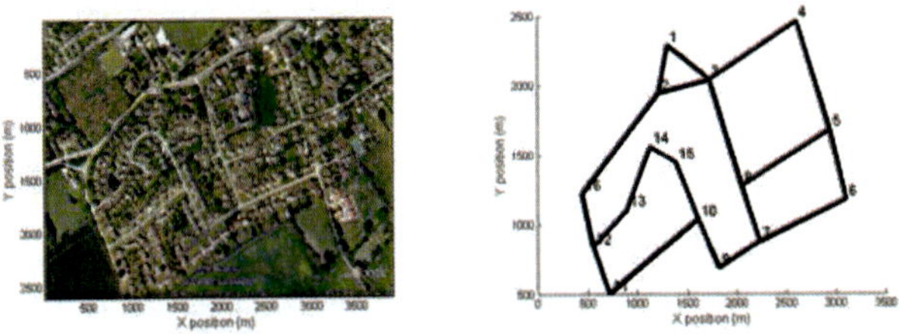

(a) Google map of village in the UK

(b) Graphic representation

Fig. 1 Road network problem

Travelling Salesman Problem (TSP): A salesman has to visit several cities (or road junction). Starting at a certain city, he wants to find a route of minimum length which traverses each of the destination cities exactly once and leads him back to his starting point.

Chinese Postman Problem (CPP): A postman has to deliver mail for a network of streets. Starting at a given point, e.g., the post office, he tries to find a route of minimum length allowing him to traverse each street at least once and leading him back to the post office.

This paper focuses on the CPP and its variation which involves constructing a tour of the road network traveling along each road with the shortest distance. Typically, the road network is mapped to an undirected graph $G = (V, E)$, and edge weights $w : E \rightarrow \boldsymbol{R}_0^+$, where the roads are represented by the edge set E and road crossings are represented by the node set V. Each edge is weighted with the length of the road or the amount of time needed to pass it. The CPP algorithm involves first constructing an even graph from the road network graph. This even graph has a set of vertices with an even number of edges attached to them. This is required as any traverse of the junction by approaching on one road and leaving on another, which means that only an even number of edges will produce an entry and exit pair for the tour. As the road network graph may have junctions with an odd number of edges, some roads are chosen for duplication in the graph. The technique chooses a set of roads with the shortest combined length to minimise duplication. The tour of the even graph is calculated by determining the Euler tour of the graph, which visits every edge exactly once or twice for duplicated edge.

3 Heuristic Algorithms for Multi-Agent Road Network Search

The CPP has a lot of variations such as the Capacitated CPP which capacitates the total edge cost they can bear, the Rural CPP which visits certain roads but not necessarily all of them, the Windy CPP which has different value for the same edge according to the direction and the k-CPP which deals with the deployment of several postmen [3]. In this section, Min-Max k-CPP (MM k-CPP) algorithms are described for multi-agent road network search with the map as given in Sec. 2. MM k-CPP is a variation of k-CPP which considers the route of the similar length. This objective can be required if UAV should finish road search mission with the minimum mission completion time. The MM k-CPP problem was first mentioned by [16] and later, solved by several algorithms [2]. This study implements two heuristic algorithms for the MM k-CPP with the similar way as in the literature and compared the results.

3.1 Cluster Algorithm

This algorithm is based on the cluster first - route second approach. In other words, in the first step, the edge set E is divided into k clusters and then, a tour for each cluster is computed. This algorithm can be represented as a constructive heuristic method, and described as follows [2].

Algorithm description

1. Determine the set F of representative edges
   First of all, k representative edges $f_1, \ldots, f_k$ of cluster F_i for each vehicle are determined. Let f_1 be the edge having the maximum distance from the depot and f_2 be the edge having maximum distance from f_1. The rest of representative edges are successively determined by maximizing the minimum distance to the already existing representatives. Then, remaining edges are assigned to the cluster according to the weighted distance between e and f_i considering the distance between representative edges and depot, number of assigned edge to the cluster F_i and cost of the cluster.
2. Include edges for connectivity
   Add edges between every vertex and depot and determine minimum spanning tree [1] which includes original edges in each cluster for connection between the edges.

3. The rural Chinese Postman Problem
   Compute CPP route of required subset of edges out of total edges by using the conventional Chinese Postman algorithm explained in [17].

Figure 2 shows the result of Cluster Algorithm which is applied to the scenario given in Sec. 2 with four ground vehicles. In this figure, solid line represents the road visited once and dashed line represents duplicated path for Even graph added by the CPP algorithm. Total distance of all vehicle is 24340.7 metre and maximum distance of one vehicle is 7318.3 metre for agent 1.

(a) Agent 1 (b) Agent 2

(c) Agent 3 (d) Agent 4

Fig. 2 Search path result of the cluster algorithm. Solid line is vehicle path visited once and dashed line is duplicated vehicle path.

3.2 *First Route Algorithm*

Unlike the cluster algorithm, the First Route algorithm follows a route first - cluster second approach. In a first step, postman tour which covers all edges is computed, and then, this tour is divided by k tour segments which have the similar length. This algorithm is described as follows.

Algorithm description

1. Compute an optimal 1-postman tour C^* using the conventional Chinese Postman algorithm
2. Compute splitting nodes
   $k-1$ splitting nodes $(v_{p_1}, \ldots, v_{p_{k-1}})$ on C^* are determined in such a way that they mark tour segments of C^* approximately having the same length. Approximated tour segment length, L_j is computed by using Shortest Path Tour Lower Bound, s_{max}.

$$s_{max} = \frac{1}{2} \max_{e=\{u,v\} \in E} \{w(SP(v_1, u)) + w(e) + w(SP(v, v_1))\} \quad (1)$$

$$L_j = (\frac{j}{k})(w(C^* - 2s_{max})) + s_{max}, \; 1 \leq j \leq k-1 \quad (2)$$

where k denotes the number of vehicles, $w(a)$ represents the distance of the subtour a and SP represents the shortest path between nodes considering road network. Then, the splitting node v_{p_j} is determined as being the last node such that $w(C^*_{v_{p_j}}) \leq L_j$. $C^*_{v_n}$ is the subtour of C^* starting at the depot node and ending at v_n. The details can be found in [16].

3. k-postmen Tours
   Constructed k tours $C = \{C_1, \ldots, C_k\}$ by connecting tour segments with shortest paths to the depot node.

Figure 3 shows the result of the First Route algorithm which is applied to the scenario given in Sec. 2 with four ground vehicles. Total distance of all vehicle is 24340.7 metre and maximum distance of one vehicle is 7818.5 metre for agent 4, which is slightly longer than the Cluster algorithm. Although both of two heuristic algorithms are intuitive, easy to implement and show reasonable performance without heavy computation burden for multiple ground vehicles, they can be far from an optimal solution and difficult to consider characteristic of the vehicle.

4 Coordinated Road Network Search for Multiple UAVs

For the road network search using multiple UAVs, variation of the typical CPP algorithm is required, so that it can consider the operational and physical characteristic of the UAV in the search problem. As shown in Fig. 4, since the UAV cannot change its heading angle instantaneously due to the physical constraint, the trajectory has to meet the speed and turn limits of the UAVs. Moreover, differently from ground vehicle, the UAV does not have to fly along the road only to cover a certain edge which is not connected. This modified search problem is formulated as MMKP which is to find an optimal solution minimizing flight time. Classical MMKP is to pick up items for knapsacks for maximum total value so that the total resource required does not exceed the resource constraints of knapsack [13]. For applying MMKP to the road network search, UAVs are assumed as the knapsacks, the roads to be searched are resources and limited flight time or energy of each UAV is capacity of knapsack. MMKP formulation allows to consider the limitation of each UAV flight time and different types of roads, vehicles and minimum turning radius and get the sub-optimal solution of the coordinated road search assignment. Moreover, the Dubins path planning produces the shortest and flyable paths taking into consideration their dynamical constraints, thus the Dubins path is used to calculate the cost function of the modified search problem. The details of proposed road network search algorithm for multiple UAVs are as follows.

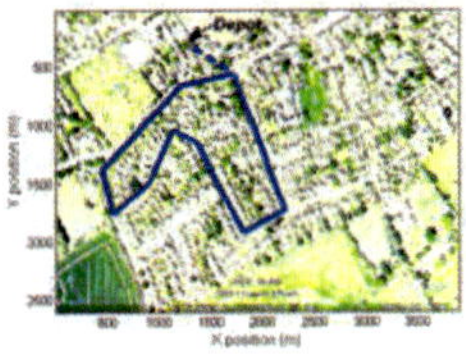

(a) Agent 1

(b) Agent 2

(c) Agent 3

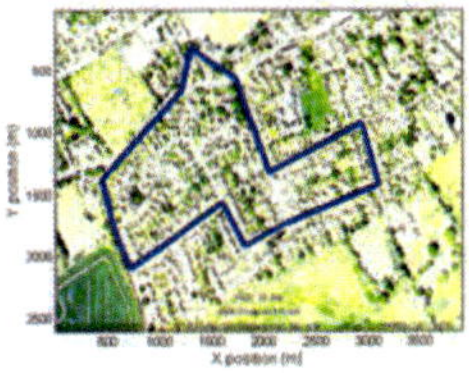

(d) Agent 4

Fig. 3 Search path result of cluster algorithm. Solid line is vehicle path visited once and dashed line is duplicated vehicle path.

4.1 Generation of the Shortest Edge Permutation

First, unordered feasible edge permutation is generated with a given petal size. Petal size means the maximum number of edge that can be visited by one UAV and it is determined by available resources of each UAV. In case that the end vertex of one edge and any vertex of next edge are not connected, connect them with the edge which has shorter distance. Then, shortest order-of-visit edge permutation among the permutations which consist the same edges is computed under the assumption that the path is line.

4.2 Dubins Path Planning

Once shortest edge permutations are determined, next step is to compute and store the cost (path length or flight time) of them. After UAVs complete their mission,

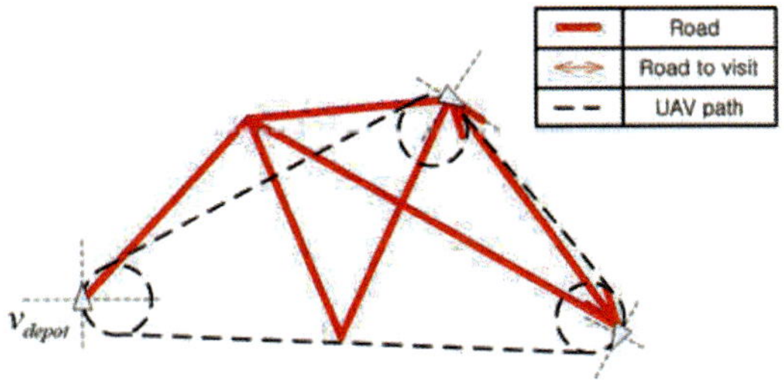

Fig. 4 Road network Search by UAV

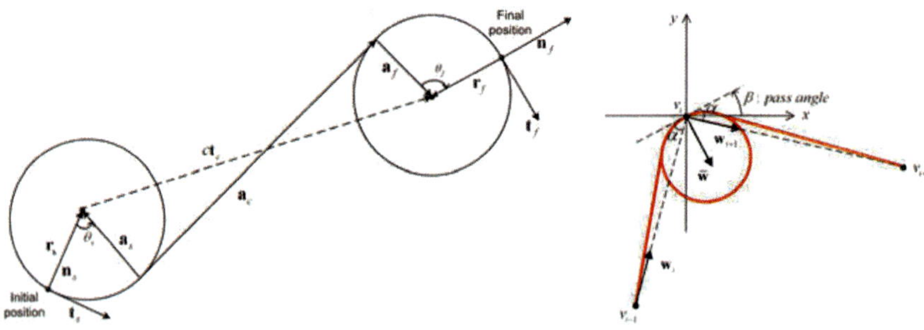

(a) The 2D Dubins arc geometry

(b) Determination of intermediate pass angle

Fig. 5 Geometry of Dubins path planning

since they should come back to the depot, line path from the end of edge permutation to the depot is connected additionally. In this step, this study uses the Dubins path which takes into account the orientation and path constraints of the UAV instead of using Euclidean distance of each edge. The Dubins path is the shortest path connecting two configuration under the constraints of a bound on curvature. It is formed either by concatenation of two circular arcs with their common tangent or by three consecutive tangential circular arcs. Using the principles of differential geometry, the Dubins path can be determined by curvature in two dimensions [18]. For the 2D manoeuvre, the initial and final tangent vectors are coplanar and the straight line manoeuvre is not uniquely defined for this case and it must be calculated. The Dubins arc is shown in Fig. 5(a). The vector sum for the position vector $\boldsymbol{p}$ which is a position of the final point $\boldsymbol{p}_f$ relative to the start point $\boldsymbol{p}_s$ in start axes is given by:

$$\boldsymbol{p} = \boldsymbol{p}_f - \boldsymbol{p}_s = \boldsymbol{r}_s - \boldsymbol{a}_s + \boldsymbol{a}_c + \boldsymbol{a}_f - \boldsymbol{r}_f \tag{3}$$

This equation can be arranged for the vector $\boldsymbol{c}$ which connects the centres of the turn circles as:

$$\boldsymbol{c} = c\boldsymbol{t}_c = \boldsymbol{p} - \boldsymbol{r}_s + \boldsymbol{r}_f = -\boldsymbol{a}_s + \boldsymbol{a}_c + \boldsymbol{a}_f \tag{4}$$

where $\boldsymbol{t}_c$ and c are the centre vector and its length, respectively. The connecting vectors $\boldsymbol{a}_s$, $\boldsymbol{a}_f$ and $\boldsymbol{a}_c$ can be written in terms of the start basis vectors. Then, the centre vector equation, Eqn. (4) now becomes:

$$\boldsymbol{c} = c\boldsymbol{t}_c = \mathbf{R}(\theta_s)' \begin{pmatrix} a \\ \frac{\pm 1}{\kappa_f} - \frac{\pm 1}{\kappa_s} \end{pmatrix} \tag{5}$$

where a is the distance of vector $\boldsymbol{a}_c$, θ_s denotes the rotation angle of the first turn, $\mathbf{R}(\theta_s)$ represents direction cosine matrix, and κ_s and κ_s are the maximum curvature of initial and final turn, respectively. If both initial and final angle are given, a set of four paths are produced and the shortest path is selected. In this paper, the pass angle for the Dubins path is determined by simple geometry. First of all, consider the edge vectors defined by the three vertices v_{i-1}, v, and v_{i+1}, and let path unit vector $\mathbf{w}_i$ and $\mathbf{w}_{i+1}$ and unit vector along the bisector of the angle formed by the three vertices $\bar{\mathbf{w}}$ as in Fig. 5(b). Then, the intermediate pass angle β which makes the path remain near the roads is given by Eqn. (6). Initial and final angle of each permutation are assumed to be zero.

$$\beta = tan^{-1}(\frac{\bar{\mathbf{w}}_y}{\bar{\mathbf{w}}_x}) \pm \frac{\pi}{2} \tag{6}$$

4.3 MMKP Formulation and MILP Optimisation

Final step of the proposed algorithm is to allocate edge permutation to each UAV to cover every edge with the minimum total time. This can be expressed MMKP formulation and given by:

$$\begin{aligned} &min\ J = \sum_{i=1}^{N_{UAV}} \sum_{j=1}^{N_{p_i}} T_j x_{ij} \\ &subject\ to \sum_{i=1}^{N_{UAV}} \sum_{j=1}^{N_{p_i}} E_{kj} x_{ij} \geq 1,\ k \in \{1, \ldots, N_{edge}\} \\ &\sum_{j=1}^{N_{p_i}} x_{ij} = 1,\ i \in \{1, \ldots, N_{UAV}\} \\ &x_{ij} \in \{0, 1\},\ i \in \{1, \ldots, N_{UAV}\},\ j \in \{1, \ldots, N_{p_i}\} \end{aligned} \tag{7}$$

where N_{UAV}, N_{edge}, and N_{p_i} denote the number of UAVs, edges to be visited and permutaions generated by the i-th UAV, respectively. T_j represents mission cost (flight time) of j-th permutation and E_{kj} represents the matrix whose k-th element of j-th permutation is 1 if edge k visited, otherwise 0 and x_{ij} is either 0, implying permutation j of the i-th vehicle is not picked, or 1 implying permutation j of the i-th UAV is picked. First constraint represents that UAVs should visit every edge once or more and second one represents the allocation of the exact one edge permutation to the each UAV. This MMKP problem is solved by SYMPHONY MILP (Mixed Integer Linear Programming) solver [19].

5 Numerical Simulations

To evaluate the performance of the proposed search algorithm, numerical simulation is performed with four UAVs and the map given in Sec. 2. UAVs have the following characteristic:

- Minimum turning radius ρ_{min}: [50 40 30 30] metre
- Maximum cruise speed $V_{c,max}$: [40 35 25 25] m/s

Maximum curvature κ_{max} of UAVs for the Dubins path can be computed by $\kappa_{max} = 1/\rho_{min}$. UAVs are assumed to have maximum cruising speed during entire

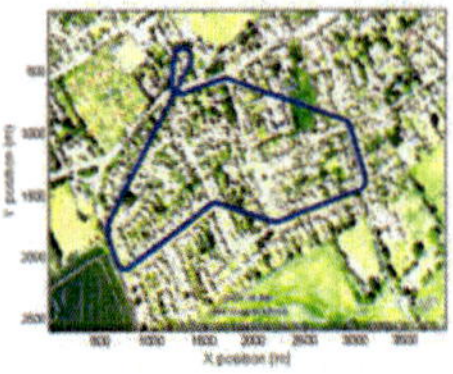

(a) UAV 1

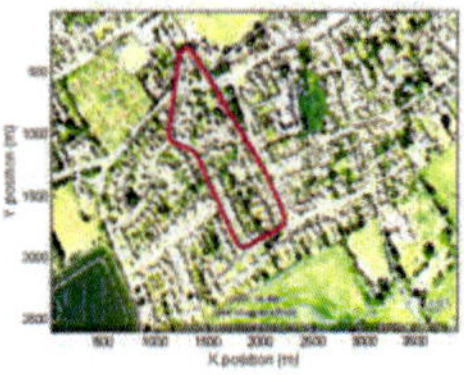

(b) UAV 2

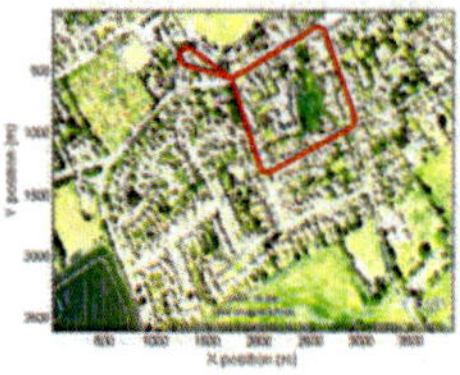

(c) UAV 3

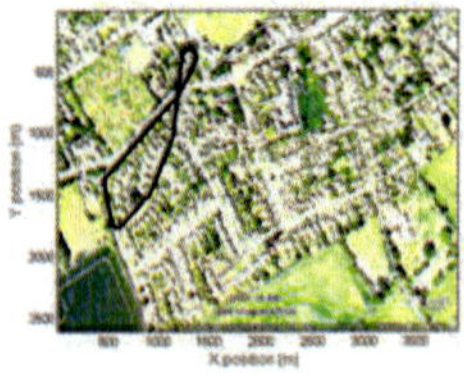

(d) UAV 4

(e) Total UAV paths

Fig. 6 Multiple UAVs road search result

simulation and maximum petal size of edge permutation is set to 4. Figure 6 shows the road network search result. It can be observed that the longest flight path is given UAV 1 whose speed is the fastest among four UAVs and flight path is smooth and flyable due to the Dubins path planning. Moreover, every roads are searched once and UAVs come back to the depot as shown in Fig. 6(e). As explained in the previous section, since the UAV does not need to fly along the road only to cover all edges for the minimum time, the road search result includes additional paths which connect some of edges or the edge and depot.

6 Conclusion

This paper described a coordinated road network search synthesis for multiple UAVs. This paper first defined a road network search problem as a graph and dealt with it using the conventional CPP algorithm. For the application to multiple UAVs, the search algorithm was modified in the consideration of the physical constraint of UAVs. The modified problem was solved by the proposed synthesis which mainly consists of the Dubins path planning and MMKP formulation for the shortest and flyable path with the minimum time solution. The performance of the proposed synthesis was verified via numerical simulation for the given map.

References

1. Gibbons, A.: Algorithmic Graph Theory. Cambridge University Press, Cambridge (1999)
2. Ahr, D.: Contributions to Multiple Postmen Problems. PhD thesis, Heidelberg University (2004)
3. Gross, J.L., Yellen, J.: Handbook of Graph Theory. CRC Press, Boca Raton (2003)

4. Bektas, T.: The multiple traveling salesman problem: An overview of formulations and solution procedures. The International Journal of Management Science 34(3), 209–219 (2006)
5. Bellingham, J., Tillerson, M., Richards, A., How, J.: Multi-Task Allocation and Path Planning for Cooperating UAVs, Cooperative Control: Models, Applications and Algorithms. Kluwer Academic Publishers, Dordrecht (2003)
6. Jin, Y., Liao, Y., Minai, A., Polycarpou, M.: Balancing search and target response in cooperative unmanned aerial vehicle(uav) teams. IEEE Transactions on Systems, Man, and Cybernetics-Part B: Cybernetics 36(3), 571–587 (2006)
7. Chandler, P.R., Pachter, M., Swaroop, D., hwlett, J.M., Rasmussen, S., Schumacher, C., Nygard, K.: Compexity in uav cooperation control. In: American Control Conference, Anchorage, AK (2002)
8. Ryan, J., Bailey, T., Moore, J., Carlton, W.: Reactive tabu search in unmanned aerial reconnaissance simulations. In: 30th Conference on Winter Simulation, Washington, DC (1998)
9. Perrier, N., Langevin, A., Campbell, J.F.: A survey of models and algorithms for winter road maintenance. Computers and Operational Research, Part IV: Vehicle Routing and Fleet Sizing for Plowing and Snow Disposal 34, 258–294 (2007)
10. Easton, K., Burdick, J.: A coverage algorithm for multi-robot boundary inspection. In: IEEE International Conference on Robotics and Automation (2005)
11. Salva, K., Frazzoli, E., Bullo, F.: Traveling salesperson problems for the dubins vehicle. IEEE Trans. on Automatic Control 53(6), 1378–1391 (2008)
12. Rathinam, S., Sengupta, R., Darbha, S.: A resource allocation algorithm for multi-vehicle systems with non holonomic constraints. IEEE Transactions on Automation Sciences and Engineering 4(1), 98–104 (2007)
13. Hifi, M., Michrafy, M., Sbihi, A.: A reactive local search-based algorithm for the multiple-choice multi-dimensional knapsack problem. Computational Optimization and Applications 33, 271–285 (2006)
14. Dubins, L.E.: On curves of minimal length with a constraint on average curvature, and with prescribed initial and terminal positions and tangents. American Journal of Mathematics 79(3), 497–516 (1957)
15. `http://maps.google.co.uk/maps`
16. Frederickson, G.N., Hecht, M.S., Kim, C.E.: Approximation algorithms for some routing problems. SIAM Journal on Computing 7, 178–193 (1978)
17. Edmonds, J., Johnson, E.L.: Matching, euler tours, and the chinese postman. Mathematical Programming 5, 88–124 (1973)
18. Shanmugavel, M., Tsourdos, A., White, B.A., Zbikowski, R.: Differential geometric path planning of multiple uavs. Journal of Dynamic Systems, Measurement and Control 129, 620–632 (2007)
19. `http://www.coin-or.org/SYMPHONY`

Eigenstructure Assignment and Robustness Improvement Using a Gradient-Based Method

Erik Karlsson, Stephan Myschik, and Florian Holzapfel

Abstract. This paper presents a gradient-based method for increasing the robustness of multivariable systems, measured by the stability margins of the broken SISO loops. The method uses iterative perturbations of the closed loop eigenvalues to minimize a certain cost function containing weighted gain and phase margins and the variation of the eigenvalues. The initial and perturbed closed loop dynamics are specified using eigenstructure assignment. The algorithm presented is used in the gain design process of the lateral part of the flight control system for a generic transport aircraft. Results obtained with this novel approach are analyzed regarding the SISO and MIMO stability margins.

1 Introduction

The eigenstructure assignment method for assigning the dynamics of multivariable systems has been widely used in the aerospace sector, especially in the design of aircraft flight control systems ([7], [11], [12]). The method offers great freedom in the eigenvalue placement via state or output feedback. Extra degrees of freedom provided by independent control inputs can be utilized to specify parts of the corresponding eigenvectors. However, with the exact assignment of eigenvalues and eigenvector components, the method features no optimization of the robustness of the closed loop system compared to other multivariable design techniques such as H_∞ loop-shaping ([10]). As the requirements on the closed loop eigenvalues

Erik Karlsson
Research Assistant, Institute of Flight System Dynamics, Technische Universität München
e-mail: erik.karlsson@tum.de

Stephan Myschik
Senior Researcher, Industrieanlagen-Betriebsgesellschaft mbH
e-mail: myschik@iabg.de

Florian Holzapfel
Director of the Institute of Flight System Dynamics, Technische Universität München
e-mail: florian.holzapfel@tum.de

are rarely defined as absolute values, a slight variation is usually allowed. This provides room for increasing the robustness using numerical algorithms.

Several approaches to a more robust eigenstructure assignment have been presented, e.g. multi-objective optimization via genetic algorithms ([5], [6]) or eigenvector projection methods ([9]).

The well-known gain and phase margins for single-input/single-output (SISO) systems have since long been used as a measure of robustness for such systems. In general, they cannot be applied straight forward on the different loops of multiple-input/multiple-output (MIMO) systems. This is due to the fact that the SISO margins only consider uncertainties present in one loop. Uncertainties may be present in multiple loops, and have cross-coupling effects not considered by the SISO margins. The SISO margins have however been extended to multivariable systems using the minimum singular value of the return difference ([2], [3]) and further on using structured singular value analysis ([10], [13]). In flight control applications, stability margin requirements are however given as gain and phase margin requirements on the SISO loops.

This paper presents a first order gradient-based algorithm aiming at optimizing the SISO stability margins. This is done by an iterative perturbation of the closed loop eigenvalues in a direction, which minimizes a cost function containing the SISO margins and the eigenvalue perturbations. The MIMO margins are however also examined in order to ensure robust stability for the entire system.

The paper consists of the following parts. First, an approach to the eigenstructure assignment for a general multivariable system is given, with a pseudo-inverse based projection to find the best achievable eigenvectors ([4]). The MIMO stability margins are presented using the structured singular value of the sensitivity function matrix. Then, a gradient-based algorithm for improving the closed loop stability margins via eigenvalue perturbations is presented. Thereafter, the presented gradient-based algorithm is used in the gain design of the lateral-directional part of a flight control system for a generic transport aircraft. The results are analyzed regarding the SISO and MIMO stability margins. In the last part, conclusions are drawn from the results.

2 Eigenstructure Assignment

The eigenstructure assignment procedure described in this section follows the one given in ([4]), other approaches can be found in ([6], [9], [11], [12]).

Consider a linear, time-invariant, multivariable system on state space form

$$\dot{\mathbf{x}}(t)=\mathbf{A}\mathbf{x}(t)+\mathbf{B}\mathbf{u}(t), \tag{1}$$

with system matrix $\mathbf{A}\in R^{n\times n}$, input matrix $\mathbf{B}\in R^{n\times m}$, state vector $\mathbf{x}\in R^{n}$ and input vector $\mathbf{u}\in R^{m}$. The system has m independent control inputs. Assuming all states are not available for feedback, output feedback is used. The output equation is given by

$$\mathbf{y}(t) = \mathbf{C}\mathbf{x}(t), \tag{2}$$

with output matrix $\mathbf{C} \in R^{r \times n}$ and output vector $\mathbf{y} \in R^r$. The control law

$$\mathbf{u}(t) = -\mathbf{K}\mathbf{y}(t) + \mathbf{u}_C(t), \tag{3}$$

with the constant feedback matrix $\mathbf{K} \in R^{m \times r}$ and commanded input $\mathbf{u}_C(t)$ leads to the closed loop dynamics

$$\dot{\mathbf{x}}(t) = (\mathbf{A} - \mathbf{BKC})\mathbf{x}(t) + \mathbf{B}\mathbf{u}_C(t) = \tilde{\mathbf{A}}\mathbf{x}(t) + \mathbf{B}\mathbf{u}_C(t). \tag{4}$$

Purpose of the control design is now to determine the feedback gain matrix $\mathbf{K}$ in such a way that the closed loop system matrix $\tilde{\mathbf{A}} = (\mathbf{A} - \mathbf{BKC})$ is given the desired dynamics. The closed loop eigenvalues λ_i and corresponding right eigenvectors $\mathbf{v}_i$ are determined by the equation (5).

$$(\mathbf{A} - \mathbf{BKC}) \cdot \mathbf{v}_i = \mathbf{v}_i \cdot \lambda_i, \; i = 1,\dots,n. \tag{5}$$

Equation (5) can be rewritten on matrix form to include the input directions $\mathbf{z}_i = \mathbf{KC}\mathbf{v}_i$. This leads to equation (6).

$$[\lambda_i \mathbf{I} - \mathbf{A} \quad \mathbf{B}] \cdot \begin{bmatrix} \mathbf{v}_i \\ \mathbf{KC}\mathbf{v}_i \end{bmatrix} = [\lambda_i \mathbf{I} - \mathbf{A} \quad \mathbf{B}] \cdot \begin{bmatrix} \mathbf{v}_i \\ \mathbf{z}_i \end{bmatrix} = \mathbf{0}. \tag{6}$$

The desired eigenvalues of the closed loop system must not be the same as those of the open loop system, i.e. the matrix $[\lambda_i \mathbf{I} - \mathbf{A} \quad \mathbf{B}]$ must have full rank (be invertible). All nontrivial solutions $\mathbf{v}_i$, $\mathbf{z}_i$ to the equation system (6) must be within the null space (the kernel) of the matrix $[\lambda_i \mathbf{I} - \mathbf{A} \quad \mathbf{B}]$. This means that any possible solution can be described as a linear combination of the base vectors $\bar{\mathbf{n}}_i$ of the null space. The null space base vectors can be divided into one upper part $\mathbf{n}_i$ related to the eigenvectors, and a lower part $\hat{\mathbf{n}}_i$ related to the input vectors.

$$\bar{\mathbf{N}}_{\lambda_i} = [\bar{\mathbf{n}}_{1,\lambda_i} \quad \cdots \quad \bar{\mathbf{n}}_{m,\lambda_i}] = \begin{bmatrix} \mathbf{n}_{1,\lambda_i} & \cdots & \mathbf{n}_{m,\lambda_i} \\ \hat{\mathbf{n}}_{1,\lambda_i} & \cdots & \hat{\mathbf{n}}_{m,\lambda_i} \end{bmatrix} = \begin{bmatrix} \mathbf{N}_{\lambda_i} \\ \hat{\mathbf{N}}_{\lambda_i} \end{bmatrix}. \tag{7}$$

Now, the achievable eigenvectors can be written as

$$\begin{bmatrix} \mathbf{v}_i \\ \mathbf{z}_i \end{bmatrix} = [\bar{\mathbf{n}}_{1,\lambda_i} \cdots \bar{\mathbf{n}}_{m,\lambda_i}] \cdot \begin{bmatrix} l_{1,\lambda_i} \\ \vdots \\ l_{m,\lambda_i} \end{bmatrix} = \bar{\mathbf{N}}_{\lambda_i} \cdot \mathbf{l}_{\lambda_i} = \begin{bmatrix} \mathbf{N}_{\lambda_i} \\ \hat{\mathbf{N}}_{\lambda_i} \end{bmatrix} \cdot \mathbf{l}_{\lambda_i}, \tag{8}$$

where $\mathbf{l}_{\lambda_i}$ is an arbitrary m-dimensional parameter vector. Sorting matrices $(\mathbf{P}_i^S)_{s \times n}$ and $(\mathbf{P}_i^U)_{(n-s) \times n}$ are introduced to sort the elements of the eigenvectors in those specified (number of specified elements is s) and those unspecified (number of unspecified elements is hence $n-s$). Elements of greater importance are specified, whereas lesser important elements remain unspecified. This leads to equation (9).

$$\begin{bmatrix} \left(\mathbf{v}_i^S\right)_{s\times 1} \\ \left(\mathbf{v}_i^U\right)_{(n-s)\times 1} \end{bmatrix} = \begin{bmatrix} \left(\mathbf{P}_i^S\right)_{s\times n} \\ \left(\mathbf{P}_i^U\right)_{(n-s)\times n} \end{bmatrix} \cdot \mathbf{v}_i \,. \tag{9}$$

Through the use of a weighting matrix $\mathbf{Q}_{\lambda_i}$ the elements of the desired eigenvector can be assigned greater or lower importance, giving the weighted desired eigenvectors

$$\overline{\mathbf{v}}_{id}^S = \mathbf{Q}_{\lambda_i} \cdot \mathbf{v}_{id}^S \,, \tag{10}$$

and the relation between the desired eigenvectors $\mathbf{v}_{id}^S$ and the parameter vector $\mathbf{l}_{\lambda_i}$, described by equation (11).

$$\left(\mathbf{Q}_{\lambda_i}\right)_{s\times s} \cdot \left(\mathbf{v}_{id}^S\right)_{s\times 1} = \left(\mathbf{Q}_{\lambda_i}\right)_{s\times s} \cdot \left(\mathbf{N}_{\lambda_i}^S\right)_{s\times m} \cdot \left(\mathbf{l}_{\lambda_i}\right)_{m\times 1} \,. \tag{11}$$

Now two different cases are possible. If the number of elements to be specified is equal to the number of independent control effectors available, i.e. $s = m$, an exact assignment of the specified eigenvector elements is possible. In this case the parameter vector is given by

$$\mathbf{l}_{\lambda_i} = \left[\mathbf{Q}_{\lambda_i} \cdot \mathbf{N}_{\lambda_i}^S\right]^{-1} \cdot \mathbf{Q}_{\lambda_i} \cdot \mathbf{v}_{id}^S \,. \tag{12}$$

If the number of elements to be specified is larger than the number of control effectors available, i.e. $s > m$, no exact assignment is possible. By using a (Moore-Penrose) pseudo-inverse based projection onto the null space, the parameter vector can be calculated according to

$$\mathbf{l}_{\lambda_i} = \left[\left(\mathbf{N}_{\lambda_i}^S\right)^H \cdot \mathbf{Q}_{\lambda_i} \cdot \mathbf{N}_{\lambda_i}^S\right]^{-1} \cdot \left(\mathbf{N}_{\lambda_i}^S\right)^H \cdot \mathbf{Q}_{\lambda_i} \cdot \mathbf{v}_{id}^S \,. \tag{13}$$

With the best achievable parameter vector known (either through exact assignment or approximated through projection on the null space), the achievable eigenvector and the corresponding input directions are given by

$$\mathbf{z}_i = \hat{\mathbf{N}}_{\lambda_i} \cdot \mathbf{l}_{\lambda_i} \,, \quad \mathbf{v}_i = \mathbf{N}_{\lambda_i} \cdot \mathbf{l}_{\lambda_i} \,. \tag{14}$$

This leads to the desired gain matrix, shown in equation (15).

$$\mathbf{K} = \left[\mathbf{z}_1 \quad \mathbf{z}_2 \quad \cdots \quad \mathbf{z}_r\right] \cdot \left(\mathbf{C} \cdot \left[\mathbf{v}_1 \quad \mathbf{v}_2 \quad \cdots \quad \mathbf{v}_r\right]\right)^{-1} . \tag{15}$$

The resulting gain matrix now places the closed loop eigenvalues as desired, and yields the specified eigenvector elements or their best approximation.

3 Robustness Enhancement Using a Gradient-Based Method

The method described in this section aims at increasing the gain and phase margins of the broken SISO loops of the closed loop system. The robustness of such a system can either be measured by the mentioned stability margins of each broken loop, assuming uncertainties only occur in one loop at a time. Or, more general, using the conservative multivariable stability margins, which consider concurrent uncertainties in multiple loops ([1]).

The multivariable stability margins are in the following given using the structured singular value of the sensitivity function matrix.

Consider a generalized multiple-input/multiple-output system with plant $\mathbf{G}$ and constant feedback matrix $\mathbf{K}$, extended by an inverse multiplicative input uncertainty matrix $\mathbf{\Delta} = diag\,\{\Delta_1,\ldots,\Delta_m\}$ containing uncertain elements $\Delta_k = r_k \cdot e^{i\phi_k}$, $k = 1,\ldots,m$ (uncertain complex numbers located within a circle of radius r_k [10]), see Fig. 1. The transfer function from the output $\mathbf{w}$ of the uncertainty block to the input $\mathbf{z}$ is then given by

$$\mathbf{z} = [\mathbf{I} + \mathbf{K} \cdot \mathbf{G}]^{-1} \cdot \mathbf{w} = \mathbf{S} \cdot \mathbf{w}\,, \tag{16}$$

where $\mathbf{S}$ is the sensitivity function at the actuator inputs. The system can now be represented by a structure containing only $\mathbf{S}$ and $\mathbf{\Delta}$, connected via $\mathbf{w}$ and $\mathbf{z}$. Assuming $\mathbf{S}$ is stable, robust stability for such a system is guaranteed (for all allowed perturbations) if and only if the Nyquist plot of $\det[\mathbf{I} - \mathbf{S\Delta}(j\omega)]$ does not encircle the origin, i.e. when $\det[\mathbf{I} - \mathbf{S\Delta}(j\omega)] \neq 0$, $\forall\omega, \forall\mathbf{\Delta}$ ([10]). The structured singular value $\mu(\mathbf{S})$ of the sensitivity function is defined as the inverse of the smallest gain c which destabilizes the system ($\mu = 1/c$), i.e. for which $\det[\mathbf{I} - c\mathbf{S\Delta}(j\omega)] = 0$. This definition of robust stability of the $\mathbf{S\Delta}$-system can now be written as

$$\mu(\mathbf{S}(j\omega)) \cdot \bar{\sigma}(\mathbf{\Delta}(j\omega)) < 1\,,\ \forall\omega\,. \tag{17}$$

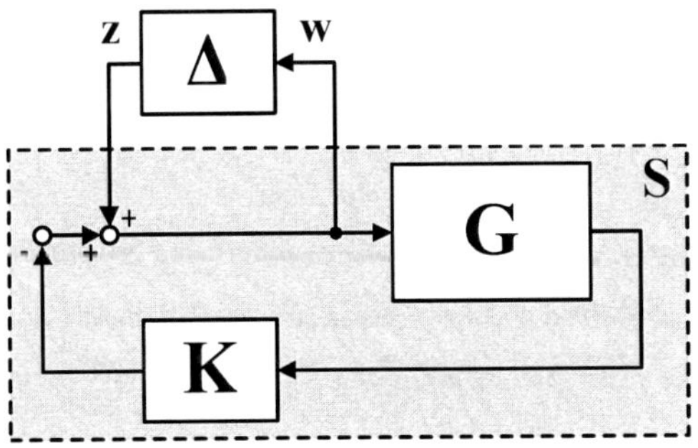

Fig. 1 Multivariable feedback system with uncertainties.

This means that robust stability is guaranteed when the product of the structured singular value of the sensitivity function and the maximum singular value of the uncertainty matrix is less than 1 for all frequencies (a "generalized small gain theorem") ([10]). Since $\mathbf{S}(j\omega)$ is a frequency dependent function matrix, robust stability is ensured only when the maximum singular value of the uncertainty matrix is smaller than the minimum value of the inverted structured singular value of the sensitivity function over all frequencies. This leads to the relation (18).

$$\bar{\sigma}(\mathbf{\Delta}) < k_{\min} = \min_{\omega \in R} \frac{1}{\mu(\mathbf{S}(j\omega))}. \tag{18}$$

With the definition of the minimum value of the inverted structured singular value of the sensitivity function matrix, $k_{\min}$, the multivariable gain and phase margins are given in ([3]). They are defined in equation (19).

$$\mathrm{GM} = \frac{1}{1 \pm k_{\min}}, \quad \mathrm{PM} = \pm\theta \text{ with } \theta = 2 \cdot \arcsin\left(\frac{k_{\min}}{2}\right). \tag{19}$$

The different loops of the system may be disturbed by a set of gains Δ satisfying the condition $1/(1+k_{\min}) < \Delta < 1/(1-k_{\min})$, or by phases θ satisfying the condition $|\theta| < 2\arcsin(k_{\min}/2)$, without leading to an unstable closed loop system.

The aim is now to perturb the closed loop eigenvalues, given by the eigenstructure assignment, in order to increase the gain and phase margins of the broken loops of the system. The loops are cut both at the actuator inputs and at the sensor outputs. When one loop is cut open, the others remain closed. Assuming the number of actuator inputs is q and the number of sensor outputs is r, the task can be formulated as a constrained optimization problem with the cost function to be minimized given by

$$\begin{aligned} E(\mathbf{p}) = -\frac{1}{2}\mathbf{v}_{GM}^T(\mathbf{p})\mathbf{W}_{GM}\mathbf{v}_{GM}(\mathbf{p}) - \frac{1}{2}\mathbf{v}_{PM}^T(\mathbf{p})\mathbf{W}_{PM}\mathbf{v}_{PM}(\mathbf{p}) + \\ + \frac{1}{2}(\mathbf{p}-\mathbf{p}_d)^T\mathbf{W}_p(\mathbf{p}-\mathbf{p}_d) \end{aligned} \tag{20}$$

with vectors $\mathbf{v}_{GM}, \mathbf{v}_{PM} \in R^{q+r}$ containing the cut loop gain and phase margins. The parameter vector $\mathbf{p} \in R^l$ contains those of the closed loop eigenvalues allowed to be varied (or some parameter given by the eigenvalues, e.g. relative damping or natural frequency, time constants etc.). The vector $\mathbf{p}_d \in R^l$ consists of the corresponding desired values. Weighting matrices $\mathbf{W}_{GM}, \mathbf{W}_{PM} \in R^{(q+r)\times(q+r)}$ consider the gain and phase margins and $\mathbf{W}_p \in R^{l\times l}$ weights the eigenvalue variations. All weighting matrices are diagonal and positive definite. The weighting matrices can either be identity matrices, or chosen to emphasize the improvement of certain gain or phase margins. One approach is to specify the weighting coefficients according to the relative distance of the corresponding margin from a certain required value. The smaller the margin, the more weighted it is. Thus, the algorithm focuses on improving the worst margins.

The constraint under which the cost function is to be minimized is that the eigenvalues remain within the boundaries defined by certain maximum and minimum values, $\mathbf{p}_{\min} \leq \mathbf{p} \leq \mathbf{p}_{\max}$. The cost function is iteratively minimized using a variation of the eigenvalue parameter vector given by the steepest descent direction,

$$\Delta\mathbf{p} = -\eta \cdot \frac{\partial E(\mathbf{p})}{\partial \mathbf{p}}, \tag{21}$$

with the step length $\eta > 0$. The step length selection is a tradeoff between a sufficient reduction of the cost function and the computation time of finding the appropriate step length. The selection method chosen is based on the Armijo sufficient decrease condition ([8]), ensuring that the step length is not chosen too small. The step size is iteratively increased from an initial value until an appropriate step length is found, fulfilling the condition

$$E(\mathbf{p} - \eta\nabla E) \leq E(\mathbf{p}) - c\eta\nabla E^T \nabla E, \tag{22}$$

for a constant $c \in [0,1]$. A maximum step length is also defined in order to limit the selection computation time. When performing a gain design over thousands of grid points, the number of iterations in each point greatly influences the total computation time. The gradient of the cost function is given by equation (23).

$$\frac{\partial E(\mathbf{p})}{\partial \mathbf{p}} = -\left[[\mathbf{W}_{GM}\mathbf{v}_{GM}]^T \cdot \left[\frac{\partial \mathbf{v}_{GM}}{\partial \mathbf{p}} \right] \right]^T - \left[[\mathbf{W}_{PM}\mathbf{v}_{PM}]^T \cdot \left[\frac{\partial \mathbf{v}_{PM}}{\partial \mathbf{p}} \right] \right]^T + \\ + \mathbf{W}_p(\mathbf{p} - \mathbf{p}_d). \tag{23}$$

The Jacobian matrices of the gain and phase margin vectors,

$$\left[\frac{\partial \mathbf{v}_{GM}}{\partial \mathbf{p}}\right] = \begin{bmatrix} \frac{\partial v_{1,GM}}{\partial p_1} & \cdots & \frac{\partial v_{1,GM}}{\partial p_l} \\ \vdots & \ddots & \vdots \\ \frac{\partial v_{q+r,GM}}{\partial p_1} & \cdots & \frac{\partial v_{q+r,GM}}{\partial p_l} \end{bmatrix}, \quad \left[\frac{\partial \mathbf{v}_{PM}}{\partial \mathbf{p}}\right] = \begin{bmatrix} \frac{\partial v_{1,PM}}{\partial p_1} & \cdots & \frac{\partial v_{1,PM}}{\partial p_l} \\ \vdots & \ddots & \vdots \\ \frac{\partial v_{q+r,PM}}{\partial p_1} & \cdots & \frac{\partial v_{q+r,PM}}{\partial p_l} \end{bmatrix}, \tag{24}$$

have to be determined through numerical differentiation, e.g. using Newton's difference quotient, given by

$$\frac{\partial v_i}{\partial p_j} = \frac{v_i(\mathbf{p} + \delta\mathbf{p}_j) - v_i(\mathbf{p})}{\delta p_j}, \quad i = 1,\ldots,q+r, \quad j = 1,\ldots,l. \tag{25}$$

Here $\delta\mathbf{p}_j$ is a vector with a small step δp_j in the j:th parameter. When the parameter vector is varied by $\Delta\mathbf{p}$, the value of the cost function is decreasing according to

$$\Delta E = \left[\frac{\partial E}{\partial \mathbf{p}}\right]^T \cdot \Delta \mathbf{p} = -\eta \cdot \left[\frac{\partial E}{\partial \mathbf{p}}\right]^T \left[\frac{\partial E}{\partial \mathbf{p}}\right] \leq 0\,. \tag{26}$$

If the calculated variation $\Delta\mathbf{p}$ leads to a parameter vector not satisfying the constraints, the elements exceeding are adjusted and set to the closest bound.

This algorithm leads to an improvement of the included stability margins under the given constraints on the eigenvalue variation.

4 Implementation and Results

The eigenstructure assignment method is implemented in the gain design of the lateral control system for a generic transport aircraft, described in detail in [4]. The model dynamics is linearized for horizontal, straight-forward flight at an altitude of 6000 m and a Mach number of 0.4. The aircraft configuration features a total mass of 110 tons. Flaps and gear are retracted.

The inherent dynamics has roll and spiral mode time constants $T_R = 0.99$ s and $T_S = 27.4$ s. The Dutch roll dynamics is given by the relative damping $\zeta_{DR} = 0.171$ and natural frequency $\omega_{0,DR} = 0.633$ rad/s.

A control allocation scheme converts the lateral control effectors aileron, rudder and roll spoilers into two independent virtual controls for the creation of rolling and yawing moments, $\mathbf{u} = [\delta_{Roll}, \delta_{Yaw}]^T$, giving the possibility to decouple the roll and yaw axis.

The initial roll mode time constant, $T_{R,d} = 1.4$ s, ensures level 1 handling qualities. With this choice, level 1 handling qualities for the given aircraft can still be achieved while obeying maximum control surface deflections in the linear gain design process. An inherently stable spiral mode ($T_{S,d} = 1.0$ s) is chosen to provide automatic wings leveling for small bank angles and bank angle resistance to disturbances. A PI-command filter, not described in this paper, provides neutral spiral stability from the pilot's point of view ([4]). The desired Dutch roll dynamics is given by the relative damping $\zeta_{DR,d} = \sqrt{2}/2$ and a natural frequency $\omega_{0,DR,d} = 2.0$ rad/s.

The linearized lateral model contains the stability axis roll and yaw rate, the sideslip and roll angle, the integrator states for the control error integration as well as the position and translational states of the actuators. The output vector used for feedback contains all but the actuator states, $y = [r_S, \beta, p_S, \Phi, \Phi_I, \beta_I]^T$. The two independent control variables make it possible to exactly assign two elements of each eigenvector.

Roll rate and angle are set to zero in the Dutch roll eigenvector, yaw rate and angle of sideslip to zero in the roll and spiral mode eigenvectors.

After the initial eigenstructure assignment, the eigenvalues are perturbed iteratively and reassigned, with the parameter vector $\mathbf{p} = [T_S, T_R, \zeta_{DR}, \omega_{0,DR}]^T$

Table 1 Stability margins before and after robustness enhancement.

		Before	After	Relative change
MIMO	GM [dB]	[-5.40, 17.21]	[-5.49, 18.46]	[+1.7%, +7.3%]
	PM [°]	51.07	52.25	+2.3%
Roll	GM [dB]	32.36	34.72	+7.3%
	PM [°]	71.15	72.35	+1.7%
Yaw	GM [dB]	26.54	27.42	+3.3%
	PM [°]	57.41	62.03	+8.1%
r_s	GM [dB]	27.50	28.18	+2.5%
	PM [°]	78.77	81.24	+3.1%
β	GM [dB]	27.44	29.63	+8.0%
	PM [°]	57.79	62.75	+8.6%
p_s	GM [dB]	32.89	35.21	+7.1%
	PM [°]	108.21	119.83	+10.7%
Φ	GM [dB]	28.04	29.79	+6.2%
	PM [°]	61.20	62.14	+1.5%

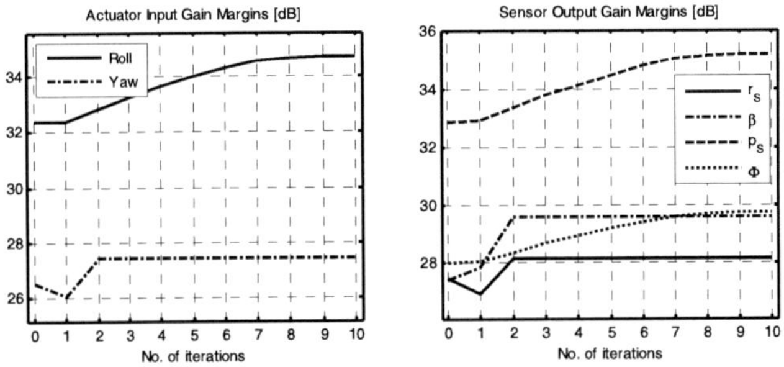

Fig. 2 Gain margins for the broken SISO loops, measured at the actuator inputs and sensor outputs.

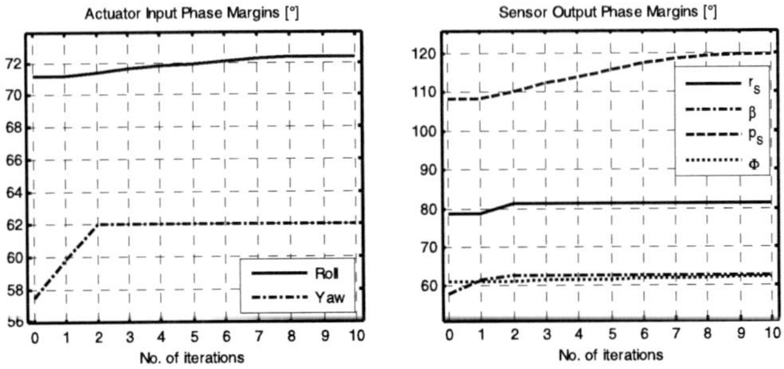

Fig. 3 Phase margins for the broken SISO loops, measured at the actuator inputs and sensor outputs.

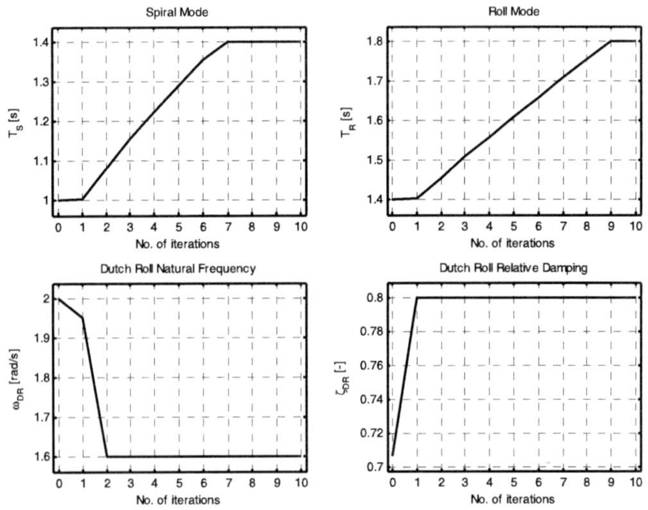

Fig. 4 Curves for the parameter vector elements.

containing the roll and spiral mode time constants and the Dutch roll dynamics. The elements of the parameter vector are bounded, constraining the adjusted closed loop dynamics to remain within specified limits, in this case $T_S \leq 1.4$ s, $T_R \leq 1.8$ s, $\zeta_{DR} \leq 0.8$ and $\omega_{0,DR} \geq 1.6$.

The SISO stability margins before and after the iterations are presented in Table 1 together with the MIMO margins. All gain and phase margins have been improved by the algorithm. Especially large improvements can be seen in the phase margins of the yaw control input and the roll rate and angle of sideslip sensor outputs. Fig. 2 shows how the gain margins at the actuator inputs and sensor outputs are improved during the iteration. Fig. 3 depicts the corresponding phase margins. It can be observed how the roll command and the roll rate and angle sensor gain margins show similar curves during the iteration. When compared to the changing closed loop dynamics described in Fig. 4, the roll gain margins are depending on the roll and spiral mode time constants. The roll command and roll rate phase margins have the same dependence in Fig. 3. Gain margins of yaw command and yaw rate sensor are at first decreasing as the Dutch roll relative damping is increasing. When the damping reaches its maximum allowed value, the yaw margins start to increase again, and keeps increasing until the Dutch roll natural frequency reaches its minimum value.

5 Conclusions

The presented gradient-based algorithm offers a simple and fast way of improving the stability margins of a multivariable system via iterative perturbations of the closed loop eigenvalues. It has been successfully implemented in combination with an eigenstructure assignment algorithm and is used in the gain design process of a flight control system for the lateral motion of a generic transport aircraft. The algorithm has been shown to converge to the best eigenvalues, minimizing the cost function under the given constraints. The stability margins of the SISO loops as well as the multivariable stability margins are increased. By examining how the stability margins vary with the parameters of the closed loop dynamics, relations between specific margins and time constants or damping/frequencies can be identified.

The focus of the presented method has been the applicability to real control design problems. The method has been proven useful on a typical problem in flight control design: vary the closed loop dynamics within limits in order to achieve improved gain and phase margins.

References

1. Bar-on, J.R., Adams, R.J.: Multivariable Gain and Phase Margin Analysis of a Fully Coupled Six-Degree-of-Freedom Guided Missile. In: Proceedings of the 1999 IEEE Intl. Conference on Control Applications, pp. 152–157 (1999)
2. Doyle, J.C., Stein, G.: Multivariable Feedback Design: Concepts for a Classical/ Modern Synthesis. IEEE Transactions on Automated Control AC-26(1), 4–16 (1981)
3. Faleiro, L., Magni, J.-F., et al.: Eigenstructure Assignment. In: Magni, et al. (eds.) Robust Flight Control: A Design Challenge. Springer, Heidelberg (1997)
4. Holzapfel, F., da Costa, O., et al.: Development of a Lateral-Directional Flight Control System for a New Transport Aircraft. In: AIAA Guidance, Navigation and Control Conference and Exhibit, Keystone, Colorado, August 21-24 (2006)
5. Liu, G.P., Patton, R.J.: Robust control design via eigenstructure assignment, genetic algorithms and gradient-based optimisation. In: IEE Proceedings on Control Theory Applications, vol. 141(3), pp. 202–208 (1994)
6. Liu, G.P., Patton, R.J.: Robust control design using eigenstructure assignment and multi-objective optimisation. International Journal of Systems Science 27(9), 871–879 (1996)
7. Mengali, G.: Role of Eigenvectors in Aircraft Dynamics Optimization. AIAA Journal of Guidance, Control and Dynamics 26(2), 340–346 (2003)
8. Nocedal, J., Wright, S.J.: Numerical Optimization, 2nd edn. Springer, Heidelberg (2006)
9. Rew, D.W., Junkins, J.L.: Robust Eigenstructure Assignment by a Projection Method: Applications Using Multiple Optimization Criteria. AIAA Journal of Guidance, Control and Dynamics 12(3), 396–403 (1988)
10. Skogestad, S., Postlethwaite, I.: Multivariable Feedback Control – Analysis and Design. John Wiley & Sons, Chichester (1996)
11. Sobel, K.M., Lallman, F.J.: Eigenstructure Assignment for the Control of Highly Augmented Aircraft. AIAA Journal of Guidance, Control and Dynamics 12(3), 318–324 (1988)
12. Sobel, K.M., Shapiro, E.Y.: Application of Eigenstructure Assignment to Flight Control Design: Some Extensions. AIAA Journal of Guidance, Control and Dynamics 10(1), 73–81 (1986)
13. Tsao, T.T., Lee, F.C., Augenstein, D.: Relationship between Robustness μ-analysis and Classical Stability Margins. In: Proceedings of the 1998 IEEE Aerospace Conference, vol. 4(1), pp. 481–486 (1998)

Helical Flight Path Trajectories for Autopilot Evaluation

Gertjan Looye

Abstract. A helical flight path trajectory (helix) involves flying exact circles over the ground while climbing or descending at a given flight path angle and speed profile. The manoeuvre is challenging to fly in windy conditions, since the path reference is inertial whereas the aircraft naturally tends to move with the air mass. Tracking a helix introduces periodical lateral and longitudinal wind shears in turn. This makes the helix an excellent manoeuvre for testing autopilot control laws, allowing to evaluate co-ordination of longitudinal and lateral modes, tracking accuracy along a curved flight path, combined tracking of inertial (flight path) and air mass-based references (airspeed), and to evaluate the trade-off between behaviour in turbulence and wind shear. Since helical flight path trajectories are not a standard option in most autopilot / flight management systems, this chapter derives generally applicable reference variables and high-level control strategies for use with typical autopilot structures. This allows the reader to fly the helix manoeuvre using his or her own autopilot design. As an example, simulation and flight test results for an autopilot developed for DLR´s test aircraft ATTAS will be discussed.

Nomenclature

m	aircraft mass	β	side slip angle	e	error
g	acceleration of gravity	χ	track angle (course)	ref	reference
h	altitude	γ	flight path angle	A	air mass referenced
n	number	μ	bank angle	AC	aircraft
r	position vector	ϕ	roll attitude angle	AP	autopilot
v	velocity vector	θ	pitch attitude angle	CAS	calibrated airspeed
y	lateral position	τ	filter time constant	I	inertial
D	distance			W	wind
N_1	engine fan shaft speed	*Abbreviations, subscripts*			
R	radius	c	commanded	*(See also Table 1)*	
V	velocity	compl	complementary		

Gertjan Looye
German Aerospace Center, DLR–Oberpfaffenhofen, Institute of Robotics and Mechatronics
e-mail: `Gertjan.Looye@dlr.de`

1 Introduction

A major challenge in the design of autopilot flight control laws is to achieve satisfactory closed-loop aircraft performance during severe atmospheric disturbances, like wind shears, turbulence and wake vortices. Especially in the case of passenger aircraft, besides tracking accuracy of velocity and flight path, careful attention has to be paid to aspects such as ride comfort, control (throttle!) activity, and safety limits.

During several projects the DLR Institute of Robotics and Mechatronics was involved in, helical flight path trajectories were found to be extremely useful to assess design aspects listed above. During a helix manoeuvre, one basically encounters a considerable amount of challenges in autopilot design: co-ordination of longitudinal and lateral modes, accurate tracking of a flight path that is continuously "bending away", and combined tracking of inertial (the flight path) and air mass-based references (the airspeed). From an atmospheric point of view, during a helix manoeuvre even little constant wind results in continuous wind shears in lateral and longitudinal directions.

The idea of performing the helix test manoeuvre came from publications by Kaminer and Lambregts. Kaminer flew small UAVs along helical flight paths for observation purposes [1], also proposing the trajectory as a benchmark. Lambregts demonstrates the ability to accurately fly circular ground tracks in windy conditions using a Control Wheel Steering mode developed in the NASA TCV program [2].

This chapter will focus on the helix manoeuvre as a benchmark trajectory for autopilot control laws. Using flight mechanics principles, generally applicable reference variables and high-level control strategies for autopilot control laws are derived, allowing the reader to make his or her autopilot to fly this manoeuvre for testing purposes. As an example, the results from simulations and a recent flight test using DLR´s ATTAS aircraft will be discussed, highlighting interesting design aspects that are typically revealed by flying the proposed benchmark trajectory.

2 The Helix Trajectory

This section geometrically describes the helix trajectory and discusses relevant flight mechanical aspects. The trajectory as used in this work is depicted in Fig. 1. It starts at the point *Init* and ends at the point *Final*. The over-all manoeuvre begins with a straight segment. This allows for easy capture by the aircraft using standard functionality in the autopilot. Geometrically, the actual helix is initiated at the point where the initial straight segment perpendicularly passes the helix centre. After the helical segment, the trajectory ends with a straight segment, providing a clear and safe point to stop the manoeuvre by making sure the aircraft is in trimmed, wings level flight. Handling the discontinuous transitions between the straight path segments and the actual helix will be discussed in Section 3.3.

The trajectory is defined by the parameters in Table 1. The objective for an autopilot is to capture the initial straight segment of the helix manoeuvre, track the inertial helix flight path at a commanded V_{CAS} (which may be varied during the task,

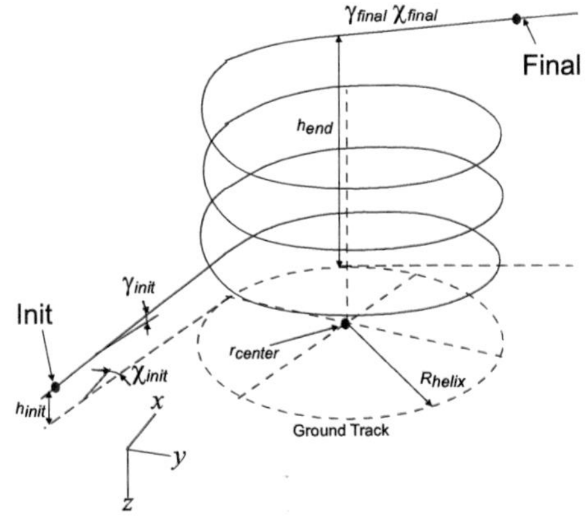

Fig. 1 Definition of the helix trajectory

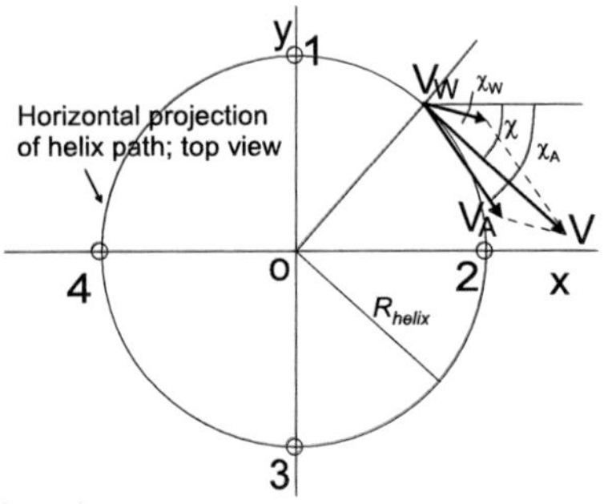

Fig. 2 Top view of helical trajectory (straight segments have been omitted)

Table 1 Parameters describing the helix trajectory in Fig. 1

Helical segment:		Initial and final straight segments:		
r_{center}	GPS position of the helix center			
R_{helix}	Helix radius			
$VCAS_{helix}$	Calibrated airspeed	$VCAS_{init}$	$VCAS_{final}$	Calibrated airspeed
h_{end}	Altitude at which helix ends	h_{init}	–	Initial altitude
n_{turn}	Number of full turns	γ_{init}	γ_{final}	Inertial flight path angle
dir	Direction (counter-) clockwise	χ_{init}	χ_{final}	Inertial track angle

of course), and return to steady symmetric flight, ending at the point marked *Final*. The challenge is to make sure the ground track is a single circle, even in case of steady wind. The control strategy to achieve this can be derived from basic flight mechanics principles, as will be discussed in the following.

Fig. 2 shows the true airspeed vector v_A (magnitude V_A), wind vector v_W (magnitude V_W) and the inertial speed vector v (magnitude V, necessarily tangent to the circle). The variables $V_W$and χ_W are given wind speed and direction respectively. Furthermore, the variables V_A and χ are the momentary airspeed and track angle on the reference trajectory respectively. It is of interest to compute V and the crab angle $\chi_{crab} = \chi_A - \chi$ (χ_A as defined in Fig. 2) that are necessary to stay on the circular path. Assuming that $cos\gamma_A \approx 1, cos\gamma_W \approx 1, cos\gamma \approx 1$ (with γ_A the air mass-referenced flight path angle), it is easy to derive the following two equations:

$$V = V_W \cos(\chi_W - \chi) + \sqrt{V_A^2 - V_W^2 \sin^2(\chi_W - \chi)} \quad (1)$$

$$\chi_{crab} = -\arcsin\left(\frac{V_w \sin(\chi_W - \chi)}{V_A}\right) \quad (2)$$

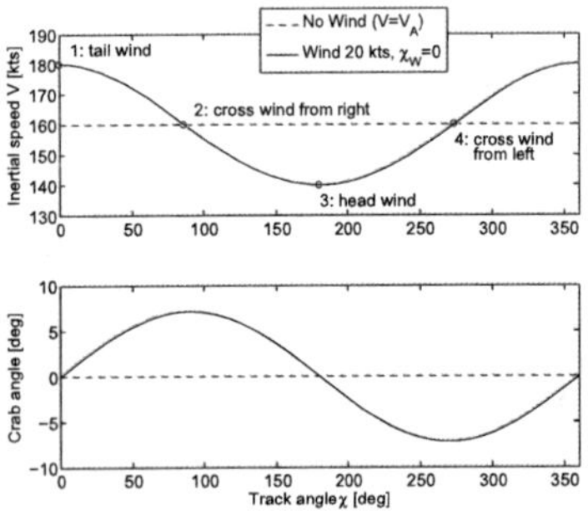

Fig. 3 Inertial speed and crab angle as a function of the track angle for fixed airspeed

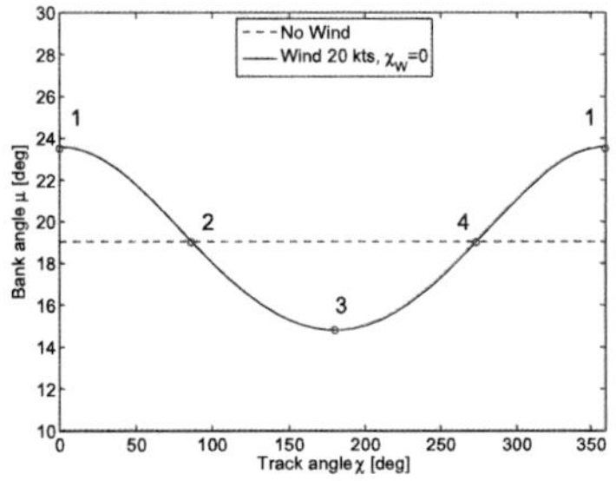

Fig. 4 Bank angle required to maintain helix radius ($V_A = 160$ kts, $R_{helix} = 2000$ m, $\gamma = 0$ deg)

The inertial speed and crab angle thus vary as a function of the angular position of the aircraft on the helix path. This is illustrated in Fig. 3, where the inertial speed and crab angle have been plotted as a function of the track angle during a full turn. The numbered wind conditions correspond with Fig. 2.

The term $V_w \sin(\chi_W - \chi)$ in equation 2 is the lateral wind component w.r.t. the inertial flight path (positive to the right). Its time derivative is the lateral wind shear that has to be compensated for:

$$-V_W \cos(\chi_W - \chi)\dot{\chi}, \quad \text{with:} \quad \dot{\chi} = \frac{V\cos\gamma}{R_{helix}} \tag{3}$$

The autopilot thus has to make sure the crab angle is continuously adapted in order to stay on the circular track. This is usually already achieved by feedback of χ and sufficient weather cock stability, e.g. provided by a good yaw damper. More important is the effect on the bank angle μ required to maintain a circular flight path. This angle is easily derived from lateral equilibrium in a co-ordinated turn [3]:

$$\frac{mV^2\cos^2\gamma}{R_{helix}} = mg\tan\mu \quad \text{so that:} \quad \tan\mu = \frac{V^2}{gR_{helix}}\cos^2\gamma = \frac{V}{g}\dot{\chi}\cos\gamma \tag{4}$$

where g is the acceleration of gravity. Since V varies as a function of the track angle, the bank angle μ must vary as well in order to keep the horizontal distance to the centre of the helix constant. An example for $\gamma = 0$ is depicted in Fig. 4.

At this point, it is interesting to make a short side step to two more scenarios.

Constant airspeed / bank angle turn

Maintaining a constant bank angle based on a constant airspeed results in turning relative to the moving air mass:

$$\tan\mu = \frac{V_A^2}{gR_{helix}}\cos^2\gamma_A \tag{5}$$

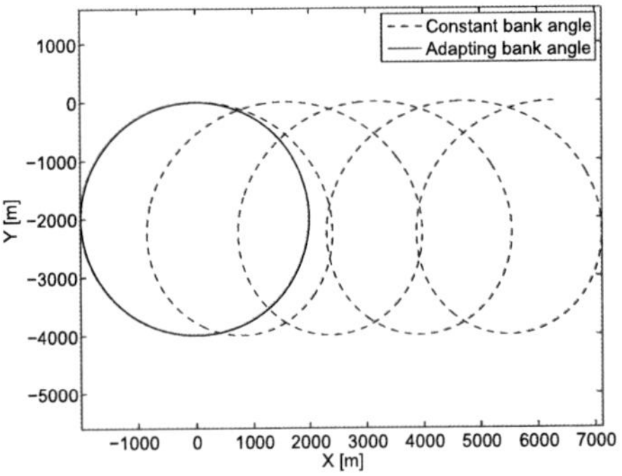

Fig. 5 Comparison between constant and ground-speed corrected bank angle (wind and flight parameters as in Fig. 4)

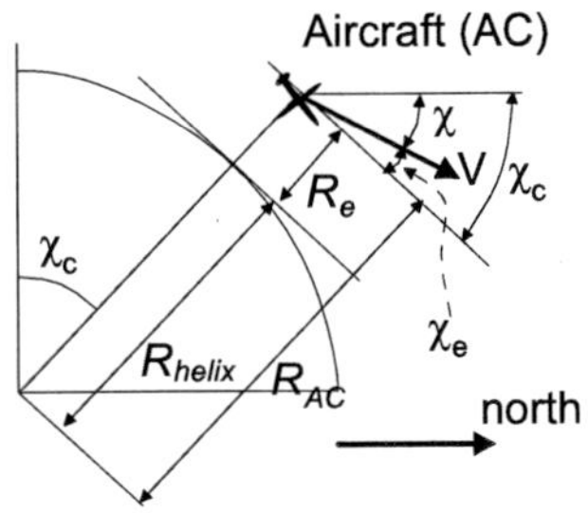

Fig. 6 Aircraft deviation from circular flight path

This equation (usually with $\gamma_A = 0$) is incorrectly used in some text books. The constant bank angle causes a superposition of the moving air mass due to wind and the turning aircraft, causing it to circle away from the helix centre. Fig. 5 compares the trajectories resulting from continuously adapting and constant bank angles.

Constant ground speed turn
Returning to eqn. 4, it is obvious that in case a constant bank angle is desired while staying on track, this can only be achieved by maintaining a constant ground speed. This is not common practice, but it was for example used by the autopilot of the Lockheed SR-71 Blackbird, since reconnaissance sensors required (near) constant bank angle turns for focusing on their specific targets on the ground [4].

For a more general discussion on flight mechanical aspects of manoeuvring under significant wind conditions, the reader is also referred to the work of Rysdyk [5].

3 Automatic Helix Tracking

Obviously, the helix trajectory is not a standard mode in autopilot and flight management systems (FMS). In order to to allow the manoeuvre to be flown by a given autopilot, appropriate command signals have to be provided. These signals will be derived from the flight mechanical considerations discussed above.

A typical autopilot control law structure as used in most commercial transport aircraft is depicted in Fig. 7. Only relevant connections have been labelled. Most structures consist of inner loops for stability and command augmentation (SCA), middle loops for flight path and speed tracking, and guidance loops for lateral (LNAV) and vertical navigation (VNAV). The signal processing block prepares feedback signals for use by the autopilot loops, including computation or estimation of variables not measured directly, smoothing airdata signals by complementary filtering with appropriate inertial measurements, etc.

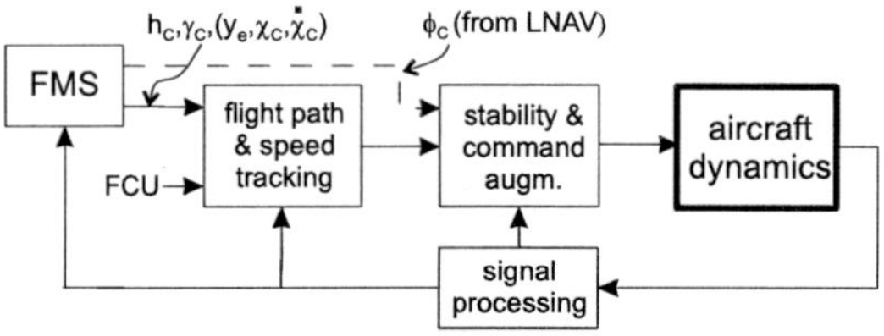

Fig. 7 Typical autopilot / FMS control law structure

The VNAV and LNAV functions are usually provided by the FMS; the flight path and speed tracking modes may also be directly controlled by the pilot via the Flight Control Unit (FCU). Most Flight Management Systems use the available autopilot altitude and climb modes for vertical navigation by commanding appropriate altitude (h_c) and flight path angle (γ_c) or rate of climb commands ($\dot{h}_c$). Lateral navigation is usually done using an internal LNAV control law that generates roll angle commands (ϕ_c) for the autopilot SCA loops. Alternatively, the FMS may provide lateral path error (y_e), turn rate ($\dot{\chi}_c$) and track angle (χ_c) commands to an autopilot LNAV mode.

3.1 Lateral Control Strategy

Fig. 6 sketches the a top view of the situation in which the aircraft has drifted away from the reference circular flight path. The intention is to maintain constant calibrated airspeed (V_{CAS}), without imposing time constraints on the navigation along the helix path. Therefore, as a reference for computing position errors and track angle commands, always the momentary radial line is used. This line is computed in local geodetic co-ordinates from the aircraft and helix centre GPS positions, using algebra and parameter values that come with the WGS-84 standard [6]. From the radial line, the currently desired track angle χ_c can be easily determined and provided to the autopilot. The radius error $R_e = R_{AC} - R_{helix}$ is provided as the lateral path error (y_e, see Fig. 7). The commanded turn rate is V_{ground}/R_{helix} (eqn. 3), where V_{ground} is the momentary ground speed ($V_{ground} = V cos\gamma$). The nominal bank angle command is to be computed from eqn. 4 (not eqn. 5!). Most autopilot SCA systems track roll angle (ϕ_c) rather than bank angle (μ_c) command signals, since measured ϕ is directly available from most Inertial Reference Systems (IRS)[1]. For transport aircraft usually $\mu \approx \phi$, since pitch attitude angles (θ) are relatively small in most flight phases. This in turn allows $\phi_c \approx \mu_c$.

3.2 Longitudinal Control Strategy

The actual helical flight path starts at an altitude $h_{init,helix}$ and ends at h_{end}. The reference altitude depends on the angular position and the number of the turn (see Fig. 1):

[1] For an explanation of the fundamental difference between bank and roll angles, see e.g. [3].

$$h_c = \left(\frac{h_{end} - h_{init,helix}}{2\pi n_{turn} R_{helix}} \right) (2\pi(i_{turn} - 1) + \chi_c - \chi_{init}) R_{helix} + h_{init,helix} \quad (6)$$

where i_{turn} is the number of the current turn and $h_{init,helix}$ is the altitude at which the helix starts. The first term is the tangents of the desired flight path angle:

$$\tan \gamma_c = \left(\frac{h_{end} - h_{init,helix}}{2\pi n_{turn} R_{helix}} \right) \quad (7)$$

The command signals h_c and γ_c may be directly provided to the autopilot control laws (Fig. 7).

3.3 Transition to the Helix Trajectory

The complete trajectory as depicted in Fig. 1 contains two important transitions: (1) from the initial straight segment onto the actual helix and (2) from the helix onto the final straight segment. These transitions should be well timed and smooth in order to prevent large path deviations and flight crew discomfort. For vertical path tracking, most autopilots provide such functionality by early switching to altitude and flight path angle commands from the next segment (see for example [3]). For lateral path tracking, a similar technique is used. The functionality is briefly described, since it is not provided by most autopilots.

Fig. 8 (left) shows a top view of the aircraft about to capture the helix. Even though the aircraft is still on the straight segment, the geometric angular and position errors w.r.t. the helix may already be computed. Obviously, the track angle error would make an autopilot turn the aircraft to the left ($\mu_{c,track} < 0$). The radial error would make it turn right ($\mu_{c,radial} > 0$). The reference bank angle, computed from eqn. 4, is to the right ($\mu_{c,ref} > 0$). The best point to capture the helix is where: $|\mu_{c,track}| = |\mu_{c,ref} + \mu_{c,radial}|$, initially resulting in zero bank angle command. With the aircraft approaching the helix, $\mu_{c,track}$ will usually decrease more rapidly than $\mu_{c,radial}$ so that the aircraft will slowly bank to the right and nicely capture the circular flight path. Leaving the helix and capturing the final straight segment can be done using the very same principles.

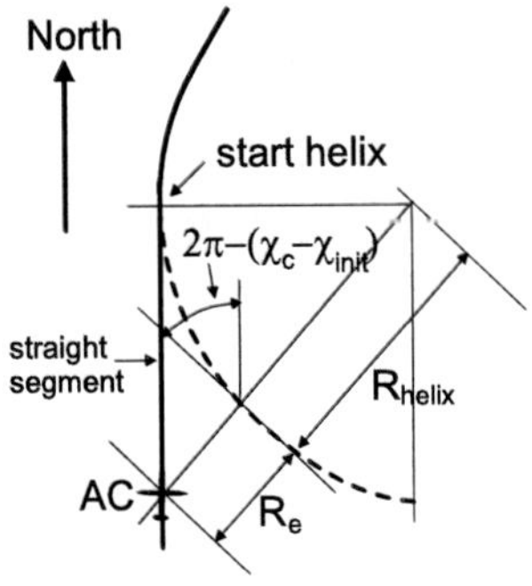

Fig. 8 Capturing the first helix turn

4 Flight Test and Results

The helix manoeuvre was performed three times during a flight test using DLR´s fly-by-wire test bed ATTAS (Advanced Technologies Testing Aircraft System) on June 10, 2009. ATTAS is a highly modified VFW-614 small passenger aircraft with two turbofan engines [7]). The purpose of the flight test was to evaluate autopilot control laws developed at the DLR Institute of Robotics and Mechatronics regarding tracking accuracy of flight path and calibrated airspeed, with particular attention to aspects like piloting, mode switching, comfort, and control activity during atmospheric disturbances. The autopilot control laws are partly described in [8]; the SCA and lateral path tracking functions are based on Nonlinear Dynamic Inversion (NDI), vertical path and speed tracking is based on the Total Energy Control System (TECS, [9]).

The autopilot command signals were computed as described in the previous section and has been implemented in a dedicated software routine. The autopilot LNAV mode was used. The helix manoeuvre was integrated in landing approaches to runway 26 of Braunschweig airport by setting the parameters in Table 1 appropriately, see Fig. 9[2]. The manoeuvre is initiated in an altitude of 7500 ft (h_{init}) in the direction of the runway, with $\gamma_{init} = 0\,\text{deg}$. The helix centre ($r_{helix}$) is located at a distance $D_{helix} = 3200\ m$ before the runway threshold. The latter point is also the reference for computing local x,y,z co-ordinates. After exactly three full turns the manoeuvre ends with the final straight trajectory, which matches the standard ILS approach path. Shortly before ending the final circle, the ILS signal is checked. If the aircraft is within the cone, the autopilot changes into its ILS tracking mode. Otherwise, flight continues in runway direction, but at constant altitude h_{end}. The calibrated airspeed is held constant at 160 kts (selected by the pilot on the FCU). During the final quarter of the last turn, it is reduced to the adopted final approach speed of 140 kts. All approaches were followed by a go-around; automatic landing capability had already been successfully flight tested before [7].

The trajectories as resulting from the flight test are depicted in Fig. 10. The helical approaches are hardly distinguishable and the ground tracks are nearly perfect circles. This is very encouraging, since the weather conditions were quite challenging for achieving accurate tracking of the helical flight path. At the airport winds between 15 and 20 kts from direction 255 deg (runway heading 265 deg) were measured. At higher altitudes winds up to 28 kts were encountered, see Fig. 11. Between and below clouds (broken cumulus, base 4000 ft, top 7000 ft) thermal turbulence was encountered. In the clouds, turbulence levels were considerably higher.

Since the wind conditions were strongest during the first approach, the discussion in the following will focus on this part of the flight test. The path tracking accuracy is confirmed by Fig. 12. Here the time along the helix path has been stretched along the horizontal-axis while the lateral deviation is represented by the vertical axis.

---

[2] This allowed for the organisation of an add-on noise measurement campaign on the ground, evaluating helical approaches as an alternative potential noise abatement procedure, see [10]. For this reason, also standard Instrument Landing System (ILS) based and steep landing approaches were flown to provide noise reference values.

Fig. 9 The helix trajectory during flight test

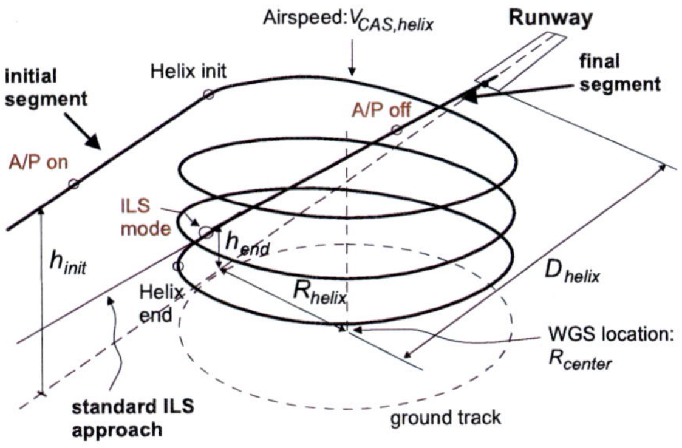

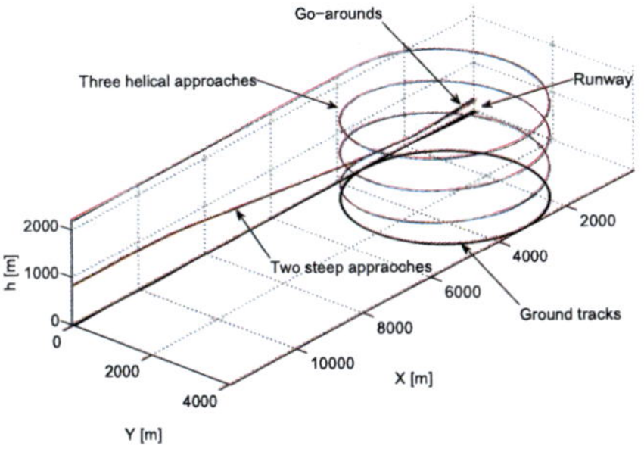

Fig. 10 Trajectories plotted from flight data

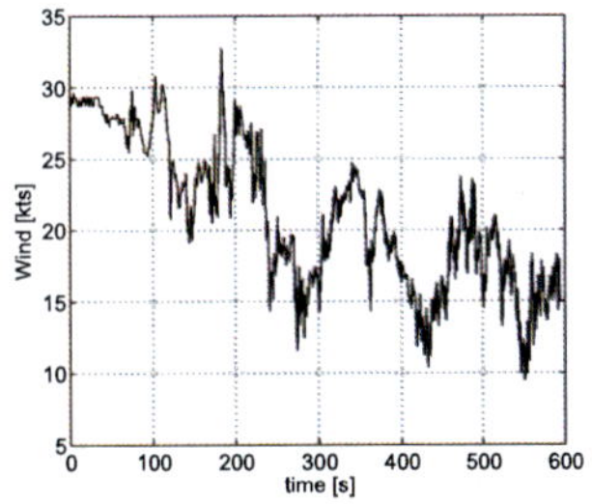

Fig. 11 Estimated wind speeds during first helical approach

Most of the time the lateral path deviation is well within the wing span of the aircraft. Two peaks occur: when capturing the helix (left) and when capturing the ILS path (right). They arise at the moment of switching to the helix path and to the ILS modes respectively. The error results from the early activation for smooth capture as discussed in Section 3.3 and nicely demonstrates disturbance rejection performance.

The next prime tracking variable is the calibrated airspeed. A first challenging aspect of the helix trajectory is the artificial creation of wind shears relative to the flight path. This can clearly be seen from Fig. 13, showing the calibrated airspeed and inertial speed for the first helical approach. Clearly, during the helix the autopilot manages to track the calibrated airspeed reference quite well. As expected, the ground speed varies considerably (see also Fig. 3). This situation is typical for wind shear: while holding the airspeed approximately constant, the ground speed gets a strong gradient.

As briefly addressed in Section 2, it is common practice to smooth air data signals using complementary filtering. For acceleration feedback, the following signal is formed:

$$\dot{V}_{compl} = \frac{s}{\tau s+1} V_{CAS} + \frac{\tau s}{\tau s+1} \dot{V} \tag{8}$$

where the time constant τ determines the complementary frequency content of both measurements. During the helix, the ground speed varies periodically, causing periodical variations in its derivatives as well. This causes continuously nonzero gradients of the acceleration signal, in turn causing $\dot{V}_{compl}$ to continuously deviate from $\dot{V}_{CAS}$. This explains the periodical error in the calibrated airspeed signal in Fig. 13.

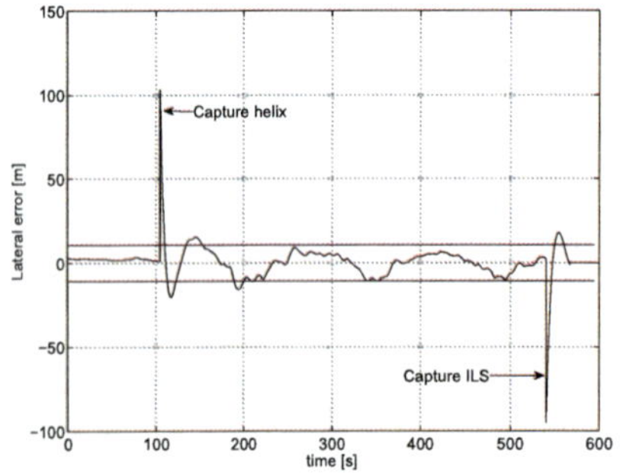

Fig. 12 Lateral path deviation during first helix approach (the distance between two solid horizontal lines indicates the aircraft wing span)

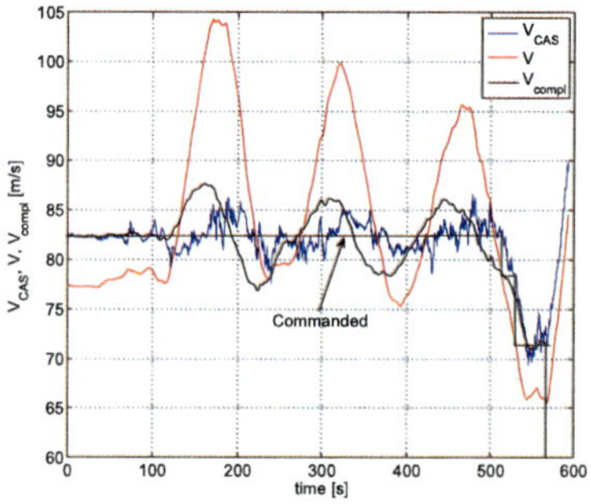

Fig. 13 Calibrated airspeed, ground speed and complementary speed

The bank angle during the first helix approach is depicted in Fig. 14. As expected, it varies strongly as a function of time (compare with Fig. 4) and accurately follows the values commanded by the autopilot. The side slip angle is quite noisy and not directly available to the control laws. It therefore is estimated instead [8]. The resulting smoothed signal (black) remains within 1 deg during the manoeuvre, indicating excellent weathercock stability and turn co-ordination.

The crab angle is depicted in Fig. 15. During the initial straight segment (to the left) a crab angle of two degrees is already necessary due to a small wind

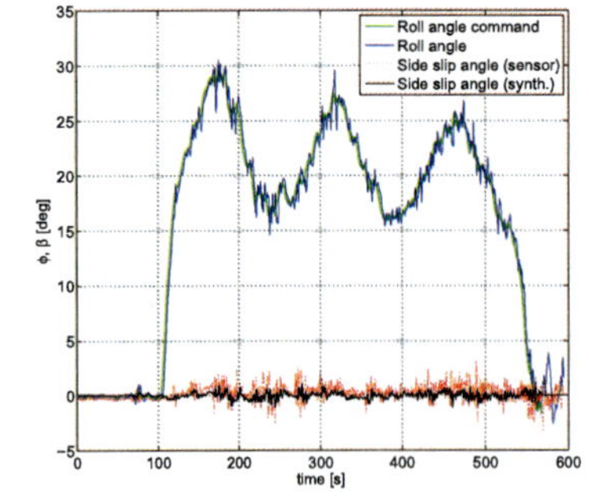

Fig. 14 Bank, side slip, and estimated side slip angles

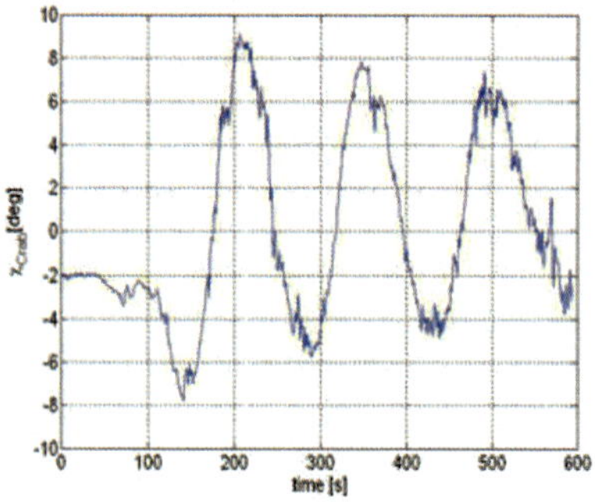

Fig. 15 Crab angle

component from the left. Fig. 16 depicts the altitude profile, comparing measured and commanded values. The initial error arises at the moment the helix is initiated and is quickly reduced. The error remains within a 5 m band most of the time. Larger peaks occur between 100s and 300s, where relatively strong turbulence was encountered (now shown). The slight undulation is caused by the varying ground speed.

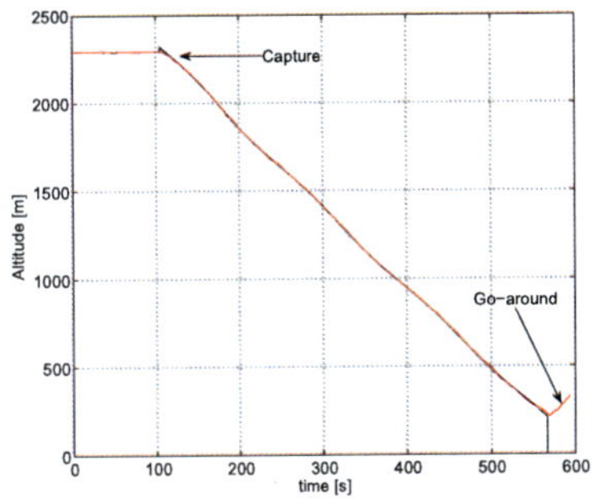

Fig. 16 Commanded and actual barometric altitudes

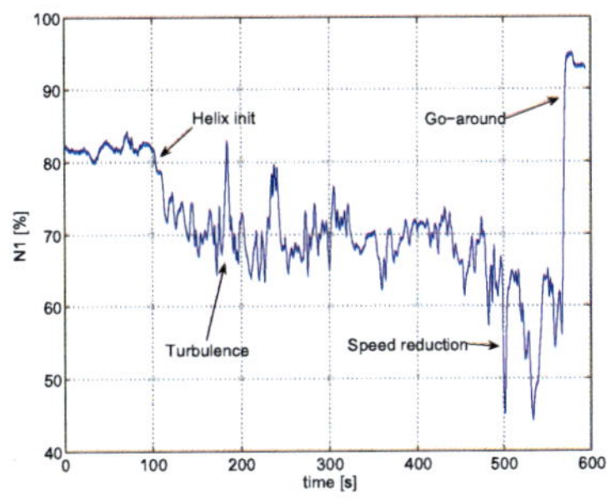

Fig. 17 Mean fan shaft speed N_1 of both engines

Finally, the mean fan shaft speed of both engines is depicted in Fig. 17. Providing good tracking performance in considerable turbulence, thrust control activity by the autopilot was low, very much pleasing the flight crew. This important feature is attributed to the Total Energy Control System.

5 Summary

Helix manoeuvres are interesting trajectories for evaluating and validating autopilot flight control law performance. The main objectives of this chapter have been to discuss relevant flight mechanical aspects and to derive appropriate command signals that allow such a manoeuvre to be flown (in simulation or flight test) by a given autopilot. As an example, flight test results of an example autopilot have been discussed, addressing helix-related performance challenges such as path and speed tracking accuracy, influence of wind shears, engine throttle activity, and path segment capture behaviour. Detailed metrics for these performance aspects are under development and will allow for quantitative comparison of different autopilot designs.

Acknowledgements

The author would like to thank the DLR flight test department for all their efforts related to the ATTAS flight test and their willingness to carry out the somewhat exotic approach trajectory. The author also would like to thank Dirk Leißling from the Institute of Flight Systems for implementing the control laws on ATTAS and for all preparatory work in the simulator.

References

1. Kaminer, I., Pascoal, A.M., Hallberg, E., Silvestre, C.: Trajectory Tracking for Autonomous Vehicles: An Integrated Approach to Guidance and Control. AIAA Journal of Guidance, Control and Dynamics 21(1), 29–38 (1998)
2. Lambregts, A.A., Cannon, D.G.: Development of a control wheel steering mode and suitable displays that reduce pilot work load and improve efficiency and safety of operation in the terminal area and in wind shear. AIAA-79-1887 (1979)
3. Brockhaus, R.: Flugregelung. Springer, Heidelberg (1994)
4. Graham Richard, H.: SR-71 revealed: the untold story. Zenith Press, Minneapolis (1996)
5. Rolf, R.: Course and Heading Changes in Significant Wind. AIAA Journal of Guidance, Control and Dynamics 30(4) (2007); Erratum published in 33(4) (2010)
6. National Imagery and Mapping Agency, 3rd edn. Department of Defence World Geodetic System, NIMA TR 8350.2 (2000)
7. Bauschat, M., Mönnich, W., Willemsen, D., Looye, G.: Flight Testing Robust Autoland Control Laws. AIAA-2001-4208 (2001)
8. Looye, G.: An integrated approach to aircraft modelling and flight control law design. PhD. thesis, TU-Delft (2008)
9. Lambregts, A.A.: Vertical flight path and speed control autopilot design using total energy principles. AIAA-83-2239 (1983)
10. Bertsch, L., Looye, G., Eckhard, A., Schwanke, S.: Flyover Noise Measurements of a Spiralling Noise Abatement Approach Procedure. In: 48th AIAA Aerospace Sciences Meeting Including the New Horizons Forum and Aerospace Exposition, Orlando, Florida, USA (2010)

Maneuver Envelope Determination through Reachability Analysis

E.R. van Oort, Q.P. Chu, and J.A. Mulder

Abstract. Knowledge of the safe maneuvering envelope is of vital importance to prevent loss of control aircraft accidents. In this paper, determination of the safe maneuvering envelope is addressed in a reachability framework. The forwards and backwards reachable sets for a set of initial trim conditions are obtained by solving a Hamilton Jacobi partial differential equation through a semi-Lagriangian method. Results obtained using this approach are presented for a nonlinear, high-fidelity, F-16 aircraft model.

1 Introduction

During the last decades adaptive control in its many forms has received a lot of attention within the flight control community. These control algorithms are able to deal with changes in the system's dynamics due to possible system component faults and failures. A question that still remains unanswered is which parts of the state space are safe to operate, often even when the dynamics of the system are completely understood or assumed known. This question is of fundamental importance in the safety verification of control systems and system validation. Recent statistics show that the majority of accidents in aviation nowadays are due to Loss-of-Control (LOC) [1, 2]. This also shows that LOC is not a phenomenon solely attributed to military aircraft, but is equally a real problem for commercial aircraft and general aviation.

The relevance of knowledge of the flight envelope is emphasized by means of an accident which ultimately is the result of a violation of the safe flight envelope. On October 4, 1992, a Boeing 747 cargo plane crashed into two apartment

E.R. van Oort
e-mail: E.R.vanOort@tudelft.nl

Q.P. Chu · J.A. Mulder
Delft University of Technology, Faculty of Aerospace Engineering, Kluyverweg 1, P.O. Box 5058, 2600 GB, Delft, The Netherlands

buildings in the Bijlmermeer neighborhood of Amsterdam, near Schiphol Airport. Engine number three separated from the right wing of the aircraft shortly after take-off, damaging the wing flaps, and struck engine number four which also separated. Analysis showed that the aircraft still had marginal controllability left in a severely restricted flight envelope [3]. Simulator experiments using various different fault tolerant flight control approaches have shown that landing the aircraft safely was possible [4, 5, 6]. Additional simulations showed that a very experienced pilot was able to land the aircraft using the standard control system when he was informed about the severely restricted flight envelope. However, the Boeing 747 aircraft does not have such a fault tolerant control system and, more importantly, the pilots did not have any knowledge about the restricted flight envelope of the aircraft. When the crew tried to reduce the speed for landing the aircraft banked sharply to the right without chance of recovery: control of the aircraft was completely lost with disastrous results.

In this paper a semi-Lagrangian level set approach is taken to obtain the safe maneuvering envelope for aircraft dynamics that are assumed known. The safe maneuvering set is defined as the intersection between the forwards and backwards reachable sets for a given set of a-priori known safe states. This approach has been used to solve fluid flow problems, for example in [7] and [8]. The novelty in this work is the application of the semi-Lagrangian approach to systems with control and disturbance inputs, and its application in higher dimensions on kd-tree grids. The method is applied to a high-fidelity nonlinear F-16 model at different flight conditions and configurations to demonstrate their effects on maneuverability. The safe maneuvering sets obtained can be used for trajectory generation, path planning and controller synthesis, even in post-failure conditions such that safety is improved.

The paper is organized as follows. First, in section 2 the concept of safe maneuver envelope is defined, and how it can be obtained through reachability analysis. Then, in section 3 the level set method and the Semi-Lagrangian approach are discussed. The longitudinal maneuver envelope for an high-fidelity model of an F-16 aircraft is evaluated in section 4 at different flight conditions. Finally, section 5 draws conclusions from the work presented, and states future research directions.

2 Maneuver Envelope and Reachability

The conventional definition of the flight envelope is "[the flight envelope] describes the area of altitude and airspeed where an airplane is constrained to operate." [9]. The flight envelope boundaries are defined by various limitations on the performance of the airplane, for example available engine power, stalling and buffet characteristics, structural considerations and requirements on maximum noise production. A common way to present the flight envelope is the doghouse diagram which relates the altitude, velocity and possibly other variables at which the aircraft can safely fly.

2.1 Flight Envelope Protection

As noted in [10] one of the most promising techniques to prevent LOC-related accidents is envelope protection. The technique tries to prevent the aircraft from exceeding the safe maneuvering envelope. A prerequisite for flight-envelope protection systems is a fly-by-wire (FBW) system. In the FBW, the pilot's inputs are sent to a computer which then calculates the desired commands, i.e. there is no direct link between the pilot and the controls. Such systems have existed for over 30 years but are currently only used in military aircraft, several commercial aircraft, and a very limited amount of general aviation aircraft. A prerequisite for protection is accurate knowledge of the maneuvering envelope, both to safeguard against excursions and LOC events, and not place too conservative constraints on performance of the aircraft. Furthermore, when the envelope is known, the maneuvering space can be presented to the pilot to increase situation awareness.

2.2 Safe Maneuvering Envelope

The boundaries defined on the flight envelope in the doghouse plot are adequate during normal operation of aircraft. The main problem with the conventional definition of flight envelope is that only constraints on quasi-stationary aircraft states are taken into account, for example during coordinated turns and cruise flight. Additionally, constraints posed on the aircraft state by the environment are not part of the conventional definition. The aircraft's dynamic behavior can pose additional constraints on the flight envelope, for example due to inertia coupling effects. Such constraints would especially be important for military and acrobatic aircraft, aircraft having experienced upset, and aircraft with airframe and/or actuator damage or malfunctions. Thus, an extended definition of the flight envelope is required which will be called the safe maneuver envelope.

Definition 1 (Safe Maneuver Envelope). The *safe maneuver envelope* is the part of the state space for which safe operation of the aircraft and its cargo can be guaranteed and external constraints will not be violated.

The safe maneuver envelope is defined by the intersection of three envelopes:

- *Dynamic Envelope:* Constraints posed on the envelope by the dynamic behavior of the aircraft, due to its aerodynamics and kinematics.
- *Structural and Comfort Envelope:* Constraints posed by the airframe, pilot, passengers and cargo. These constraints are defined through maximum accelerations and loads.
- *Environmental Envelope:* Constraints due to the environment in which the aircraft operates.

The last two envelopes pose *external* constraints on the flight envelope, constraints which are generally well-known and can be quantified easily. Examples of such external constraints are the terrain and no-fly zones around the aircraft, and the maximum load-factor the airframe can sustain before breaking. The example given in the introduction is considered to be a violation of the dynamic flight envelope. The focus in this paper is on the first type, i.e. the flight envelope directly related to the aircraft's dynamic behavior. A more formal definition of the *dynamic flight envelope* is given below.

Definition 2 (Dynamic Flight Envelope). The region of the aircraft's state space in which the aircraft can be safely controlled and no loss-of-control events can occur.

Constraints posed on the aircraft by the dynamic flight envelope are for example a maximum roll-rate at high angle of attack in order to prevent the aircraft from entering a potentially hazardous inertia coupling phenomenon or spin.

2.3 Reachable Sets

Reachable set analysis is an extremely useful tool in safety verification of systems. The reachable set describes the set that can be reached from a given initial set within a certain amount of time, or the set of states that can reach a given target set with a certain time. The dynamics of the system can be evolved backwards and forwards in time resulting in the backwards and forwards reachable sets respectively. The difference between these two sets is illustrated in figure 1. For a forwards reachable set, the initial conditions are specified and the set of all states that can be reached along trajectories that start in the initial set are determined. For the backwards reachable sets, a set of target states is defined, and a set of states from which trajectories start that can reach that target set are determined.

Assume that the dynamics of the system are given by

$$\dot{x} = f(x, u, d) \tag{1}$$

where $x \in \mathbb{R}^n$ is the state of the system, $u \in \mathscr{U} \subset \mathbb{R}^m$ is the control input, and $d \in \mathscr{D} \subset \mathbb{R}^q$ a disturbance input. Then, the formal definition of the backwards and forwards reachable sets is given by definitions 3 and 4 respectively.

Definition 3 (Backwards Reachable Set). The backwards reachable set $\mathscr{S}(\tau)$ at time $\tau(0 \leq \tau \leq t_f)$, of the system (1) starting from the target set $\mathscr{T}_0$, is the set of all states $x(\tau)$, such that there exists a control input $u(t) \in \mathscr{U}(\tau \leq t \leq t_f)$, for all disturbance inputs $d(t) \in \mathscr{D}(\tau \leq t \leq t_f)$, for which some $x(t_f) \in \mathscr{T}_0$ are reachable from $x(\tau)$ along a trajectory satisfying (1).

Definition 4 (Forwards Reachable Set). The forwards reachable set $\mathscr{V}(\tau)$ at time $\tau(0 < \tau \leq t_f)$ of the system (1) starting from the initial set $\mathscr{S}_0$, is the set of all states $x(\tau)$, such that there exists a control input $u(t) \in \mathscr{U}(\tau \leq t \leq t_f)$, for all disturbance inputs $d(t) \in \mathscr{D}(\tau \leq t \leq t_f)$, for which $x(\tau)$ is reachable from some $x(0) \in \mathscr{S}_0$ along a trajectory satisfying (1).

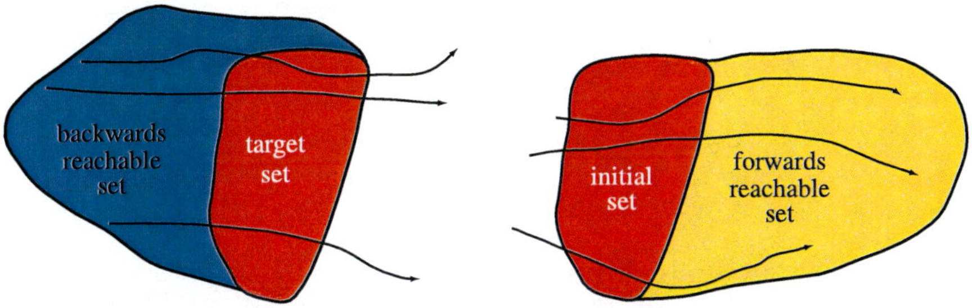

(a) Backwards Reachable Set (b) Forwards Reachable Set

Fig. 1 Backwards and forwards reachable set definitions.

2.4 Safe Maneuver Envelope through Reachable Set Analysis

Now, the safe envelope for a given system with a preliminary set of *safe* states can be found by the intersection of the forwards and backwards reachable set of this safe set. States that are part of both these sets can be reached from the safe set, and can reach the safe set within a certain time. Therefore, if the initial/target set is known to be safe, then all states that are part of both the forwards and backwards reachable sets can be considered *safe* as well. This is illustrated by figure 2. For example, an aircraft can enter a spin starting from a certain initial condition, and the spin trajectory would then be included in the forwards reachable set. If recovery to a safe flight condition from the spin is possible, the spin trajectory, or part of it, is also included in the backwards reachable set. A similar example is deep stall of the aircraft.

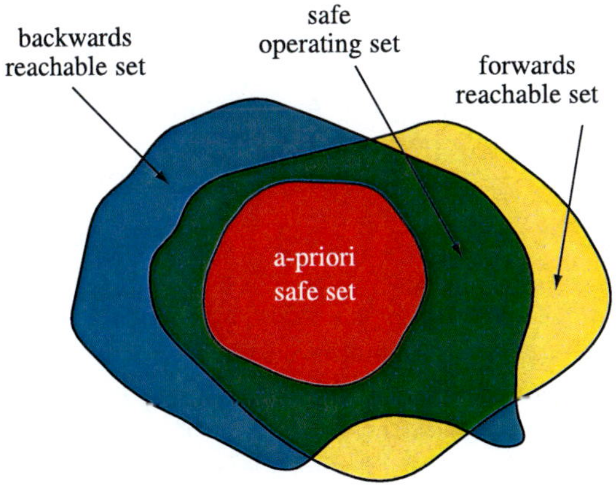

Fig. 2 The safe envelope for a known safe set is defined by the intersection of the forwards and backwards reachable sets.

3 Semi-lagrangian Reachable Set Analysis

Sets can be represented either explicitly by enumerating all parts belonging to the set, or implicitly as the level set of some function. This function can be evolved in time such that the reachable set can be tracked through time by tracking the level set of the function in time. In this section different solution methods to track the interface evolution methods over time are discussed.

3.1 Time Evolution

Explicit and implicit surface descriptions only give a representation of a set. To obtain the reachable set the interface has to be evolved in time. Suppose that the velocity of each point on the interface is given by an external velocity field $f(x,t)$. The simplest method to move the interface with the velocity field is by solving the ordinary differential equation

$$\frac{dx}{dt} = f(x,t) \tag{2}$$

for every point x on the interface, a *Lagrangian* formulation of the interface evolution equation. To avoid the problems with instabilities, and deformation of the interface elements, an implicit function φ is used both to represent the interface and to evolve it. A simple convection partial differential equation

$$\varphi_t + \nabla\varphi \cdot f(x,t) = 0, \tag{3}$$

where the subscript t denotes a temporal partial derivative in the time variable t, and ∇ is the gradient operator. This is an *Eulerian* formulation of the interface evolution, since the interface is captured by the implicit function φ as opposed to being *tracked* by interface elements as was done in the Lagrangian formulation [11, 12]. This equation can be solved using an upwind schemes combined with forward Euler integration. Stability of this approximation has to be enforced using the Courant-Friedrichs-Lewy (CFL) condition[13], stating that numerical waves should propagate at least as fast as the physical waves. This leads to the CFL time-step restriction of

$$\Delta t \sum_{i=1}^{n} \frac{|f_i(x,t)|}{\Delta x_i} < \alpha \tag{4}$$

where $0 \leq \alpha \leq 1$ is a safety factor usually taken to be 0.9. This condition clearly poses a stringent constraint on the allowed time-step for high resolution grids in combination with fast dynamical systems.

The time restriction posed by the CFL condition can be eliminated by allowing unbounded stencils [14]. The time-step can be decoupled from the CFL-condition by using an explicit, unconditionally stable time-stepping scheme. These schemes can be interpreted as *semi-Lagrangian* schemes. The CFL condition is satisfied for large time-steps by shifting the stencil. The first order hyperbolic partial differential equation (3) propagates the values φ along the characteristic curves $s(t)$ defined by

$$\dot{s} = f(s,t). \tag{5}$$

Thus, the value of φ at any time t can be determined by finding the characteristic curve passing through (x,t) and tracing it backwards in time to a point (x_0,t_0) where the value of φ is known, then $\varphi(x,t) = \varphi(x_0,t_0)$. This observation is the basis of the backwards characteristic, or the Courant-Isaacson-Rees (CIR) scheme [15], the simplest semi-Lagrangian scheme. Given φ at time t^k, the CIR-scheme approximates the $\varphi(x,t^{k+1})$ at any point x at time $t^{k+1} = t^k + \Delta t$ by evaluating the velocity $f(x,t^k)$ and approximating the backwards characteristic through x by a straight line

$$x - \Delta t f(x,t^k) \approx s(t^k). \tag{6}$$

Then, $\varphi(x,t^{k+1})$ is set equal to the interpolated value at the location $s(t^k)$. For linear PDEs, the Lax-Richtmyer equivalence theorem guarantees that CIR converges to the exact solution when $\Delta t, \Delta x \to 0$ if the discretization is stable and consistent[14].

Semi-Lagrangian schemes combine the regular mesh of an Eulerian scheme with the unconditional stability of a Lagrangian scheme. The difference between the Eulerian, Lagrangian and semi-Lagrangian schemes is shown in figure 3.

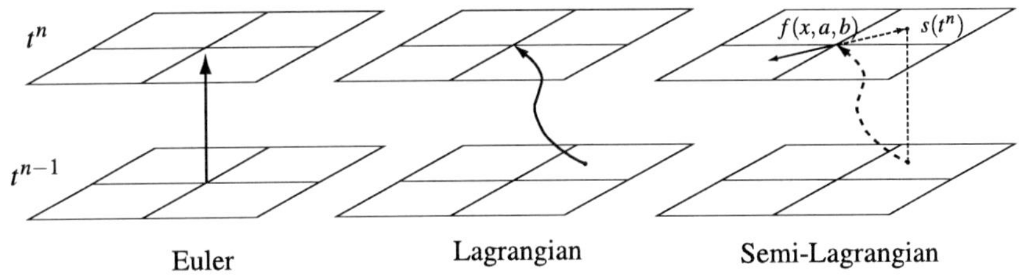

Fig. 3 Euler, Lagrangian, and semi-Lagrangian schemes for the Level Set equation.

3.2 *Hamilton-Jacobi Partial Differential Equation*

For systems with control and disturbance inputs the level set equation (3) needs to be reformulated as an Hamilton-Jacobi partial differential equation. One input, marked as input b, will try to keep the system away from the target or to initial set, the other, input a, will try to drive it towards the target or away from initial set. The reachable set can then be obtained as the viscosity solution of a time dependent Hamilton-Jacobi-Isaacs equation [16, 17].

The dynamic programming approach yields the backwards reachable set as the viscosity solution of the terminal value problem, with subscript b indicating backwards,

$$\frac{\partial V}{\partial t} + \min\left[0, H_b(x, \frac{\partial V}{\partial x})\right] = 0,\ V(x,0) = T(x) \tag{7}$$

where the Hamiltonian is defined as

$$H_b(x, \frac{\partial V}{\partial x}) = \min_u \max_d \left(\frac{\partial V}{\partial x}\right)^T f(x,u,d). \tag{8}$$

Similarly, the forwards reachable set can be found as the viscosity solution of the initial value problem, with subscript f indicating forwards,

$$\frac{\partial V}{\partial t} + \max\left[0, H_f(x, \frac{\partial V}{\partial x})\right] = 0,\ V(x,0) = S(x) \tag{9}$$

where

$$H_f(x, \frac{\partial V}{\partial x}) = \max_u \min_d \left(\frac{\partial V}{\partial x}\right)^T f(x,u,d). \tag{10}$$

$T(x)$ and $S(x)$ are functions for which the level 0 describes the target and initial sets $\mathcal{T}$ and $\mathcal{S}$ respectively. The comparison with 0 in (7) and (9) is added such that the reachable set is only allowed to grow over time [16].

4 F-16 Longitudinal Maneuver Envelope

In this section the longitudinal maneuver envelope for a high-fidelity nonlinear model of an F-16 aircraft is determined at two different flight conditions.

4.1 F-16 Model

When the full maneuvering capabilities of an aircraft are considered the reachable set calculations have to be run in 8 dimensional space, if the altitude is considered to be fixed. Three states related to the airspeed, three rotational rates, and two states defining the relevant attitude. In this example only the longitudinal dynamics are considered to reduce the computational load, and simplify representation of the maneuver set. Therefore, only four states are considered: the airspeed V_T, the angle of attack α, the pitch rate q_B, and the pitch attitude defined through a quaternion component q_2. The engine thrust and stabilizer deflections are considered as control inputs. In this particular case, no disturbance inputs were considered. However, it is possible to include uncertainty on the aerodynamic parameters, aircraft parameters, and wind as disturbance inputs and obtain the worst-case safe maneuvering envelope.

The dynamics for the F-16 model are given by

$$\begin{aligned}
\dot{V}_T &= \tfrac{1}{m}\left(-D(\alpha, q_B, \delta_h) + T\cos\alpha + mg_1\right)\\
\dot{\alpha} &= q_B + \tfrac{1}{mV_T}\left(-L(\alpha, q_B, \delta_h) - T\sin\alpha + mg_3\right)\\
\dot{q}_B &= \tfrac{1}{I_{yy}}\bar{M}(\alpha, q_B, \delta_h)\\
\dot{q}_2 &= \tfrac{1}{2}q_B q_0
\end{aligned}$$

where

$$q_0 = \sqrt{1 - q_2^2},$$
$$g_1 = \left[-2q_0q_2\cos\alpha + (q_0^2 - q_2^2)\sin\alpha\right]g,$$
$$g_3 = \left[2q_0q_2\sin\alpha + (q_0^2 - q_2^2)\cos\alpha\right]g,$$

with q_0 a quaternion component, and g the gravitational acceleration. The horizontal stabilizer deflection δ_h is constrained to ± 25 degrees, and the engine thrust T is limited between 0 and 75000 Newton. The lift, drag and pitching moment are computed through lookup-tables obtained from [18]. In order to calculate the optimal control inputs to evolve the level set, a linear approximation in the stabilizer deflection of the drag, lift and pitching moment coefficients was made as, for example

$$C_{m_T}(\alpha, \delta_h) \approx C_{m_0}(\alpha) + C_{m_{\delta_h}}(\alpha)\delta_h \tag{11}$$

The true coefficients were used to propagate the implicit function over time. Another option is to pose additional constraints on the elevator deflection such that at its maximally allowed deflection the minimal/maximal aerodynamic effects are generated. The leading-edge flap is not deflected in any of the simulations.

4.2 Scenarios and Trim Set Determination

Before the reachable set calculations can be performed, first the safe set has to be defined. In this case the aircraft was trimmed for straight and level flight at 0m and 10000m altitude for a center of gravity location of 0.30% of the mean aerodynamic chord. The result is a trim curve relating the airspeed and the angle of attack for each of the flight conditions and configurations described above. Figure 4(a) shows the trim curves for each of the considered scenarios.

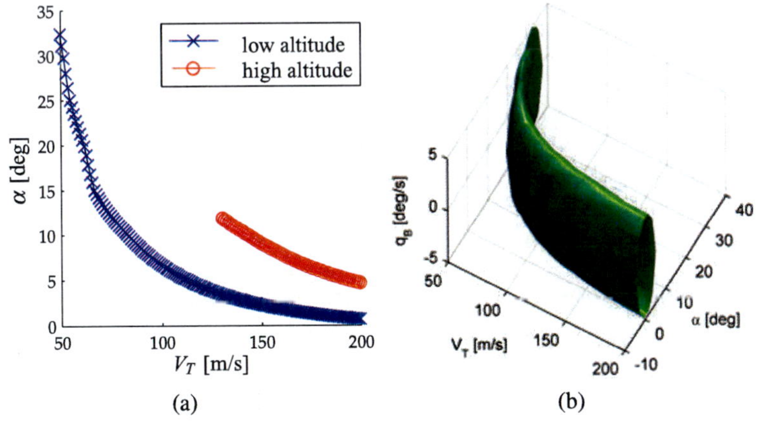

Fig. 4 Trim curve of the F-16 aircraft model at low and high altitude (a). Extended trim curve (b).

The trim curves have to be translated into implicit surfaces before the reachable set calculation can be performed. This can be done by calculating the weighted distance to the nearest trim point for each point in the domain of interest. Implicit surface methods require that at least one grid node is within the initial or target set such that the location of the interface can be found. Therefore, the converted trim curve is not sufficient to initialize the level set calculations. Hence, a region within a certain weighted distance of the trim curve is considered to be safe, where the weighted distance defining the band should be larger than the smallest allowed grid cell. A different option would be to propagate a narrow band on a high resolution grid over a small timestep, and then start the calculations with the resulting reachable sets as initial set on a coarser grid. Another option would be to extend the trim set to include non-stationary, non-level flight conditions and create an implicit set description from that set. This would however require a very large number of trim points.

In order to limit the computational load and clearly show that the forwards and reachable set yield different reached portions of the state space, the evolution time is set to 1 second. This time was reached by evolving the initial/target set forwards or backwards in time respectively, by 100 steps of 0.01 seconds. The computation time for the F-16 application was between 1 and 4 hours on a single core of an Intel Xeon X5500 processor running at 2.13 Ghz and equipped with 12 GB of RAM.

4.3 Airspeed Comparison

First of all a comparison is made between the maneuverability of the F-16 aircraft at low altitude for different airspeeds. Figures 5(a), and 5(b) show the reachable sets and safe envelope at airspeed of 60 and 150 m/s respectively. Clearly, with increasing dynamic pressure, the aircraft becomes more maneuverable as can be observed from the increased size of the safe maneuver set. Furthermore, the expected relations between the angle of attack and the pitch attitude, the angle of attack and pitch rate, and the pitch attitude and pitch rate can all be observed from the plots.

4.4 Altitude Comparison

There exists a large difference between the trim curves of the aircraft at low and high altitude as can be observed from figure 4(a). The lookup tables of the used F-16 model do not depend on Mach number. The only difference between the two flight conditions is therefore the dynamic pressure caused by a difference in air density. The resulting effect on the safe maneuver set can be observed by comparing Fig. 5(b) and Fig. 6(b). The dynamic pressure at 10000m and 150 m/s is about 75% of flying at 0m and 100 m/s. The same ratio can be observed from the comparison of the maneuver sets shown in figure 6(a) and figure 6(b).

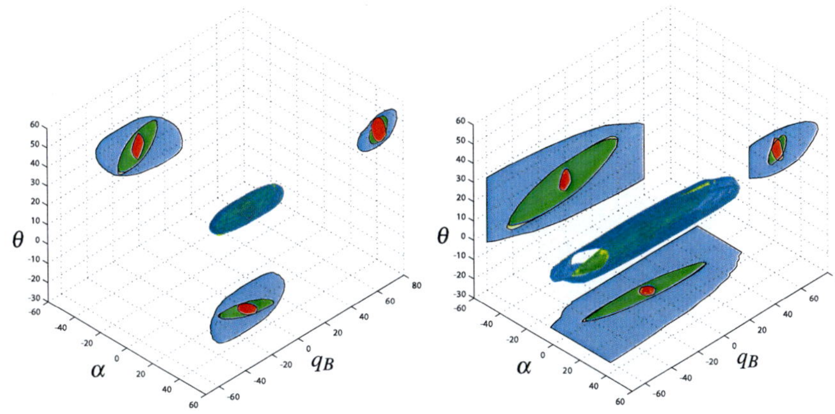

(a) $V_T = 60$ m/s, altitude = 0m (b) $V_T = 150$ m/s, altitude = 0m

Fig. 5 Safe maneuver envelopes for two flight conditions with different dynamic pressure.

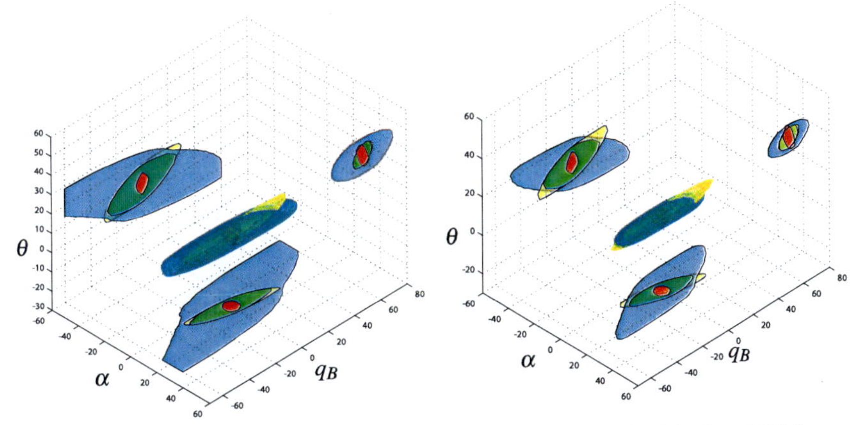

(a) $V_T = 100$ m/s, altitude = 0m (b) $V_T = 150$ m/s, altitude = 10000m

Fig. 6 Safe maneuver envelopes for two flight conditions with dynamic pressure of 6125 Nm^{-2} and 4643 Nm^{-2}.

5 Conclusions and Recommendations

In this paper a definition of maneuvering and dynamic flight envelope was given based on the dynamics of an aircraft. Furthermore, a method to determine this envelope based on reachable sets was formulated and applied to a nonlinear high-fidelity model of an F-16 aircraft. The safe maneuvering envelope results agree with what is expected from flight dynamics knowledge. Especially for general aviation and commercial aircraft, it would be interesting to investigate whether the full envelope determination problem can be split into fast, and slow dynamics by means of time-scale separation arguments. This would simplify the dynamic envelope problem into five and three dimensional subproblems which are more computationally tractable than the original problem.

References

1. Ranter, H.: Airliner accident statistics 2006. Tech. rep., Aviation Safety Network (2007), `http://aviation-safety.net/`
2. Boeing Company, Statistical summary of commercial jet airplane accidents: Worldwide operations since 1959. Tech. rep., Boeing Commercial Airplanes (2009), `http://www.boeing.com/news/techissues/pdf/statsum.pdf`
3. Smaili, M.H., Mulder, J.A.: AIAA Modeling and Simulation Technologies Conference and Exhibit, Washington, DC, AIAA-2000-4586 (2000)
4. Lombaerts, T.J.J., van Oort, E.R., Chu, Q.P., Mulder, J.A., Joosten, D.A.: Journal of Guidance, Control and Dynamics 33(3), 707 (2010)
5. Lombaerts, T.J.J., Smaili, M.H., Stroosma, O., Chu, Q.P., Mulder, J.A.: Journal of Guidance, Control and Dynamics 32(6), 1747 (2009)
6. Alwi, H., Edwards, C., Stroosma, O., Mulder, J.A.: Journal of Guidance, Control and Dynamics 31(5), 1186 (2008), doi:10.2514/1.35066
7. Losasso, F., Gibou, F., Fedkiw, R.: SIGGRAPH 2004: ACM SIGGRAPH 2004 Papers, pp. 457–462. ACM, New York (2004), doi:http://doi.acm.org/10.1145/1186562.1015745
8. Wang, Z., Wang, Z.J.: Proceedings of the 44th AIAA Aerospace Sciences Meeting and Exhibit, AIAA-2006-887 (2006)
9. Ruijgrok, G.J.J.: Elements of Airplane Performance. Delft University Press, The Netherlands (1996)
10. Lambregts, A.A., Nesemeier, G., Wilborn, J.E., Newman, R.L.: AIAA Modeling and Simulation Technologies Conference and Exhibit, Honolulu, Hawaii, AIAA-2008-6867 (2008)
11. Osher, S., Fedkiw, R.: Level Set Methods and Dynamic Implicit Surfaces. Springer, Heidelberg (2003)
12. Neuman, S.P.: Journal of Computational Physics 41(2), 270 (1981), doi:10.1016/0021-9991(81)90097-8
13. Courant, R., Friedrichs, K., Lewy, H.: Mathematische Annalen 100(1), 32 (1928)
14. Strain, J.: Journal of Computational Physics 151, 498 (1999)
15. Courant, R., Isaacson, E., Rees, M.: Communications on Pure and Applied Mathematics 5, 243 (1952)
16. Mitchell, I.M.: Application of level set methods to control and reachability problems in continuous and hybrid systems. Ph.D. thesis, Stanford University (2002)
17. Kurzhanski, A.B., Varaiya, P.: Journal of Optimization Theory and Applications 108(2), 227 (2001)
18. Nguyen, L.T., Ogburn, M.E., Gilbert, W.P., Kibler, K.S., Brown, P.W., Deal, P.L.: Simulator study of stall/post-stall characteristics of a fighter airplane with relaxed longitudinal static stability. Tech. Rep. 1538, NASA Langley Research Center (1979)

Modelica Landing Gear Modelling and On-Ground Trajectory Tracking with Sliding Mode Control

Fabrizio Re

Abstract. A control system for an aircraft taxiing on ground based on sliding mode has been developed. The controller is capable to track the trajectory assigned in terms of longitudinal velocity and yaw rate and to drive an aircraft equipped with electric motors in the main gear as well as conventional brakes and nose gear steering. In addition, it can successfully handle saturation of the actuators. The algorithm is shown to be robust against parameter uncertainties (e.g. aircraft mass) as well as low friction coefficients at the interface tyre-ground. In order to test the tracking controller, an accurate virtual aircraft model has been designed in Modelica, with particular attention to the landing gears.

1 Introduction

In recent years, a great deal of attention in aircraft transportation research has focused on reduction of noise and pollution in airports. One specific research topic involves the possibility of driving the aircraft by electric motors integrated in the landing gears instead of the jet engines while taxiing and manoeuvring on the ground. This configuration should be complemented by an on-ground control system capable of driving the aircraft autonomously by tracking an assigned trajectory. In fact, while automated flight controls are an established reality and advanced functions keep being investigated and marketed, the aircraft on ground is still controlled manually by the pilot to a large extent; automated ground controls, if at all present, offer very limited features. Indeed, a robust ground speed and heading control system can offer interesting benefits, like for instance insensitivity to wind gusts while taxiing. In a broader perspective, automated trajectory planning and tracking through an on-ground autopilot function could free the pilot from driving and let him

Fabrizio Re
DLR German Aerospace Center, Institute of Robotics and Mechatronics, Oberpfaffenhofen, Germany
e-mail: fabrizio.re@dlr.de

concentrate on pre-flight or post-flight activity (tower communication, preliminary controls, etc.); even more importantly, such a system could become part of a fully automated airport ground control, which might improve airport capacity and safety.

Research on the topic of aircraft-on-ground dynamics and control has not been extensive so far. Rankin et al. [13, 14, 15] have studied the lateral dynamics of a mid-size commercial aircraft comprehensively; they carried out bifurcation analyses to investigate the aircraft cornering behaviour in relation to longitudinal speed, steering angle at the nose wheel, steering rate, tyre-ground friction and position of the centre of gravity. Roos et al. [16, 17, 18] and Biannic et al. [1] propose control algorithms based on nonlinear or linear parameter-variable (LPV) aircraft models and anti-wind-up compensators. Duprez et al. [3] showed that nonlinear inversion and linear control techniques are not sufficient for yaw control at very low speeds because the system is highly nonlinear due to high sideslips and secondary tyre effects; moreover, errors become important if compared to measured or estimated quantities (e.g. lateral velocity), causing a need for robust control.

On the other side, the problem faced in this paper is not really different from a vehicle dynamics problem, hence it is appropriate to review the literature on this field. Research on autonomous road vehicles is way more advanced, reaching as far as to predictive control applications [4] and optimal control of overactuated systems through control allocation [2]. Solea and Nunes [21] propose a trajectory planning and tracking system for an autonomous vehicle based on sliding mode control.

This paper shows an approach to designing an automatic aircraft-on-ground trajectory tracking control based on sliding mode. The attention will focus on the inner loop controller receiving kinematic inputs (e.g. velocities and/or accelerations) and commanding the actuators accordingly. Such a system can be operated by the pilot himself or can be coupled with an outer-loop trajectory planning controller that receives trajectory data and calculates the required kinematics. First of all, a model of the real aircraft has been realised as a virtual test-rig for the control validation. The landing gear and particularly the tyres have been modelled precisely to enhance the realism of this virtual aircraft. The development of this model with the Modelica commercial compiler *Dymola* will be described in section 2. Section 3 illustrates the control system architecture and design. Some simulation results showing the behaviour of the controlled aircraft will be presented in section 4. Finally, conclusions and the outlook of this work will be discussed in section 5.

2 Virtual Aircraft

Since 1995, the DLR Institute of Robotics and Mechatronics has constantly been developing a Modelica Flight Dynamics Library (FDL) [9] that allows aircraft modelling in a realistic world environment at different levels of detail. The FDL makes use of the open modelling language Modelica [6] and has already been used in a number of internal and international research projects (e.g. [10], [8], [19], [5]).

In the first versions of the FDL, the landing gear was modelled in a rather simple way, by means of an interaction of basic forces between aircraft and ground. This part has been improved in the present work by modelling a proper landing gear with rolling wheels.

A wheel model has been used together with elements of the Modelica Multibody Library to build a model of a whole landing gear. Three variants have been realised:

- steerable two-wheel bogie without brakes, for use as nose gear;
- fixed (non-steerable) two-wheel bogie with brakes, for use as main gear on small and medium aircraft;
- fixed four-wheel bogie with brakes, for use as main gear on large aircraft.

The vertical top end of the gear structure is connected to the airframe model through Modelica Multibody interfaces, which represent physical connections between tridimensional objects with exchange of forces and torques. A suspension and a damper are also modelled between the gear structure and the body interface, with a simplified stopper to limit the suspension travel within the allowed range. All dimensional parameters can be adjusted to model landing gears of different sizes. A *Bearing Friction* block from the standard Modelica Rotational Library [6] is connected to each wheel to model the frictional effects in the axle supports. In addition, there is a connector to the avionic bus for data exchange with the avionic layer.

The steerable gear features a rotational joint and an *Impressed Position* element to rotate the whole bogie around the vertical axis. The steering command given by the pilot is processed through a first-order lowpass filter to reproduce the delays of real steering systems. This part may be replaced with a model of a real steering actuator for a more accurate dynamic simulation. In the brakeable gears instead, brake elements from the standard Modelica Rotational Library are provided for each wheel. In addition, a simple ABS model was developed, appropriately limiting the braking force when the longitudinal slip exceeds 15% to avoid wheel blocking.

For the work described in this paper, the midsize aircraft configuration with two-wheel main gears has been used.

A Modelica tyre model package [23] already available at DLR was adapted to this work. This package features a semi-physical tyre model based on parametrized friction coefficients and their dependency on the slip velocities. The longitudinal and lateral forces are a function of the tyre slips in both longitudinal and lateral direction as well as the camber angle and the normal force at the contact patch. Those functions can be varied through several external parameters, such as e.g. friction coefficients in different predefined conditions (e.g. at zero longitudinal slip, at large longitudinal slip, etc.). The tyre model has been tuned by means of comprehensive experimental data on aircraft tyres provided by a tyre manufacturer. For this purpose, a test rig was modelled in Modelica to replicate those experiments, e.g. rolling wheel with given speed and slip angle imposed. The tyre model parameters were varied to match the simulation results with the experimental data available as closely as possible.

The parameters used in the wheel model include masses, inertias, dimensions as well as tyre model parameters and are arranged in data sheets. The gear model included in the FDL already contains some real data sets available by default, like nose and main gear wheels of a mid-range and long-range aircraft respectively. The user can easily change the parameters to those of any other aircraft by compiling a blank data sheet available as template in the library.

The terrain model of the FDL was enhanced with a basic ground model. It is possible to define a distance from the EGM96 geoid surface and a normal vector to define the ground slope. In the basic version, these quantities are constant parameters which are set at the beginning of the simulation, resulting in a simple, smooth ground with constant slope.

As regards the ground driving system, a simple model of an electric motor with an ideal characteristic was designed. The motor produces a constant torque below a certain rotational speed, and works at constant power above that speed. The input is a real number in the range $[0;1]$ which proportionally controls the amount of torque produced, being 0 = no torque and 1 = maximum torque available at the current speed. One motor model has been connected to each of the main gear wheels; thus the whole aircraft model features four motor models which are independent of each other. A driving/braking controller is provided for each gear. It receives the commanded driving/braking moments from the on-ground control system described further on and regulates the electric moments and the brakes of the two gear wheels. The driving/braking effort is equally divided between the two wheels of each gear.

The steering system has been modelled in a simple way as an integrator of the rate of steering angle commanded by the on-ground controller, as the standard input of a real steering system is assumed to be the steering rate (commanding the position of the servovalve, which in turn regulates the hydraulic flow rate, hence the steering rate). This model also features a first-order delay to account for the delayed response of the real system.

3 On-Ground Controller

The control system shall be capable to handle critical on-ground situations, since it will be used as global control system for the complete ground mission (landing, turn-off, taxiing, U-turning, turn-on, take-off). While it can be argued that a simple linear model should suffice for standard taxiing, situations occur in which linear approximations no longer hold. This is the case at low speeds or large steering angles [3]; also, it can be expected in runway high-speed manoeuvres (e.g. turn-on, turn-off) in poor weather conditions. In addition, the behaviour of the aircraft on the ground is influenced by external factors (e.g. aerodynamics) and a number of parameters that are variable even during the same mission (e.g. tyre behaviour, taxiway surface, variable aircraft mass), therefore the control system must be robust against both unmodelled dynamics and parameter uncertainties.

For these reasons, and also in order to keep the computational effort at a low level, a control architecture has been chosen based on a feedforward nonlinear controller and a feedback sliding mode control. This kind of control is known for offering high robustness in spite of a quite simple mathematical structure ([20], [12]). The control architecture is shown in Fig. 1. It features a feedforward controller based on a nonlinear aircraft model, a feedback sliding mode control that adds robustness

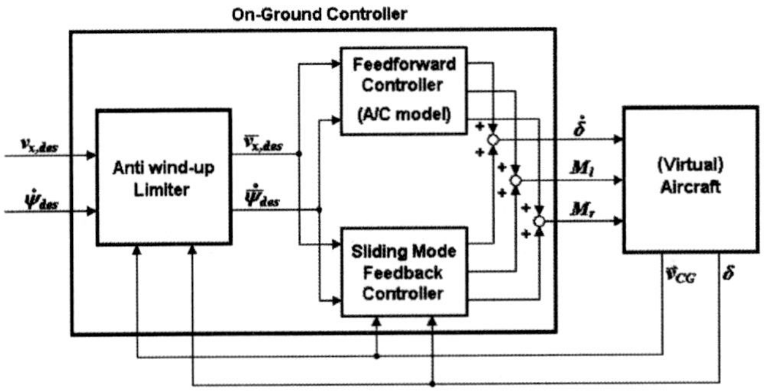

Fig. 1 On-Ground Control Architecture

against uncertainties, and an anti-windup limiter that handles actuator or tyre saturation. The system inputs are the requested longitudinal velocity $v_{x,des}$ and yaw rate $\dot{\psi}_{des}$. The outputs of the controller are the two driving/braking moments M_l, M_r for left and right wheel, and the steering angle rate $\dot{\delta}$ at the front gear. The actual steering angle δ and the velocity vector $\mathbf{v}_{\mathbf{CG}}$ referred to the centre of gravity (CoG) and containing v_x, v_y, $\dot{\psi}$ are then fed back from the virtual aircraft to the controller.

3.1 Feedforward Controller

The feedforward controller is based on an inverse nonlinear aircraft model. Vertical dynamics and aerodynamics are neglected here; their effect will be compensated by the feedback control. Since the wheels of each gear are relatively near to each other, it is reasonable to assume that their behaviour will be approximately the same. Hence only one wheel is modelled in each landing gear. The whole model will then have three wheels, one in the front and two in the rear. The dynamic equations of the aircraft on ground are:

$$\begin{aligned} m\left(\dot{v}_x - v_y\dot{\psi}\right) &= F_{x,CG} = F_{xf}\cos\delta - F_{yf}\sin\delta + F_{xl} + F_{xr} \\ m\left(\dot{v}_y + v_x\dot{\psi}\right) &= F_{y,CG} = F_{xf}\sin\delta + F_{yf}\cos\delta + F_{yl} + F_{yr} \\ J_z\ddot{\psi} &= M_{z,CG} = b_f\left(F_{xf}\sin\delta + F_{yf}\cos\delta\right) + \\ &\qquad -b_r\left(F_{yl} + F_{yr}\right) + a\left(F_{xr} - F_{xl}\right) \end{aligned} \tag{1}$$

where $v_x, v_y, \dot{\psi}$ are the longitudinal and lateral velocity referred to the CoG and the yaw rate in an aircraft-fixed reference system; $F_{x,CG}$, $F_{y,CG}$, $M_{z,CG}$ are the aircraft longitudinal resp. lateral force on the CoG and the yaw moment; F_{xf}, F_{yf} are the longitudinal resp. lateral force on the front gear wheel; the same applies for the left main wheel (with index l) and right main wheel (with index r). The mass m and the moment of inertia J_z around the vertical axis are assumed known. b_f and b_r are the longitudinal distances of the nose gear resp. main gears to the CoG; a is the lateral distance of each main gear to the CoG. In addition to eq. (1), a tyre model with

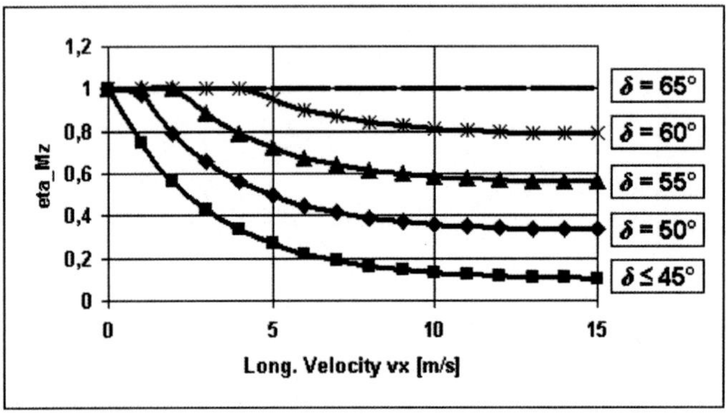

Fig. 2 Coefficient η_{M_z} as function of v_x and δ

relationships between tyre forces and kinematics is needed to close the equation system. A simplified version of Pacejka's Magic Formula [11] has been used here.

The aircraft longitudinal speed v_x and yaw rate $\dot{\psi}$ are the inputs of the feedforward controller and coincide with $v_{x,des}$ and $\dot{\psi}_{des}$. Recall that, in a road vehicle with only one steering axle, the absolute velocity and the yaw rate are coupled [7], so v_y is already determined here. As they are imposed externally, it follows that their derivatives in eq. (1) are known. The outputs of the feedforward controller are the two driving (braking) moments M_l, M_r and the variation of front steering angle $\dot{\delta}$.

Since the system is overactuated, additional equations are needed. The required yaw moment $M_{z,CG}$, resulting directly from $\ddot{\psi}$, can be generated through both the steering system and differential driving moments on the gear wheels, so a strategy must be defined to split it among the actuators. It is firstly noted that increasing the steering angle δ has a little effect on the yaw moment at low longitudinal speeds. As the longitudinal speed decreases, the yaw moment is more efficiently generated by differential driving moments between left and right wheel. Secondly, the greater the steering angle δ already is, the less effective an additional increase of δ is in generating yaw moment. This is obvious from the yaw equation of the on-ground vehicle (1), where the term $F_{yf} \cos \delta$ decreases for increasing δ while F_{yf} will eventually remain constant or even decrease because tyre saturation has occurred. Finally, the achievable steering angle is limited, therefore an increase of δ in the feedforward controller should be prevented once the maximum steering angle has been reached, as only differential moments can produce additional yaw moment in this condition. Based on these considerations, an appropriate continuous function $\eta_{M_z} = f(v_x, \delta)$ is defined, with $0 < \eta_{M_z} \leq 1$. It expresses which part of the needed yaw moment will be generated by differential moments; the part generated by the steering system is then $1 - \eta_{M_z}$. As shown in Fig. 2, η_{M_z} is 1 for $v_x = 0$ and decreases monotonically to 0.1 for $v_x \to \infty$ when $|\delta|$ is smaller than 45°. For $|\delta| > 45°$, η_{M_z} increases; it becomes 1 regardless of v_x for the assumed maximum steering angle of 65°.

Starting from $\eta_{M_z} \cdot M_{z,CG}$ and considering the wheel radius and the distances of the main gears to the CoG, the differential motor moment M_{diff} that generates the yaw moment is calculated. The requested longitudinal force is completely generated by the wheel driving moments; the total sum of the driving moments $2 \cdot M_{long}$ is calculated by dividing the longitudinal force by the wheel radius. Finally, the motor moment on each wheel is:

$$
\begin{aligned}
M_l &= M_{long} + M_{diff} \\
M_r &= M_{long} - M_{diff}
\end{aligned} \tag{2}
$$

The feedforward controller is implemented as a Modelica object. The Modelica parser automatically solves the system for the unknowns δ, M_l, M_r without need to rearrange the equations manually; $\dot{\delta}$ is then directly available by differentiating δ.

3.2 Sliding Mode Feedback Controller

Let $\tilde{v}_x(t) = v_x - v_{x,des}$ and $\dot{\tilde{\psi}}(t) = \dot{\psi} - \dot{\psi}_{des}$ be the tracking errors of the state variables to be controlled. We define two surfaces for both variables (see also [20]):

$$
\begin{aligned}
s_{v_x} &= \dot{\tilde{v}}_x + \lambda_{v_x} \tilde{v}_x = 0 \\
s_{\dot{\psi}} &= \ddot{\tilde{\psi}} + \lambda_{\dot{\psi}} \dot{\tilde{\psi}} = 0
\end{aligned} \tag{3}
$$

Assuming the initial conditions $\tilde{v}_x(0), \dot{\tilde{\psi}}(0) = 0$ and $\dot{\tilde{v}}_x(0), \ddot{\tilde{\psi}}(0) = 0$, it is straightforward to show that $\tilde{v}_x(t), \dot{\tilde{\psi}}(t) = 0$ is the unique solution of eq. (3). The tracking problem is then equivalent to that of finding a control law such that $s_{v_x}, s_{\dot{\psi}} = 0 \;\forall\; t > 0$. The situation where $s_{v_x}, s_{\dot{\psi}} = 0$ is called the sliding condition. However, the real plant could differ from the internal model due to modelling errors (e.g. aerodynamic effects) and parametric uncertainties (e.g. different mass). Hence the control law must first provide that $s \to 0$, that is, that all trajectories must point towards the sliding surface s; once $s = 0$ is reached, that $\dot{s} = 0$. The former requirement can be formalized as follows [20]:

$$
\frac{1}{2}\frac{d}{dt}s^2 \leq -k\,|s| \tag{4}
$$

with the strictly positive constant k. A control law of the form $u = -k\,\mathrm{sgn}\,(s)$ guarantees that the sliding condition is reached in a finite time $t \leq s\,(t=0)\,/k$; this can be proved by integrating eq. 4 in time. Once the sliding surface has been reached ($s = 0$), the dynamics is given by eq. 3, that is, the tracking errors will tend to zero exponentially with the time constants $\lambda_{v_x}^{-1}, \lambda_{\dot{\psi}}^{-1}$. However, such a control law would result in practice in high-frequency chattering around the the sliding condition. Therefore, the sign function is replaced with the following saturation function:

$$
\mathrm{sat}\,(s/\Phi) = \begin{cases} s/\Phi & if\,|s/\Phi| \leq 1 \\ \mathrm{sgn}\,(s/\Phi) & \text{otherwise} \end{cases}
$$

where Φ is a strictly positive constant. The control law guarantees now that the boundary layer $B(t) = \{\mathbf{x}\colon |s(\mathbf{x},t)| \leq \Phi\}$ is reached in a finite time smaller than $s\,(t=0)\,/k$, still allowing good tracking within a predefined tolerance for sufficiently small Φ; more precisely, $|\tilde{x}| \leq 2\Phi \quad \forall t \geq 0$ holds for any trajectory starting inside $B(t)$, where $\tilde{x}$ is the tracking error of the controlled state variable.

In the present problem, two constants Φ_{v_x} and $\Phi_{\dot\psi}$ must be defined for the two variables tracked. For this application, they have both been set as 0.1. While such a precision requirement is reasonable on the longitudinal velocity error where the absolute values are in an order of magnitude of 1–10, it might arguably be too large for the yaw rate. However, the simulation results show that very good tracking can nonetheless be achieved; in fact, a smaller $\Phi_{\dot\psi}$ would affect the computational time seriously without adding any noticeable benefit. Both $\lambda_{v_x}, \lambda_{\dot\psi}$ have been set as 3. This means a time constant of 0.33 s for both tracked variables, which is deemed appropriate considering the system characteristics.

The control laws of the feedback part are then:

$$
\begin{aligned}
M_{long} &= -k_M \operatorname{sat}\left(s_{v_x}\right) \\
M_{diff} &= -\eta_{M_z} k_M \operatorname{sat}\left(s_{\dot\psi}\right) \\
\dot\delta &= -\left(1-\eta_{M_z}\right) k_{\dot\delta} \operatorname{sat}\left(s_{\dot\psi}\right)
\end{aligned}
\tag{5}
$$

where η_{M_z} is the coefficient in Fig. 2. $k_M, k_{\dot\delta}$ are the gains of the control laws. As explained previously, they determine the time in which the sliding condition is reached. It should be noted that the saturation functions cannot be greater than 1 in modulus, so it is not significant to set gains much greater than the actuator saturation values; on the other hand, the feedback control must be able to correct a possibly inadequate feedforward control, which might be at saturation in the opposite direction in the worst case. For this reason, the first two control laws, which command a motor moment, have the same $k_M = 7000$ which is twice the maximum motor moment assumed of 3500 Nm; the third control law, responsible for the front steering, has $k_{\dot\delta} = 3$ being 1.5 rad/s the assumed maximum turning rate of the steering system.

The left and right driving moments resulting from the first two control laws are determined according to eq. (2).

3.3 Anti Wind-Up Limiter

With the control configuration described so far, the derivatives of the control inputs v_x and $\dot\psi$ should be limited through an appropriate rate limiter for two reasons. On one hand, the control system is still not able to deal with the saturation of the actuators. Modelica can only solve the feedforward controller equations as a closed equation system; actuator saturation cannot be implemented there because the equation system would have no exact solution for certain inputs. In the configuration without rate limiting, inputs that would be impossible in reality because of actuator saturation are still valid inputs of the feedforward controller and cause its states to diverge significantly from the real aircraft ones. This, together with the logic of the sliding mode control commanding large controls to compensate that difference, causes an effect analogue to wind-up in integrators ([22], [20]). This is problematic when the control input is not monotone, because a sign change in the control input rate has to bring the model states back to the real states first and has therefore only a delayed effect on the real system. On the other hand, it is important to ensure that the controlled system always remains in a stable region. This means in particular that the tyre slip must never be greater than the slip corresponding to the maximum force in both longitudinal and lateral direction. A solution is to limit the rate, i.e. the derivative, of the control inputs if the inputs themselves diverge from the system states beyond a certain threshold. In this way, the model states can match the real ones as closely as possible and no unrealistic inputs can be passed to the control system.

The threshold variable for the longitudinal direction is chosen to be the difference between real and model longitudinal speed v_x. This is immediately available as measurement (ground speed) in the real system. For the lateral direction, we define:

$$q_{lat} = \text{sgn}\,(\delta)\;\sigma_{y,f} + \begin{cases} 0 & if\,|\delta| < \delta_0 \\ 2.5\,(\delta - \delta_0\;\text{sgn}\;\delta) & \text{otherwise} \end{cases}$$

where δ is the actual steering angle and $\sigma_{y,f}$ is the lateral slip at the front tire calculated as the ratio of tyre lateral velocity to tyre absolute velocity: $\sigma_{y,f} = v_{y,f}/\sqrt{v_{x,f}^2 + v_{y,f}^2}$. These velocities are available through either measurements or estimates. It should be noted that this anti wind-up limiter only needs to look at the slip conditions at the front tyre to ensure permanence in the stable region. If the rear tyres should lose grip, the real yaw rate would become quickly greater than the one in the controller model, and the sliding mode feedback controller would immediately contrast this by countersteering and commanding a moment difference in the opposite direction, much like an ESP system in a road vehicle.

The anti wind-up limiter regulates the derivative of the outputs (which will be the commanded states of the feedforward controller) as a function of the inputs, their derivatives and their respective threshold variable. If the threshold variable exceeds the allowed value, the output rate is decreased and follows the rate of the real system, which is supposed to be the maximum feasible rate since the model and the real system were diverging. If the threshold variable is below the allowed value, the output rate is equal to the input rate to guarantee tracking precision; if there is a difference between feedforward controller states and real system states due to prior rate limitation, the rate limiter no longer intervenes, and if necessary, the output rate is additionally increased resp. decreased to reduce the state difference to zero.

This anti wind-up arrangement results in limiting the aircraft longitudinal force resp. the yawing moment whenever the real system cannot physically achieve them. However, if the pilot or the outer-loop controller commands a different velocity or yaw rate whose derivatives are smaller in modulus or change sign, the anti wind-up limiter no longer intervenes. In this sense, this limiter does not bring additional risk of pilot induced oscillations, nor does it limit or delay the pilot inputs in any way if no saturation has occurred.

4 Simulation Results

A virtual aircraft model of a mid-size aircraft has been built in Dymola 7.1 using the FDL (section 2) and electrical motors have been added to each of the four rear wheels. The tracking controller composed of a nonlinear feedforward controller

Table 1 Different features of the simulations carried out

Simulation	Features
#1	Mass = 56900 kg, friction coeff. = 0.8 ∗ nominal
#2	Mass = 66900 kg, friction coeff. = 1 ∗ nominal
#3	Mass = 76900 kg, friction coeff. = 0.6 ∗ nominal

and a sliding mode feedback controller has been connected appropriately and three different simulations have been carried out. To test the controller robustness, the aircraft mass and the tyre-ground friction coefficients stored in the virtual aircraft model were different in each simulation, while the aircraft model in the feedforward controller remained unchanged (see table 1). The inputs are predefined time-dependent profiles of longitudinal velocity and yaw rate. The simulations were run in Dymola using the DASSL algorithm with variable time step.

It can be seen from the longitudinal velocity diagram (Fig. 3) that the controller accelerates the aircraft as much as possible to reach the requested v_x; the acceleration differences are due to the different mass in each case. The braking phases track the requested deceleration precisely in all cases since the brakes are never saturated.

The yaw rate is tracked with very good precision in all simulations. The curves in the diagram (Fig. 4) mostly overlap with the requested yaw rate. During the last corner, the simulation with the highest mass and the smallest friction coefficients shows a clear delay in reaching the requested yaw rate because the front wheels are saturated and yaw rate limiting occurs to keep the aircraft in the condition of maximum lateral force. This indeed assures a stable operation; on the other side, it affects the precision of the trajectory tracking because the cornering radius is reached later than requested. This circumstance shall be handled at the outer control loop level which assigns the requested inputs, that is the pilot or the autopilot system; they must

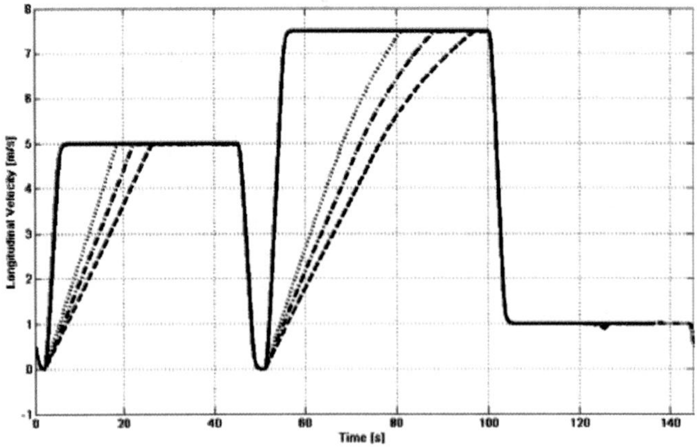

Fig. 3 Longitudinal Velocity - solid line = input velocity; dotted line = simulation #1; dashed-dotted line = simulation #2; dashed line = simulation #3

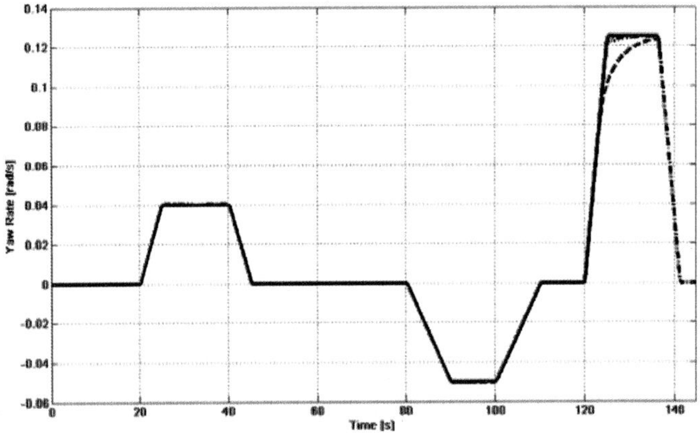

Fig. 4 Yaw Rate - solid line = input velocity; dotted line = simulation #1; dashed-dotted line = simulation #2; dashed line = simulation #3

recognize (by feedback from the rate limiters) that the physical limits of the system have been reached and recalculate new inputs to be able to follow the trajectory.

5 Conclusions

An aircraft on-ground control system based on sliding mode has been designed in Modelica. The controller allows precise tracking of assigned longitudinal velocity and yaw rate by controlling the steering angle, the braking system and electric motors connected to the main gear wheels; at the same time, it takes actuator saturation and physical limits of the system into account. It has been shown through simulation results that the controller is robust against different aircraft mass and varying tyre-ground friction coefficients.

During this work, new Modelica landing gear and tyre-ground interaction models have been developed as an extension to the DLR Flight Dynamics Library. The new parts allow detailed simulation of an aircraft moving on ground with particular regard to the dynamic aspect. The new models can be used as sub-components in larger Modelica models, thus allowing global, multi-disciplinary aircraft simulation. For instance, the landing shock and its influence on the aircraft structure can be analysed in a flexible aircraft model; also, side forces acting on the landing gears can be quantified, and their components can be dimensioned accordingly.

References

1. Biannic, J.M., Marcos, A., Jeanneau, M., Roos, C.: Nonlinear simplified LFT modelling of an aircraft on ground. In: 2006 IEEE International Conference on Control Applications (2006)

2. Bünte, T., Andreasson, J.: Global chassis control based on inverse vehicle dynamics models. In: 19th IAVSD Symposium Supplement to Vehicle System Dynamics (44) (2005)
3. Duprez, J., Mora-Camino, F., Villaumé, F.: Aircraft-On-Ground Lateral Control for Low Speed Manoeuvers. In: Proceedings of the 16th IFAC Symposium on Automatic Control in Aerospace, St. Petersburg, Russia (June 2004)
4. Falcone, P., Borrelli, F., Asgari, J., Tseng, H., Hrovat, D.: Predictive Active Steering Control for Autonomous Vehicle Systems. IEEE Transactions on Control Systems Technology 15 (2007)
5. Fielding, C., Varga, A., Bennani, S., Selier, M.: Advanced techniques for clearance of flight control laws. LNCIS, vol. 283. Springer, Heidelberg (2002)
6. Fritzson, P., Bunus, P.: Modelica-a general object-oriented language for continuous and discrete-event system modeling and simulation. In: Simulation Symposium, Annual 0, 0365 (2002), doi:http://doi.ieeecomputersociety.org/10.1109/SIMSYM.2002.1000174
7. Isermann, R.: Fahrdynamik-Regelung. ATZ/MTZ-Fachbuch. Vieweg+Teubner (2006)
8. Looye, G.: Integrated flight mechanics and aeroelastic aircraft modeling using object-oriented modeling techniques. In: Proceedings of the AIAA Modeling and Simulation Technologies Conference, Portland, USA (1999)
9. Looye, G.: The new DLR flight dynamics library. In: Bachmann, B. (ed.) Proceedings of the 6th International Modelica Conference, vol. 1, pp. 193–202. The Modelica Association, Bielefeld (2008), `http://www.modelica.org`
10. Magni, J.F., Bennani, S., Terlouw, J.: Robust flight control - a design challenge. LNCIS, vol. 224. Springer, London (1997)
11. Pacejka, H.B.: Tyre and Vehicle Dynamics. Butterworth-Heinemann, Butterworths, London (2006)
12. Perruquetti, W., Barbot, J.P.: Sliding Mode Control in Engineering. Marcel Dekker Inc., New York (2002)
13. Rankin, J., Coetzee, E., Krauskopf, B., Lowenberg, M.: Bifurcation and Stability Analysis of Aircraft Turning on the Ground. Journal of Guidance, Control, and Dynamics 32(2) (March-April 2009)
14. Rankin, J., Krauskopf, B., Lowenberg, M., Coetzee, E.: Operational Parameter Study of an Aircraft Turning on the Ground. In: Progress in Industrial Mathematics at ECMI 2008, vol. 15 (2010)
15. Rankin, J., Krauskopf, B., Lowenberg, M., Coetzee, E.: Nonlinear Analysis of Lateral Loading During Taxiway Turns. Journal of Guidance, Control and Dynamics 33(6) (November-December 2010)
16. Roos, C., Biannic, J.M.: Aircraft-on-gound Lateral Control by an Adaptive LFT-based Anti-windup Approach. In: 2006 IEEE Conference on Control Applications (2006)
17. Roos, C., Biannic, J.M., Tarbouriech, S., Prieur, C.: On-ground Aircraft Control Design Using an LPV Anti-windup Approach, vol. 365, pp. 117–145. Springer, Heidelberg (2007)
18. Roos, C., Biannic, J.M., Tarbouriech, S., Prieur, C.M.J.: On-ground Aircraft Control Design Using a Parameter-Varying Anti-windup Approach. Aerospace Science and Technology 14 (2010)
19. Rouwhorst, W.F.J.A.: Robust and Efficient Autopilots Control Laws Design, Demonstrating the Use of Modern Robust Control Design Methodologies in the Autoland System Design Process - the REAL Project. In: Proocedings of the Aeronautic Days 2001, Hamburg, Germany (2001)

20. Slotine, J.J.E., Li, W.: Applied Nonlinear Control. Prentice Hall, Englewood Cliffs (1991)
21. Solea, R., Nunes, U.: Trajectory planning and sliding-mode control based trajectory-tracking for cybercars. Integrated Computer-Aided Engineering 14, 33–47 (2007)
22. Yokoyoma, M., Kim, G.N., Tsuchiya, M.: Integral sliding mode control with anti-windup compensation and its application to a power assist system. Journal of Vibration and Control 16(4), 503–512 (2010)
23. Zimmer, D., Otter, M.: Real-time models for wheels and tyres in an object-oriented modelling framework. Vehicle System Dynamics 48, 189–216 (2010)

Obstacle Avoidance Strategy for Micro Aerial Vehicle

Cezary Kownacki

Abstract. Obstacle avoidance of Micro Aerial Vehicle (MAV) in urban environment is the most complex, difficult and essential part of the autonomous flight problems. The paper presents a simple ad-hoc strategy using a pair of miniature laser rangefinders (i.e. MLR100) and two PIDs cooperating with an obstacle avoidance controller. The strategy can be realized as an additional routine integrated with the autopilot's firmware (i.e. MP2128^{HELI}). The main advantage of the proposed strategy is simplicity of its implementation in small-sized MAVs and its power efficiency. All previous works, especially the vision-based ones require high performance microprocessors which is an important limitation when applying on real MAVs. On the other hand, the autonomous controller, which is based on optic flow sensors, is easy to implement on even tiny MAVs, but optic flow sensors require applicable level of contrast variation, so their performance is strongly sensitive to weather conditions. The proposed idea of the autonomous obstacle avoidance system in urban environment was simulated using MATLAB – SIMULINK software. In the real flight all computations and controls will be realized by the advanced autopilot, hence the rest of autonomous control and complex flight dynamics are not included in the simulation. The assumption allows to spot a more focused attention on the obstacle avoidance problem and a simpler model of the MAV can be used in the simulation. The results presenting the 2D trajectories confirm that the effectiveness and safety of the proposed strategy of obstacle avoidance is attainable during the real flight in streets' canyons.

Keywords: obstacle avoidance, autonomous flight, autopilot, streets' canyons, miniature aerial vehicle laser rangefinder.

1 Introduction

The main purpose of research on autonomous MAVs is to achieve the best autonomy and flight performance in unknown environments as much as possible.

Cezary Kownacki
Bialystok Technical University, ul. Wiejska 45C, 15-351 Białystok, Poland
e-mail: cezarkw@poczta.onet.pl

The urban environment is the most difficult and uncertain area where MAVs can perform flights. The assurance and rapid obstacle detection are highly expected because of the significant variability and variety of objects appearing around the flight path. Flying over an urban environment also requires the waypoint navigation in order to planning paths which fulfills the goals of the flight and it is essential to the diverse missions of MAVs. Therefore the robustness of the autonomous flight is depended on navigation algorithms and autonomous control laws.

Many researchers work on improving the effectiveness of the autonomous systems used in MAVs. Most of them focused on the vision-based systems and digital image processing [1, 3, 5]. Vision-based systems allow observing a visible area around the MAVs and it makes possible to create a 3D or 2D map. The map can be further utilized by the path planners [12].But the obvious disadvantage lies in the complexity of the image processing routines, which require efficient DSPs with high power consumption. In turns it shortens the flight duration. Increasing the capacity of batteries also increases the weight of the whole system. Moreover, the image stabilization is an additional problem. Although a mechanical stabilizer can help, it means more electronics and more computations. Vision-based systems are the future of autonomous flight control systems, but today's technology still fails to meet the specifications of the tiny MAVs.

Some of the recent researches on the autonomous systems presented the control strategy based on optic flow sensors and it gives satisfying results. However [10, 11, 12], most demonstrations required special patterns as landmarks in order to guarantee sufficient visual features to estimate safe flight path. If the variation of the contrast is too low the control strategy based on optic flow will be useless and collisions will be unavoidable. That's why optic flow sensors performance is strongly sensitive to weather conditions and illumination.

Designing of miniature fixed-wing or delta-wing aerial vehicles requires to be light-weighted and low-power consumption. Hence it requires a power efficiency of the on-board equipment realizing autonomous flight control. On the other hand, complex algorithms of the autonomous guidance and obstacle avoidance must involve high performance microprocessors, which are necessary in rapid computing for a safe and sound flight paths. Most of the developed algorithms are only offline experiments, computer simulations or online experiments carried out in selected environments without verifications in different flight scenarios or weather conditions [1-11]. So the challenge is to design an effectively autonomous flight control system which will guarantee a reliable autonomous flight in various real world scenarios and it could be realized by available equipment [12].

We propose an strategy of autonomous obstacle avoidance, while micro aerial vehicle flies in streets' canyons, which can be realized on readily available equipment: two miniature laser rangefinders MRL100 (from Aerius Photonics) and advanced autopilot MP2128^{heli} (from MicroPilot). We couldn't use laser scanner, because it is too heavy to place them in our airframe (wingspan: 1245 mm and total weight: 1200 gram). Some strategies of obstacle avoidance use a single laser rangefinder, but such configuration disables possibility of flight in streets' canyons

[12]. So we decided to use configuration, which enables autonomous flight in canyons, but instead of optic flow sensors we choose laser rangefinders [12].All necessary computations can be realized by the autopilot, so any additional electronics would not be needed. The manufacturer of the autopilot delivers a development tool called XTENDER, which allows modifying the autopilot firmware with additional user code. Although the strategy presented in this paper is validated only in simulations, it proves that there is a great chance to introduce it into real MAV. And it will be the next step of our research.

2 Obstacle Avoidance Strategy

2.1 Hardware and Configuration

Two miniature laser rangefinders (LRF) are the base of proposed autonomous obstacle avoidance strategy (Figure 1). They will measure ranges between MAV and objects lying on both sides of the vehicle track. Controlling these ranges by PIDs' loops and obstacle avoidance controller will allow avoiding obstacles and continuing flight in street canyon [12]. The great advantages of miniature laser rangefinder are its features predisposing it as perfect sensor which can be used in miniature aerial vehicle. Features of the miniature laser rangefinder are presented in Table 1.

Table 1 Features of miniature laser rangefinder MRL100

Feature	Value
Weight	26 g
Power rating	<400mW
Size	1,25”x1,5”x1,6”
Pulse repetition	500Hz
Resolution	<0,2 m
Range	~0.1m to >100m
Divergence angle	10x10 mrad
Output	UART
Filters	averaging and median filters

The most important features are small size and weight, power efficiency, high repetition rate and UART serial port which is indispensable for measured data transmission. Also detectable maximum range fits specific conditions of obstacle avoidance during flight in streets' canyons.

Of course there are many other sensors which can be used in obstacle avoidance system, but their possibilities make them losing out with the laser rangefinder. Miniature low resolution cameras are smaller and lighter, but image processing routines running on energy efficient microprocessors, are still insufficient to reduce influence of contrast and obstacle illumination conditions. The extremely case, when always cameras fail, is a night flight. Also optic flow sensors are sensitive to contrast and weather conditions. For instance they are unable to detect range to obstacle properly while MAV is flying nearby uniform building wall with no applicable contrast differences. Such situation can happen in urban environment regularly. The last kind of sensor, which is frequently used in robotics, is an ultrasonic sonar. Its outdoor version has similar weight and dimensions as the laser rangefinder. The main advantage of ultrasonic sensors is a wide beam cone, so the scanned area is much bigger than the laser rangefinder measurement at single point. In the other hand scanning range is limited up to a few meters against more than 100 meters achieved by the laser rangefinder. If MAV flies fast, a few meters scanning range would not be enough to ensure space for turn. Hence in our opinion the laser rangefinders are the best sensors for our purpose at the moment which fulfill requirements of obstacle detection in urban environment.

Fig. 1 Miniature laser rangefinder MRL100 from AERIUS.

Configuration of miniature laser rangefinders is a significant aspect of the system reliability and usability. In our opinion the sensors configuration, which enables flight possibility in streets' canyons, is as follows: both laser beams are placed tangent to the MAV plane and they are forward looking, but swept back from nose by specified angle. So two laser beams will create "V" shape (Figure 2).

The configuration satisfies both aspects of autonomous flight in urban environment: flight in streets' canyons and obstacle avoidance, because laser rangefinders scan not only area at MAV front but also on both sides in directions of potential flight. Angle between laser beams can be evaluated from formula:

$$\alpha = 2 \cdot \arctan\left(\frac{R}{\sqrt{3R}}\right) = \frac{\pi}{3}[rad] \tag{1}$$

where: R- minimum turn radius.

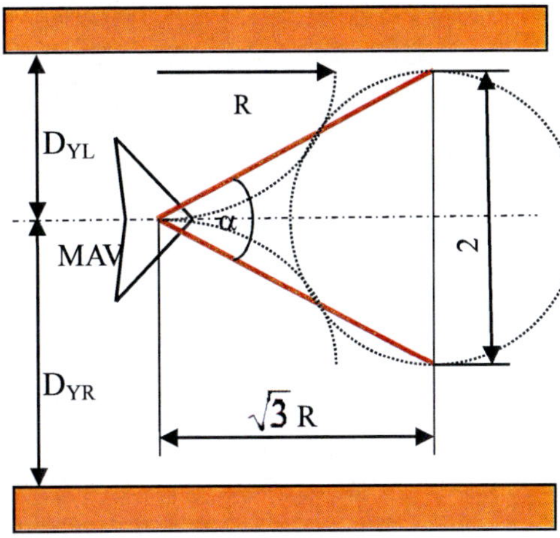

Fig. 2 The "V" shape created by laser beams of miniature laser rangefinders: α - angle between laser beams, R – minimum turn radius.

Thanks to the chosen configuration, the α angle is independent from flight parameters such as speed or minimum turn radius.

Typical use of ultrasonic or laser rangefinders is scanning and sweeping the environment while the mobile robot is moving. Such strategy allows rendering the environment occupancy grid map and thus it is possible to generate the safe trajectory. While occupancy grid map calculations are simple for 2D environments, things goes worse for 3D environments, which are typical for aerial vehicles. Dimensions and sizes of occupancy grid map are increasing since MAV 3D operating zone is much bigger than 2D zone typical for mobile robot. Even if the MAV flight zone could be scanned only in 2D, the laser sensor must be stabilized by special gimbals' platform, so beam targeting also won't be easy.

That's why laser rangefinders will be fixed to the MAV body. Thus measured ranges should be independent from MAV body frame orientation, what is essential for determining true range to obstacle. Hence we have to transform coordinates

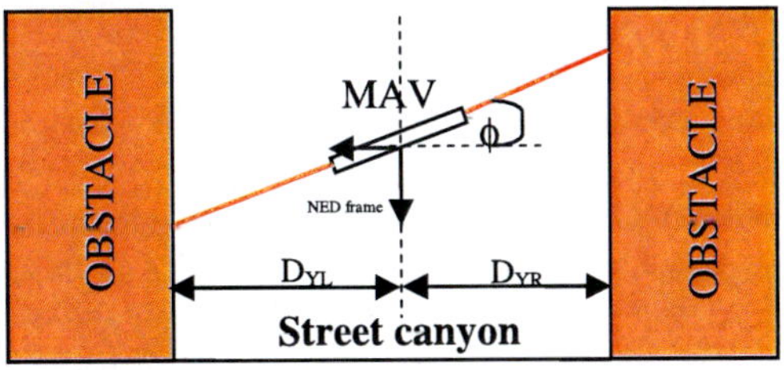

Fig. 3 MAV flight in street canyon, ϕ - roll angle, D_{YL} – NED frame *y* coordinate of obstacle on the left, D_{YR} – NED frame *y* coordinate of obstacle on the right.

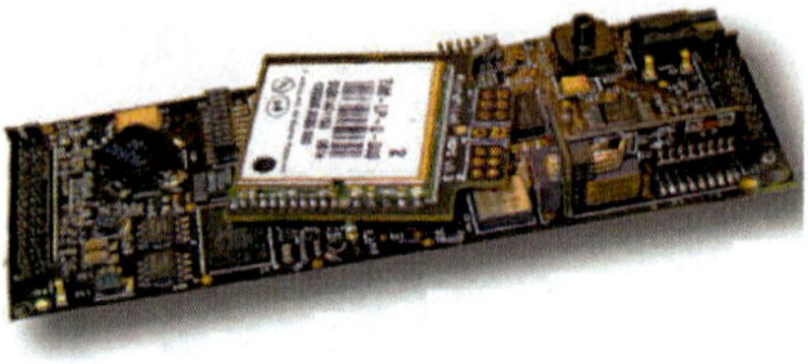

Fig. 4 MP2128heli from MicroPilot

Table 2 Main features of autopilot system MP2128^{heli}

Feature	Value
Weight with GPS	28 g
Power rating	140 mA @ 6.5V
Size	100x40x15 mm
selectable inner loop update rates	30/60/180 Hz
user definable PID feedback loops (for camera stabilization etc)	8
user definable table lookup functions	8
autonomous takeoff and landing supported by AGL	yes
change altitude at waypoint	yes
change airspeed at waypoint	yes
user definable holding patterns	yes
user definable error handlers (loss of GPS, low battery etc.)	yes
servo resolution	11bit
servo update rate	50 - 200 Hz
altimeter maximum altitude	12,000m
3 axis accelerometers	2g
maximum angular rate	150° per second
attitude update rate	200 Hz
12 state Kalman filter	yes
GPS update rate	4Hz
supports DGPS accuracy	yes
waypoints	1000

of obstacle location from MAV body frame to NED frame (North East Down) (Figure 3). Required values of actual roll and actual pitch angles are available from autopilot gyros.

Reliability of proposed autonomous obstacle avoidance strategy strongly relies on predictable flight control, so the high quality autopilot system should be used. Creating individual autopilot for our purpose seems be pointless, because probably it would take much time and final result couldn't be as good as available commercial autopilot from MicroPilot or Procerus. Table 2 presents main features of MP2128^{heli} (Figure 4), the most advanced autopilot system from MicroPilot, which will be suitable enough for our purpose.

2.2 *Diagram of Obstacle Avoidance Control System*

Usage of laser rangefinders in the autonomous obstacle avoidance strategy requires introducing adequate control algorithm. We propose the algorithm, which is based on two PIDs' loops cooperating with obstacle avoidance controller. The routine of the algorithm will be integrated with the autopilot software responsible for standard control laws and waypoint navigation. Obstacle avoidance controller will decide about actual priority of each control task i.e. waypoint navigation and obstacle avoidance including flight in streets' canyons. Obstacle avoidance controller will be able to override standard autopilot controls.

The diagram of designed autonomous obstacle avoidance system is presented in Figure 5.

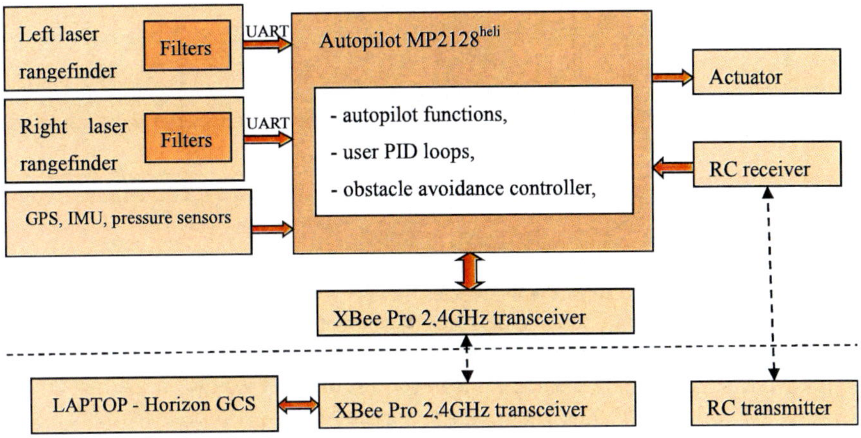

Fig. 5 The diagram of designed autonomous obstacle avoidance system

GPS and IMU sensors are internal parts of the autopilot and they are used in standard control functions and waypoint navigation. GCS (ground control station) is communicating with thc autopilot via 2,4 GHz radio modem. MIAV can be also controlled manually using standard RC equipment. The strategy presented in the paper will be only an extension of standard functions of MP2128^{Heli}.

Laser rangefinders should measure range between MAV and buildings during a flight in streets' canyons. Of course there are many smaller objects i.e. street lights, traffic lights, trees, road signs which can't be properly detected by the obstacle avoidance strategy. This is an important disadvantage. Any small obstacle detected by laser rangefinders will introduce peaks in ranges sequence similar to random disturbances or noise. So it could probably affect on MAV behavior negatively making it unstable. We decided to filter out these peaks as useless information. It can be achieved by enabling median and averaging filters which are standard features of MLR100 laser range finder. Hence it requires making an assumption that the MAV is able to fly in streets' canyons at patrol altitude.

2.3 Signal Flow and Processing

The diagram of signal flow and signal processing is presented in Figure 6.

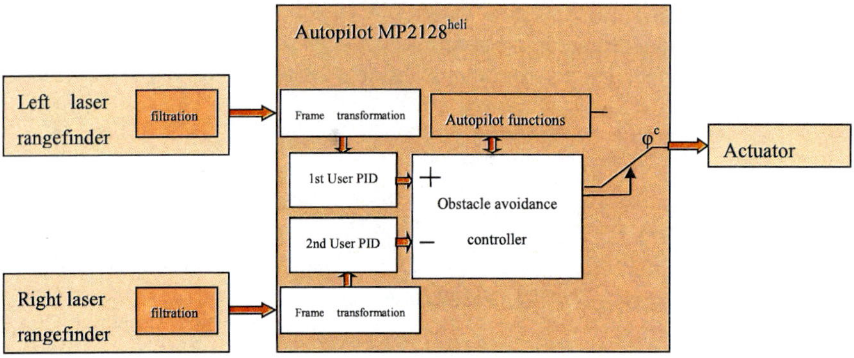

Fig. 6 The diagram of signal flow

The initial step of signal processing is filtration essential for eliminating range peaks generated by small obstacles, wind or other disturbances. It will be done by internal filters of laser rangefinder MLR100 (median filter and averaging filter). MLR100 transmits filtered range values over a serial port, so they are almost ready for use. If the MLR100 internal signal processing become to be inadequate, there will be possibility to introduce software filtration realized by the autopilot. But we have to keep in mind fact that additional computations are able to slow down autopilot operation.

The next step of signal processing is coordinates frame transformation. Obstacle coordinates expressed in body frame can be derived from range value when α angle is known (Equation 1).

$$\begin{pmatrix} x_{BF} \\ y_{BF} \end{pmatrix} = \begin{pmatrix} \cos\left(\frac{\alpha}{2}\right) \\ \sin\left(\frac{\alpha}{2}\right) \end{pmatrix} \cdot D^{*}{}_{BF} \tag{2}$$

where: D^{*}_{BF} – measured range (body frame),
α – angle between laser beams,
x_{BF}, y_{BF} – obstacle coordinates in body frame.

Next the body frame obstacle coordinates must be transformed to NED (North-East-Down) frame. It allows to determine true range between obstacle and MAV. Formula of transformation is given by equation:

$$\begin{pmatrix} x_{NED} \\ y_{NED} \\ z_{NED} \end{pmatrix} = \begin{pmatrix} 1 & 0 & 0 \\ 0 & \cos\phi & \sin\phi \\ 0 & -\sin\phi & \cos\phi \end{pmatrix} \cdot \begin{pmatrix} \cos\theta & 0 & \sin\vartheta \\ 0 & 1 & 0 \\ -\sin\theta & 0 & \cos\theta \end{pmatrix} \cdot \begin{pmatrix} x_{BF} \\ y_{BF} \\ 0 \end{pmatrix} \tag{3}$$

where: x_{BF}, y_{BF} – body frame obstacle coordinates,
x_{NED}, y_{NED} – NED frame obstacle coordinates,
ϕ – actual roll angle,
θ – *actual* pitch angle.

NED frame coordinates can be used to compute range D^{*}_{NED} expressed in NED frame.

$$D^{*}_{NED} = \sqrt{\left(x_{NED}^{2} + y_{NED}^{2}\right)} \tag{4}$$

The next step is computing derivative of D^{*}_{NED} and use it as control signal, what makes obstacle avoidance system sensitive to velocity and direction of approaching MAV to obstacle. Because if $(D^{*}_{NED})/dt$ equals zero, it would mean that the MAV is flying parallel to the obstacle. The maximum value of $(D^{*}_{NED})/dt$ presents situation when the MAV is flying towards to obstacle, so collision probability increases.

The final step of signal processing is setting up thresholds which allow ignoring obstacles placed further than specified safe range value D_{safe}. It also allows ignoring obstacles, which MAV are receding from ($(D^{*}_{NED})/dt$ is a positive value (Equation 5)). Thresholds ensures that relevant component of PID's input signal will equal zero (Equation 6). The PID's input signal PID_{in} is defined as follows:

$$\begin{cases} E_D = D^{*}_{NED} - D_{safe} & D^{*}_{NED} - D_{safe} < 0 \\ E_D = 0 & D^{*}_{NED} - D_{safe} \geq 0 \\ E^{*}_D = \dfrac{d(D^{*}_{NED})}{dt} & \dfrac{d(D^{*}_{NED})}{dt} < 0 \\ E^{*}_D = 0 & \dfrac{d(D^{*}_{NED})}{dt} \geq 0 \\ PID_{in} = E_D + E^{*}_D \end{cases} \tag{5}$$

Where: E_D – first component of PID input signal,
D_{safe} – safe range to obstacle,
D^*_{NED} – range to obstacle expressed in NED frame,
E_D^* – the MAV to obstacle approaching rate- second component of PID input signal,
PID_{in} – input signal of PID controller.

Presented definition of PID's input signal makes the obstacle avoidance strategy activate only when the MAV is approaching towards or too close to obstacle. Because MAV to obstacle approaching rate E_D^* is proportional to the MAV speed, the obstacle avoidance strategy is indirect function of the MAV speed, Whereas PIDs' outputs has the sense of desired roll angle with opposite sign for each PID (left PID sign + and right PID sign -).

2.4 Obstacle Avoidance Controller

The role of the obstacle avoidance controller is to specify what flight control task should be considered actually: waypoint navigation or obstacle avoidance. Its input signals are outputs of PIDs and actual values of both D^*_{NED} ranges. The obstacle avoidance controller works as a kind of commuting element which switches control of desired roll angle between PIDs and autopilot waypoint navigation dependently on actual obstacle position. The desired heading angle of delta-wing MAV is a function of desired roll angle ϕ^C and desired velocity V^C. Then having constant desired velocity V^C the strategy of desired roll angle control realized by the obstacle avoidance controller is described by Equation 6 (Figure7).

$$\phi^C = \begin{cases} \phi_{autopilot} & ;\ \phi_{PIDright} = 0 \cap \phi_{PIDleft} = 0 \\ \phi_{PIDright} + \phi_{PIDleft} & ;\ \phi_{PIDright} \neq 0 \cap \phi_{PIDleft} \neq 0 \cap \left|\phi_{PIDright}\right| \neq \left|\phi_{PIDleft}\right| \\ \phi_{PIDright} & ;\ \left(\phi_{PIDright} \neq 0 \cap \phi_{PIDleft} = 0\right) \cup \\ & \left(\begin{matrix} \phi_{PIDright} \neq 0 \cap \phi_{PIDleft} \neq 0 \cap \left|\phi_{PIDright}\right| = \left|\phi_{PIDleft}\right| \cap \\ D^{**}_{NEDleft} > \mathrm{D}^{**}_{\mathrm{NEDright}} \end{matrix}\right) \\ \phi_{PIDleft} & ;\ \left(\phi_{PIDright} = 0 \cap \phi_{PIDleft} \neq 0\right) \cup \\ & \left(\begin{matrix} \phi_{PIDright} \neq 0 \cap \phi_{PIDleft} \neq 0 \cap \left|\phi_{PIDright}\right| = \left|\phi_{PIDleft}\right| \cap \\ D^{**}_{NEDleft} \leq \mathrm{D}^{**}_{\mathrm{NEDright}} \end{matrix}\right) \end{cases} \tag{6}$$

where: ϕ^C – the MAV desired roll angle,
$\phi_{autopilot}$ – desired roll angle from waypoint navigation,
$\phi_{PIDleft}$ – desired roll angle from left PID,
$\phi_{PIDright}$ – desired roll angle from right PID,

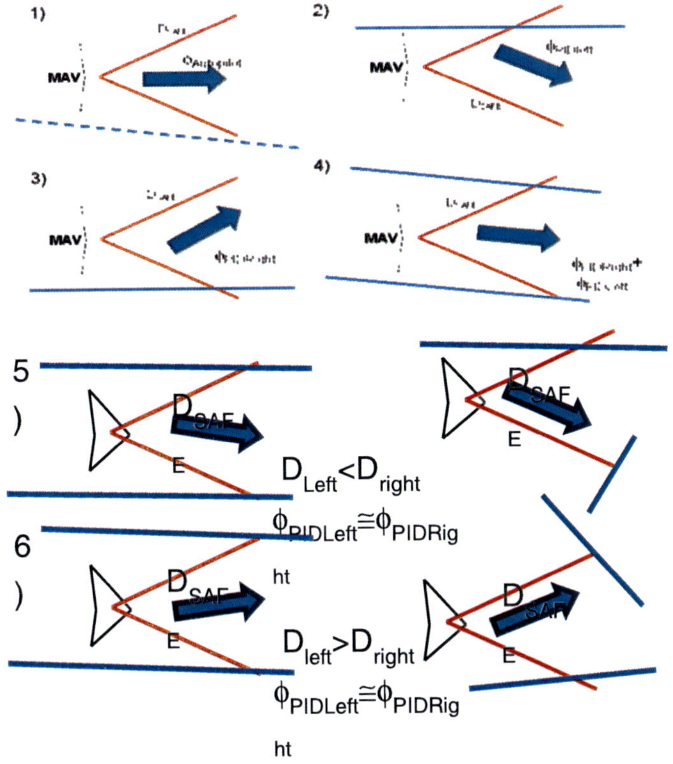

Fig. 7 Control strategy of desired roll angle realized by obstacle avoidance controller

If outputs of both PIDs are zero, waypoint navigation will control the desired roll angle (Figure 7-1). When both of them have two different non-zero absolute values, desired roll angle will be calculated as a sum of both PIDs' outputs (Figure 7-4). But at the moment it's necessary to mention that $\phi_{PIDleft}$ is positive desired roll angle (right wing down) and $\phi_{PIDright}$ is negative desired roll angle (right wing up). So $\phi_{PIDleft}$ makes the MAV turning right and $\phi_{PIDright}$ makes the MAV turning left. Another situation is occurring when only the left PID output is zero, or when both PIDs outputs have the same non-zero absolute values and D^*_{NED} value from left sensor is greater than D^*_{NED} from right sensor (Figure 7-3 and Figure 7-6). In the case the right PID output will control the desired roll angleϕ^C. The last one case is opposite to the pervious. The left PID output will control the desired roll angleϕ^C, when only the right PID output is zero, or both PIDs outputs have the same non-zero absolute values and the D^*_{NED} value from left sensor is smaller than the D^*_{NED} from right sensor(Figure 7-2 and Figure 7-5).

In summary the obstacle avoidance controller controls the desired roll angle accordingly to actual state of surrounding obstacles and coordinates of the next waypoint.

3 Simulation

3.1 Delta Wing MAV Model

The simulations of proposed obstacle avoidance strategy were performed in MATLAB – SIMULINK environment, because at the moment there is no possibility to involve light simulator, like it's possible for Kestrel autopilot [12]. Hence adequate model of MAV is required for deriving simulated flight paths. We used simple delta-wing MAV model which is defined by following equations [6]:

$$
\begin{aligned}
\dot{x} &= V \cdot \cos\psi, \\
\dot{y} &= V \cdot \sin\psi, \\
\psi &= \frac{g}{V} \cdot \tan\phi, \\
\dot{V} &= \alpha_V (V^C - V), \\
\phi &= \alpha_\phi (\phi^C - \phi).
\end{aligned}
\qquad (7)
$$

where: x, y – MAV actual coordinates,
V– the MAV actual velocity,
V^C– the desired MAV velocity,
ψ– the MAV actual heading angle,
ϕ– the actual roll angle,
ϕ– the desired roll angle,
α_ϕ, α_V– time constants of MAV dynamics,
g – the gravitational constant.

The model is simple, but it's sufficient for our simulations, which main purpose is to present trajectory of MAV flight in different scenarios of surrounding obstacles. The model excludes several parameters of MAV flight in the real world, which are more dependent on autopilot performance than rather on the obstacle avoidance strategy. Creating the individual autopilot and its control software is not the purpose of our research. We decided to use commercial MP2128^{Heli} autopilot, so we can afford to introduce presented simplifications. But the important part of our future flight tests will be precise autopilot tuning to ensure good flight performance in different weather conditions.

Our simulations assume that the flight path is controlled by two parameters: flight velocity V^C and desired roll angle ϕ^C. We also assumed that the flight velocity V^C is a constant and only desired roll angle ϕ^C is controlled by the obstacle avoidance strategy.

3.2 Autopilot Model

A autopilot model is also required for good simulation performance. Because the main purpose of the simulation is presenting only 2D flight paths, autopilot model functions were restricted to the waypoint navigation realized by heading angle control. The autopilot was modeled by a single PID controller which minimizes the track angle error between the actual heading angle and the actual bearing angle pointing on the next waypoint. So the model can be formulated as follows:

$$\begin{aligned} \psi_{err} &= \psi_{WP} - \psi, \\ \psi_{WP} &= \arctan\left(\frac{y_{WP} - y}{x_{WP} - x}\right) \\ \phi_{autopilot} &= PID(\psi_{err}) \end{aligned} \tag{8}$$

where: ψ_{err} – the track angle error,
ψ – the actual heading angle,
ψ_{WP} – the actual bearing angle pointing on the next waypoint WP,
x, y – actual coordinates of MAV,
x_{WP}, y_{WP} – coordinates of the next waypoint WP,
PID – the transitions function of PID,
$\phi_{autopilot}$ – the desired roll angle.

The desired roll angle $\phi_{autopilot}$ is the model output and it is applied when the actual range to the nearest obstacle is greater than safe range D_{safe} (Figure 7-1).

The whole simulation model is presented in figure below.

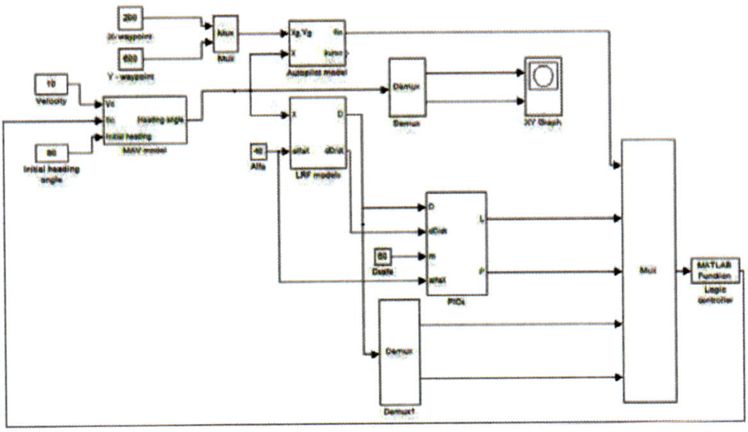

Fig. 8 The simulation model of proposed autonomous obstacle avoidance strategy.

Four synthetic maps representing streets' canyons and obstacles were used in flight simulations.

4 Results

Four maps of environment characterizing different flight scenarios were prepared to present the control strategy possibilities. Two of them are concerning the problem of flight in streets' canyons. Next two present the problem of obstacle avoidance. In each simulation the level flight with initial heading angle was started at the point marked as SP. The level flight target was the next waypoint marked as WP. Results

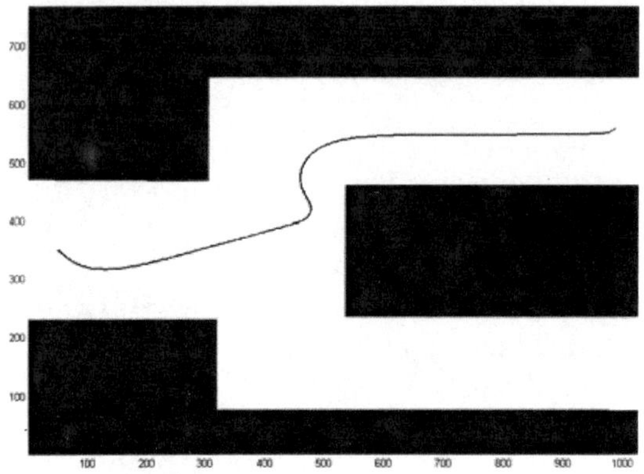

Fig. 9 The MAV flight path in the first simulation: fight in street canyon. Initial heading angle –0^0, starting point SP – x=50,y=350, the next waypoint WP – x=1000,y=550.

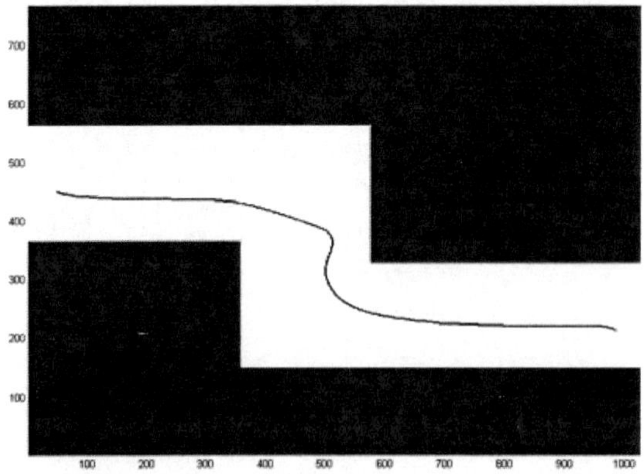

Fig. 10 The MAV flight path in the second simulation: fight in street canyon. Initial heading angle – 0^0, starting point SP – x=50,y=450, the next waypoint WP – x=1000,y=200.

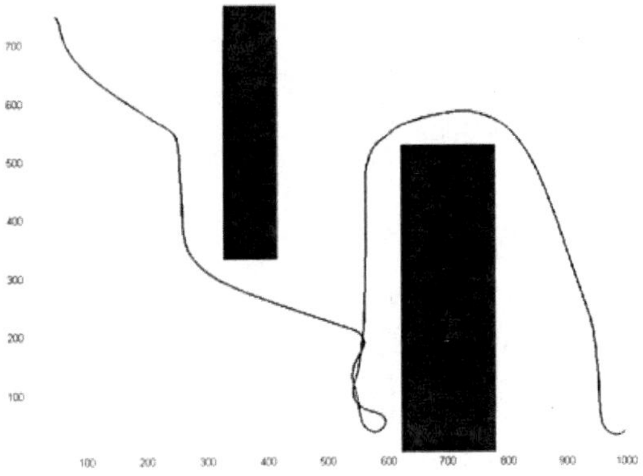

Fig. 11 The MAV flight path in the first simulation of obstacle avoidance. Initial heading angle – 0^0, starting point SP – x=50,y=750, the next waypoint WP – x=1000,y=50.

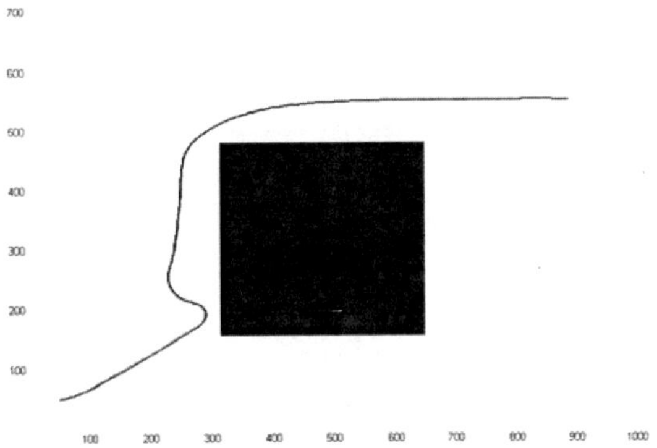

Fig. 12 The MAV flight path in the second simulation of obstacle avoidance. Initial heading angle – 50^0, starting point SP – x=50,y=50, the next waypoint WP – x=900,y=600.

of simulations are presented in Figures 9-12. Obstacles are represented by shapes filled with black color. The maps boundaries were treated as obstacles, so MAV should also avoid them. Following values of simulations' and models' parameters were used: V^C=12 m/s, D_{safe}=60 m, α=30^0, α_V=0.5, α_ϕ=0.5. It also was assumed that there was no wind or other external disturbances during simulated flight.

If we take a look on results presenting flight paths, we will notice that the MAV always flies holding a safe range to obstacle. Of course sizes of all obstacles are much bigger than MAV size. So the MAV position is represented by single points creating the flight paths in all figures. That's why some turns seems to be too sharp to perform by delta-wing MAVs. However more details of the obstacle avoidance strategy behavior can be determined by observation and analysis of obstacle avoidance controller inputs and outputs: D^*_{NED}, ϕ^C, $\phi_{autoplot}$ and ψ. Plots of specified signals are presented in figures 13-15. They are concerning the simulated scenario of obstacle avoidance from figure 11.

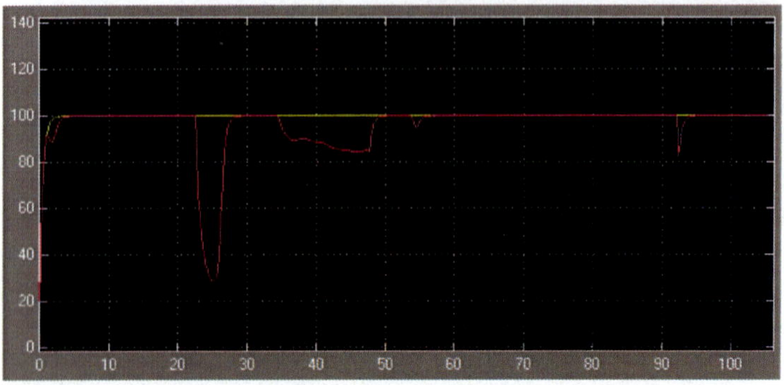

Fig. 13 Simulated range signals D^{*}_{NED}. Yellow – left laser rangefinder, pink – right laser rangefinder.

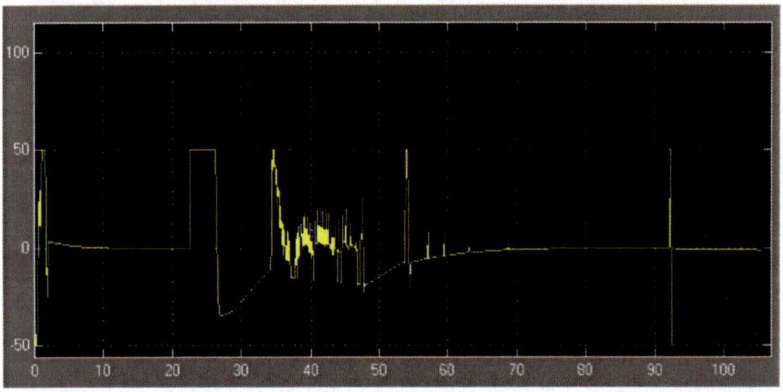

Fig. 14 Simulated ϕ^{C} signals – desired roll angle.

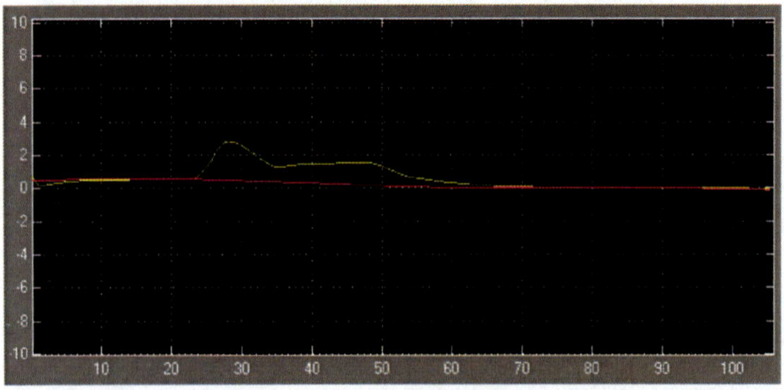

Fig. 15 Simulated $\phi_{autopilot}$ signals – desired roll angle from waypoint navigation (yellow) and ψ– actual heading angle (pink).

Simulated flight path are not the shortest path of possible MAV flight (for example in Figure 11), because in our research we skipped the problem of path optimization. Path planner requires computations which can't be handled by the autopilot without performance drop. In some research path planner was design as a separate PC application, and computed optimal patch was uploaded to the autopilot as waypoints list [12]. So presented path planner can't be treated as a part of autonomous system. Absence of path planning function is some disadvantage of presented obstacle avoidance strategy. But the main research purpose is to present functional idea of obstacle avoidance strategy which would be based on available equipment and fit requirements of micro aerial vehicles.

5 Testing Airframe

To verify received simulation results we start to prepare a testing airframe based on delta wing RC model made with ELAPOR. ELAPOR is a material which gives great possibility of introducing any required modifications in airframe body. It's also resistant to damages in the case of eventual crash and easy to repair.

The MP2128^{heli} autopilot, speed controller and standard 35 MHz RC receiver were mounted inside the airframe cockpit which is closed with plastic cover protecting onboard electronics. XBee Pro 2,4 GHz radio modem, GPS antenna and batteries were placed in cavities located in both wings symmetrically. We also designed special balsa enclosures for MLR100 laser rangefinders which will be fixed to the wings surface. The laser rangefinders are communicating with the autopilot via two separate serial ports. These serial ports were configured by modifying assignment of unused digital outputs and inputs of MP2128heli onboard processor. The autopilot firmware was expanded with routine of obstacle avoidance.

Fig. 16 The testing airframe – delta wing MAV

Actually we are preparing to perform first flights which will present the airframe performance and give required data for the autopilot and the obstacle avoidance strategy tuning. When all configuration procedures are completed, the airframe will be fully functional and ready to perform autonomous flights, which verifies obstacle avoidance strategy in different collision scenarios.

6 Conclusion

Figures 9-12 present results of four different simulations which routed flight paths of MAV being under control of the obstacle avoidance strategy. They are clearly proving possibilities of the strategy and allow expecting its effectiveness in real experiments. In each case the MAV flight succeeded and the next waypoint was reached without collision with surrounding obstacles. Results presented in Figures 9-10indicate also possibilities of the MAV flight in streets' canyons. This is the essential feature for the MAV autonomous flight in urban environment. The obstacle avoidance strategy meets also requirements of autonomous obstacle avoidance what was shown in Figures 9-10.

In summary it can be underscored that the aim of the paper was achieved and the research can be continued. Basing on these results it can be predicted that the proposed obstacle avoidance strategy will effective and can be handled by commercial autopilots like a MP2128heli and micro aerial vehicles. The advantage of the algorithm would be fast response thanks to real-time signal processing. Others methods involving path planners or image processing introduce some delays by a reason of much computation time. The obstacle avoidance strategy verification will be the next step of our research.

Acknowledgment

The research was funded by finances of Polish Ministry of Science and Higher Education for science in 2008-2010 as developmental project No. O R00 0059 06.

References

1. Frew, E., Sengupta, R.: Obstacle Avoidance with Sensor Uncertainty for Small Unmanned Aircraft. In: 43rd IEEE Conference on Decision and Control, Paradise Island, Bahamas (December 2004)
2. Frew, E., Spry, S., Howell, A., Hedrick, J., Sengupta, R.: Flight Demonstrations of Self-Directed Collaborative Navigation of Small Unmanned Aircraft. In: AIAA 3rd Unmanned Unlimited Technical Conference, Workshop & Exhibit, Chicago, IL (September 2004)
3. Frew, E., McGee, T., Kim, Z., Xiao, X., Jackson, S., Morimoto, M., Rathinam, S., Padial, J., Sengupta, R.: Vision-Based Road Following Using a Small Autonomous Aircraft. In: IEEE Aerospace Conference, Big Sky, MT (March 2004)
4. Frew, E., Langelan, J.: Receding Time Horizon Control for Passive, Non-cooperative UAV See-and-Avoid. In: IEEE International Conference on Robotics and Automation, Barcelona, Spain (April 2005)

5. He, Z., Venkataraman Iyer, R., Chandler, P.R.: Vision-based UAV flight control and obstacle avoidance. In: IEEE Automatic Control Conference (2006)
6. Saunders, J.B., Call, B., Curtis, A., Beard, R.W., McLain, T.W.: Static and Dynamic Obstacle Avoidance in Miniature Air Vehicles. In: Infotech@Aerospace, AIAA, Arlington, Virginia, September 26 - 29 (2005)
7. Anderson, E.: Extremal control and unmanned air vehicle trajectory generation. Master's thesis, Brigham Young University (April 2002)
8. Dong, T., Liao, X.H., Zhang, R., Sun, Z., Song, Y.D.: Path Tracking and Obstacle Avoidance of UAVs - Fuzzy Logic Approach. In: FUZZ 2005 The 14th IEEE International Conference on Fuzzy Systems, Reno, pp. 43–48 (2005)
9. Scherer, S.: Flying Fast and Low Among Obstacles: Methodology and Experiments. The International Journal of Robotics Research 27(5), 549–574 (2008)
10. Zufferey, J.-C., Klaptocz, A., Beyeler, A., Nicoud, J.-D., Floreano, D.: A 10-gram Vision-based Flying Robot. Advanced Robotics 21(14), 1671–1684 (2007)
11. Zufferey, J.-C., Beyeler, A., Floreano, D.: Optic Flow to Steer and Avoid Collisions in 3D. In: Flying Insects and Robots. Springer, Berlin (2009)
12. Beyeler, A., Zufferey, J.-C., Floreano, D.: OptiPilot: control of take-off and landing using optic flow. In: EMAV 2009, Delft, Netherland (2009)
13. Griffiths, S., Saunders, J., Curtis, A., Barber, B., McLain, T., Beard, R.: Obstacle and Terrain Avoidance for Miniature Aerial Vehicles. In: Advances in Umanned Aerial Vehicles, State of Art and the Road to Autonomy, pp. 213–244. Springer, Heidelberg (2007)

Post-Optimal Sensitivities of Flight Trajectories with Respect to Selected Parameters

C. Büskens, F. Fisch, and F. Holzapfel

Abstract. In the paper at hand, the sensitivities of optimized flight trajectories with respect to selected parameters are to be investigated. Especially atmospheric influences like wind or deviations from the standard atmosphere shall be regarded since those parameters strongly influence the respective optimal trajectory.

1 Introduction

In many guidance or control applications, an aircraft, e.g. an UAV, ought to follow a given reference trajectory. Often, the reference trajectory therefore is obtained by a high-fidelity offline-optimization of the specific aircraft mission before the flight mission is conducted. Then, this reference trajectory together with an appropriate guidance or control law are stored onboard. For the offline-optimization, certain parameters concerning the environmental conditions are required. Usually, the values of those parameters are not known exactly a priori or might even be subject to change during the flight mission. Consequently, the offline-computed reference trajectory may neither be valid nor optimal any more. In most cases, an online re-optimization of the flight-mission utilizing the high-fidelity offline-optimization algorithm is by far too time-consuming. One possibility would be to use a simplified optimization algorithm that is real-time capable and can thus be implemented onboard. For example, Yakimenko et al. ([8], [9]) utilize high-order polynomials as reference functions to describe an aircraft trajectory, thereby reducing the optimization problem to the determination of a few parameters. The solution of the reduced optimization problem can then be accomplished online. In this paper, another possibility is introduced where valid, sub-optimal reference trajectories can be computed in real-time onboard (see e.g. [6], [7]) without any time-consuming online re-optimization of the aircraft trajectory. At this, post-optimal sensitivities of the optimized trajectory with respect to selected parameters are calculated offline. The sensitivities can then be utilized to restore in real-time a valid, high-quality close-optimal reference trajectory for the guidance of aerial vehicles if perturbations in the environmental parameters occur. In this case, the tracking of the guidance trajectories by any flight control system is facilitated since the guidance trajectories can be corrected with respect to changed environmental influences.

2 Aircraft Simulation Model

The flight system dynamics are represented by a 3-DoF point-mass simulation model. In the following, the equations of motion are depicted and the proper inclusion of the wind influence within the point-mass simulation model is explained. Furthermore, an atmospheric model is given that allows to take into account temperature deviations from the standard atmosphere.

The position equations of motions are specified in the *NED*-Frame. The *NED*-Frame is a coordinate system with its origin located at the aircraft's center of gravity. The x-axis of the *NED*-Frame is pointing into northward, the y-axis into eastward and the z-axis into downward direction. The position equations of motion are:

$$\begin{pmatrix} \dot{x} \\ \dot{y} \\ \dot{z} \end{pmatrix}_O^E = \begin{pmatrix} V_K^G \cdot \cos\chi_K^G \cdot \cos\gamma_K^G \\ V_K^G \cdot \sin\chi_K^G \cdot \cos\gamma_K^G \\ -V_K^G \cdot \sin\gamma_K^G \end{pmatrix}_O \tag{1}$$

where V_K is the kinematic velocity, χ_K the kinematic course angle and γ_K the kinematic climb angle. Assuming flight over flat, non-rotating earth, the following simplified translation equations of motion result:

$$\dot{V}_K^G = \frac{\sum \left(F_x^G\right)_K}{m} \tag{2}$$

$$\dot{\chi}_K^G = \frac{\sum \left(F_y^G\right)_K}{m \cdot V_K^G \cdot \cos\gamma_K^G} \tag{3}$$

$$\dot{\gamma}_K^G = -\frac{\sum \left(F_z^G\right)_K}{m \cdot V_K^G} \tag{4}$$

Here, m represents the aircraft mass and is assumed to be constant. The total sum of external forces $\Sigma\,\vec{\mathbf{F}}$ given in the Kinematic Reference Frame K is made up of the gravitational force $\vec{\mathbf{F}}_G$, the aerodynamic force $\vec{\mathbf{F}}_A$ and the propulsive force $\vec{\mathbf{F}}_P$:

$$\sum \left(\vec{\mathbf{F}}^G\right)_K = \left(\vec{\mathbf{F}}_A^G\right)_K + \left(\vec{\mathbf{F}}_P^G\right)_K + \left(\vec{\mathbf{F}}_G^G\right)_K \tag{5}$$

The gravitational force $\vec{\mathbf{F}}_G$ is given by

$$\left(\vec{\mathbf{F}}_G^G\right)_K = \mathbf{M}_{KO} \cdot \left(\vec{\mathbf{F}}_G^G\right)_O = \mathbf{M}_{KO} \cdot \begin{pmatrix} 0 \\ 0 \\ mg \end{pmatrix}_O \tag{6}$$

where g denotes the gravitational constant. The transformation matrix $\mathbf{M}_{KO}$ is given by Eq. (18). The aerodynamic force $\vec{\mathbf{F}}_A$ with respect to the center of gravity G of the aircraft is computed by the following formula:

$$\left(\vec{\mathbf{F}}_A^G\right)_K = \mathbf{M}_{KA}\left(\vec{\mathbf{F}}_A^G\right)_A = \mathbf{M}_{KA}\begin{pmatrix} -D \\ Q \\ -L \end{pmatrix}_A = \mathbf{M}_{KA}\begin{pmatrix} -\bar{q}\cdot S\cdot C_D \\ \bar{q}\cdot S\cdot C_Q \\ -\bar{q}\cdot S\cdot C_L \end{pmatrix}_A \tag{7}$$

where D denotes the drag, Q the aerodynamic force in the direction of the y-axis of the Aerodynamic Frame A and L the lift. The transformation $\mathbf{M}_{KA}$ is given by Eq. (17). The dynamic pressure $\bar{q}$ is calculated from the air density ρ and the aerodynamic velocity V_A:

$$\bar{q} = \frac{1}{2}\cdot\rho\cdot\left(V_A^G\right)^2 \tag{8}$$

The aerodynamic coefficients are given by the following equations, where a linear lift curve and a quadratic drag polar are implemented:

$$C_D = C_{D0} + k\cdot\left(C_L - C_{L0,C_D}\right)^2 + C_{D\dot{p}}\cdot\dot{p}^2 \tag{9}$$

$$C_Q = C_{Q\beta}\cdot\beta_{A,CMD} \tag{10}$$

$$C_L = C_{L0} + C_{L\alpha}\cdot\alpha_{A,CMD} \tag{11}$$

Here, α_A is the commanded angle of attack and β_A the commanded sideslip angle. The propulsion force $\vec{\mathbf{F}}_P$ is assumed to be aligned with the x-axis of the aircraft's kinematic velocity:

$$\left(\vec{\mathbf{F}}_P^G\right)_K = \begin{pmatrix} T \\ 0 \\ 0 \end{pmatrix}_K \tag{12}$$

The following relationship for the thrust force T is implemented that applies to propeller-powered aircraft:

$$T = T_{\max}\cdot\frac{V_{ref}}{V_A^G}\cdot\left(\frac{\rho}{\rho_{ref}}\right)^{n_\rho}\cdot\delta_{T,CMD} \tag{13}$$

where T_{max} is the maximum thrust, V_{ref} the reference speed, ρ_{ref} the reference air density and n_ρ the density exponent. δ_T is the commanded thrust lever position. For the computation of the air density ρ, the atmospheric model given by Eq. (14) is utilized where the air density is a function of the aircraft's downward position z given in the *NED*-Reference Frame:

$$\rho = \rho_{ref}\left[1 + \frac{\gamma_{ref}}{T_{ref}}\cdot(-z)\right]^{\left(-\frac{g}{R\cdot\gamma_{ref}}-1\right)} \tag{14}$$

Here, T_{ref} is the reference temperature, γ_{ref} the reference temperature gradient and R the specific gas constant. Eq. (14) is valid in the troposphere layer, i.e. it is valid for altitudes between -2km and 11km. The reference values involved in the atmosphere model are taken from the International Standard Atmosphere DIN ISO 2533 [5]. Temperature deviations from the norm standard atmosphere can be taken into account by adjusting the reference temperature T_{ref} accordingly:

$$T^*_{ref} = T_{ref} + \Delta T_{ISA} \tag{15}$$

Furthermore, a differential equation for the aerodynamic bank angle μ_A is added to the simulation model:

$$\dot{\mu}^G_A = p_{CMD} \tag{16}$$

with p being the commanded roll rate. The aerodynamic force $\vec{\mathbf{F}}_A$ acting on the aircraft's center of gravity is computed in the Aerodynamic Frame A and then transformed into the Kinematic Reference Frame K with the help of the transformation matrix $\mathbf{M}_{KA}$ between the Aerodynamic Frame A and the Kinematic Reference Frame K (see Eq. (7)). The transformation matrix $\mathbf{M}_{KA}$ between the Aerodynamic Frame A and the Kinematic Frame K is given by:

$$\mathbf{M}_{KA} = \mathbf{M}_{KO} \cdot \mathbf{M}_{O\bar{A}} \cdot \mathbf{M}_{\bar{A}A} \tag{17}$$

$$\mathbf{M}_{KO} = \begin{bmatrix} \cos\chi^G_K \cdot \cos\gamma^G_K & \sin\chi^G_K \cdot \cos\gamma^G_K & -\sin\gamma^G_K \\ -\sin\chi^G_K & \cos\chi^G_K & 0 \\ \cos\chi^G_K \cdot \sin\gamma^G_K & \sin\chi^G_K \cdot \sin\gamma^G_K & \cos\gamma^G_K \end{bmatrix} \tag{18}$$

$$\mathbf{M}_{O\bar{A}} = \begin{bmatrix} \cos\chi^G_A \cdot \cos\gamma^G_A & -\sin\chi^G_A & \cos\chi^G_A \cdot \sin\gamma^G_A \\ \sin\chi^G_A \cdot \cos\gamma^G_A & \cos\chi^G_A & \sin\chi^G_A \cdot \sin\gamma^G_A \\ -\sin\gamma^G_A & 0 & \cos\gamma^G_A \end{bmatrix} \tag{19}$$

$$\mathbf{M}_{\bar{A}A} = \begin{bmatrix} 1 & 0 & 0 \\ 0 & \cos\mu^G_A & -\sin\mu^G_A \\ 0 & \sin\mu^G_A & \cos\mu^G_A \end{bmatrix} \tag{20}$$

where χ_A represents the aerodynamic course angle and γ_A the aerodynamic climb angle. The aerodynamic course angle and the aerodynamic climb angle have to be restored utilizing the respective kinematic relationships, taking into account the influence of the wind. Therefore, the aerodynamic velocity of the aircraft's center of gravity with its components denoted in the *NED*-Frame is first obtained by:

$$\left(\vec{\mathbf{V}}^G_A\right)^E_O = \left(\vec{\mathbf{V}}^G_K\right)^E_O - \left(\vec{\mathbf{V}}^G_W\right)^E_O \tag{21}$$

Here, $\vec{\mathbf{V}}_W$ represents the wind field given in the *NED*-Reference Frame with its components u_w, v_w and w_w, where u_w points into the northward, v_w into the eastward and w_w into the downward direction. With the help of the aerodynamic velocity $\vec{\mathbf{V}}_A$, the aerodynamic course angle χ_A, the aerodynamic climb angle γ_A and the absolute aerodynamic velocity V_A can be restored:

$$\left(\vec{\mathbf{V}}_A^G\right)_O^E = \begin{pmatrix} u_A^G \\ v_A^G \\ w_A^G \end{pmatrix}_O^E = \begin{pmatrix} V_A^G \cdot \cos\chi_A^G \cdot \cos\gamma_A^G \\ V_A^G \cdot \sin\chi_A^G \cdot \cos\gamma_A^G \\ -V_A^G \cdot \sin\gamma_A^G \end{pmatrix}_O^E \tag{22}$$

$$V_A^G = \left\|\left(\vec{\mathbf{V}}_A^G\right)_O^E\right\|_2 = \sqrt{\left[\left(u_A^G\right)_O^E\right]^2 + \left[\left(v_A^G\right)_O^E\right]^2 + \left[\left(w_A^G\right)_O^E\right]^2} \tag{23}$$

$$\chi_A^G = \arctan\left(\frac{\left(v_A^G\right)_B^E}{\left(u_A^G\right)_B^E}\right) \tag{24}$$

$$\gamma_A^G = \arctan\left(\frac{\left(w_A^G\right)_B^E}{\sqrt{\left[\left(u_A^G\right)_B^E\right]^2 + \left[\left(v_A^G\right)_B^E\right]^2}}\right) \tag{25}$$

In total, the aircraft simulation model comprises seven states that are the northward position x, the eastward position y, the downward position z, the kinematic velocity V_K, the kinematic course angle χ_K, the kinematic climb angle γ_K and the aerodynamic bank angle μ_A. The command inputs are the aerodynamic angle of attack α_A, the aerodynamic sideslip angle β_A, the roll rate p and the commanded thrust lever position δ_T. For the solution of the air race trajectory optimization problem described in paragraph 3, the commanded sideslip angle β_A has been set to zero and the commanded thrust lever position δ_T has been set to 1.

3 Air Race Trajectory Optimization Problem

The task of the air race trajectory optimization problem is to find the minimum possible race time t_f for a given race track. The optimal control problem can be stated as follows: Determine the optimal control history

$$\mathbf{u}_{opt}(t) \in \mathbb{R}^m \tag{26}$$

and the corresponding optimal state trajectory

$$\mathbf{x}_{opt}(t) \in \mathbb{R}^n \tag{27}$$

that minimize the Bolza cost functional

$$J = e\left(\mathbf{x}(t_f), t_f\right) + \int_{t_0}^{t_f} L\left(\mathbf{x}(t), \mathbf{u}(t), t\right) dt \tag{28}$$

subject to the state dynamics

$$\dot{\mathbf{x}}(t) = \mathbf{f}(\mathbf{x}(t), \mathbf{u}(t), t) \tag{29}$$

the initial and final boundary conditions

$$\boldsymbol{\psi}_0(\mathbf{x}(t_0), t_0) = 0 \qquad \boldsymbol{\psi}_0 \in \mathbb{R}^q \qquad q \leq m + n \tag{30}$$

$$\boldsymbol{\psi}_f(\mathbf{x}(t_f), t_f) = 0 \qquad \boldsymbol{\psi}_f \in \mathbb{R}^p \qquad p \leq m + n \tag{31}$$

the interior point conditions

$$\mathbf{r}(\mathbf{x}(t_i), \mathbf{u}(t_i), t_i) = 0 \qquad \mathbf{r} \in \mathbb{R}^k \tag{32}$$

and the equality and inequality conditions

$$\mathbf{C}_{eq}(\mathbf{x}(t), \mathbf{u}(t), t) = 0 \qquad \mathbf{C}_{eq} \in \mathbb{R}^r \tag{33}$$

$$\mathbf{C}_{ineq}(\mathbf{x}(t), \mathbf{u}(t), t) \leq 0 \qquad \mathbf{C}_{ineq} \in \mathbb{R}^s \tag{34}$$

For the air race trajectory optimization problem, the Bolza cost functional reduces to a Mayer functional since the only objective is to minimize the final time:

$$J = t_f \tag{35}$$

The state vector **x**, the control vector **u** and the according state dynamics were given in paragraph 2. The initial and final boundary conditions as well as the interior point conditions are defined by the positions respectively the type of the various air race gates. While no equality path constraints are present, inequality path constraints arise from safety regulations or from aircraft performance limits. First of all, a certain ground clearance has to be respected by the pilots:

$$z_{\min} - z(t) \leq 0 \tag{36}$$

Furthermore, the safety regulations require that an upper limit and a lower limit of the load factor $n_{z,B}$ in the direction of the z-axis of the Body Fixed Reference Frame B is never exceeded:

$$\begin{aligned} n_{Z,\min} - n_Z(t) &\leq 0 \\ n_Z(t) - n_{Z,\max} &\leq 0 \end{aligned} \tag{37}$$

Besides the never-exceed speed V_{max} given by the safety regulations, the velocity V_A of the aircraft must not go below the stall speed V_{stall} of the aircraft:

$$\begin{aligned} V_{stall} - V_A(t) &\leq 0 \\ V_A(t) - V_{\max} &\leq 0 \end{aligned} \tag{38}$$

Additional inequality path constraints are due to aircraft performance limitations with respect to the minimum and maximum angle of attack α_A as well as the minimum and maximum roll rate p_K:

$$\begin{aligned} \alpha_{A,\min} - \alpha_A(t) &\le 0 \\ \alpha_A(t) - \alpha_{A,\max} &\le 0 \end{aligned} \tag{39}$$

$$\begin{aligned} p_{K,\min} - p_K(t) &\le 0 \\ p_K(t) - p_{K,\max} &\le 0 \end{aligned} \tag{40}$$

The corresponding values for the various bounds of the inequality path constraints are listed in Table 1.

Table 1 Path constraints specifications

Path constraints specifications		
Altitude	z_{min} [m]	7.5
load factor	$n_{z,min}$ [-]	-2.0
	$n_{z,max}$ [-]	12.0
Velocity	V_{stall} [m/s]	25.0
	V_{max} [m/s]	102.9
angle of attack	$\alpha_{A,min}$ [rad]/[°]	-0.35/-20.02
	$\alpha_{A,max}$ [rad]/[°]	0.35/20.02
roll rate	$p_{K,min}$ [rad/s]/[°/s]	-7.33/-420.0
	$p_{K,max}$ [rad/s]/[°/s]	7.33/420.0

4 Parametric Optimal Control Problems with Control and State Constraints

Problem (1) to (40) defines a parametric nonlinear control problem subject to control and state constraints of the following form:

Parametric control problem *OCP(p)*:

$$\begin{array}{llcl} \textit{Minimize} & F(x,u,p) & = & g(x(t_f),p) \\ \textit{subject to} & \dot{x}(t) & = & f(x(t),u(t),p), \\ & x(t_0) & = & \varphi(p), \quad \psi(x(t_f),p) = 0, \\ & C(x(t),u(t),p) & \le & 0, \quad t \in [t_0,t_f]. \end{array} \tag{41}$$

Here $x(t) \in \mathbb{R}^n$ denotes the state of a system and $u(t) \in \mathbb{R}^m$ the control in a given time interval $[t_0, t_f]$. Data perturbations in the system are modeled by a parameter $p \in P := \mathbb{R}^q$. The functions $g:\mathbb{R}^n \times P \longrightarrow \mathbb{R}$, $f:\mathbb{R}^{n+m} \times P \longrightarrow \mathbb{R}^n$, $\varphi:P \longrightarrow \mathbb{R}^n$, $\psi: \mathbb{R}^n \times P \longrightarrow \mathbb{R}^r$, $0 \le r \le n$ and $C:\mathbb{R}^{n+m} \times P \longrightarrow \mathbb{R}^k$ are assumed to be sufficiently smooth on appropriate open sets. The admissible class of control functions is that of piecewise continuous controls. The final time t_f is assumed to be free.

Note, that the Bolza cost functional in (28) can easily be transformed to the objective functional in (41) by simply introducing an additional differential equation for the integral term in (28) and initial value zero.

5 Numerical Solution of Optimal Control Problems by Nonlinear Programming Techniques

Direct optimization methods for solving the optimal control problem *OCP(p)* are based on a suitable discretization of problem (41). In this section we describe an approach where for simplicity the discussion will be restricted to Euler's method. It is mentioned that the sensitivity analysis that is utilized for the computation of the sub-optimal reference trajectories is not limited to the collocation method described below. Moreover, any other direct optimization method like for example a multiple shooting method could constitute the basis for the sensitivity analysis outlined in paragraph 6.

Choose a natural number N and let $\tau_i \in [t_0, t_f]$, $i = 1, \ldots, N$, be mesh or grid points with

$$t_0 = \tau_1 < \ldots < \tau_{N-1} < \tau_N = t_f . \tag{42}$$

For notational simplicity we assume that the discretization in (42) is equidistant:

$$h := \frac{t_f - t_0}{N-1}, \qquad \tau_i = t_0 + (i-1) \cdot h, \quad i = 1, \ldots, N . \tag{43}$$

Let the vectors $x^i \in \mathbb{R}^n$ respectively $u^i \in \mathbb{R}^m$, $i = 1, \ldots, N$, be approximations of the state variable $x(\tau_i)$, respectively control variables $u(\tau_i)$ at the grid points. Then Euler's approximation applied to the differential equation in (41) yields

$$x^{i+1} = x^i + h \cdot f(x^i, u^i, p), \quad i = 1, \ldots, N-1 . \tag{44}$$

We treat the control variables together with t_f as the only optimization variables while the state variables are obtained recursively from the state equation (44). Therefore, we consider the optimization variable

$$z := (u^1, u^2, \ldots, u^{N-1}, u^N, t_f) \in \mathbb{R}^{N_z}, \qquad N_z := m \cdot N + 1, \tag{45}$$

and compute recursively the state variables from (44),

$$x^i = x^i(z, p) := x^i(u^1, .., u^{i-1}, t_f, p), \tag{46}$$

as functions of the control variables with initial condition $x^1 = \varphi(p)$ given in (41). This leads to the following discretized optimal control problem

Discretized optimal control problem *DOCP(p)*:

$$\begin{aligned} &\textit{Minimize} \quad F(z,p) := g(x^N(z,p),p) \\ &\textit{subject to} \quad \psi(x^N(z,p),p) = 0, \\ &\qquad\qquad\quad C(x^i(z,p),u^i,p) \le 0, \quad i = 1,\dots,N. \end{aligned} \tag{47}$$

Let $G(z) = (G_1(z), \dots, G_{N_c}(z))$ denote the collection of functions defining the equality and inequality constraints of (47). Then the number of optimization variables $N_z = mN + 1$ and constraints $M = kN + r$ results in a dense structure in the Hessian of the Lagrangian (49), whereas about 50% of the elements in the Jacobian of the constraints are zero. Setting the dimensions $N_e = r$, $N_c = N_e + kN$, the discretized control problem *DOCP*(*p*) defines a *NLP*-problem of the form

***NLP*(*p*):**

$$\begin{aligned} &\textit{Minimize} && F(z,p) \\ &\textit{subject to} && G_i(z,p) = 0, \quad i = 1,\dots,N_e, \\ & && G_i(z,p) \le 0, \quad i = N_e+1,\dots,N_c. \end{aligned} \tag{48}$$

Problems of the form (48) can be solved efficiently using sequential quadratic programming (SQP) methods; compare, e.g., the code WORHP of Büskens and Gerdts. Instead of Euler's method incorporated into the relations (44) or (46) one can use higher order integration methods combined with a higher order control approximation. The convergence of solutions discretized via Euler's method to solutions of the continuous control problem has been proved in Malanowski, Büskens and Maurer [4].

The Lagrangian function for problem *NLP*(*p*) is

$$L(z,\mu,p) = F(z,p) + \mu^T G(z,p), \quad \mu \in \mathbb{R}^{N_c}. \tag{49}$$

Let $\bar{z}$ be an optimal solution for *NLP*(*p*) with associated Lagrange multiplier $\bar{\mu}$ satisfying first order necessary optimality conditions. Consider the set of active indices defined by $I_a(p) := \{i \in \{1, \dots, N_c\} \mid G_i(\bar{z}, p) = 0\}$ and let $m_a := \#I_a(p)$. Denote the active constraints by $G^a := (G_i)_{i \in I_a(p)}$ and the multiplier corresponding to active constraints by $\bar{\mu}^a \in \mathbb{R}^{ma}$. Let G_z^a be the Jacobian with dimension $m_a \times N_z$.

The following strong second order sufficient conditions are well known, cf., Fiacco [3].

Strong Second Order Sufficient Conditions

Assume that

a) F and G are twice continuously differentiable with respect to z and p,

b) the gradients in G_z^a are linearly independent, i.e. rank($G_z^a(\bar{z}, p)$) = m_a,

c) strict complementarity $\bar{\mu}^a > 0$ of the Lagrange multipliers holds,

d) the Hessian of the Lagrangian is positive definite on Ker($G_z^a(\bar{z}, p)$),

$$v^T L_{zz}(\bar{z},\bar{\mu},p)v > 0, \quad \forall \quad v \in Ker(G_z^a(\bar{z},p)), \quad v \neq 0. \tag{50}$$

Then $\bar{z}$ *is a local minimum for NLP(p).*

A numerical check of the SSC (50) consists in evaluating the projected Hessian on $Ker(G_z^a(\bar{z}, p))$ and in verifying that its eigenvalues are positive, cf. Büskens and Maurer [1].

6 Parametric Sensitivity Analysis

A thorough treatment of parametric sensitivity analysis can be found in Fiacco [3], a compact overview on parametric sensitivity analysis for discretized optimal control problems is given in Büskens and Maurer [1].

Solution Differentiability for *NLP(p)*:

Suppose that the optimal solution (z_0, μ_0)*, satisfies the strong second order sufficient conditions for the nominal problem* $NLP(p_0)$*. Then for p near to* p_0 *the unperturbed solution* (z_0, μ_0) *can be embedded into a* C^1*-family of perturbed optimal solutions* $(z(p), \mu(p))$ *for NLP(p) with* $(z(p_0), \mu(p_0)) = (z_0, \mu_0)$*. The active sets* $I_a(p)$ *coincide with* $I_a(p_0)$ *and hence it follows that* $\mu_i(p) = 0$ *for* $i \notin I_a(p)$*. The sensitivity differentials of the optimal solutions and Lagrange multipliers are given by the formula*

$$\begin{pmatrix} \dfrac{dz}{dp}(p_0) \\ \dfrac{d\mu^a}{dp}(p_0) \end{pmatrix} = -\begin{pmatrix} L_{zz}(z_0,\mu_0,p_0) & G_z^a(z_0,p_0)^T \\ G_z^a(z_0,p_0) & 0 \end{pmatrix}^{-1} \begin{pmatrix} L_{zp}(z_0,\mu_0,p_0) \\ G_p^a(z_0,p_0) \end{pmatrix}. \tag{51}$$

Moreover, the sensitivity of the objective function is obtained from

$$\frac{dF}{dp}(z(p_0),p_0) = L_p(z_0,\mu_0,p_0). \tag{52}$$

Note that the so-called Kuhn-Tucker matrix on the right hand side of (51) is *regular* since second order sufficient conditions (50) are assumed to hold. One possibility of evaluating the Kuhn-Tucker matrix and the incorporated Hessian L_{zz} consists in using BFGS-updates which are obtained in the process of computing the nominal solution. However, these BFGS-updates usually give rather inaccurate approximations. Hence, in our approach the Hessian L_{zz} and the Kuhn-Tucker matrix will be evaluated after computing the optimal solution (z_0, μ_0), which yields much more accurate results.

Eq. (52) can be used in a first order Taylor expansion of the objective functional,

$$F(z(p),p) \approx F(z,p_0) + \frac{dF}{dp}[p_0](p-p_0), \tag{53}$$

where the notation $[p_0]$ stands for all nominal arguments. The computation of $dz/dp(p_0)$ via (51) yields an approximation for the sensitivity differentials of the perturbed optimal control solutions at the mesh points, i.e., for the quantities

$$\frac{\partial u}{\partial p}(\tau_i, p_0) \approx \frac{du^i}{dp}(p_0), \quad i = 1, \ldots, N. \tag{54}$$

Since a free terminal time t_f is handled as an additional optimization variable in $DOCP(p_0)$, its sensitivity differential $dt_f/dp(p_0)$ can likewise be calculated from equation (51).

Eq. (54) grants also the sensitivity differentials of the *state* with respect to the parameter, if we take into account the recursive formulation (44) together with the relation (46). The state sensitivities $dx/dp(\tau_i, p_0)$ are obtained from

$$\frac{\partial x}{\partial p}(\tau_i, p_0) \approx \frac{dx^i}{dp}[p_0] = \frac{\partial x^i}{\partial z}(z_0, p_0)\frac{dz}{dp}(p_0) + \frac{\partial x^i}{\partial p}(z_0, p_0). \tag{55}$$

7 Result

The method proposed for the real-time update of optimized aircraft trajectories is applicable to a large variety of aircraft trajectory problems if variations in environmental parameters occur. The main practical relevance of the wind example may be for civil airliners or other aircraft flying on prescribed routes. However, the air race example chosen as the underlying trajectory optimization problem is very challenging since moment dynamics is no longer negligible, the trajectory is highly curved in all dimensions, states and controls frequently hit saturations and the aircraft is flying at a high dynamic bandwidth for a longer period of time. By choosing the air race example it was intended to demonstrate the capability and potential of the approach assuming that showing its applicability for this example includes the applicability for problems with lower dynamics.

In Fig. 1, the optimized race trajectory for the air race that took place in San Diego in 2009 is shown. The optimal air race trajectory has been obtained utilizing a direct optimization method. The colorization of the trajectory indicates the sensitivity of the optimized trajectory with respect to wind blowing along the y-axis of the depicted coordinate system. At this, the following measure for the sensitivity of the trajectory with respect to the selected parameter has been utilized:

$$\frac{ds^i}{dp}(p_0) = \sqrt{\left(\frac{dx^i}{dp}(p_0)\right)^2 + \left(\frac{dy^i}{dp}(p_0)\right)^2 + \left(\frac{dz^i}{dp}(p_0)\right)^2} \tag{56}$$

where x denotes the northward, y the eastward and z the downward position of the aircraft. As can be seen from Fig. 1, especially at the Quadro (i.e. at the 270° degree turn) where the aircraft pulls up the trajectory is very sensitive with respect to the considered wind influence.

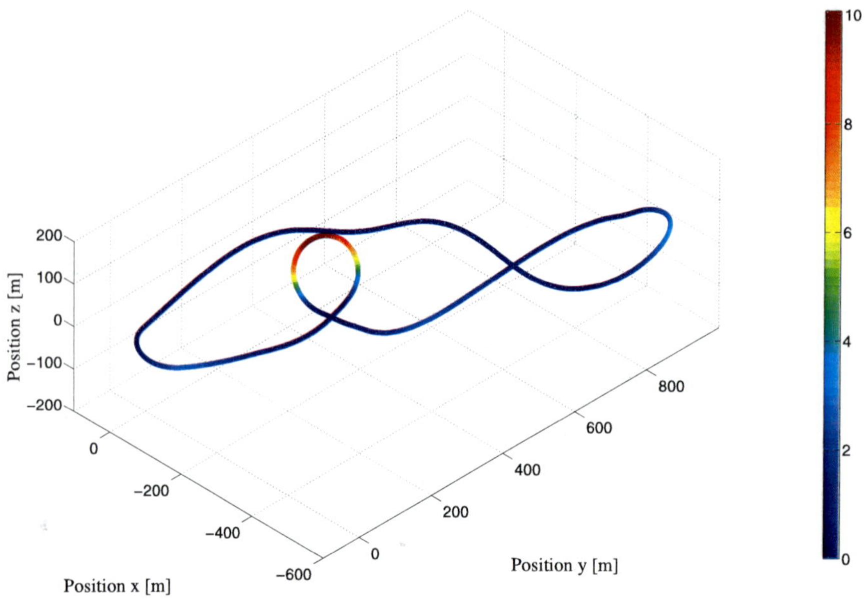

Fig. 1 Sensitivity of the optimized trajectory with respect to wind blowing along the y-axis of the depicted coordinate system

8 Conclusions

In the paper at hand, the parametric sensitivities of air race trajectories with respect to wind are investigated. Therefore, a simulation model is implemented that takes into account wind influences. The air race trajectory optimization problem is formulated and the discretization approach that transforms the optimal control problem into a nonlinear programming problem is introduced. For the optimal race trajectory, the sensitivities with respect to the selected parameters are computed. It has been found that the trajectory is quite sensitive with respect to wind if an aerobatic maneuver is flown that requires the aircraft to pull up. By the proposed method, offline-optimized reference trajectories utilized for the guidance of flight vehicles can be updated in real-time onboard the aircraft to give valid, close-optimal reference trajectories taking into account deviations of environmental parameters. At this, the tracking of the guidance trajectories in the presence of perturbed environmental parameters can be facilitated.

References

1. Büskens, C., Maurer, H.: SQP-methods for Solving Optimal Control Problems with Control and State Constraints: Adjoint Variables, Sensitivity Analysis and Real-Time Control. Journal of Computational and Applied Mathematics 120, 85–108 (2000)

2. Büskens, C., Maurer, H.: Sensitivity Analysis and Real-Time Control of Nonlinear Optimal Control Systems via Nonlinear Programming Methods. In: International Series of Numerical Mathematics, vol. 124. Birkhäuser Verlag, Basel (1998)
3. Fiacco, A.V.: Introduction to Sensitivity and Stability Analysis in Nonlinear Programming. Mathematics in Science and Engineering, vol. 165. Academic Press, New York (1983)
4. Malanowski, K., Büskens, C., Maurer, H.: Convergence of approximations to nonlinear optimal control problems. In: Fiacco, A.V. (ed.) Mathematical Programming with Data Perturbations. Lecture Notes in Pure and Applied Mathematics, vol. 195, pp. 253–284. Marcel Dekker, Inc., New York (1997)
5. N.N.: The Standard Atmosphere, DIN ISO 2533, International Organization for Standardization, Genf, Schweiz (1975)
6. Grimm, W., Hiltmann, P.: Direct and Indirect Approach for Real-Time Optimization of Flight Paths. LNCIS, pp. 190–206. Springer, Heidelberg (1987)
7. Twigg, S., Calise, A., Johnson, E.: On-line Trajectory Optimization for Autonomous Air Vehicles. In: AIAA Guidance, Navigation and Control Conference, Austin, Texas, AIAA 2003-5522 (2003)
8. Yakimenko, O.: Direct Method for Real-Time Prototyping of Optimal Control. In: Proceedings of the International Conference "Control 2006", Glasgow, Scotland, August 30 - September 1 (2006)
9. Bevilacqua, R., Romano, M., Yakimenko, O.: Online Generation of Quasi-Optimal Spacecraft Rendezvous Trajectories. Acta Astronautica (2008)

Simple Control Law Structure for the Control of Airplanes by Means of Their Engines

Nicolas Fezans

Abstract. In this paper a simple control law structure is presented for the control of airplanes using only the engines' thrust. For the design of such a propulsion controlled aircraft control law, the approach followed in this work is to look for the right level of performance in order to avoid both excessive engines activity and reduction of robustness properties. Another goal is to keep the control law and its tuning as simple as possible: for this a control law structure whose terms can easily be interpreted is proposed. The capability of the proposed control law to permit safe landing was shown by simulator tests as well as flight tests.

1 Introduction

One of the objectives of the research conducted at the DLR (German Aerospace Center) Institute of Flight Systems is to improve the safety of all types of aircraft. This objective takes a central part in many projects concerning human-machine interaction (e.g. situation awareness, training of pilots), detection of adverse conditions (e.g. gust, wake-vortices), fault detection and isolation, and reconfiguration of control laws.

In the past, several incidents and accidents were caused by partial or total loss of aircraft primary control systems. Even though a total loss of aircraft primary control systems is theoretically extremely rare, it did happen several times [1] with an estimate of more than 1200 casualties made in 1996 [2]. The fact that such events are extremely rare must be demonstrated during the certification process. Indeed, looking at the past as well as at accidents that happened more recently (e.g. DHL A300 in Baghdad in 2003), it appears clearly that the loss of primary control devices is generally a consequence of other failures with some complex relations between

Nicolas Fezans
DLR (German Aerospace Center) Institute of Flight Systems, Lilienthalplatz 7, 38108 Braunschweig, Germany
e-mail: nicolas.fezans@dlr.de

them and in some cases related to maintenance errors or to external causes. The interactions between the faults and failures in such systems can be extremely complex. The author's point of view is that even with very serious work on safety analysis a significant number of elements of the fault tree of such complex systems will be omitted. The level of redundancy and the extensive use of computers and real tests in the design of modern aircraft allowed to reach very high safety standards. However, human errors in the maintenance or the design of the aircraft can still happen as well as adverse weather conditions, wake vortex encounters, or terrorist actions. When technically and economically possible, design and implementation of emergency systems capable of dealing with a wide spectrum of improbable situations should be made. The propulsion-based emergency control law presented in this paper is worth being integrated in modern transport aircraft as such an emergency system.

Several studies have been conducted at NASA in the 1990's to prove the feasibility of controlling the aircraft and landing using only the engines [1, 2, 3]. It appeared clearly at that time that a system assisting the pilot is required and that with such a system probability of casualties would drastically be reduced. Implementation of such a system in aircraft equipped with an autothrottle would not require any new hardware (and thus there would be no increase of weight) and for many modern aircraft only software modifications would be required to include this new functionality. However, more than 15 years later still no civil transport aircraft possesses such an emergency system. Even aircraft that were entirely designed after this technology was demonstrated in flight are not equipped!

In this paper, the case of total loss of primary control devices is considered. For such a deteriorated airplane a control law based on engine thrust is designed with the approach of keeping a very simple structure and using the lowest gains possible. Longitudinal control is achieved using symmetrical variations of thrust and lateral control using asymmetrical variations. This control strategy is similar to the one studied at NASA in the 1990's under the name of "Propulsion Controlled Aircraft" (PCA). As the name PCA properly describes the content of the work presented in this paper, it will also be used. At the DLR Institute of Flight Systems, experiments on Fault-Tolerant Control (FTC) are made using the ATTAS research aircraft [4, 5, 6]: some of them leading to the control of the ATTAS using only the engines. In particular, approaches based on Model Predictive Control (MPC) were tested with the aim of reaching very high performance by taking directly into account the limitations of the airplane and in particular of its actuators and engines. In the current work a low-gain approach is preferred.

The paper is organized as follows: Section 2 presents the motivation for a structured control law with a low-gain tuning. Section 3 discusses the properties of the aircraft with PCA control law that are required to permit safe landings. Section 4 presents the control law: its structure, tuning, and protections. Section 5 presents a summary of the results obtained during a flight test that took place in November 2009 with the presented propulsion-based control law.

2 Motivation for a Structured Control Law with a Low-Gain Tuning

Failures are not always extremely severe and in some cases it makes sense to try to integrate the deteriorated aircraft into the regular air traffic. The failures leading to the use of a control law based only on the engines are very severe. In such cases, an emergency would be declared and the real objective for the design of an emergency system capable of controlling the aircraft by means of the engines is to permit an acceptable landing. An acceptable landing for such an emergency system is, to the author's mind, a landing avoiding both the loss of human lifes and major injuries. Relatively low vertical speed and normal attitude (slightly positive pitch, small roll) are major criteria for such a landing.

Besides, in such a severe failure case, quite strong differences in terms of dynamical behavior between a nominal model of the airplane and the real deteriorated airplane can be expected. For instance, the Boeing 747 of the Japan Airlines flight 123 had lost almost the entire vertical stabilizer and a significant part of the left wing was missing in the case of the Airbus A300 of DHL in Baghdad. These differences lead to the need for strong robustness properties of the system, either by intrinsic robustness of the control law or by its adaptation. The robustness analysis of airplanes with a propulsion-based control law is challenging due to the slow and highly nonlinear responses of the engines. Their dynamics result from a combination of saturations, nonlinear dynamics and state-dependent rate-limiters. The use of typical robustness metrics for design purposes in such a case is very difficult. For this reason, it has been decided to focus in this work on the definition of a simple control law whose components can easily been interpreted. Thus, the approach here is to compensate the lack of practically usable rigorous mathematical tools for such problems by the use of physical interpretations. Precise evaluations of the obtained robustness properties could also be made by means of Monte-Carlo methods but have not yet been considered in this study. The only evaluations that have been made until now rely on several hundreds of simulations. It should be noticed that stability of the controller-augmented airplane is not required as pilots will be part of the closed-loop. Moreover, a stability criterion does not permit to estimate whether a pilot is really able to reach the airport and land or not.

However, to the best of the author's knowledge no classical handling qualities criterion is applicable to a propulsion controlled aircraft, therefore there is a need for a new evaluation of handling qualities for such an airplane. An objective of this research is thus to understand which design criteria are really important with pilots in the loop and which criteria can be withdrawn. The purpose of this paper is not to address this question, but to present the simple control law structure that has been designed in order to ease future simulator and flight-test studies that will address it. With an adequate tuning of this control law the pilots must be able to land successfully in a wide spectrum of situations. Later on, once the required performance criteria for such a propulsion controlled aircraft will be properly defined and understood, the design and the validation of such an emergency system will be eased.

A single set of parameters has been used on the entire flight domain and for all configurations. The only scheduling made takes place in the inner loop which controls the engines: the proportional gain depends on the rotational speed of the fan (N1), see section 4.5. Such a choice would often be suboptimal, in particular if we want to force the closed-loop to exhibit some defined behavior. But for such a system the exact behavior of the controller-augmented aircraft is not very important: what matters is that pilots succeed in controlling it and finally succeed in landing the airplane. In our experiments and in the work presented here this single set of parameters with a low-gain tuning allowed pilots to control the aircraft and to follow both glideslope and localizer signals with a good precision. This shows that both high-gain solutions and control strategies based on reference models are not required for this application.

3 Requirements

In this section, the main requirements for successful approach and landing by means of a PCA system are discussed with focus on how desirable they are, how difficult it will be to reach them, and which trade-off between the performance criteria should be made. Obviously, classical handling qualities criteria are not applicable for an aircraft having a propulsion-based control law.

3.1 Longitudinal Control

The longitudinal motion is mainly composed of the phugoid mode and the short-period mode. The period of the phugoid is generally between 30 and 60 seconds for transport airplanes. The frequency of the short-period mode depends on the aircraft and its center of gravity location, but would typically lie between 1.5 and 3 rad/s.

Increasing the total thrust of the engines leads to an increase of the energy rate of the aircraft. To really know the effect of this additional thrust on the movement of the aircraft, the pitch equation as well as the aerodynamics and mass characteristics of the aircraft must be known. For typical configurations, a simplified reasoning can be expressed as follows: a constant additional total thrust $\Delta T_t = \sum_i \Delta T_i > 0$ leads to a positive variation of the flight path angle γ and vice-versa, i.e. $\Delta T_i > 0 \Rightarrow \Delta\gamma(t \to \infty) > 0$ and $\Delta T_i < 0 \Rightarrow \Delta\gamma(t \to \infty) < 0$. This makes it possible to control the trajectory of the aircraft in the vertical plane.

With the typical frequencies and damping ratios of the short-period mode and of the phugoid as well as the typical dynamics of engines, no real challenge is expected in designing and tuning a control law assisting the pilots in the control of the flight path angle. Such a control law will basically consist of controlling the phugoid (acceptable response time, good damping, and no static error on the flight path) while avoiding unnecessary excitation of the short-period mode.

3.2 Lateral Control

The lateral dynamics of an aircraft are composed of:

- the Dutch roll mode exhibiting a pair of complex conjugate and stable poles with very low damping,
- the roll mode which is aperiodic and stable,
- the spiral mode which is slow and quite often slightly unstable.

As for the short-period mode, the Dutch roll mode and the roll mode are generally too fast to be significantly modified by means of the engines, in particular in the low-thrust domain that will generally be required for descent and approach. However, a control law based on thrust can easily modify the spiral mode in order to ease its control by a human pilot. For this, the coupling between yaw and roll is used: the pilot controls only the roll motion and the control law generates a yaw motion by means of asymmetric thrust allowing to get the induced roll corresponding to the pilot's commands. In previous studies PCA control laws were designed to follow a reference bank angle ϕ_{ref} that was provided by the pilot. During the current research activities several other possibilities have been investigated in the ATTAS ground simulator. In particular a rate-command attitude-hold and a combination of roll rate and bank angle commands are being tested. They are both based on the control law presented hereafter: the difference is the way pilots provide the references to the system.

Some reasonable goals for the lateral part of the control law are: to permit enough maneuverability, to reduce pilot workload by damping the lateral dynamics, and to ensure acceptable disturbance rejection without any action of the pilot.

4 Propulsion-Based Control Law

This section presents the propulsion-based control law that was used for the simulator tests and flight tests that are shown in section 5. The global structure of the control law is first presented (section 4.1). After that all elements constituing this structure are presented separately in sections 4.2 - 4.5. The easiness of the physical interpretation of each component will appear clearly.

4.1 Global Control Structure

The global control structure is presented in figure 1. This structure is based on a cascade control strategy with inner loops controlling the engines through commands in terms of Power Lever Angle (PLA) and outer loops controlling the longitudinal and lateral motions through symmetric and asymmetric thrust. The inner loops have a crossfeeding in case of saturation as detailed in section 4.5 (see signals $\mathrm{PLA}_{sat}\mathrm{L}$ and $\mathrm{PLA}_{sat}\mathrm{R}$). A block labeled "Mixing priorities & Protections" connects the outer loops to the inner loops by allocating longitudinal ($\overline{\mathrm{N1}}_{cmd}$) and lateral ($\Delta\mathrm{N1}_{cmd}$) control actions to the two engines while satifying the limits for each engine.

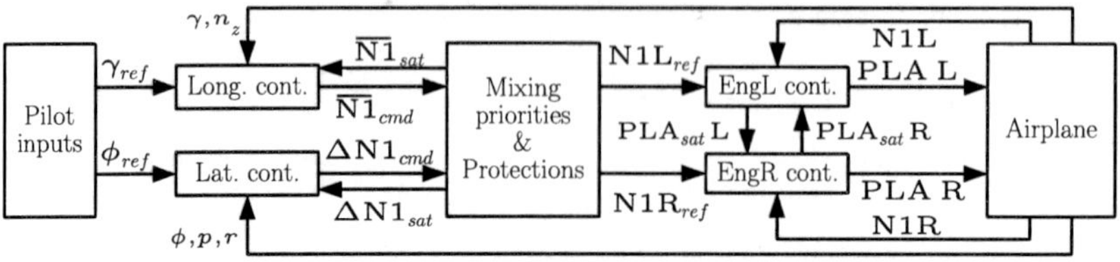

Fig. 1 Global architecture of the propulsion-based control law

This leads to the two references $\mathrm{N1L}_{ref}$ and $\mathrm{N1R}_{ref}$ that are provided to the inner loops. Although this does not appear very explicitly in figure 1 the "Mixing priorities & Protections" block also connects the two outer loops by means of the anti-windup feedback signals $\overline{\mathrm{N1}}_{sat}$ and $\Delta\mathrm{N1}_{sat}$. This particular point is more detailed in section 4.4.

On one side imposing a particular structure on the controller generally restricts the maximum achievable performance and robustness. On the other side it allows to guarantee both, the physical meaning of the controller's parameters and the simplicity of the future implementation on on-board computers. In this work the trade-off between reachable performance, engine activity, and robustness is investigated under the assumption that a simple and comprehensible structure is required. Using such a structure is therefore not a choice made during this work but a constraint directly included in the addressed problem itself.

4.2 Longitudinal Controller

The longitudinal controller is presented in figure 2a. It uses a PID structure with a filter ($K_{n_z}(s)$) on the derivative part and two feedforward elements: a dynamical one $K_{FF\int\gamma}(s)$ and a static one $K_{FF\gamma}$. The filter $K_{n_z}(s)$ is required to remove the medium and high-frequencies measured by the accelerometer and therefore remove undesirable control activity due to gusts and turbulence. As the phugoid mode is very slow with respect to the typical frequency content of these types of atmospheric disturbances, a simple first-order low-pass filter with a bandwidth higher than the frequency of the phugoid mode was used.

To get the desired reference tracking of the flight path angle γ, the static gain $K_{FF\ \int\gamma}(\omega=0)$ must be equal to 1. The degrees of freedom provided by these transfer functions allow seperate tuning of the reference tracking dynamics and of the disturbance rejection properties. For instance, defining $K_{FF\ \int\gamma}(s)$ as a low-pass filter permits to use a relatively high value for $K_{\int\gamma}$ without getting high overshoots of the reference tracking response and without requiring high values of the derivative gain. Such a tuning is interesting because it gives the pilot a control input that is not excessively sensible in the medium to high frequency domain while permitting quite efficient rejections of disturbances by the feedback without any action of the pilot. As the obtained performance was already satisfying with a simple gain for $K_{FF\ \gamma}$, this part of the feedforward has been kept static. If finer tuning of the

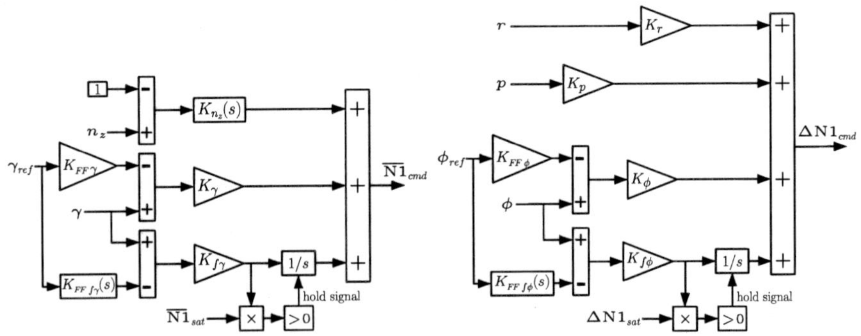

(a) Longitudinal: γ-mode

(b) Lateral: ϕ-mode

Fig. 2 Longitudinal and lateral controllers

reference tracking response is required some additional degrees of freedom could be obtained by using a transfer function instead of the $K_{FF\,\gamma}$ gain.

4.3 Lateral Controller

The lateral controller is based on the same principle as the longitudinal controller and therefore has a similar structure. However, the types of dynamics to be controlled by these two controllers are quite different. In the longitudinal case, the static response to a symmetric variation of thrust is a variation of the flight path angle γ, which is directly the variable that must be regulated. In the lateral case, the static response to an asymmetric variation of thrust is a variation of the sideslip angle β, which is related to the second time derivative of the regulated variable ϕ (bank angle).

This is a logical consequence of the fact that there is no roll authority and roll is now controlled by means of the yaw-roll coupling. Therefore, the chosen controller structure (figure 2b) is not only based on the bank angle ϕ and the roll rate p, but also on the yaw rate r. Feedback based on the sideslip angle β would probably also help getting good closed-loop properties, but β is generally not used in flight control systems for various practical reasons. This fact has been taken as a constraint and is not discussed. The feedback of p and r should ideally be restricted to high frequencies in order to avoid counteracting normal turns. Indeed the feedback $K_{\int\phi}$ ensures rejection of such side effects and simple gains K_p and K_r can be taken instead of some high-pass filters.

During the landing task, the required corrections for the lateral movement are much more demanding than for the longitudinal motion. This leads to a more difficult tuning of the lateral controller. The balance between good enough handling qualities and high activity of engines is more difficult to find, in particular when considering uncertainties of engines' dynamics. As mentioned in the introduction, the analysis tools for this type of nonlinear system are difficult to put into practice and with a pilot in the loop stability is neither a sufficient nor a required condition. Thanks to the physical interpretations of each part of the controller, a good set of parameters can be found in a short time. However, a more systematic and rigorous methodology would be desirable.

4.4 Mixing Priorities and Outer Loop Protections

Authority on both longitudinal and lateral movements is severely restricted. The same actuators (i.e. engines) are used for both and thus some mixing priorities should be defined. As the controlled dynamics of the lateral motion are faster than the controlled dynamics of the longitudinal motion, it seems logical to give priority to the lateral control action over the longitudinal control actions in the case both cannot be satisfied simultaneously. Moreover, a poor control of the spiral mode will strongly disturb the control of the flight path angle, whereas the opposite coupling can be neglected in most cases (e.g. without stall, overspeed, etc.). This priority can be implemented as shown in figure 3 where the absolute value of the lateral control signal is used in the definition of limits applied to the longitudinal control signal. This permits to ensure that the sum of both control signals will respect the constraints of each engine and to give priority to the thrust difference between them over the mean thrust.

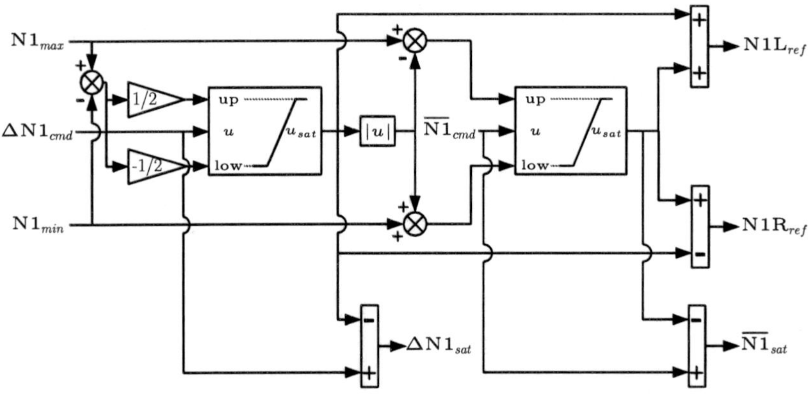

Fig. 3 Mixing priorities and outer-loops protections

In figure 3, signals $\overline{\mathrm{N1}}_{sat}$ and $\Delta\mathrm{N1}_{sat}$ are respectively the differences between input and output of longitudinal and lateral control saturations. They are used for the "integrator hold" antiwindup strategy that is implemented in both controllers (figures 2a-2b). After giving priority to the thrust difference over the mean thrust, the signs of these signals are compared to the sign of the signals entering the integrators in figures 2a and 2b. If their signs are identical, then the corresponding integrator would be winding and this is prevented by holding the integrator at its previous value. This way of preventing integrator windup does not allow to desaturate the command in order to keep some authority, but in the considered application there is no need for a more complex antiwindup strategy.

4.5 Inner Loops for Engines Control

The structure of the inner loop controllers is depicted in figure 4. In this figure the controller of the left engine is shown: the letters L and R are use to distinguish left and right. For the right engine, the letters L and R must be inverted in this figure. The controller is basically a PI controller with a static feedforward on the proportional part and an "integrator hold" antiwindup strategy. In addition, the saturation signal of the other engine (here $PLA_{sat}R$) is crossfed in the output in accordance with the priorities defined in the previous section.

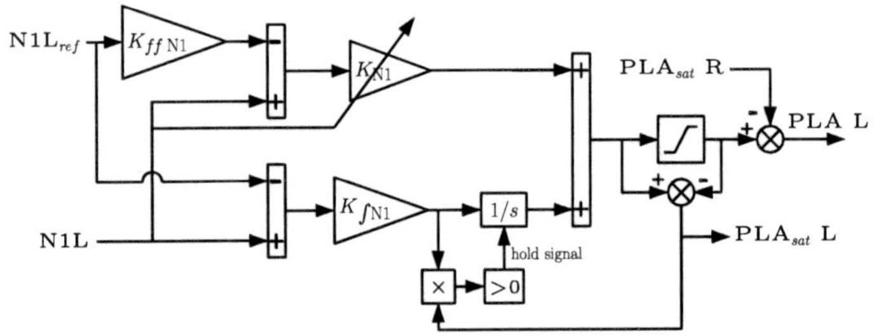

Fig. 4 Inner loop for the left engine control. Invert L and R for the right engine.

During our tests (both in simulator and in flight), the aforementioned antiwindup and crossfeed in these inner loops did not appear to be absolutely required as outer loop gains were intentionally chosen relatively low. For more performance-driven tuning or in the presence of uncertainties these elements are likely to be of the greatest importance. As they result in negligible increases of controller complexity, they should be kept for all implementations of such controllers.

Note that in our applications, the parameters of these loops were systematically taken equal for both engines even though significant differences of the left and right engines were known and can be observed at medium to high frequencies (with respect to closed-loop bandwidth). Note that engines may also differ significantly during their life and the control law must be able to deal with such variations. This can be achieved through the robustness of the control law or by means of adaptive techniques. Adaptive techniques have not been considered at all during this work.

5 Summary of Flight Test Results

After several simulator tests of the system, the system was tested on Nov. 20, 2009 with the DLR ATTAS (VFW-614) research airplane. Goal of this flight test was to verify and test the handling qualities provided by this propulsion-based control law for navigation, ILS intercept, approach, and go-around. During previous simulator tests pilots were asked to rate several possible tuning options for the control law parameters and to express their preferences. These preferences were also put into perspective with their foreseeable drawbacks in terms of robustness, disturbance rejection properties, engine activity. During this analysis, rough assessments of the closed loop at 11 points of the flight domain were made. The considered points are based on 3 different flap configurations, 4 different speeds, 3 different altitudes, and landing gear extended or retracted (note that not all combinations of these parameters were considered). A 200 ms delay was introduced at engine inputs and for each flight condition the frequency of the Dutch roll of the ATTAS model was artificially shifted until the closed loop was destabilized. Besides, simulations with demanding inputs (e.g. successions of maximal stick inputs at various frequencies around the bandwidth of the stick to output transfer function) as well as simulations with initial conditions were performed with the aim of analyzing both the input-output behavior (including its nonlinearities) and the internal dynamics of the closed loop. The objective was not to compute robustness margins, but rather to obtain a qualitative evaluation of each set of parameters regarding robustness and disturbance rejection properties. More complete and precise evaluations will be performed later for validation purposes and not for design purposes. This analysis lead to test two controller parameter sets (called tuning A and tuning B hereafter) during this flight, even though a clear preference for tuning A had already been identified before the flight. It corresponds to a faster but less damped behavior for the lateral motion.

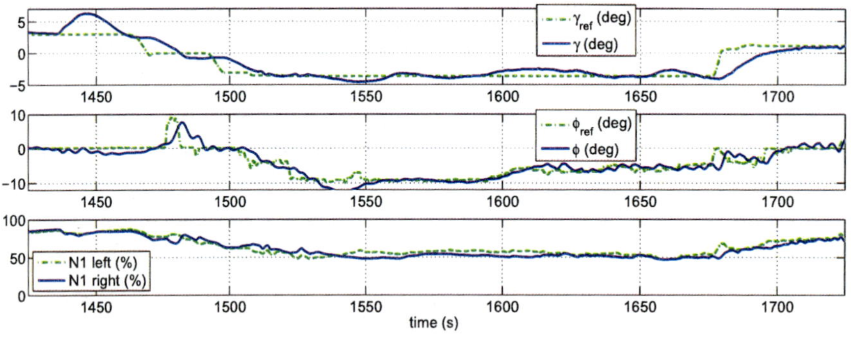

Fig. 5 Reference tracking with the propulsion-based control law.

Weather was sunny and cloudless. Wind at 2500 ft was 40 to 50 kt from 220° and wind on ground was 5 kt from south-east. Light turbulence. After take-off and activation of the experimental systems, some maneuvers were made by the test pilot in order to check the dynamical behavior of the aircraft with this control law and to provide data with strong enough inputs for future analysis. A short sequence of these maneuvers is shown in figure 5. It shows how well the references were followed as well as how the deviations caused by the very light turbulence are rejected in spite

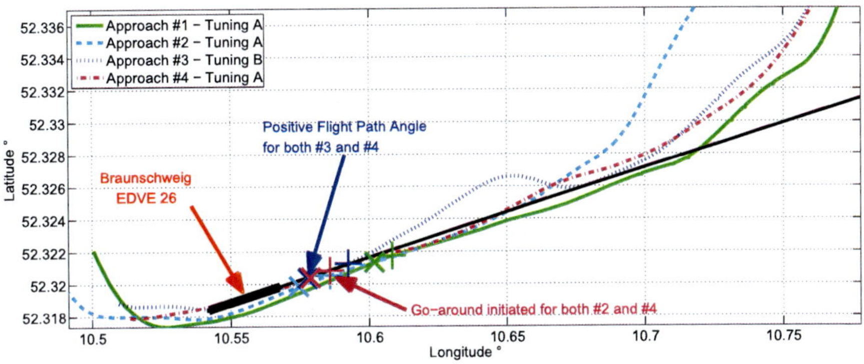

Fig. 6 Ground tracks of approaches performed with the PCA control law.

of low engine activity. After that the entire sequence "navigation, ILS intercept, approach, and go-around" has been flown four times.

The preference for tuning A was rapidly confirmed before attempting to land and therefore three of the four "intercept, approach, and go-around" sequences flown were performed using the tuning A. The ground tracks of the sequences are shown in figure 6. The longitudinal control was identical in both parameter sets and the glide slope could be very well followed (results are not shown here). During the four approaches, both localizer and glide-slope deviations could easily be controlled with less than a half dot precision. In figure 6, the $+$ symbols show the point where the go-around was initiated and the $\times$ symbols where the flight path angle became positive.

Comments of pilots and flight engineers made in the flight test report were that "the airplane can very well be flown with careful inputs", "desired heading can be exactly controlled", "during height changes maximal overshoot was 100 ft", and "there is no concern about pushing the experiment until landing on a long and wide runway in calm air and without lateral wind". Besides, during the flight and the post-analysis, it appeared that dynamics of the engines at very high thrust were significantly faster than predicted by the model used. This lead to a limit cycle in the lateral control (about $+/-$ $3°$ of β) loop during max climb maneuvers. After having analyzed the flight-test data, the correction of this limit cycle was straightforward by adjusting the gains on the inner loops at high N1 values. As low N1 values are required to descent, this gain modification has no consequence on handling qualities during approach and landing.

6 Conclusion

To conclude, a simple structure for a propulsion-based control law has been proposed. Simulator as well as flight tests have shown that performance obtained with this structured control law with a low-gain tuning is clearly sufficient for the purpose of landing. Further tests are required and planned in the next months which includes actual landings with this system. Some of the typical autopilot modes are being developed as outer loops generating the references ϕ_{ref} and γ_{ref} given as input of the control law that is presented in this paper. In particular an autoland function is currently under development.The evaluation of the pros and cons of each mode depending on damage cases while taking into account the variability of pilot behaviors and weather conditions will be performed in the near future.

This work has been applied to the VFW-614 ATTAS airplane in both flight tests and simulator tests. Within the next months, it will also be adapted and pursued using two simulators of the DLR Institute of Flight Systems: the Airbus A320 ATRA (Advanced Technology Research Aircraft) simulator and the "Future Military Transport Aircraft" simulator.

References

1. Burcham Jr., F.W., Burken, J.J., Maine, T.A., Gordon Fullerton, C.: Development and Flight Test of an Emergency Flight Control System Using Only Engine Thrust on an MD-11 Transport Airplane. Technical Paper: NASA/TP-97-206217. NASA Dryden
2. Burcham Jr., F.W., Maine, T.A., Gordon Fullerton, C., Webb, L.D.: Development and Flight Evaluation of an Emergency Digital Flight Control System Using Only Engine Thrust on an F-15 Airplane. Technical Paper: NASA/TP-3627. NASA Dryden (1997)
3. Bull, J., Mah, R., Hardy, G., Sullivan, B., Jones, J., Williams, D., Soukup, P., Winters, J.: Piloted Simulation Tests of Propulsion Control as Backup to Loss of Primary Flight Control for a B747-400 Jet Transport. Technical Memorandum: NASA TM-112191. NASA Ames
4. de Almeida, F.: Waypoint Navigation Using Constrained Infinite Horizon Model Predictive Control. In: AIAA-2008-6462. AIAA Guidance, Navigation and Control Conference and Exhibit, Honolulu, Hawaii, USA, August 18-21 (2008)
5. de Almeida, F., Leißling, D.: Fault-Tolerant Model Predictive Control with Flight Test Results on ATTAS. In: AIAA-2009-5621. AIAA Guidance, Navigation, and Control Conference, Chicago, Illinois, USA, August 10-13 (2009)
6. de Almeida, F., Leißling, D.: Fault-Tolerant Model Predictive Control with Flight-Test Results. Journal of Guidance, Control and Dynamics 0731-5090 33(2) (2010)

The Development of Perspective Displays for Highly Precise Tracking Tasks

Alexander Efremov and Mikhail Tjaglik

Abstract. A generic technique is presented for optimization of the display-aircraft system. The technique includes mathematical modeling based on the modified Hess model and selection of pilot model parameters by minimization of the variance of predicted path angle error. The prediction time is selected through minimization of variance of error characterizing the goal of the investigated piloting task. These tasks are refueling, terrain following and landing. The experiments fulfilled on a ground based simulator with Head up Display (HUD) demonstrated the tunnel image with preliminary selected sizes and projection of predicted path angle on a surface sliding in the tunnel and located on appropriately selected distance ahead. The results demonstrated that the developed technique allowed considerable improving accuracy in each piloting task.

Symbols:

e(t) - error:

$e(t) = \Delta\varepsilon_\gamma = i(t) - \varepsilon_\gamma(t)$ - error in single loop compensatory task;

$e(t) = \Delta H(t) = H - H_i(t)$ - error in landing, terrain following task;

$e(t) = \Delta\varepsilon_\theta = \varepsilon_\theta(t) - H_i / L$ - error in refueling task;

$i(t),\ H_i(t)$ - input signals in different task;

L_{pr} - predicted distance;

$n_\alpha = \partial n / \partial \alpha$;

n - normal acceleration;

$T_{pr} = L_{pr} / V$ - prediction time;

$Z^\alpha = Z_W W$;

$Z_W = \partial Z / \partial W \cdot 1/m$;

W - vertical component of perturbed translation velocity of aircraft;

Alexander Efremov
Moscow Aviation Institute, Russia
e-mail: pvl@mai.ru

Mikhail Tjaglik
Moscow Aviation Institute, Russia
e-mail: pvl@mai.ru

ξ_{sp}, ω_{sp} - damping ratio, frequency in short period motion;

$\gamma_{pr} = \dot{\gamma} T_{pr} / 2 + \gamma$ - predicted path angle;

$\varepsilon_\gamma = \gamma_{pr} + H / L$;

$\varepsilon_\theta = \theta + H / L$ - aiming angle;

H - altitude;

γ - flight path angle;

L - distance between probe and drogue, distance between pilot and surface were predictor is projected;

$\Delta\bar{Z}$, $\Delta\bar{H}$ - sizes of tunnel.

1 Introduction

Number of piloting tasks such as terrain following, refueling, spot landing (including carrier landing are precise tracking tasks). All of them are characterized by the necessity to control the aircraft flight path. The aircraft dynamics has second and third order pole at the origin in these cases. It requires additional information (climb rate, pitch, yaw angles etc) which pilot uses to close additional feedbacks to ensure system stability and accuracy. To ease piloting process in some tasks (landing, terrain following) they use director indicator generating a director signal improving the dynamics of display-aircraft system. However, such director indicators do not change the compensatory type of the task. Recently a number of researches [1,2] has been made to develop a new type of displays capable of generating 3D elements (like tunnels) and additional signals with information indicating future position of aircraft. The results of investigations demonstrated that displays of this type allow improving accuracy considerably. Below is a generic approach towards the display design for some piloting tasks.

2 Approach

Three precise tracking piloting tasks: were considered landing, terrain following, refueling. When landing and terrain following the longitudinal control is characterized by tracking of program trajectory $H_i(t)$. When landing such trajectory is a glide path. In the case of terrain following it might by defined preliminary on the basis of required proximity to the terrain. Its definition has to take into account limits of accelerations, path angles, climb rate and minimum altitude. The pilot has to minimize error $e(t)$ between $H_i(t)$ and aircraft altitude $H(t)$,

$$e(t) = \Delta H = H_i(t) - H(t) \tag{1}$$

in both tasks.

The transfer function of aircraft in these tasks

$$\frac{H(s)}{\delta_e(s)} = \frac{K_c n^\alpha V}{s^2 (s^2 + 2\xi_{sp}\omega_{sp} s + \omega_{sp}^2)} \tag{2}$$

has the second order pole at the origin. It is well known that precise control in dynamics of this kind is very difficult.

As far as refueling is concerned, the controlled element element dynamics is the angle ("aiming error")

$$\varepsilon_\theta(t) = \theta(t) + \frac{H(t)}{L}$$

where L is a distance between the pilot and the drogue (basket) (Fig 1). When refueling the pilot has to minimize the error $e(t)$, where

$$e(t) = \Delta\varepsilon_\theta = \varepsilon_\theta(t) - \frac{H_i(t)}{L} \qquad (3)$$

And $H_i(t)$ is the basket displacement.

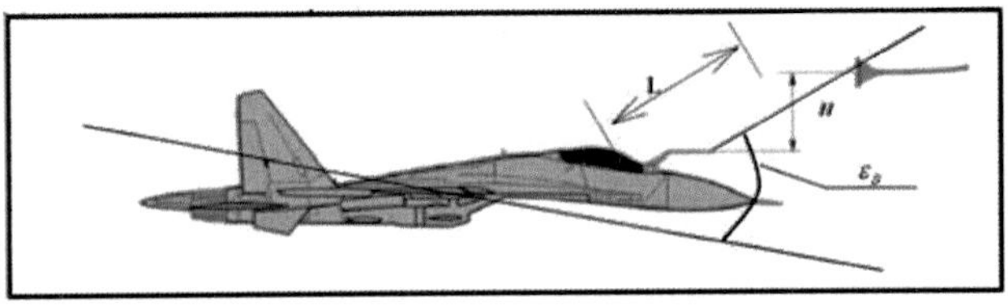

Fig. 1 Kinematics of refueling

The transfer function $\frac{\varepsilon_\theta(s)}{\delta_e(s)}$ of controlled element dynamics is the following

$$\frac{\varepsilon_\theta(s)}{\delta_e(s)} = \frac{K_c(s^2 - Z^\alpha s - Z^\alpha V/L)}{s^2(s^2 + 2\xi_{sp}\omega_{sp}s + \omega_{sp}{}^2)} \qquad (4)$$

In case of short distance to the basket V/L>>1 this transfer function becomes close to

$$\frac{\varepsilon_\vartheta(s)}{\delta_e(s)} \cong \frac{K_c(-Z^\alpha)V/L}{s^2(s^2 + 2\xi_{sp}\omega_{sp}s + \omega_{sp}{}^2)}$$

This transfer function is similar to (2). It means that the pilot compensation in this task has to be considerable too. The gain coefficient in this transfer function ($K_c\overline{Z}^\alpha V/L$) for short distance L is considerable too. For this reason the pilot changes the strategy to suppress the instability in pilot-aircraft system. The pilot starts adjusting the aircraft altitude rather than the angle ε_θ [3]. Such dynamics is not simple one, but it is characterized by a constant gain coefficient.

Controlled element dynamics can be improved in a number of ways. Among them is employment of a flight control system capable of altering parameters of denominator *D(s)* of aircraft transfer function $W_c = \frac{N(s)}{D(s)}$. The other way is alteration of numerator *N(s)* of transfer function $W_c = \frac{N(s)}{D(s)}$ for any considered piloting task. It can be realized by the image of predicted path angle $\gamma_{pr}(t)$ on HUD and its projection on the surface locating at a distance L_{pr} and sliding in front of the plane with the velocity corresponding to the aircraft velocity. The value of $L_{pr\ opt}$ (or $T_{pr\ opt}=L_{pr\ opt}/V$) is defined through the simulation on workstation or mathematical modeling by calculation of $\sigma_e^2 = f(T_{pr})$ and definition of its minimum. The error e(t) is defined by (1) for landing and terrain following tasks or by (3) for refueling task. Efficiency of information on the future position of the aircraft was verified through the fixed-based simulation, where a computer-generated visual system (CGVS) created an image close to the investigated piloting task. Moreover, CGVS also generated HUD images of a predictor and a tunnel. By virtue of such tunnel pilot can see the aircraft position in the space and change his or her compensatory behavior to the preview type. The efficiency of such transformation has been shown in a number of studies [4].

3 Analysis of Prediction Distance L_{pr} on Pilot Aircraft System Variables

To optimize L_{pr} or corresponding prediction time ($T_{pr}=L_{pr}/V$) it is necessary to take into account its influence on the following task variables: controlled element dynamics $W_c(s)$, transfer function W_c^* defining the relation between variable characterizing the task performance (*H(s)* in landing and terrain following, and $\varepsilon_\theta(s)$ in refueling) and controlled element dynamics output signal ε_γ, input signal. They are defined below these variables and influence of L_{pr} on them for each investigated piloting task. To define controlled element dynamics it is necessary to know predicted path angle $\gamma_{pr}(t)$. The Fig. 2 shows that $\gamma_{pr}(t) = \dot{\gamma}\frac{T_{pr}}{2} + \gamma$, where $T_{pr}=L_{pr}/V$.

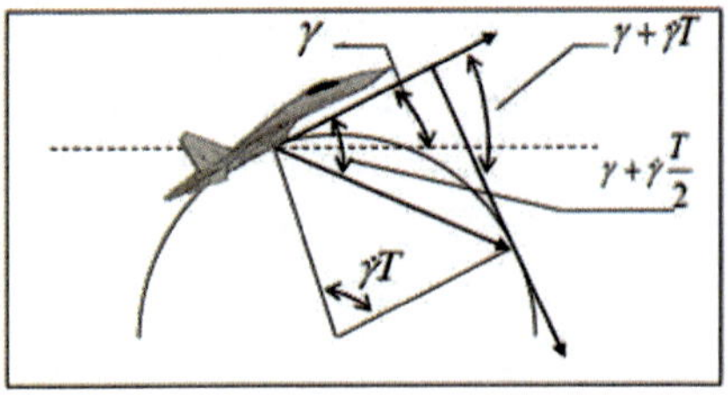

Fig. 2 Definition of predicted path angle

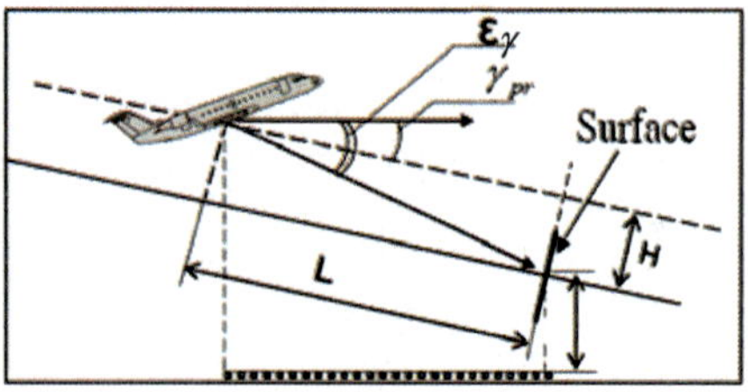

Fig. 3 Projection of predicted path angle

For all piloting tasks under consideration change of altitude influences on the location of projection of prediction path angle $\gamma_{pr}(t)$ on surface at distance L_{pr} (fig 3). It influences controlled element dynamics output $\varepsilon_\gamma = \gamma_{pr} + H / L_{pr}$.

The block-diagram of the considered task (Fig 4) allows to obtain the transfer function of controlled element dynamics

$$W_c = \frac{\varepsilon_\gamma(s)}{X_e(s)} = \frac{K_c(T_{pr}s^2 + 2s + 2/T_{pr})}{2s^2(s^2 + 2\xi_{sp}\omega_{sp}s + \omega_{sp}^2)} \tag{5}$$

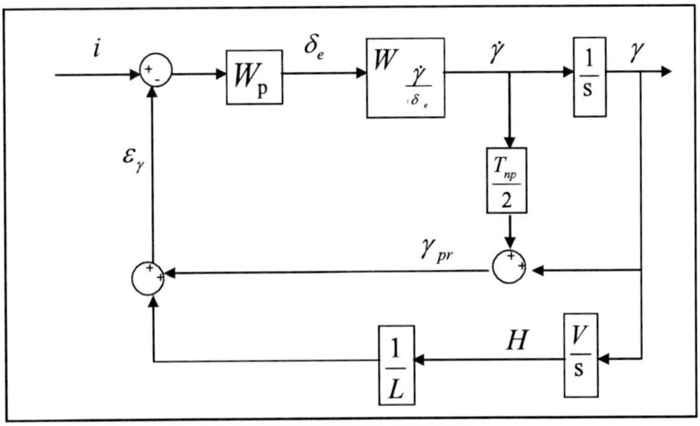

Fig. 4 Pilot-aircraft system for ε_γ angle control.

In general the dynamics (5) with the constant T_{pr} (or L_{pr}) provides better flying qualities in comparison with the dynamics characterized by the variable parameter L_{pr}.(or T_{pr}). The parameter L_{pr} influences on the transfer function W_c^* .

Below is the transfer function in landing and terrain following tasks:

$$W_c^* = \frac{H(s)}{\varepsilon_\gamma(s)} = \frac{2V}{T_{pr}s^2 + 2s + 2/T_{pr}} \tag{6}$$

In case of decrease of T_{pr}, equation (6) can be simplified as $W_c^* \cong \frac{H(s)}{\varepsilon_\gamma(s)} = \frac{V}{s + 1/T_{pr}}$, which does not contain the first order pole at the origin. The increase of T_{pr} leads to increase of the delay of altitude response in relation to ε_γ .

For example, when T>>1

$$W_c^* = \frac{H(s)}{\varepsilon_\theta(s)} \cong \frac{V}{s(T_{pr}/2s + 1)} .$$

Deterioration of harmonization between variables *H(s)* and $\varepsilon_\gamma(s)$ can lead to deterioration of tracking accuracy *H(t)*, in spite of improvement of tracking accuracy ε_γ. This disadvantage can be avoided by corresponding selection of parameter T_{pr} and additional information on current aircraft altitude position. The latter can be intensively used by a pilot, which means that he or she closes additional visual feedback loop. As an example of such information, 3-D corridor (tunnel) was generated on HUD during experiments carried out on a fixed based simulator. The sizes of such tunnel $\Delta\bar{H}$, $\Delta\bar{Z}$ allow the pilot to estimate current displacement relatively to the program trajectory.

For refueling task the output signal is $\varepsilon_\theta(s)$ and W_c^* is the following

$$W_c^* = \frac{\varepsilon_\theta(s)}{\varepsilon_\gamma(s)} = \frac{s^2 - Z^\alpha s - Z^\alpha V / L}{s^2 T_{pr}/2 + s + 1/T_{pr}} \tag{7}$$

For short distances to the basket and the high values T_{pr} equation (7) transforms to $W_c^* \cong \dfrac{-Z^\alpha V / L}{s(sT_{pr}/2+1)}$.

It seems that response ε_θ is characterized by the delay in comparison with ε_γ in that case. The harmonization between $\varepsilon_\theta(s)$ and $\varepsilon_\gamma(s)$ can be achieved by corresponding selection of parameter T_{pr} and demonstration of the tunnel on HUD with preliminary selected boundaries $\Delta\bar{Z}$ and $\Delta\bar{H}$.

A projection of angle ε_γ on the surface at a distance L_{pr} in front of the pilot causes dependence of input signal *i(t)* from L_{pr}, $i(t) = H / L_{pr}$. The increase of T_{pr} lessens variance of input signal perceived by pilot $\sigma_i^2 = \dfrac{\sigma_{\Delta H}^2}{(VT_{pr})^2}$.

A high L_{pr} affects visual accuracy threshold. Therefore, T_{pr} should be of optimum value for which the variance of altitude error ΔH (1) (in landing or terrain following task) or error $\Delta\varepsilon_\theta$ (3) (refueling task) is minimum. The procedure for its definition is considered below.

4 Technique of Research

The optimum value of T_{pr} is defined by mathematical modeling and experimental investigations.

4.1 Mathematical Modeling

Mathematical modeling has been carried out in the following steps:

1. Modification of pilot model
2. Determination of input signal
3. Determination of optimal value T_{pr}

4.1.1 Modification of Pilot Model

The structure of pilot model used in the research is shown on Fig. 5:

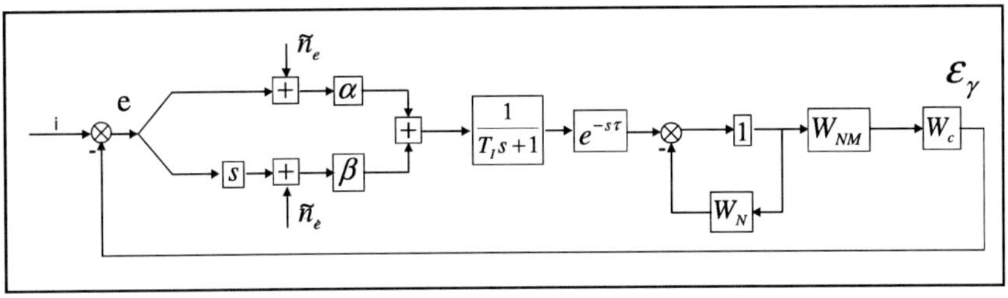

Fig. 5 Modified Hess pilot model

The models of spectral density of $\tilde{n}_e$ and $\tilde{n}_{\dot{e}}$ take into account the influence of the visual threshold (Δ): $S_{\bar{n}_e \bar{n}_e} = \dfrac{0.01\pi}{K^2_{\bar{n}_e}}$; $S_{\bar{n}_{\dot{e}} \bar{n}_{\dot{e}}} = \dfrac{0.01\pi}{K^2_{\bar{n}_{\dot{e}}}}$ where stochastic gain coefficients are $K_{\bar{n}_e} = erf(\Delta / \sigma_e)$, $K_{\bar{n}_{\dot{e}}} = erf(\Delta / \sigma_{\dot{e}})$. The transfer functions $W_{KN} = \dfrac{K_N s^2}{T_N^2 s^2 + 3\xi T_N s + 1}$ and $W_{NM} = \dfrac{\omega_{NM}^2}{s^2 + 3\xi_{NM}\,\omega_{NM}\, s + \omega_{NM}^2}$ (ξ_{NM} =0.1, ω_{NM} =12 1/sec) characterize the pilot's adaptation to kinesthetic information and neuromuscular dynamics correspondingly. The selection of pilot variable parameters ($K_L = \alpha$, $T_L = \alpha / \beta$, K_N, T_N) was carried out by minimization of variance of error σ_e^2. Here *e(t)* is the error signal predicted by pilot (see Fig. 5). The procedure for such parametric optimization is presented in [3.7]. This model is a modification of the well known Hess model [5]. The proposed modification allowed to obtain a better agreement with the experimental results in low and high frequency ranges in comparison with the basic model.

4.1.2 Input Signal

An input signal in terrain following task is a program trajectory. It is a basket motion in refueling task and aircraft response on atmosphere turbulence in landing task. In the mathematical modeling a general form of spectral density spectral $S_{ii} = k^2 / (\omega^2 + \omega_i^2)^2$ has been used. The ω_i parameter is defined from the requirement of agreement of this spectral density to the real input signal spectral density corresponding to the investigated task. The gain coefficient k is selected from condition that variance of input signal located at distance L_{pr} was $\sigma_i^2 = \sigma^{*2} / (L_{pr})^2$, where σ_i^{*2} variance of input signal.

4.1.3 Optimum Prediction Time

The mathematical modeling of pilot-aircraft system corresponding to each piloting task was fulfilled for different values T_{pr}. For each of them it was necessary to define pilot model parameters (K_L, T_L, …) providing minimum $\sigma^2_{\Delta\varepsilon_\gamma}$ (where according to Fig. 5 $e(t) = i(t) - \varepsilon_\gamma(t) = \Delta\varepsilon_\gamma$), value of this minimum, variances $\sigma^2_{\Delta H}$ or $\sigma^2_{\Delta\varepsilon_\theta}$ and their normalized variances $\bar{\sigma}^2_{\Delta H} = \sigma^2_{\Delta H} / \sigma^2_{\Delta H\,\max}$, $\bar{\sigma}^2_{\Delta\varepsilon_\theta} = \sigma^2_{\Delta\varepsilon_\theta} / \sigma^2_{\Delta\varepsilon_\theta\,\max}$, $\bar{\sigma}^2_{\Delta\varepsilon_\gamma} = \sigma^2_{\Delta\varepsilon_\gamma} / \sigma^2_{\Delta\varepsilon_\gamma\,\max}$. Here $\sigma^2_{\Delta H\,\max}$, $\sigma^2_{\Delta\varepsilon_\theta\,\max}$, $\sigma^2_{e\,\max}$ are maximum values of variances calculated for all T_{pr}. As a result, dependence $\bar{\sigma}^2_{\Delta\varepsilon_\gamma} = f(T_{pr})$ is obtained which demonstrated that an increase of prediction time (up to T_{pr}=1÷2 sec) leads to a decrease of variance of error $\sigma^2_{\Delta\varepsilon_\gamma}$. Such result is common for all investigated piloting tasks. The result of calculation of $\bar{\sigma}^2_{\Delta H}$ for landing task on Fig. (6) shows that the optimum value T_{pr} is close to 0.7 sec.

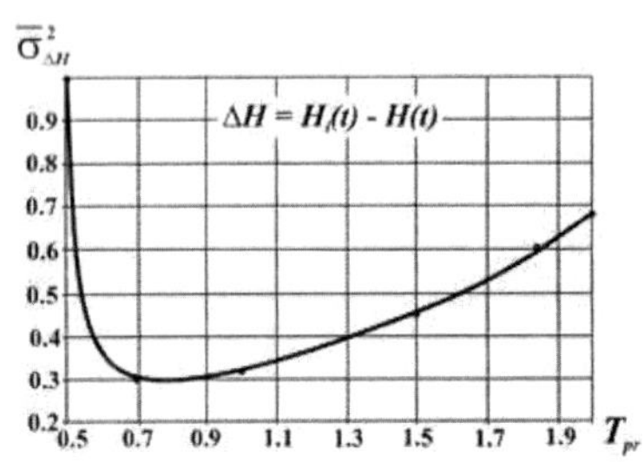

Fig. 6 Optimal prediction time for landing and terrain following tasks

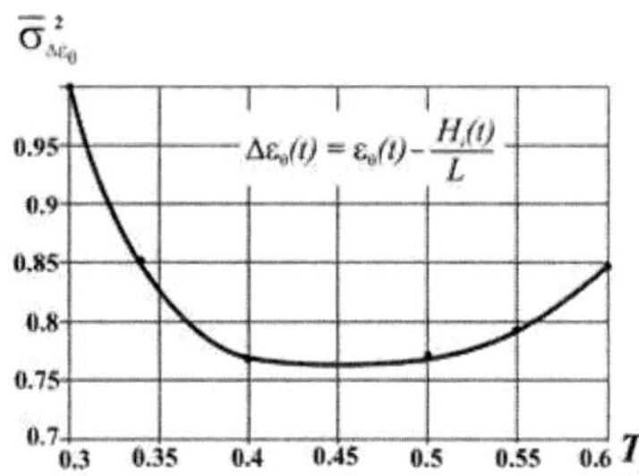

Fig. 7 Optimal Prediction time for re fueling task

It was shown that influence of input signal bandwidth ω_i on T_{pr} is insignificant. As for the refueling task the dependence is calculated as $\bar{\sigma}^2_{\Delta\varepsilon_\theta} = f(T_{pr})$ which allowed obtaining optimum value T_{pr}. It can be seen (Fig. 7) that optimum value is T_{pr} = 0.4÷0.5 sec. The influence of input spectral density parameter ω_i does not influence practically on this value either.

4.2 *Experimental Investigations*

The results of mathematical modeling were verified by experimentally on computer and ground based simulator. The workstation simulations were fulfilled for situation when pilot carried out compensatory task with aircraft dynamics corresponding to the equation (5) with constant T_{pr}. Pilot task was the minimization of error

$e(t) = \Delta\varepsilon_\gamma = i(t) - \varepsilon_\gamma(t)$. Upon each experiment pilot and pilot-aircraft frequency response characteristics and variances were defined. All of them were close to the same characteristics calculated in mathematical modeling. The typical results of workstation simulation (normalized variances $\overline{\sigma}^2_{\Delta\varepsilon_\gamma} = \sigma^2_{\Delta\varepsilon_\gamma} / \sigma^2_{\Delta\varepsilon_\gamma \max} = f(T_{pr})$ and $\overline{\sigma}^2_{\Delta H} = \sigma^2_{\Delta H} / \sigma^2_{\Delta H \max} = f(T_{pr})$) are shown on Fig. 8.

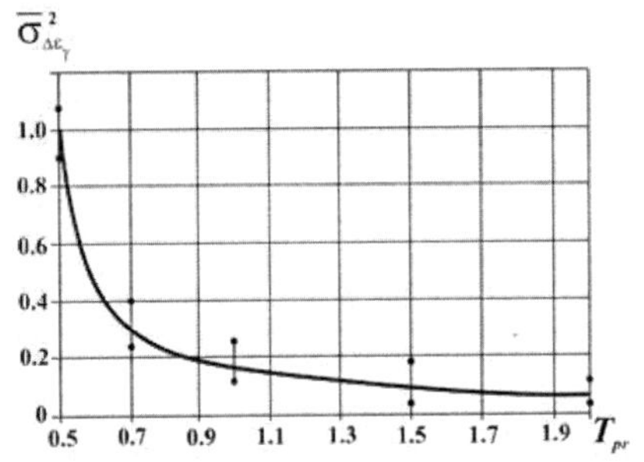

Fig. 8a Experimental results of computer-aided simulation modeling

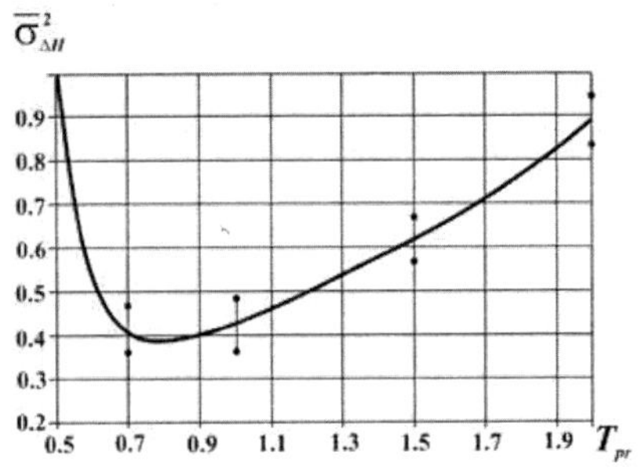

Fig. 8b Results of computer-aided simulation

These results substantiated the mathematical modeling that an increase of prediction time T_{pr} leads to decrease of normalized variance $\overline{\sigma}^2_{\Delta\varepsilon_\gamma}$ and optimum value of T_{pr} providing the minimum of variance $\overline{\sigma}^2_{\Delta H}$ is close to $T_{pr\,opt}$=0.7÷0.8 sec.

The efficiency of prediction information was checked through experiments on ground based simulator. This simulator has a computer generated visual system which demonstrates an image of outside world corresponding to each piloting task. Besides, a HUD image was projected on the screen of this visual system. The intention was to estimate:

- pilot subjective Cooper-Harper rating
- mean of error $m_x(t_i)$
- mean square error $\sigma_x(t_i)$,

where x is path variables (H, Y, $\dot{H}$) calculated experimentally.

For terrain following task two versions of HUD were investigated:

- HUD with information shown on fig 9
- HUD with the same image without the tunnel.

The desired and accepted task performances were defined by a method developed in [6]. The task performances were defined as requirements to aircraft path position to stay at least 87% and 70% correspondingly of total flight time inside the 3D tunnel.

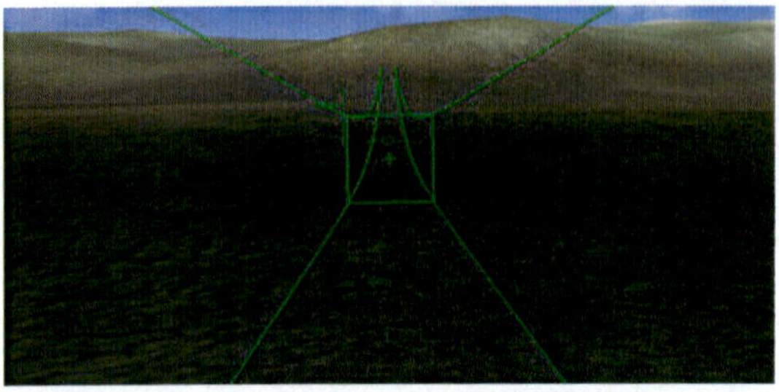

Fig. 9 Head up Display for terrain following task

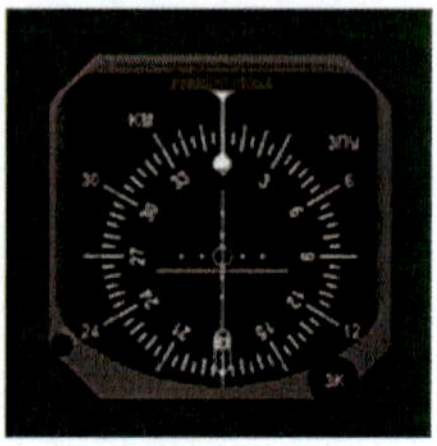

Fig. 10 Director indicator used for the landing task

Totally 80 experiments were carried out. Data reduction allowed to calculate the variance of altitude errors $H(t)=H(t_i)-H_i(t_i)$, where $H_i(t)$ is the input signal, in each moment t_i:

$$\sigma^2_{\Delta H}(t_i) = \frac{\sum_N \sqrt{(e(t_i)-m_e(t_i))^2}}{N-1}$$; N - number of experiments, $m_{\Delta H}(t_i)$ - mean of error in moment t_i. After that average mean square error for all flight was defined.

$$M[\sigma_{\Delta H}] = \sum_{i=1}^{k} \sigma_{\Delta H}(t_i)/k$$; where k is a total number of moments t_i.

During experiments the full six-degree of freedom aircraft mathematical model was simulated in a real time mode.

Experimental investigations demonstrated that any considered HUD allowed carrying out the task very accurately (Fig. 11). Moreover, employment of HUD with images of predictor and tunnel allowed decreasing the mean square error of altitude up to 20% in comparison with cases when only image of tunnel was generated by HUD.

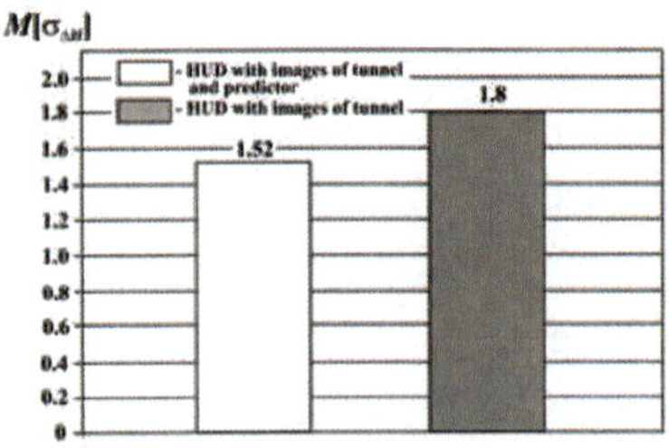

Fig. 11 Results of experiments (terrain following)

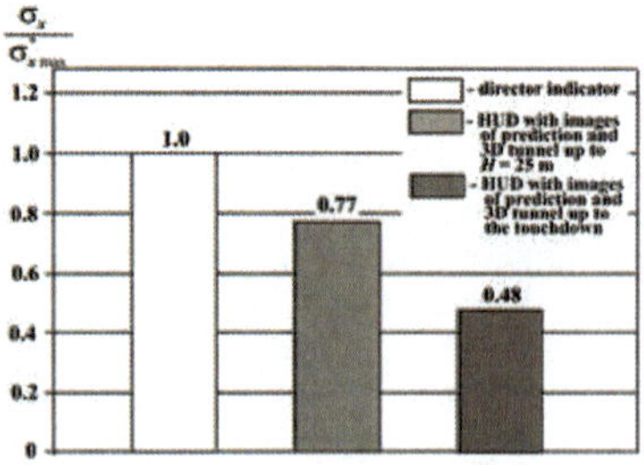

Fig. 12 Results of experiments (landing)

Landings were fulfilled under the following conditions:

- Instrumental landing with display shown on Fig 10. The pilot also used other indicators. A computer generated visual system gave pilot an image corresponding to the landing task.
- HUD display with images of predictor with 3D tunnel and sliding surface at distance L_{pr} where the predictor and 3D tunnel were demonstrated up to H=25m. After that pilot carried out flare using the visual information.
- HUD display demonstrated the same information during the approach and flare up to the touchdown.

During the flare the tunnel transforms in curved tunnel according to the flare trajectory. The trajectory of flare was defined from requirement that initial (when H=25 m) and touchdown climb rates corresponds to the pre-defined values.

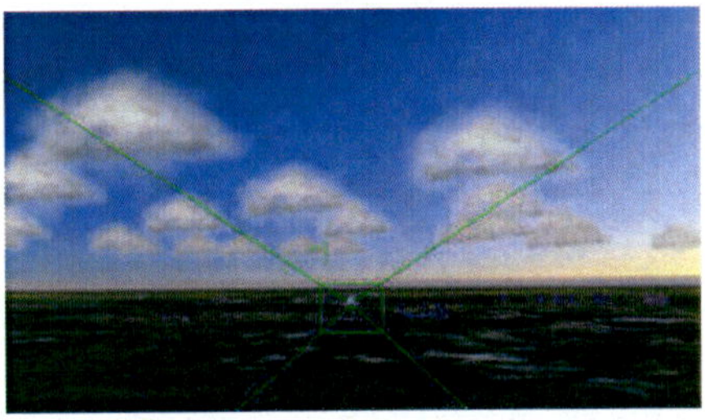

Fig. 13 HUD used for landing

Fig. 14 HUD for refueling

Almost 90 landings were fulfilled in ground based simulator. For each of them the altitude and side position during approach, touchdown climb rate, touchdown point coordinates were recorded. The reduction of data demonstrated that additional information allowed decreasing the variability of touchdown point. Employment of HUD with prediction information up to H=25 m allowed decreasing the normalized mean square error in longitudinal channel up to 23% and up to 52% for HUD with images of predictor and 3D tunnel up to touchdown. Additionally, the accuracy of landing in lateral channel increases up to 4 times.

Besides, vertical rate at touchdown point decreases in two times. Employment of HUD display when approaching allowed decreasing altitude deviation in two times.

Refueling task was investigated in two options:

- without HUD display
- with HUD shown on fig 14

Totally about 70 experiments totally was carried out. The colors of the tunnel changed with the function of velocity of approach V_a. Red color means the case with V_a<0.8 m/sec, green color, 0.8≥ V_a ≤1.2 m/sec, yellow color, V_a>1.2 m/sec. These colors inform pilot about aircraft speed which allows pilot to control it.

With a display of this type it becomes possible to decrease the mean square error of point of contact 1.5 times and ensure green color of tunnel in all experiments.

5 Conclusions

The method of employment of future HUD is offered for some piloting tasks requiring high accuracy. The method is based on principles of the general pilot aircraft system theory stated in [7]. The advanced versions of HUD allowed to improve the accuracy of task fulfillment 1.5÷2.5 times and to enhance flight safety.

References

1. Sachs, G., Sperl, R., Sturhan, I.: Curved and steep approach flight tests of a low – cost 3D-Display for general aviation aircraft. In: Proceedings of 25 ICAS Germany (September 2006)
2. Efremov, A.V., Koshelenko, A. V., Tjaglik, M. S. : Means for Flying Qualities Improvement in Piloting Tasks Required Extremely High Accuracy. In: Proceeding of AFM AIAA Conference, Chicago (2009)
3. Efremov, A.V.: Development of criteria for prediction of handing qualities of new generation of aircraft TR on contract SPC-96-4073 MAI – Wright Patterson USA AF Base, Moscow (1997)
4. Read, L.D., Drewell, N.: A pilot model for tracking with preview. In: The Eight NASA – University Annual Conference on Manual Control, pp. 191–204 (May 1972)
5. Hess, R.: Structural model of the adaptive human pilot. J. of Guidance and Control 3(5) (1979)
6. Efremov, A.V., Ogloblin, A.V., Koshelenko, A.V.: Evaluation and prediction of flying qualities. Journal "Flight" (3), 28–34 (1999) (in Russian)
7. Efremov, A.V., et al.: Pilot as a dynamic system. Mashinostroenie M. Russia (1992)

UAV Lab, Open Research Platform for Unmanned Aerial Vehicles

Péter Bauer, Paw Yew Chai, Luigi Iannelli, Rohit Pandita, Gergely Regula, Bálint Vanek, Gary J. Balas, Luigi Glielmo, and József Bokor

Abstract. This paper summarizes some of the joint research efforts of University of Minnesota, University of Sannio, Computer and Automation Research Institute (SZTAKI) and Budapest University of Technology and Economics (BME) to develop an Unmanned Aerial System. The future aim of the research collaboration is to develop and test advanced navigation, control and fault detection algorithms running over an easily accessible, customizable platform in real flights. This collaboration resulted in a flight research platform which can be easily configured with different guidance, navigation and control algorithms including an easy way of code development and testing. The paper focuses on the fine-tuning of the system components implemented by SZTAKI and BME. The related work includes sensor calibration, modification of autopilot code and testing of simple PID control algorithms. The latest flight test results are included in the paper.

Abbreviations and Notations

BME	Budapest University of Technology and Economics
CEO	Chief Executive Officer
COTS	Commercial Off The Shelf

Péter Bauer · Gergely Regula · Bálint Vanek
Computer and Automation research Institute of HAS, Kende u. 13-17, Budapest, Hungary
e-mail: bauer.peter@sztaki.hu

József Bokor
Budapest University of Technology and Economics, Műegyetem rkp. 3-9, Computer and Automation Research Institute of HAS, Kende u. 13-17, Budapest, Hungary
e-mail: bokor@sztaki.hu

Luigi Iannelli · Luigi Glielmo
University of Sannio, piazza Roma 21, Benevento, Italy
e-mail: glielmo@unisannio.it

Paw Yew Chai · Rohit Pandita · Gary J. Balas
University of Minnesota, 110 Union St SE, Minneapolis, USA
e-mail: paw@aem.umn.edu

DAQ	Data Acquisition
EKF	Extended Kalman Filter
GPS	Global Positioning System
HIL	Hardware-in-the-loop
IAS	Indicated Airspeed
IMU	Inertial Measurement Unit
PID	Proportional Integral Derivative (control)
PPM	Pulse Position Modulated (signal)
PWM	Pulse Width Modulated (signal)
RC	Radio Controlled
SIL	Software-in-the-loop
SZTAKI	Computer and Automation Research Institute, Hungarian Academy of Sciences
UAS	Unmanned Aerial System
UAV	Unmanned Aerial Vehicle
UoM	University of Minnesota, Department of Aerospace Engineering and Mechanics
UoS	University of Sannio, Department of Engineering
$\mathbf{a}$	acceleration
h	altitude
$\mathbf{H}$	magnetic vector
k	time index
$\mathbf{V}$	vector of measured voltage values
V	aircraft indicated airspeed
superscripts:	
a	acceleration related values
ω, ω_0	angular rate related values
h, H	altitude or magnetic vector related values respectively
V	IAS related values
subscripts:	
meas	measured quantity
0	quantity related to zero output

1 Introduction

Recently, FedEX CEO F. Smith told the media that they would switch their fleet to UAV's as soon as possible (Ref. [1]). However, civil, low-cost UAV's need several pre-requisites to render them viable, cost effective and regulated alternatives to existing resources. One key aspect of these requirements is to have reliable flight control systems onboard. It means that advanced and fault tolerant algorithms must be used to guarantee the safety of UAV missions.

A joint research project was initiated between UoM, UoS, SZTAKI and BME to develop a UAS that has easily accessible and customizable development environment which makes possible significantly shorter development cycles (see Ref. [7]). The long-term aim of this collaboration is to develop and flight test advanced, reliable and fault tolerant algorithms onboard the UAV. The collaboration originally started between UoM and UoS. The aim was to create an open-source project with COTS hardware components. SZTAKI and BME later joined in and their task was to test the developed UAS including all its components. This means exchange of

Fig. 1 UAS Ultrastick aircaft

documentation, component lists and software codes. This is a vital task to test and prove whether the available open-source documentation is complete or not. During this work, continuous communication and bug information exchange was conducted and several improvements were applied from which the main results are summarized here.

The outline of the paper is as follows: Sec. 2 introduces the open-source UAS, Sec. 3 summarizes the work related to the sensor unit, Sec. 4 presents the developments in the autopilot code, Sec. 5 and 6 summarize the flight test results. And Sec. 7 draws conclusions.

2 The Open-Source UAS with COTS Components

The selected aircraft is the E-flite Ultrastick 25e. It is a small RC airplane but has enough space in the fuselage to integrate all the required hardware components (see Fig. 1). The aircraft has 1.27 m wingspan and about 1 m fuselage length with 2.1 kg take-off weight (including all additional hardware).

The general architecture that is the onboard avionics together with the ground components is shown in Fig. 2. Besides the conventional RC components, the onboard system consists of the MPC555 microcomputer, the μNAV sensor, a fail-safe multiplexer board and a wireless communication modem (see Ref. [8]). The main steps of the code development for the control system are given in Fig. 3. The software-in-the-loop (SIL) test environment shown in the figure was fully implemented in Matlab. While the hardware-in-the-loop (processor-in-loop) environment uses a nonlinear Matlab simulation model of the aircraft that is connected to a hardware architecture similar to the one shown in Fig. 2. In case of HIL, the aircraft and

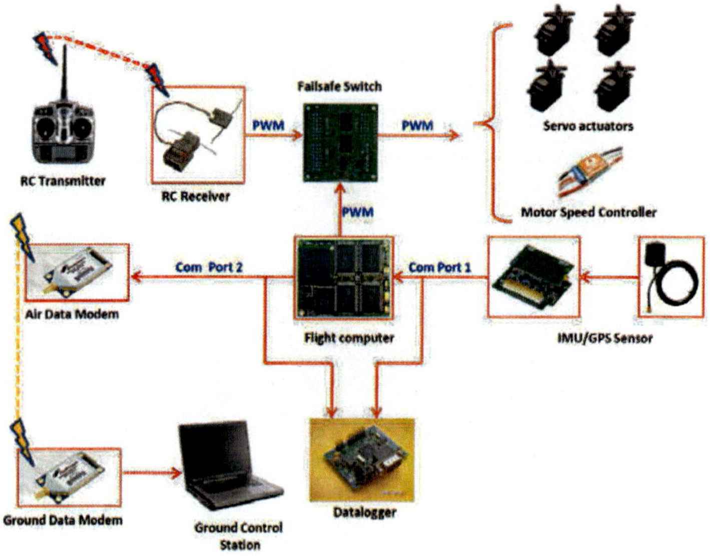

Fig. 2 General architecture for autonomous flight

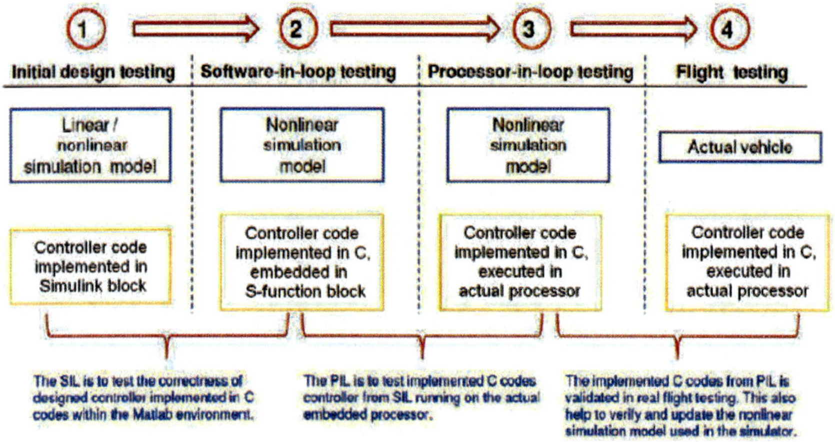

Fig. 3 UAV control system code development steps

μNAV sensor are replaced by their simulation models; with all the other components (i. e., microcontroller, multiplexer board, RC receiver) being identical to those appearing in the real system.

After introducing the main system components and explaining the steps of control code development, we focus our attention to the various modifications during system testing.

3 Modifications Related to μNAV Sensor

The first, and most important component of the system is the IMU. The Crossbow μNAV sensor is used in the UAS. It can measure three axis accelerations, angular rates, magnetic values, static and pitot pressures, temperature and GPS signals. It can convert RC PPM signal into the separate PWM signals. It is a compact, lightweight and cost effective device with variable output frequency and packet types (see Ref. [5]).

3.1 Experiences with Software and Its Modifications

Crossbow provides a software – called Micro-View – with which the sensor can be calibrated and the sensor data can be collected. Experience with this software indicates that it is unreliable as it loses connection with the sensor device if a broken data packet arrives, furthermore it does not always set the sensor into the desired mode. For this reason a handling program was written using LabWindows CVI with similar but extended capabilities, particularly with respect to packet error tolerance.

The calibration capability provided by Micro-View is not satisfactory as it considers only the diagonal elements of the calibration matrices for rate sensors, accelerometers and magnetometers; furthermore it does not take temperature dependence into consideration. However, the off-diagonal elements of these matrices should be taken into consideration, and it has turned out that the acceleration and rate sensors have significant temperature dependence. $0.85\,g$ acceleration was measured at $5°C$ instead of $1.0g$ (measured at $25°C$). This means $(0.075m/s^2)/K$ temperature dependence, therefore temperature calibration is a vital task.

μNAV can send out either sensor voltage data or scaled data. If one makes a custom calibration procedure, then it is advisable to use the former data format. But then neither PWM, nor GPS signals are sent out in the data packet (see Ref. [5]). As a consequence, the scaled mode packet should be modified to include sensor voltages (instead of scaled values). Two other modifications were also required.

1. one of the unused timers of the microcontroller on μNAV was used for checking whether the data frequency is correct or not.
2. a packet counter was set to make the detection of packet losses possible.

3.2 Sensor Calibration and Data Acquisition

The next step of the development was sensor calibration. The calibration formulae are as follows. For the explanation of notations see abbreviations and notations.

$$\mathbf{a} = K^a\left(\mathbf{V}^a_{meas} - \mathbf{V}^a_0\right) \tag{1a}$$

$$\mathbf{V}^\omega_0 = \mathbf{V}^{\omega_0}_0 + K^{\omega_0}\mathbf{a} \tag{1b}$$

$$\omega = K^\omega\left(\mathbf{V}^\omega_{meas} - \mathbf{V}^\omega_0\right) \tag{1c}$$

$$\mathbf{H} = K^H\left(\mathbf{V}^H_{meas} - \mathbf{V}^H_0\right) \tag{1d}$$

$$h_{k+1} = h_k + K^h \left(\mathbf{V}^h_{meas}(k+1) - \mathbf{V}^h_{meas}(k) \right) \tag{1e}$$

$$V_{k+1} = \sqrt{V_k^2 + K^V \left(\mathbf{V}^V_{meas}(k+1) - \mathbf{V}^V_{meas}(k) \right)} \tag{1f}$$

Equation 1 shows that bias vector and scaling matrices were calculated for the acceleration, angular rate and magnetic vector, while for the static and pitot pressure a scaling factor and the formulae were derived. The unknown parameters were determined using least-squares technique. As it turned out, higher order terms were not required as the errors of calibrated quantities were sufficiently small. The maximum/mean absolute errors are summarized in Table 1.

Table 1 Maximum and mean absolute fit errors

quantity	max error	mean error
acceleration	4%	1.5%
angular rate	3%	0.8%
magnetic data	5%	1%
altitude	27%	2.6%
IAS	100%	30.3%

As can be seen from Table 1 only the IAS errors are serious, but it is well known that its measurement with pitot tube is inaccurate. Therefore a better fit could not be achieved.

The bias vector of the angular rate (in Eq. 1b) was calculated from the acceleration with bias and scaling in a manner proposed by Ref. [4]. The temperature dependent parts of the formulae are shown in Eq. 2. It soon became evident that the magnetic and pressure data did not need temperature calibration.

$$\mathbf{V}_0^a = \mathbf{V}_{01}^a t_V + \mathbf{V}_{02}^a; \quad K^a = K_1^a t_V + K_2^a \tag{2a}$$

$$\mathbf{V}_0^{\omega_0} = \mathbf{V}_{01}^{\omega_0} t_V + \mathbf{V}_{02}^{\omega_0}; \quad K^{\omega_0} = K_1^{\omega_0} t_V + K_2^{\omega_0} \tag{2b}$$

$$K^{\omega} = K_1^{\omega} t_V + K_2^{\omega} \tag{2c}$$

Here, t_V is the output voltage of temperature sensor. The calibration formulae in Eq. 2 relate to the linear fitting on the elements of the calibration matrices and vectors in Eq. 1a–1d. The quality of temperature calibration can be tested comparing the measured vertical accelerations when the aircraft is on the ground. In this case, the pitch angle of aircraft is about 10° so, the vertical acceleration should be about $\cos(10^\circ) \cdot 1g = 0.9848g$. In accelerometer calibration the following sign convention was used: $+1$ g is output if gravity acts along the sensor's axis. The measured and calibrated a_z accelerations from flight tests - in May 2009 (with $20-25^\circ C$ air temperature), October 2009 (with $15-20^\circ C$ air temperature) and December 2009 (with $5-10^\circ C$ air temperature) in Budapest Hungary - are plotted in Fig. 4. The minimum and maximum differences from the required value were -0.57% and 0.73%,

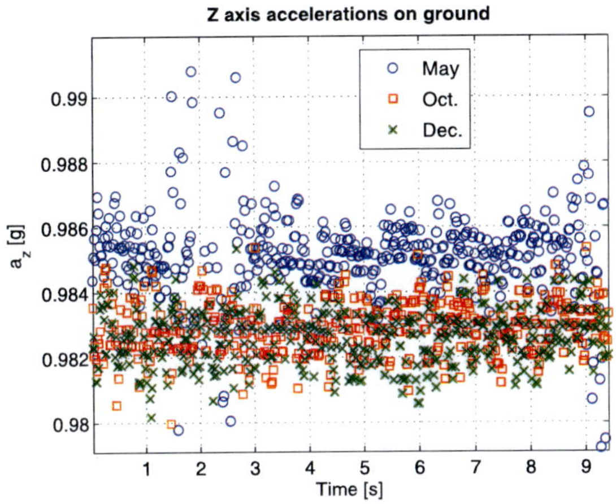

Fig. 4 Temperature-calibrated vertical acceleration in different weather conditions

respectively. It means that the calibrated measurement was carried out with very high accuracy.

Sensor data could have been collected with the ground station software, but the following two problems arose:

1. The data speed of the uplink thread was too low. After some consideration, it was set to 10Hz. With this setup, the data speed on the ground was about 8Hz.
2. The unreliability of the wireless connection caused data losses (and/or erroneous data packets).

For the aforementioned two problems, the system identification could not be performed from ground station data. However, system identification can be carried out based on the data provided onboard by *μNAV* at 50Hz, as this frequency is large enough to cover all the important modes of a small UAV.

For this purpose an SD card-based data acquisition (DAQ) device was developed by SZTAKI. The developed DAQ device is shown in Fig. 5. It collects and stores data on a standard SD card in text files. These files can be simply read and copied

Fig. 5 The SD card-based data acquisition device (developed by SZTAKI)

by any operating system. A separate program was written to convert the content of these files into μNAV packet format.

4 Control Code Developments

After completing sensor calibration and DAQ device construction as development tasks, flight tests were carried out with the Ultrastick UAV. Firstly, the system identification tests were performed. Then, tests with different levels of autonomy were flown.

During these tests the attitude estimation extended Kalman filter (EKF) was somewhat inaccurate. It was because it considered the measured acceleration vector as the instantaneous gravity vector thereby neglecting the inertial effects. Another problem was the proper handling of global variables in the different mutually exclusive threads as this led to certain instabilities in the control. For more details on thread handling see Ref. [2].

To overcome the aforementioned problems, a new and more accurate multi-mode EKF was developed and tested. It uses not only acceleration and magnetic vectors, but relies on GPS speed measurements as well. This development is presented in detail in Ref. [3].

The problem with mutual exclusions (mutex's) to handle global variables was solved by associating a global variable with one thread only, and using its local copies in every other thread. This way, mutex's are used only for a short time (to ensure copying the global values into the locals) thereby minimizing the locked times of the threads.

At the final stage of the development, flight tests with the modified MPC555 code were conducted. The developed control code has two levels: the low level implements the roll and pitch angle tracking PID controllers, while the high level control code is responsible for altitude and IAS hold and waypoint tracking control (PI and other controllers). The waypoint tracking solution is described in Ref. [6]. For more details see Ref. [8].

5 Low Level Control: Tracking of Roll and Pitch Doublet References

The low level control tracks roll and/or pitch Euler angle references with PID loops. It enables straight and level flight or turning of the aircraft. At first, the tracking of roll then the tracking of pitch doublets was tested by switching the aircraft between tracking modes. SIL, HIL and real flight tests were carried out following the individual development steps. Real flight results are plotted in Figs. 6 and 7. Tests were done on 17th June 2010 with 20-25$°C$ air temperature and moderate wind. The tracking error for roll angle was about 2.5–7.5%, while for pitch angle it was about 3.5–6.5%

6 High Level Control: Altitude and IAS Hold, Waypoint Tracking

After testing the low level controllers, we turned our attention to the higher level controllers. Four different autopilot versions were tested in SIL, HIL and real flight. The first one saves the instantaneous IAS and altitude at the time of switching and tracks them while flying straight and level (applying PI controllers). The second autopilot version tracks an IAS step function while holding the altitude (saved at the time of switching) and flying circles with constant $-30°$ roll angle. The third one tracks an altitude step function while holding the IAS (saved at the time of switching) and flying circles with constant $-30°$ roll angle. The fourth one is the full version of waypoint tracking autopilot. This latter flies between GPS coordinates while holds the altitude and IAS.

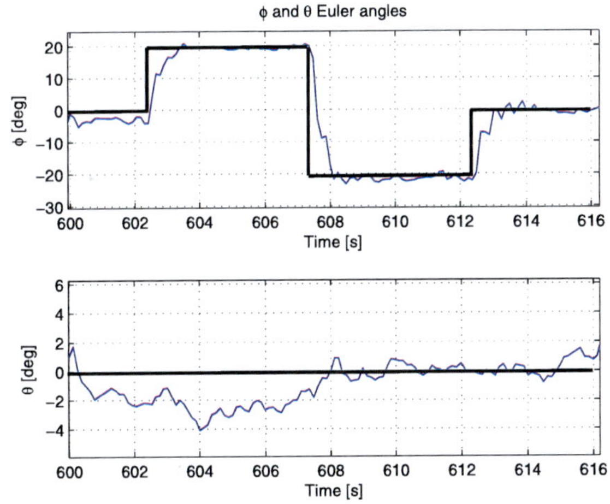

Fig. 6 Tracking of a roll doublet reference signal in real flight conditions

The results were acceptable for all the autopilot versions. Tests were done on 29th June 2010. with 25-30°C air temperature and high crosswind. The tracking results for IAS and altitude references can be seen in Fig. 8. After some transient, the IAS reference was held with $\pm 2\ m/s, \pm 9\%$, while the altitude with $+10m/-15\ m, +8\%/-12\%$ maximum error.

The tracking of waypoints can be seen in Fig. 9. The crosses show the part of flight path when the aircraft was flown by the autopilot. The small circles with the crosses show the waypoints to be tracked, while the large circles show the tolerances (20 m) around the waypoints. The figure shows that the waypoints were tracked with good accuracy (7m and 0.97m errors) so, the crosswind was well compensated. Unfortunately, the flight had to be ended earlier due to the diminished battery power.

The overall experiences show that low level controllers work well, while the high level controllers need to be tuned further to decrease the tracking errors. The IAS

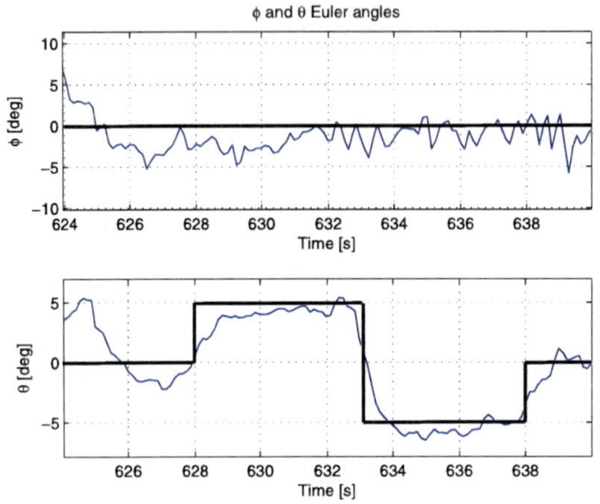

Fig. 7 Tracking of a pitch doublet reference signal in real flight conditions

tracking part (with elevator) and altitude tracking part (with throttle) has a tendency to work against each other. It is because they have very different bandwidths (the IAS tracking part is faster then the altitude tracking one) that should be tuned and harmonized.

We experienced some problems concerning the reception of GPS signals. The antenna provided with *μNAV* frequently loses reception which makes waypoint tracking impossible. Therefore, another better antenna must be purchased to improve flight performance.

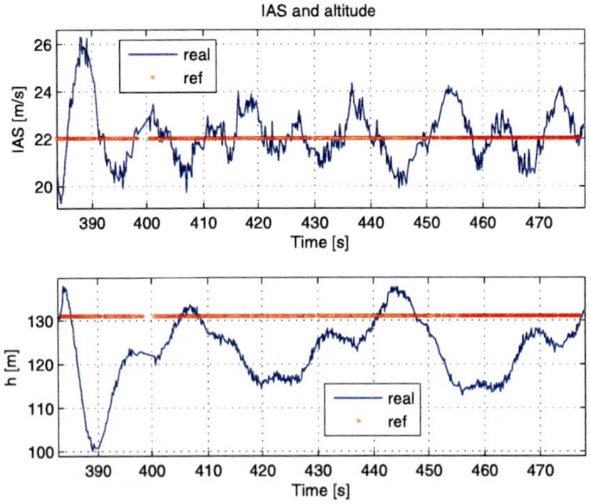

Fig. 8 Tracking of IAS and altitude references in real flight conditions

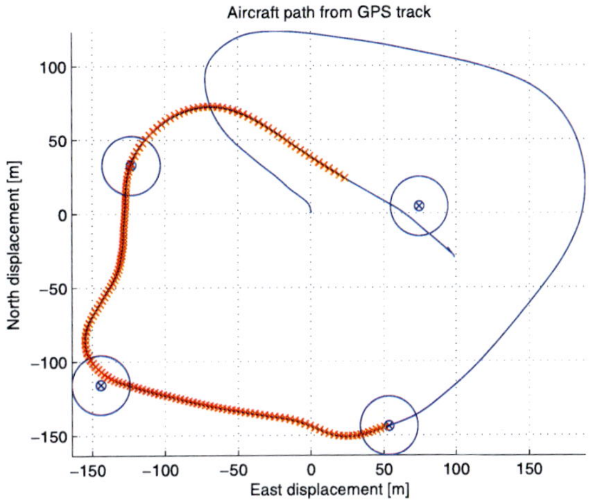

Fig. 9 Tracking of a route with four corner points in real flight conditions

7 Conclusions

This paper reports on some of the joint research efforts of UoM, UoS, SZTAKI and BME. The goal of this research was to create a UAS from COTS components with development tools which enable to rapidly synthesize, implement, analyze and validate a candidate controller design using iterative development cycles. Another goal was to use open-source philosophy. This paper described the tests performed with the developed UAS. This included building and testing the complete system starting from the documentation, component list and software codes. Several HIL tests and flights were performed. The corresponding data were plotted and evaluated.

As a result, the software codes related to μNAV sensor were modified to achieve better functionality. The sensor unit was calibrated including temperature dependence. A suitable EKF was developed for attitude estimation to improve global controller performance. Finally, the thread handling of the multithreaded autopilot code was corrected and optimized to achieve a better performance.

Flight test results show the success of the project. A UAS was developed and it serves as a platform to implement and test different autopilot codes. The participants gained hands on experience with the platform.

The future plans include system identification, implementation and testing of advanced tracking and fault detection and isolation control algorithms. Presently, fault detection and isolation is a hot topic in aerospace research and it is expected to remain so, in the coming years. The developed UAS can be used for developing, implementing and testing such control solutions.

Acknowledgements. The authors gratefully acknowledge to the support by the Hungarian National Scientific Research Fund (OTKA CNK 78168), by the TRUCKDAS project

(TECH_08-A2/2-2008-0088) and by the Control Engineering Research Group of HAS at Budapest University of Technology and Economics.

References

1. Anderson, C.: Fred Smith: FedEx wants UAV's. DIY Drones (2009), http://diydrones.com/profiles/blogs/fred-smith-fedex-wants-UAV's (Cited July 29, 2010)
2. Barney, B.: POSIX Threads Programming. Lawrence Livermore National Laboratory (2010), https://computing.llnl.gov/tutorials/pthreads/ (Cited August 18, 2010)
3. Bauer, P., Bokor, J.: Development and hardware-in-the-loop testing of an Extended Kalman Filter for attitude estimation. In: Proceedings of 11th IEEE International Symposium on Computational Intelligence and Informatics, Budapest, Hungary (2010)
4. Kis, L., Prohászka, Z., Regula, G.: Calibration and testing issues of the vision, inertial measurement and control system of an autonomous indoor quadrotor helicopter. In: Proceedings of RAAD 2008, Ancona, Italy (2008)
5. MNAV 100CA User's Manual. Crossbow Technology Inc., San Jose, USA (2005)
6. Niculescu, M.: Lateral track control law for Aerosonde UAV. In: Proceedings of 39th AIAA Aerospace Sciences Meeting and Exhibit, Reno, NV, USA (2001)
7. Yew Chai, P.: UAV Research Group. University of Minnesota (2006), http://www.aem.umn.edu/~uav/index.html (Cited August 31, 2010)
8. Yew Chai, P.: Synthesis and validation of flight control for UAV. PhD Thesis University of Minnesota, Minneapolis, USA (2009)

Visibility Cues for Communication Aware Guidance in Cluttered Environments

H. Claus Christmann and Eric N. Johnson

Abstract. This paper presents the usage of visibility based guidance cues in order to find waypoints useful for maintaining communication in a multi UAV (Uninhabited Aerial Vehicle), single operator system. Based upon the overlay of visibility graphs (for radio communication) and Voronoi diagrams (for maximum clearance motion paths), the paper presents simulations of three staged methods, allowing the computation of waypoints suitable for establishing a potential multi-hop connection between an operator and a primary UAV in an urban or otherwise cluttered environment. The methods present generic solutions for 2D planes, ensuring applicability for indoor, outdoor, and other structured environments through a potential interconnection of several non-coplanar 2D planes. The presented methods increase in computational complexity as they are capable of handling more complex scenarios. However, the presented methods are overall still deemed computationally acceptable and present themselves as good candidates for onboard implementation on vehicles with limited computational power.

1 Introduction and Motivation

Tactical Uninhabited Aerial Systems (UAS) often utilize a single Uninhabited Aerial Vehicle (UAV) tele-operated by a single control station operator. Though higher level control, i.e. the use of preprogrammed waypoints or whole trajectories, is sometimes possible, the remote operators often pilot the UAV directly via a first-person video feed, providing them with immediate sensor data and allowing them to perform tasks such as obstacle detection and classification, collision avoidance, and path planning.[1] These first-person video streams in combination with

H. Claus Christmann
Georgia Institute of Technology, School of Aerospace Engineering, Atlanta, GA 30332-0150
e-mail: hcc@gatech.edu

Eric N. Johnson
Georgia Institute of Technology, School of Aerospace Engineering, Atlanta, GA 30332-0150
e-mail: eric.johnson@ae.gatech.edu

stability augmentation systems for remote piloting allow for increased ease of operation, high situational awareness of the operator, and direct availability of the primary sensor data, the video feed. All that is achievable with relatively modest training requirements - including the control station operation and the actual remote-piloting.

However, this single-operator-single-vehicle setup limits the operational range of such a UAS to essentially the range of the utilized communication link. Furthermore, given, for example, the height of urban high rise buildings, positioning UAVs "above and behind" (as required by Line-Of-Sight (LOS) constraints) might not always allow suitable sensor access to the back side of Radio Frequency (RF) obstructing objects. Using a relay can mitigate those limits. More complex UAS, e.g. current High-Altitde-Long-Endurance (HALE) or Medium-Altitude-Long-Endurance (MALE) systems, can utilize indirect communication via communication relay nodes to overcome this LOS limit, most often at a cost of link delay and the addition of an operator dedicated to payload related activities.

For tactical scale UAS the use of satellites as relays is prohibitive, not only due to the introduction of high latency, but foremost for infeasibility of the implementation of related required avionics. Relying on potentially available HALE or MALE systems to act as relays is also challenging. Not only would the local tactical UAS operator have to coordinate with a different UAS to negotiate operational areas and coverage, but also would the link to the relay HALE or MALE UAV have to be robust to shadowing and/or multi-path effects in cluttered urban environments.

Instead of external pseudolites, other local UAVs from within the same tactical UAS could be used as communication relay nodes, effectively establishing a local multi-hop network within the UAS.

However, if operated under the same principle of remote-piloting, introducing additional UAVs as relays comes at the cost of drastically increased workload. Each additional relay UAV would require a similar amount of work as the primary UAV, mainly work related to collision avoidance and path planning. For those secondary UAVs, path planning is furthermore complicated by the dual task of getting from one location to another as well as maintaining LOS to the primary UAV and to the GCS or other intermediaries, respectively.

This work proposes visibility based cues that could allow secondary UAVs to conduct these relay tasks without major operator intervention, combining operational advantages of smaller scale tactical UAS with the benefits of swarm-enabling, higher-level automation in the background.

1.1 Limiting the Operational Zone of the Secondary UAVs

Starting from the HALE and MALE analogy, an initial replication of such a setup seems suitable. The system would deploy a single supportive UAV as a relay, this UAV would position itself "high and behind" any potential obstacle, Fig. 1(a), and as such establish a dual-hop link. An extension to this would be the usage of two supportive UAVs, positioned high enough directly above the control station and the primary UAV, respectively, to establish a three-hop link, Fig. 1(b). These setups

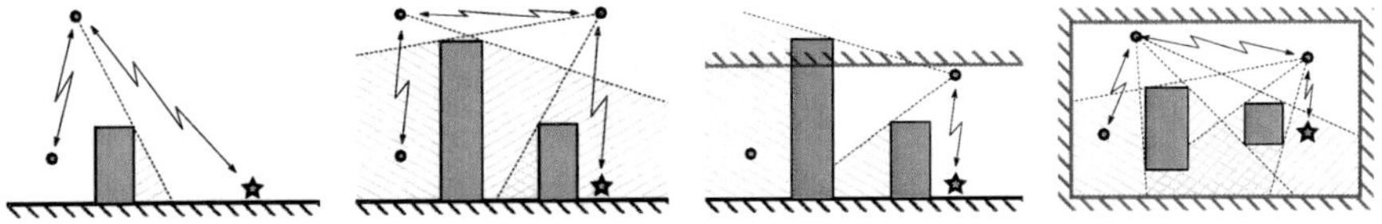

(a) Placing a single supportive UAV high and behind obstacles in cases of few obstacles.

(b) Supportive UAVs high and directly over other nodes helps in denser environments.

(c) The vertical approach could fail if going high is not an option for some reason.

(d) Utilizing a horizontal setup could enable otherwise infeasible solutions.

Fig. 1 Open environments pose no special problems as LOS is essentially guaranteed. In the presence of RF obstacles, using a vertical 2D plane to create multi-hop links between the control station and the primary UAV provides for conceptually identical solutions. If the scenario does not allow for such a positioning, using a horizontal 2D plane can expand the solution space and enable previously not possible setups.

could conceptually be called *vertical*, as the task involves the placement of supportive UAVs in an essentially vertical plane determined by the position of the control station, the primary UAV, and the "up"-direction.

However, "going up" might not always be an option. In certain scenarios the airspace could be closed above a given altitude or there is no LOS between a primary UAV and the space above, potentially due to the Area of Interest (AoI) being in a tunnel, under a large bridge, indoors, or under ground, Fig. 1(c). To include such scenarios, the solution would have to include the *horizontal* component of the environment, Fig. 1(d). However, as the problem is still a 2D problem, the proposed generic processes can also solve this setup.

Fig. 2 In complex structured 3D environments, the generality of the solution for the 2D plane case allows for an intersecting of planes to capture the environment. In the displayed example (stylizing a staircase), one horizontal plane captures each floor (B,C,D), a single vertical plane in the staircase joins them together (A). Requiring a UAV to be located at each plane intersection allows inter-plane communication based on the processes presented for the generic 2D planar case.

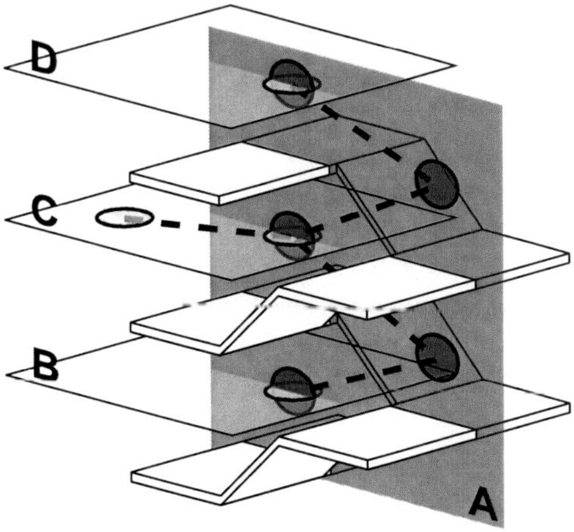

In both scenarios, the *vertical* as well as the *horizontal* one, a large part of the problem can be captured in a, respectively, vertical or horizontal 2D plane. As a lot of human created environments tend to be "2.5D" - two dimensional complexes or mazes extruded in the "up"-direction and then stacked on top of each other - the operational zone of the supportive UAVs has been chosen to be limited to spaces representable[1] by a 2D plane. This provides a generic solution for the *vertical* and the *horizontal* setup and allows an extension into structured 3D environments by dissecting the environment into a set of mutually intersecting 2D planes, Fig. 2.

1.2 Urban First-Responder Scenario

To further motivate the application, an urban first-responder scenario is proposed.[2] In this scenario, urban first-responders are assumed to have access to a tactical UAS to support their mission. The first-responders would be accompanied by a UAS operator that manages the UAS, gathers mission relevant information through it, and distributes the gathered data to the affected members of the team.

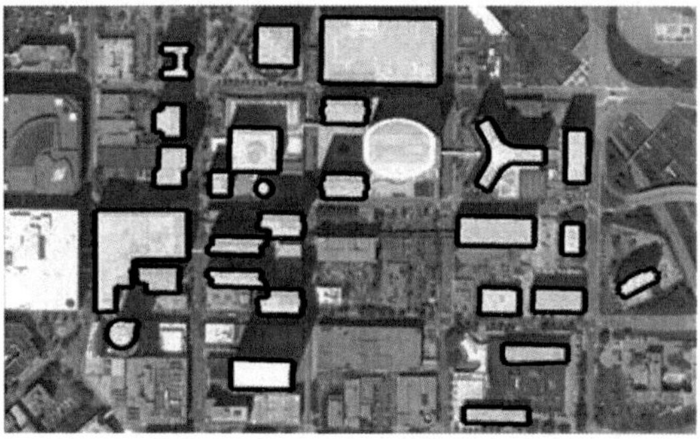

Fig. 3 An urban first-responder scenario: the UAS operator is tasked with an reconnaissance type mission on the target building (white border), i.e. a fire on the 13th floor. Relevant obstacles are also highlighted (black borders). (Aerial Image: Google)

In the scenario a designated primary UAV would be under complete operator control at all times, providing the before mentioned benefits. The (additional) secondary UAVs (acting as relay stations) would be fully autonomous. The operational zone of the secondary UAVs would be a horizontal 2D-plane at a predefined altitude, assuming that the UAVs can not go high enough to clear the buildings. This also supports the system's predictability for the operator, reducing operator workload by eliminating questions like "What is it doing?" and "Why is it doing that?" In Fig. 3 a possible control station screen is presented. The mission target is highlighted and has a white border, physical obstacles protruding the operational zone are also highlighted and have a black border. The task at hand would be to gather visual information from all sides of the target area of interest, in this case the white outlined target building.

[1] In order to be allowable, the utilized projection has to maintain the visibility property of the mapped points, i.e. a simple top-down view is only permissible in the absence of larger hills, etc.

[2] For more details on the scenario and the motivation see [2].

The scenario assumes *a priori* availability of a map as this is required for the proposed processes. Sec. 2.1 elaborates how this fits into the overall operational scenario.

2 Motion Map

Motion planning is intricately connected to the map that is used and there is extensive research on the topics of map generation as well as motion planning with urban environments being a focus area not only since the DARPA Urban Challenge. In his dissertation [3], Wooden comments on König and Likhachev [4] by stating that *"optimal planning is outweighed by the need for a "good" plan now."*

Looking for a "good plan now", ease of computation is a major driver in choosing methods and algorithms and, as mentioned in Sec. 1.1, limiting the operational area from a full 3D environment to a (vertical or horizontal) 2D plane is a big benefactor. Also, as the operational envelope of the (tele-operated) primary UAV is not at all affected through this constraint, the applicability of the proposed methods to urban, indoor, or otherwise structured environments is maintained, particularly if modeling techniques as outlined in Fig. 2 are utilized.

Limiting other driving requirements to the most fundamental one, collision free motion, generalized Voronoi diagrams present themselves as an immediate candidate to cover the motion aspect in 2D scenarios.

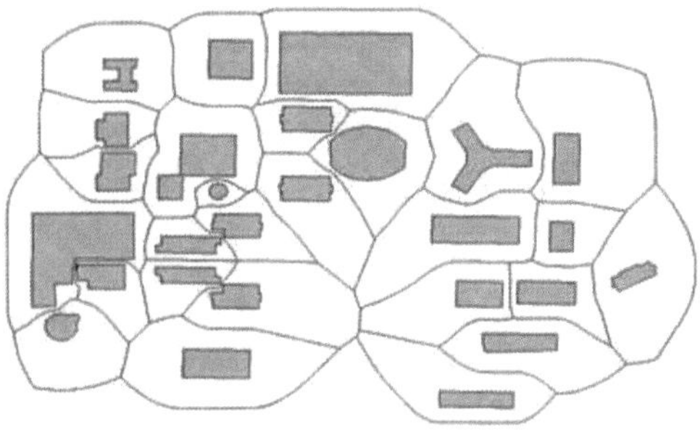

Fig. 4 Voronoi maximum clearance paths through the environment. The graph has been stripped of leaf nodes, resulting in a purely cyclic graph that segments the environment in one connected cell per obstacle. No dead ends allow for easier usage by non-hover capable aircraft, such as fixed wing Micro-UAVs.

Given a *free/occupied* classification of the environment, e.g. Fig. 3, a Voronoi diagram providing maximum clearance paths through this environment can be easily computed. Held's VRONI [5] provides a computational efficient algorithm to generate Voronoi paths through polygonal environments and the authors believe this algorithm to be suitable for UAV onboard implementation and use.

In order to generate a basic map of permitted paths for the supportive UAVs, the Voronoi graph is stripped of leaf nodes, leaving a dead end free completely cyclic graph, Fig. 4, of presumably collision free paths. The graph also segments the environment into regions associated with each motion obstacle. The borders of the cell encloseing the AoI will later be called (Voronoi) perimeter.

2.1 A Priori Data

The proposed processes conceptually fit into the *guidance* category, leaving *navigation* (and the related mapping), and *control* to other systems. As a result from that stems the requirement for *a priori* availability of a map of the environment, i.e. a *free/occupied* classification. For first-responders in metro areas, these maps are assumed to be made available or, if not, created by the operator during ingress.[3] Furthermore, the processes assume the availability of onboard collision avoidance mechanisms which could report a mismatch between the *a priori* given map and the sensed environment and trigger a conflict resolution procedure to synchronize the map with the sensor information and react appropriately.

As such, the underlying assumption is that the navigation and related mapping problem has been solved, either conventionally through a GPS corrected INS solution or, for example, through a SLAM based approach as in [6].

3 Guidance Cues

Assuming that the supportive UAVs are fully autonomous, the scenario poses a guidance problem: where to send the supportive UAVs to and how to get them there. The guidance task is to propose waypoints which are beneficial for the establishment of a Mobile Ad-hoc NETwork (MANET) in a cluttered environment.

Several researchers have presented results on how to form and maintain MANETs with UAVs (e.g. [7, 8, 9]), though one of the basic assumptions in the presented research is a free space assumption under which the establishment of a link between two nodes depends mainly on the distance of the nodes.[4]

Starting from an identical initial task - getting the primary UAV on the far side of a building - three methods to obtain guidance cues for where to position secondary UAVs are proposed.

Without any relay nodes and the assumption that the control station operator is being stationary during an active use of the primary UAV, the operational range of the primary UAV is limited to an area that has a direct LOS to the control station and is within range of the communication equipment used. Fig. 5 shows this area for an arbitrary position in the environment introduced above.

In order to complete the reconnaissance task given through the urban first-responder scenario, the operator at some point has to position the primary UAV on the far side of the target building in order to gather detailed data. Under the assumptions of this work, a conventional tactical UAS would be incapable of achieving this

[3] Given readily available geo-referenced aerial images, the operator could use a familiarity with the area to quickly "click together" a polygonal 2D *free/occupied* classification of the presumed operational area. Bounds on how close to approach these obstacles could be made conservative.

[4] In [10] the authors have proposed the usage of UAS reference scenarios to evaluate the performance of MANET protocols in a free-space situation.

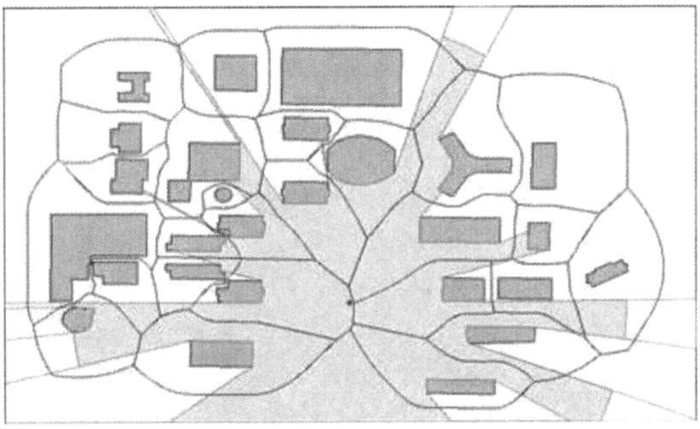

Fig. 5 Given a position of the control station operator (circle in the middle of the bottom half) the single-hop operational area is limited by range and LOS for the given position, represented by the lightly shaded area. Not all of the mission target's faces (curved building in the middle of the top half, compare Fig. 3) are visible.

as the communication between UAV and operator would be interrupted when the primary UAV leaves the direct LOS area of the GCS operator (Fig. 5).

3.1 Dual-Hop Scenario

The first proposed method to obtain cues is based on intersecting visibility polygons. Using Obermeyer's VisiLibity ([11, 12]), the visibility polygons for the current position of the GCS and the intended position of the primary UAV are computed based upon the environment data also used for the computation of the Voronoi paths.

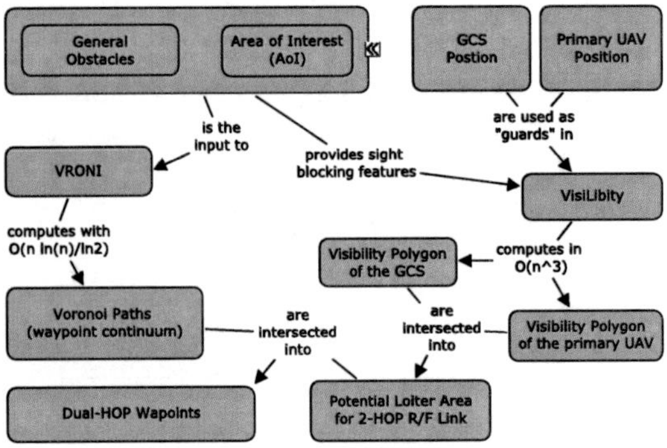

Fig. 6 The process to finding visibility cues for a dual-hop scenario. The Voronoi diagram and the visibility related computations are independent up to the last step of the process. *Guards* is a general term for the seeing entities in VisiLibity, here they represent the (supportive) UAVs.

Intersecting these two polygons and the Voronoi paths produces possible waypoints which are reachable via translations on the Voronoi paths and also fulfill the requirement to have a direct LOS to the primary UAV as well as the GCS. Fig. 7(a) shows the result of the process outlined in Fig. 6. If applicable, this method only requires one supportive UAV in addition to the tele-operate primary UAV.

3.2 Perimeter Scenario

There are cases in which the dual-hop process proposed in Sec. 3.1 either completely fails, i.e. there is no intersection of the visibility polygons and the Voronoi paths, or

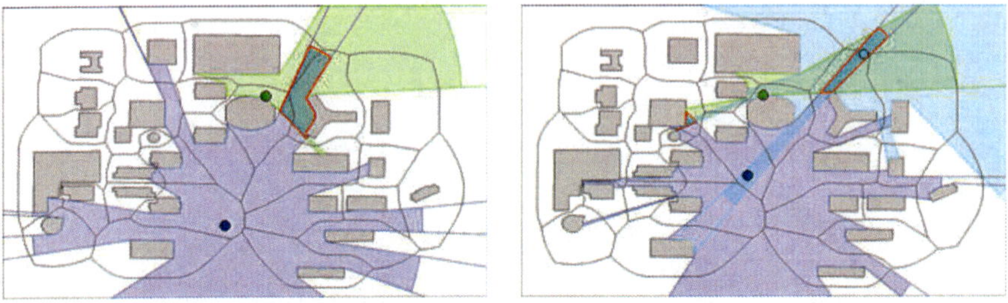

(a) Outcome of the process: waypoints on the Voronoi paths inside the intersection of the GCS visibility polygons (shaded blue) and the primary UAV's visibility polygon (shaded green) could be used as guidance cues.

(b) A feasible solution that provides very little robustness to motion of the GCS. Indicators are the slender shape of the intersection and the distance of the GCS to the edges of a secondary UAV's visibility polygon (shaded cyan).

Fig. 7 Graphical representation of the dual-hop process results. The outcome of the process is not guaranteed to be usable. Even if the intersection of the visibility polygons and the Voronoi paths are non-empty, the solution might not be robust to movements of either the GCS or the primary UAV.

the resulting cues are not very robust,[5] e.g. as shown in Fig. 7(b). Adapting the dual-hop process (Sec. 3.1) for a larger hop count (i.e. several supportive UAVs) as a main approach to counteract these disadvantages results in a computational load that might not be justifiable as the underlying method seems best suited for single-relay scenarios.

Using the AoI Voronoi perimeter, i.e. the edges of the Voronoi graph that form the segment containing the AoI (see Sec. 2), allows for the computation of a solution that is essentially independent of the positions of the GCS or the primary UAV and provides conceptually a different approach, presumably better suited for multiple relays.

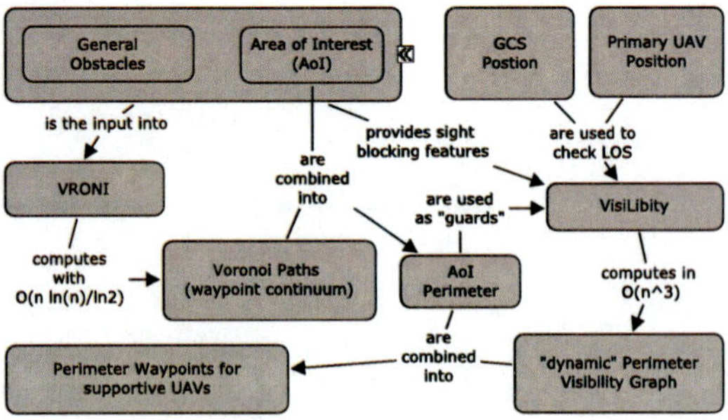

Fig. 8 The process to finding visibility cues for a perimeter scenario. The edges of the Voronoi diagram that also form the edges of the cell containing the mission target form the AoI perimeter.

[5] For computation of visibility robustness see, for example, [13].

In the process outlined in Fig. 8, the problem of finding cues is translated into the well known "Art Gallery Problem": find the minimum number of guards necessary to observe all walls of an art gallery.

In the adapted problem, the faces of the target building (the AoI) have to be completely observed and the supportice UAVs (the "guards") can only be located on the Voronoi paths forming the edges of the cell containing the AoI. Additionally, the supportive UAVs, the GCS, and the primary UAV have to be connected in the visibility graph.

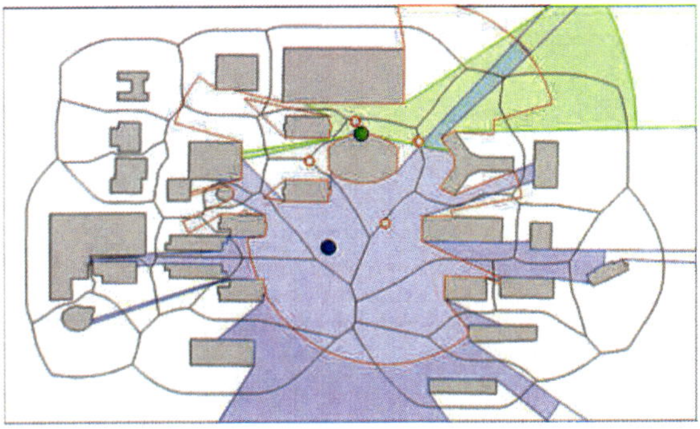

Fig. 9 Graphical representation of the perimeter process results: four guards (red outlined dots) are positioned on the AoI Voronoi perimeter. The guards are connected in the visibility graph and their combined visibility polygon is shown (red outline). As long as the GCS (blue dot) and the primary UAV (green dot) do not leave this polygon, connectedness is ensured.

Fig. 9 shows a result of this process. As long as the GCS and the primary UAV do not leave the combined visibility area of the guards, it is ensured that there always exists a multi-hop connection between the GCS and the primary UAV, using the (stationary) guards as relays.

3.3 *Dynamic Visibility*

The perimeter process provides cues for a static placement of secondary UAVs for the duration of a mission. However, the perimeter process might also lead to unusable cues, whether due to a limited number of secondary UAVs or due to other constraints, e.g. a resulting congregation of secondary UAVs that exceeds a certain space-density and is hence deemed unsafe. As a next level, a process utilizing the fact that initially only two nodes in the scenario are actually moving could be used to reduce the computational load during a mission, potentially also making use of the *a priori* availability of a map. Fig. 10 outlines this process.

As supportive UAVs are limited to positions on the Voronoi paths, the visibility of these potential positions can be precomputed to save computational time during the mission. As the Voronoi paths essentially are a continuum of possible waypoints, a smart sampling has to be developed to reduce the computational effort while keeping a higher resolution where necessary. The edges of the Voronoi path could, for example, be sampled at an increased distance between sample points as the overall

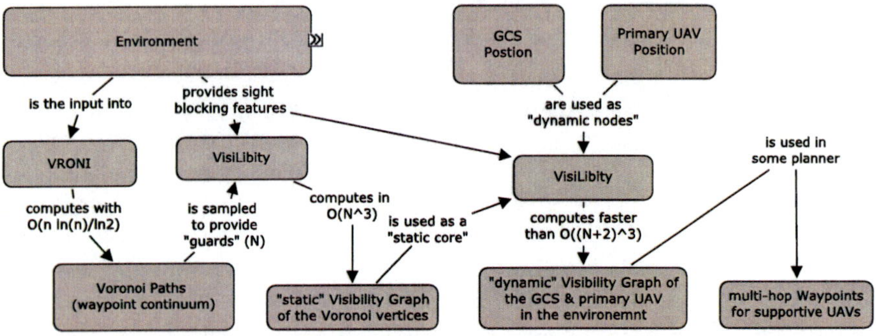

Fig. 10 The process to finding visibility cues for a dynamic scenario. Instead of recomputing the complete visibility graph of the complete environment, the visibility of select points on the Voronoi paths are precomputed (left half of the figure) and later expanded with the visibility information of the two moving nodes, the GCS and the primary UAV. This split in a static/pre-computed part and a dynamic/online-computed part fosters onboard realizability.

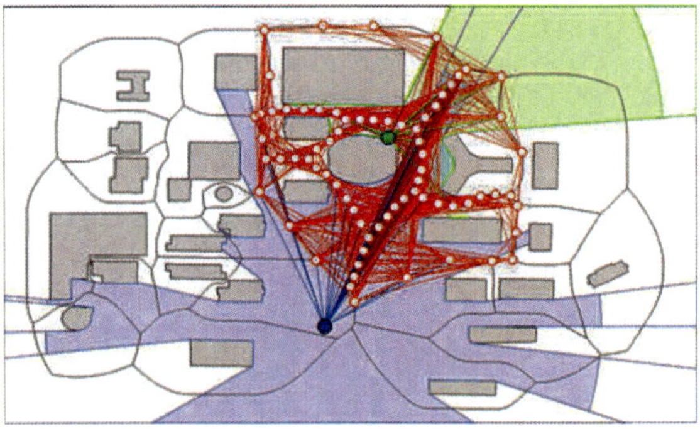

Fig. 11 Using the differentiation of static versus dynamic nodes when computing the visibility speeds up the computation. However, the computational load is still high and the results rather complex. Shown are some nodes of the static core and their respective visibility (red), as well as the visibility of the GCS (blue) and the primary UAV (green).

distance of the edge to the AoI increases. This would lead to the highest waypoint line-density on the AoI perimeter edges and to lower densities towards the outer areas of the environment.

The result of this process is not inherently geometrical and hence rather hard to visualize. Fig. 11 shows some of the possible paths given by the adjacency matrix of the undirected visibility graph. The process takes the precomputed visibility *static* adjacency matrix $S \in \mathbb{R}^{N\times N}$ of the sampled Voronoi paths and extends it with the *dynamic* adjacency matrix $D \in \mathbb{R}^{N\times 2}$ of nodes representing the GCS and the primary UAV with respect to the the static part. Hence, instead of recomputing the complete visibility graph (with a complexity of $O((N+2)^3)$), the process computes the visibility polygons[6] of the two dynamic nodes (GCS and primary UAV) and

[6] In a staged approach of trying the proposed processes in the presented order, these can be reused from the dual-hop process.

checks which of the N sampled points of the Voronoi paths are inside of them. This gives the visibility of the dynamic and the precomputed nodes and the complete adjacency matrix $A \in \mathbb{R}^{(N+2)\times(N+2)}$ can be constructed as $A = \left[\begin{array}{c|c} S & D \\ \hline D^T & \begin{array}{cc} a_{\text{pUAV}} & 0 \\ 0 & a_{\text{GCS}} \end{array} \end{array}\right]$, where $S \in \mathbb{R}^{N\times N}$ is symmetric and $D \in \mathbb{R}^{N\times 2}$.

Any preferred graph algorithm can be used to find paths from the GCS to the primary UAV in the expanded visibility matrix. In Fig. 11 a subset of the complete environment is shown. The potential positions of supportive UAVs are indicated by white circles. The corresponding visibility graph is plotted in red. This would be (part of) the static core. Dynamically computed would be the visibility of the control station (blue) and the primary UAV (green). As the visibility polygon for both has already been computed in an earlier step, the computation of the actual visibility is reduced to a checking which positions of the static core would be inside this polygon.

4 Conclusions and Remarks

The proposed methods to obtain guidance cues for communication aware UAVs in a cluttered environment are aimed at solving the problem of where to send supportive UAVs to establish a multi-hop communication network between a GCS and a primary UAV. Feasibility of the methods has been tested in simulation for non-moving vehicles, a deployment simulation or actual flight test have not yet been conducted.

The dual-hop process presented in Sec. 3.1 is computationally easy and a very good candidate for onboard implementation. The perimeter process presented in Sec. 3.2 is computationally much more complex, however, as the results are valid for the complete mission (assuming the AoI stays identical) it could be performed *a priori*. The dynamic approach outlined in Sec. 3.3 presents a method to deal with a worst case scenario by dividing potential nodes in two sets, pre-computable static nodes and moving dynamic ones. As this process would only be reached if the other presented processes fail, previously computed data can be reused to minimize the computational impact.

Although the dynamic process seems to be implementable onboard (from a computational perspective), this process also poses the most challenges in transforming the cues into actually selected target locations for secondary UAVs. As the result of the process is just an adjacency matrix (where shortest path graph algorithms can give cues for multi-hop waypoints), additional metrics have to be found or defined in order to rank the set of shortest (hop) paths between primary UAV and the GCS, assuming that fewer supportive UAVs are preferable.

The proposed methods however do not yet deal with some of the immediately imminent challenges: as the primary UAV moves unpredictably for the automation (as it is tele-operated), the automation has to anticipate its motion and pre-plan for all possibilities. This might lead to conflicting requirements for the positioning of the secondary UAV(s) when the guidance has to sort out where to send them. Future work will have to look into the challenges resulting therefrom.

Though experimentation with realistic urban scenarios seems to indicate that the dual-hop process most often leads to usable results (where the use of some notion of visibility robustness can rank the cues given by the process in order to obtain definite waypoint for the underlying guidance), more simulation - particularly of the deployment - is necessary to provide usable and implementable heuristics and algorithms.

References

1. Salas, E.: Human Factors of Remotely Operated Vehicles, Advances in Human Performance and Cognitive Engineering Research. In: Cooke, N.J., Pringle, H.L., Pedersen, H.K., Connor, O. (eds.) Advances in Human Performance and Cognitive Engineering Research, 1st edn., vol. 7, Emerald Group Publishing Limited (2006)
2. Christmann, H.C., Johnson, E.N.: AIAA Modeling and Simulation Technologies Conference and Exhibit, Toronto, Canada. AIAA-2010-8361 (2010)
3. Wooden, D.T.: Graph-based path planning for mobile robots. Ph.D. thesis, Georgia Institute of Technology (2006), http://hdl.handle.net/1853/14055
4. König, S., Likhachev, M.: IEEE Transactions on Robotics 21(3), 354 (2005), doi:10.1109/TRO.2004.838026
5. Held, M.: Computational Geometry 18(2), 95 (2001), doi:10.1016/S0925-7721(01)00003-7
6. Sobers, M.D.: Efficient ranging-sensor navigation methods for indoor aircraft. Ph.D. thesis, Georgia Institute of Technology (2010), http://hdl.handle.net/1853/34824
7. Jenkins, A., Henkel, D., Brown, T.: Proceedings on Infotech@Aerospace 2007 Conference and Exhibit. AIAA, Rohnert Park, California, United States, AIAA-2007-2747 (2007)
8. Frew, E.W., Brown, T.X.: Proceedings of the IEEE 96(12) (2008), doi:10.1109/JPROC.2008.2006127
9. Hauert, S., Leven, S., Zufferey, J.C., Floreano, D.: IEEE International Conference on Robotics and Automation (ICRA 2010), pp. 15–20 (2010), doi:10.1109/ROBOT.2010.5509421
10. Christmann, H.C., Johnson, E.N.: AIAA Modeling and Simulation Technologies Conference and Exhibit, Honolulu, HI, USA, AIAA-2008-7042 (2008)
11. Obermeyer, K.J.: Contributors. The VisiLibity library (2008), http://www.VisiLibity.org R-1
12. Obermeyer, K.J.: Visibility problems for sensor networks and unmanned air vehicles. Ph.D. thesis, Mechanical Engineering Department, University of California at Santa Barbara (2010)
13. Ben-Moshe, B., Hall-Holt, O., Katz, M.J., Mitchell, J.S.B.: SCG 2004: Proceedings of the twentieth annual symposium on Computational geometry, pp. 27–35. ACM, New York (2004), doi:10.1145/997817.997825

Part II
Guidance and Control

Adaptive Control of a High Agility Model Airplane in the Presence of Severe Structural Damage and Failures

Stephan Baur, Travis Gibson, Anuradha Annaswamy, Leonhard Höcht, Thomas Bierling, and Florian Holzapfel

Abstract. Adaptive control is a promising technology for future high-performance, safety-critical flight systems. By virtue of their ability to adjust control parameters as a function of online measurements, adaptive flight control systems offer improved performance and increased robustness. This paper addresses the adaptive control of extremely agile aircrafts in the presence of damages and failures. The FSD ExtremeStar, a modified version of the polystyrene model airplane Multiplex TwinStar II, is used as a platform for this purpose by offering a highly redundant set of control surfaces. The underlying nonlinear model, including the effect of all control inputs, is derived from first principles. A dynamic-inversion PI-error controller is proposed as the baseline controller for a model reference adaptive tracking control. The resulting performance is evaluated for aggressive maneuvers in the presence of elevator failures using the complete nonlinear model.

1 Introduction

In recent years, the interest in adaptive control for autonomous flight has been renewed. Due to their ability to self-tune their parameters using on-line measurements, control in the presence of unforeseen damages and failures is enabled through adaptation in a highly satisfactory manner. Over the past thirty years, the

Stephan Baur · Travis Gibson · Anuradha Annaswamy
Active Adaptive Control Laboratory (AAC), Massachusetts Institute of Technology (MIT), Cambridge 02139, Massachusetts, USA
e-mail: stephan.baur@mytum.de, tgibson@mit.edu, aanna@mit.edu

Leonhard Höcht · Thomas Bierling · Florian Holzapfel
Institute of Flight System Dynamics (FSD), Technische Universität München (TUM), D-85748 Garching, Germany
e-mail: leonhard.hoecht@tum.de, t.bierling@tum.de, florian.holzapfel@tum.de

field of adaptive control has laid the foundation for analysis and synthesis of such systems [1]. Stability and robustness properties of these systems are well understood. Adaptation in single-input and multiple-input systems, continuous and discrete-time systems, time-invariant and time-varying systems, as well as of linear and nonlinear systems have been studied in depth. Over the past few years, several adaptive flight controllers have flown successfully [2],[3],[4],[5],[6],[7],[8] and enabled successful flight even in the presence of severe failures and damages. The tasks in these cases have focused on way-point tracking or landing. Very few results are currently available where highly agile aerobatic maneuvers have been carried out in the presence of severe failures. This paper addresses the adaptive control in the presence of failures for a highly agile aircraft.

2 The Aircraft Model

In this section, a nonlinear model for a high agility aircraft, the FSD ExtremeStar, is derived from first principles. This model will be used for design and evaluation of adaptive controllers for performing high agile maneuvers in the presence of severe structural damage (e.g. parts of the wing or of the tail missing) or failures (e.g. reduced control surface efficiency, motors). The FSD ExtremeStar is a modified version of the polystyrene model airplane Multiplex TwinStar II. The modifications were conducted by the Akamodell München on behalf of the Institute of Flight System Dynamics of Technische Universität München for investigating flight dynamics and flight control methods (e.g. control allocation, nonlinear adaptive control) on an Unmanned Aerial System (UAS) with a large number of flight control inputs [9]. To facilitate this, the model features one aileron and one flap control surface on each wing, a canard as well as a horizontal tail with variable incidence, a rudder on the vertical tail, two wing-mounted propellers with variable vertical tilt angle and one tail-mounted propeller with variable tilt and azimuth angle (see Figure 1). Additionally, an independent control of left and right side control devices is possible. The aircraft's actuation is therefore highly redundant as the flaps or ailerons could, for instance, be used for producing a rolling as well as a pitching moment.

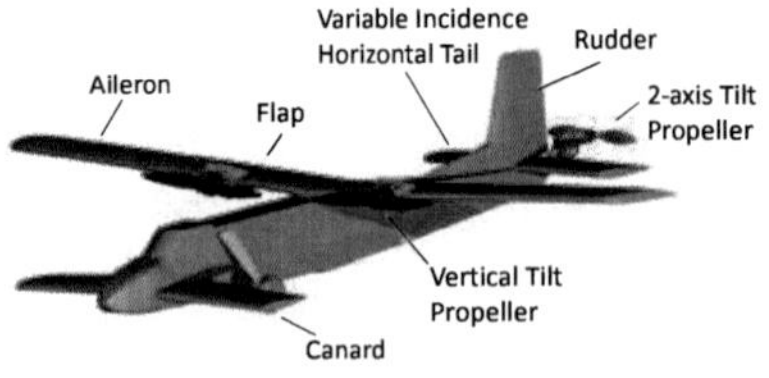

Parameter	Value
Wing span	1.40 m
Canard span	0.58 m
Horizontal tail span	0.42 m
Fuselage length	1.35 m
Left/right motor	2 x max. 500 W
Left/right propeller	CAMcarbon 11" x 6"
Tail motor	max. 75 W
Tail propeller	GWS 8" x 4.3"

Fig. 1 FSD ExtremeStar with Illustration of Multiple Control Devices and Configuration Data

The required aerodynamic terms for the nonlinear model consist of three force (C_D, C_Q, C_L) and three moment coefficients (C_l, C_m, C_n). They were calculated with the AeroTool, which was developed in cooperation between Bauhaus Luftfahrt and the Institute of Flight System Dynamics. This tool allows the calculation of aerodynamics for a composition of lifting surfaces (using potential flow theory [10]), rotors (using momentum theory [11],[12],[13],[14],[15]) and bodies (using DATCOM methods [16]) by considering the influence of these objects on each other [17] as well as for the linear and nonlinear angle of attack range. It enables the user to calculate the required aerodynamic coefficients as a function of the chosen six state variables ($\alpha, \beta, \mathrm{V}, \mathrm{p}, \mathrm{q}, \mathrm{r}$) and the 16 control inputs (9 control surfaces, 3 thrust vectors and the according tilt angles as well as the azimuth angle of the tail mounted motor). The aerodynamics were calculated for an angle of attack ranging from -27.5° to 27.5°, for an angle of sideslip from -15 to 15° and the possible control surface inputs. The results for C_L for different motor rotation speeds are shown in Figure 2 [18].

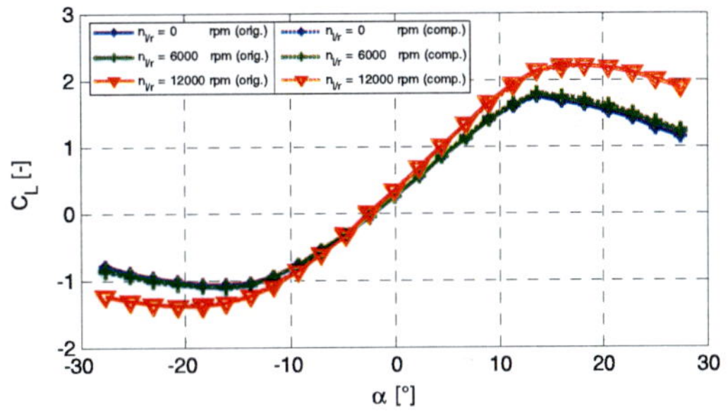

Fig. 2 Comparison of Original and Composed Model

Due to the 16 control surfaces and the considered six aerodynamic state variables it is computationally intensive to carry out a full factorial design where each input dimension is discretized on a selected range into a number of breakpoints and the output is calculated for every combination of these input values. Therefore, a simplification of the data structure, which optimally preserves the characteristics of the original numerical model, was found. The symmetry of the FSD ExtremeStar to the $\mathrm{xy}-$plane was used and thus only the forces and moments for the right side were calculated. The aerodynamics of the whole aircraft were determined by composing the calculated aerodynamic data for the symmetrical parts and by adding the calculated data for the non-symmetric components, such as fuselage and vertical tail [18],[21].

The influence of different variables on each other (e.g. angle of sideslip on the effect of an elevator deflection) was assessed and, based on this analysis, the influences of second or lower order variables (e.g. influence of canard deflection on elevator efficiency) were neglected in the calculation [18].

The described simplifications helped to reduce the computation time by orders of magnitude by still preserving the essential dependencies of the different control inputs and the overall nonlinear aerodynamic behavior of the aircraft. The simplifications were validated by comparing the results of a full aerodynamic model with those of the symmetric reduced model for arbitrarily chosen input variables for all six coefficients. The comparison of the obtained data showed very good agreement, as can be seen in Figure 2.

The actuation of the control surfaces of the FSD ExtremeStar is performed with servos, which were modeled in Matlab/Simulink using a second order lag model. The different parameters of the model as the damping, the eigenfrequency and the limits for the maximum moment, velocity and the maximum amplitude were either set according to the information as provided by the datasheets of the manufacturers (e.g. maximum moment) or determined experimentally (e.g. maximum deflection).

The FSD ExtremeStar offers three different thrust vectors, which are located on the left and the right side of the main wing and at the tail of the fuselage. For the modeling of the motor system the dynamics of the battery, of the cables and of the brushless controller are neglected and modeled as a resistance. The FSD ExtremeStar, as a high agility model airplane performing aerobatic maneuvers, will operate at nearly full motor rotation speed for most of the time, and the brushless motors are therefore modeled for simplification with a full load motor model. The forces and moments of the propeller were calculated with the previously mentioned AeroTool, whereas the moments of inertia of the rotating parts were calculated with an appropriate CatiaV5 model.

There are different types of sensors used in the aircraft such as an INS, dynamic pressure as well as magnetometer sensors, etc. For a test of the controller in presence of uncertainties like sensor errors and bias, it is necessary to simulate the real behaviour of the sensors as accurately as possible. Therefore, using the data provided by the manufacturer, models for each of the sensors were established, taking into account null shift bias, white noise, bias short-term stability, scale factor absolute value uncertainties and misalignment of the sensor axes [21].

The equations of motion for the FSD ExtremeStar are derived for a flat and non-rotating Earth, as the UAS only performs local missions at a low operation altitude. The 6-DOF equations of motion require the knowledge of the inertia tensor $I\ \left[kgm^2\right]$ relative to the reference point R. The Inertia tensor was determined with appropriate CatiaV5 models for the damaged as well as the undamaged FSD ExtremeStar and stored in appropriate data tables [19]. CatiaV5 provides the moments of Inertia for the center of gravity (CoG) of the considered body and they therefore have to be transformed to the according reference point R.

The dynamics of the FSD ExtremeStar consist of the states

$$\mathbf{x} = \left[p, q, r, q_0, q_1, q_2, q_3, V, \gamma, \chi\right]^T \tag{1}$$

where the first three entries are the rotation rates, the following four are the components of the quaternion attitude vector and the last three are the kinematic state variables. The inputs are given as

$$\mathbf{u} = \left[\eta_{cr_l}, \eta_{cr_r}, \xi_l, \xi_r, \delta_{F_l}, \delta_{F_r}, \delta_{T_l}, \sigma_l, \delta_{T_r}, \sigma_r, \eta_{ele_l}, \eta_{ele_r}, \zeta\right]^T \tag{2}$$

with the first two entries denoting the canard deflection, followed by the left and right aileron deflection, the flap deflections, the motor tilt angle and the motor rotation speed of the left and the right motor, and finally the elevator and the rudder deflections.

The resulting model is given with the well-known time derivatives of:

- *the rotation rates:*

$$\left(\dot{\vec{\boldsymbol{\omega}}}_K^{OB}\right)_B = \left(\mathbf{I}_{BB}^G\right)^{-1} \cdot \left[\sum\left(\vec{\mathbf{M}}^G\right)_B - \left(\vec{\boldsymbol{\omega}}_K^{OB}\right)_B \times \mathbf{I}_{BB}^G \cdot \left(\vec{\boldsymbol{\omega}}_K^{OB}\right)_B\right] \tag{3}$$

- *the attitude quaternion*

$$\dot{q} = \frac{1}{2}\begin{bmatrix} q_0 & -\mathbf{q}^T \\ \mathbf{q} & \mathbf{Q} \end{bmatrix}\begin{bmatrix} 0 \\ \boldsymbol{\omega} \end{bmatrix} \tag{4}$$

$$\text{with } \mathbf{q} = \begin{bmatrix} q_1 & q_2 & q_3 \end{bmatrix}^T \text{ and } \mathbf{Q} = \begin{bmatrix} q_0 & -q_3 & q_2 \\ q_3 & q_0 & -q_1 \\ -q_2 & q_1 & q_0 \end{bmatrix}$$

- *the path dynamics:*

$$\begin{pmatrix} \dot{V} \\ \dot{\chi} \\ \dot{\gamma} \end{pmatrix} = \begin{bmatrix} 1 & 0 & 0 \\ 0 & 1/(V_K \cos\gamma_K) & 0 \\ 0 & 0 & -1/V_K \end{bmatrix} \frac{1}{m} \left[\begin{pmatrix} F_{x,aero} \\ F_{y,aero} \\ F_{z,aero} \end{pmatrix}_K + mg \cdot \begin{pmatrix} -\sin\gamma_K \\ 0 \\ \cos\gamma_K \end{pmatrix}\right] \tag{5}$$

It can clearly be seen that the given model is nonlinear and provides the starting point for the dynamic inversion based controller [20],[21]. In this model, as well as in the current work, the rear motor is neglected and is a topic for future research. It can be seen that Quaternions are used for the simulation of the aircraft and will also be used for control design as they do not suffer from singularities like Euler angles.

3 Adaptive Control of the FSD ExtremeStar

In the following, an adaptive controller, which is based on the nonlinear model described in (3), (4) and (5), will be designed to accommodate for failures in the control surfaces. This controller is integrated with a dynamic inversion based baseline controller with the latter designed to ensure that the aircraft performs in a desired manner in the absence of any damages or failures. Given the large number of control inputs, two different control allocation methods are possible. In the first case, the redundant control surfaces of the FSD ExtremeStar are fully used, whereas, in the second case, the control inputs are chosen as in a conventional aircraft (only ailerons, elevators and rudder) with the remaining surfaces being fixed at their neutral positions. In the following the attention will be focused, due to space limitations, on the second method.

3.1 Dynamic Inversion Baseline Controller

A relative degree one dynamic inversion controller, based on Quaternions, for rate command tracking and attitude hold was designed as the baseline controller.

The equations and the basic principle of the implemented dynamic inversion PI-error controller can be seen in Figure 3 and 4.

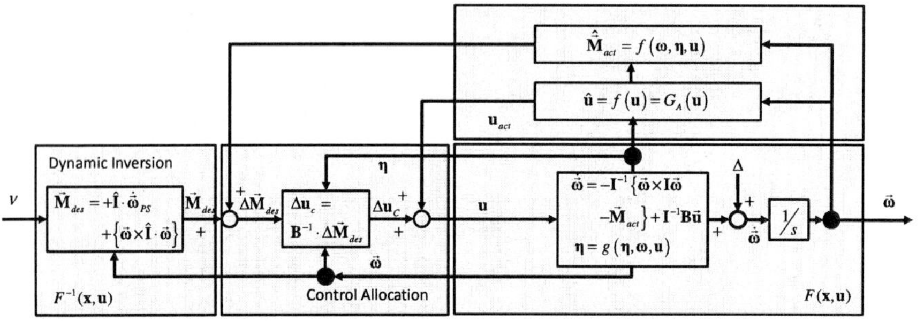

Fig. 3 Basic Layout of RD 1 Dynamic Inversion

The pseudo control input for the dynamic inversion controller is given by:

$$\mathbf{v} = \mathbf{v}_R + \mathbf{v}_{error} = \mathbf{v}_R + \mathbf{K}_{P,\omega} \cdot \left(\vec{\omega}_{RM} - \vec{\omega}\right) + \mathbf{K}_{I,\omega} \cdot \int \left(\vec{\omega}_{RM} - \vec{\omega}\right) dt \tag{6}$$

where $\mathbf{v}_{error}$ denotes the output of the PI-error controller with the proportional gain matrix $\mathbf{K}_{P,\omega}$ and the integral gain matrix $\mathbf{K}_{I,\omega}$ and $\mathbf{v}_R$ is calculated based on the commanded $\vec{\omega}_{CMD,outer}$ and the reference rotation rates $\vec{\omega}_{RM}$ by a first order linear reference model according to [21],[22]:

$$\mathbf{v}_R = \dot{\vec{\omega}}_{RM} = \mathbf{K}_{RM} \left(\vec{\omega}_{CMD,outer} - \vec{\omega}_{RM}\right) \tag{7}$$

with $\mathbf{K}_{RM}$ specifying the dynamics of (7).

The control allocation of the FSD ExtremeStar uses a constrained minimization, with Equation (8) serving as the cost function, where the control surface deflection being necessary for a desired moment is minimized [21].

$$J = \frac{1}{2} \Delta\mathbf{u}^T \mathbf{W} \Delta\mathbf{u} + \boldsymbol{\lambda}^T \left(\mathbf{B}\Delta\mathbf{u} - \Delta\mathbf{M}_{des}\right) \tag{8}$$

where $\Delta\mathbf{M}_{des}$ is the desired moment increment as provided by the dynamic inversion, and based on the current state variables and moments of the airplane, and $\Delta\mathbf{u}$ is the corresponding incremental control input being necessary to achieve $\Delta\mathbf{M}_{des}$.

The weighting matrix $\mathbf{W}$ allows prioritizing or neglecting specific control surfaces. Therefore, it allows making use of the full redundant control surfaces of the FSD ExtremeStar or to treat it like a conventional aircraft by only using the ailerons, elevators and the rudder.

The constraint minimization is performed online for each simulation step respectively for each sampling/calculation of the real controller. This allows obtaining an optimal control surface deflection for a desired commanded rotation rate and attitude. The dynamic inversion as well as the control allocation assumes an undamaged aircraft since, in the case of failures, no information about the type of severe structural damage or failures are available to the baseline controller. Therefore, the baseline controller will be augmented by an adaptive part to account for this.

3.2 MRAC Architecture for Adaptive Tracking

By applying the dynamic inversion controller to the plant, assuming the aircraft dynamics are perfectly known, the overall closed loop system has a linear behaviour [20],[21].

$$\mathbf{y} = \frac{1}{s} \cdot \mathbf{v} \tag{9}$$

where $\mathbf{y}$ is the output of the system and $\mathbf{v}$ is the pseudo command input from (6). For simplification only a single input single output (SISO) system will be considered in the following. The input for the SISO system is the reference pitch rate $q_{CMD,outer}$, as commanded to the reference model, and the output is the pitch rate q_{MEAS}, as achieved by the feedback linearized plant. For the proposed MRAC architecture an explicit direct MRAC tracking approach as shown in Figure 4 will be chosen.

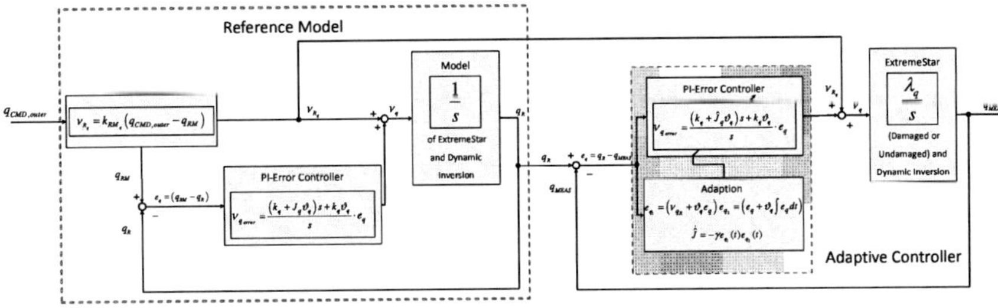

Fig. 4 Adaptive Control Architecture for Tracking in the Pitch Axis

The time derivative of the pitch rate is given with (3) to (5):

$$\dot{q}_{MEAS} = \underbrace{\frac{(I_z - I_x)\,pr + I_{xz}\left(p^2 - r^2\right)}{I_y}}_{b(\mathbf{x})} + \underbrace{\frac{1}{I_y}\cdot M(\mathbf{x},u)}_{\hat{a}(\mathbf{x})\cdot u = a\cdot B(\mathbf{x})\cdot u_{deflection}} \tag{10}$$

p and r can be assumed to be approximately zero, since these axes are stabilized by according controllers. Thus $b(\mathbf{x})$ vanishes and Equation (10) can be written as:

$$\dot{q}_{MEAS} = \hat{a}(\mathbf{x})\cdot u = a\cdot B(\mathbf{x})\cdot u_{deflection} \tag{11}$$

where $B(\mathbf{x})\cdot u_{deflection} = \frac{\partial M(\mathbf{x})}{\partial \eta_{ele}}\cdot \eta_{ele}$ for the left and the right elevator being deflected symmetrical. η_{ele} denotes the elevator deflection and $\frac{\partial M(\mathbf{x})}{\partial \eta_{ele}}$ the pitching moment due to an elevator deflection.

In the presence of damage we can represent the corresponding pitching moment derivative due to an elevator deflection as $\lambda_q \cdot \frac{\partial M(\mathbf{x})}{\partial \eta_{ele}}$, where λ_q denotes the remaining elevator efficiency with $0 \le \lambda_q \le 1$ and $\lambda_q = 1$ specifies the undamaged case. The corresponding damaged plant dynamics for the pitch axis is thus given by:

$$\dot{q}_{MEAS} = a\cdot \lambda_q \cdot B(\mathbf{x})\cdot u_{deflection} \tag{12}$$

Since the dynamic inversion controller still assumes an undamaged plant the same control input, as without damage, is used and given by:

$$u_{deflection} = \left(B(\mathbf{x})\right)^{-1}\cdot a^{-1}\cdot \nu_q \tag{13}$$

It can be shown that the plant with the applied dynamic inversion results in:

$$\dot{q}_{MEAS} = \lambda_q \cdot \nu_q \tag{14}$$

For the considered case and, if the roll and the yaw rate are kept to zero, the output q_{MEAS} is given in the frequency domain by:

$$q_{MEAS} = \frac{\lambda_q}{s}\cdot \nu_q \tag{15}$$

The PI-Error controller, as presented in (6), can be rewritten for each of the three aircraft axis (roll $i = p$, pitch $i = q$ and yaw $i = r$) as:

$$\nu_{i_{error}} = \frac{(k_i + J_i\vartheta_i)s + k_i\vartheta_i}{s} \tag{16}$$

with the fixed gains k_i, ϑ_i and $J_i = 1/\lambda_i$, where $(1-\lambda_i)$ is the failure of the single axis. In the considered case, since only the elevator is facing reduced efficiency, we can chose $J_p = J_r = 1$ and $J_q = 1/\lambda_q$.

The control input for each of the single axis can be written, using (16) and considering the first order reference model for the rates, as given in (7), as:

$$\nu_i = J_i\left(\nu_{i_R} + \vartheta_i e_i\right) + k_i\left(e_i + \vartheta_i \int e_i dt\right) \tag{17}$$

with the error $e_i = \left(\omega_{i_{RM}} - \omega_{i_{MEAS}}\right)$.

While (18) can be implemented for the roll and yaw axis, this is not possible for the pitch axis, as J_q is unknown.

Denoting $e_{q_1} = \left(\nu_{q_R} + \vartheta_q e_q\right)$ and $e_{q_2} = \left(e_q + \vartheta_q \int e_q dt\right)$ an adaptive control input is generated for the pitch axis as:

$$\nu_q = \hat{J} e_{q_1}(t) + k_2 \cdot e_{q_2}(t) \tag{18}$$

It can be shown that V in Equation (19) is a Lyapunov function:

$$V = \frac{1}{2}\left(J e_{q_2}^2 + \frac{1}{\gamma}\tilde{J}^2 \right) \tag{19}$$

with $\tilde{J} = \hat{J} - J$. Since $\dot{e}_{q_2} = \frac{1}{J}(-\tilde{J})e_{q_1}(t) - \frac{1}{J}k_q e_{q_2}(t)$ the time derivative of V is given by:

$$\dot{V} = \underbrace{-\underbrace{k_q e_{q_2}^2(t)}_{>0}}_{<0} \underbrace{-\tilde{J} e_{q_1}(t) e_{q_2}(t) + \frac{1}{\gamma}\tilde{J}\dot{\tilde{J}}}_{\overset{!}{=}0} \tag{20}$$

It can be seen that the first part of (20) will always be smaller than zero. The second part will be designed in a way that it is always equal to zero and yields the adaptive update law. If $\dot{\hat{J}}$, the adaptive parameter in (18), is adjusted as:

$$\dot{\hat{J}} = -\gamma e_{q_1}(t) e_{q_2}(t) \tag{21}$$

the stability for the chosen adaptive law is shown [1],[21],[23] while the learning rate of the adaptive law can be adjusted with the constant parameter γ.

4 Validation of Adaptive Controller and Conclusion

In the following the derived adaptive law is assessed for the control of the FSD ExtremeStar with several aggressive pitch rate commands of 60 [deg/s], where each is commanded for a duration of 6 seconds, and with an elevator efficiency of $\lambda_q = 0.2$.

Our assumption here is that only the pitch-axis loop requires adaption as the failure is in the elevator. A non-adaptive PI controller, as given in (16), is used for the roll and the yaw axis, while the adaptive controller given by (18) and (21) will be used for the pitch-axis. The assumption is that the fixed controllers will ensure that p as well as r remain near zero and the adaptive controller focuses on the damaged pitch-axis.

It can be seen that over the first few cycles, both the PI-error controller without adaption (cyan) as well as with adaption (red), exhibit similar performance. However as time proceeds, the adaptive controller adjusts its parameters and due to its learned behaviour it results in a much better performance than the non-adaptive PI-controller. Moreover it is also able to reestablish the desired nominal performance.

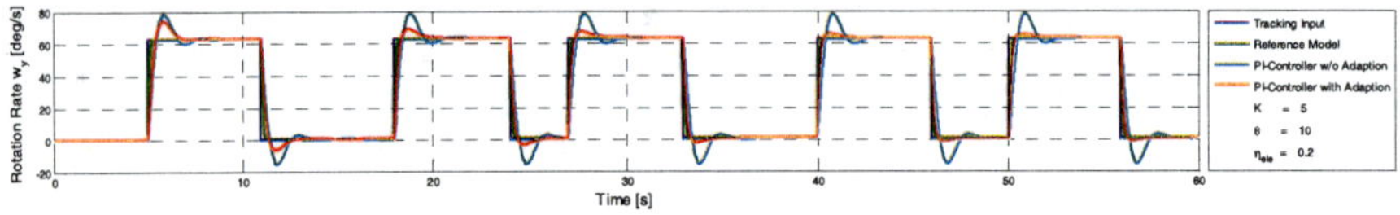

Fig. 5 Implemented Design for Adaptive Tracking

The results show that adaptive control is a promising approach for the control of high agility aircraft in the presence of severe structural damage and failures. This approach will be extended and analysed in depth in subsequent publications.

References

1. Narendra, K.S., Annaswamy, A.M.: Stable Adaptive Systems. Dover Publication, Inc., Mineola (1989)
2. Sharma, M., Lavretsky, E., Wise, K.A.: Application and Flight Testing of an Adaptive Autopilot on Precision Guided Munitions. In: AIAA Guidance, Navigation and Control Conference, Keystone, Colorado (2006)
3. Vos, D.: Five Steps to facilitating the convergence of manned and unmanned aviation. Rockwell Collins (2009), `http://www.webfulfillment.com/cedargraphics/CF/RC/LP/106/images/UAS_eBook09_5mb.pdf` (accessed September 5, 2010)
4. Michini, B.: Modeling and Adaptive Control of Indoor Unmanned Aerial Vehicles. Master Thesis, Massachusetts Institute of Technology, Cambridge, Massachusetts (2009)

5. Johnson, E.N., Calise, A.J., Blauwe, H.D.: In Flight Validation of Adaptive Flight Control Methods. In: AIAA Guidance, Navigation and Control Conference, Honolulu, Hawaii (2008)
6. Chowdhary, G., Johnson, E.N.: Flight Test Validation of a Neural Network based Long Term Learning Adaptive Flight Controller. In: AIAA Guidance, Navigation and Control Conference, Chicago, Illinois (2009)
7. Johnson, E.N., Chowdhary, G.: Guidance and Control of an Airplane under Severe Structural Damage. In: AIAA Infotech@Aerospace, Atlanta, Georgia (2010)
8. Chowdhary, G., Johnson, E.N., Kimbrell, M.S., Chandramohan, R., Calise, A.: Flight Test Results of Adaptive Controllers in Presence of Severe Structural Damage. In: AIAA Guidance, Navigation and Control Conference, Toronto, Ontario, Canada (2010)
9. Wiedenmann, R.: 13 auf einen Streich, Modellflug zu Forschungszwecken. Modellflug-Praxis, Modellflieger, Issue 01/2010, Hamburg, Germany (2010)
10. Hunsaker, D., Snyder, D.: A Lifting-Line Approach to Estimating Propeller/Wing Interactions. In: 24th Applied Aerodynamics Conference, San Francisco, California (2006)
11. Johnson, W.: Helicopter Theory. Dover Publications, Inc., Mineola, N.Y (1980)
12. Dreier, M.E.: Introduction to Helicopter and Tiltrotor Flight Simulation. AIAA, Reston, VA (2007)
13. Phillips, W.F.: Mechanics of Flight. John Wiley & Sons Inc., Hoboken (2004)
14. Hunsaker, D.F.: A Numerical Blade Element Approach to Estimating Propeller Flowfields. In: AIAA Aerospace Sciences Meeting and Exhibit, Reno, Nevada (2007)
15. Hess, J.L., Valarezo, W.O.: Calculation of Steady Flow About Propellers using a Surface Panel Method. Journal of Propulsion and Power 1, 470–476 (1985)
16. Fink, R.: USAF Stability and Control DATCOM (1978)
17. Witkowski, D.P., Lee, A.K., Sullivan, J.P.: Aerodynamic Interaction between Propellers and Wings. Journal of Aircraft 26, 829–836 (1989)
18. Steiner, H.J., Baur, S., Hornung, M., Holzapfel, F.: Modeling of Propeller-Wing Aerodynamics Considering Large Number of Control Inputs. In: DGLR Conference, Hamburg (2010)
19. Blauwe, H.D., Johnson, E.N.: Nonlinear Modeling and Simulation of Small Unmanned Aerial Vehicle with Substantial Damage. In: AIAA Modeling and Simulation Technologies Conference, Chicago, Illinois (2009)
20. Holzapfel, F.: Nichtlineare adaptive Regelung eines unbemannten Fluggerätes. Ph.D. Thesis, Technische Universität München, Munich (2004)
21. Baur, S.: Simulation and Adaptive Control of a High Agile Aircraft in the Presence of Severe Structural Damage and Failures. Diplomarbeit, Technische Universität München, Munich and Massachusetts Institute of Technology, Cambridge, Massachusetts (2010)
22. Johnson, E.N.: Limited Authority Adaptive Flight Control. Ph.D. Thesis, Georgia Institute of Technology, Atlanta, Georgia (2000)
23. Bierling, T., Höcht, L., Holzapfel, F., Maier, R., Wildschek, A.: Comparative Analysis of MRAC Architectures in a Unified Framework. In: AIAA Guidance, Navigation and Control Conference, Toronto, Ontario Canada (2010)

Adaptive Control of Non-minimum Phase Modal Systems Using Residual Mode Filters: Part I

Mark J. Balas and Susan A. Frost

Abstract. Many dynamic systems containing a large number of modes can benefit from adaptive control techniques, which are well suited to applications that have unknown parameters and poorly known operating conditions. In this paper, we focus on a direct adaptive control approach that has been extended to handle adaptive rejection of persistent disturbances. We extend this adaptive control theory to accommodate problematic modal subsystems of a plant that inhibit the adaptive controller by causing the open-loop plant to be non-minimum phase. We will modify the adaptive controller with a Residual Mode Filter (RMF) to compensate for problematic modal subsystems, thereby allowing the system to satisfy the requirements for the adaptive controller to have guaranteed convergence and bounded gains. This paper will be divided into two parts. Here in Part I we will review the basic adaptive control approach and introduce the primary ideas. In Part II, we will present the RMF methodology and complete the proofs of all our results. Also, we will apply the above theoretical results to a simple flexible structure example to illustrate the behavior with and without the residual mode filter.

1 Introduction

Applications of control theory to flexible aerospace structures have been many and varied. The survey [13] provides a foundation for structure control with many control approaches and examples. This was based upon a distributed parameter approach to control of flexible structures and other very large-scale systems [14]. Later work created the idea of a Residual Mode Filter (RMF) to offset the destabilizing effect of unmodeled modes in a feedback control environment [15]-[17]. This RMF-based structure control theory has been applied to the complex control

Mark J. Balas
Department of Electrical and Computer Engineering, University of Wyoming, Laramie, WY USA

Susan A. Frost
Intelligent Systems Division, NASA Ames Research Center, Moffett Field, CA USA

issues for large horizontal-axis utility-sized wind turbines [18]-[21], and is beginning to be applied to aeronautic problems that currently use notch filters, eg for flutter, also we are applying the theory to aircraft control where there are flexible modes in the pilot bandwidth, e.g. large civil tilt rotor.

In this paper, we extend our adaptive control theory [1]-[4], [7] to accommodate modal subsystems of a plant that inhibit the adaptive controller, in particular those residual modes that interfere with the almost strict positive real condition. The systems we consider will be large dimensioned, linear time invariant ones which can be diagonalized or placed into modal form. This will include linear flexible structures of many types. Our adaptive Control approach allows for large dimensioned systems through a foundational use of Ideal Trajectories so that the adaptive controller is of much lower dimension than the plant.

The modification will use the idea of Residual Mode Filters (RMF) introduced for fixed gain controllers in [6]. In this paper the RMF will be used to eliminate the effect of modes that prevent the almost strict positive realness of the overall system by being non-minimum phase. This is a new use of the RMF idea; in previous non-adaptive work the purpose of the RMF was to eliminate or mitigate the destabilizing effect of modes unmodeled in the control system design, whereas here the RMF is applied to reinstate the minimum phase nature of the plant under adaptive control.

Here in Part I we will review the basic adaptive control approach and introduce the primary ideas. In Part II, we will present the RMF methodology and complete the proofs of all our results using results from [8]. Also, we will apply the above theoretical results to a simple flexible structure example to illustrate the behavior with and without the residual mode filter.

2 Rejection of Persistent Disturbances

The Plant used in this theory section of the paper will be modeled by the linear, time-invariant, finite-dimensional system:

$$\begin{cases} \dot{\mathbf{x}}_p = \mathbf{A}\mathbf{x}_p + \mathbf{B}\mathbf{u}_p + \mathbf{\Gamma}\mathbf{u}_D \\ \mathbf{y}_p = \mathbf{C}\mathbf{x}_p;\ \ \mathbf{x}_p(0) = \mathbf{x}_0 \end{cases} \tag{1}$$

where the *plant state* $\mathbf{x}_p(t)$, is an N_p-dimensional vector, the *control input vector,* $\mathbf{u}_p(t)$, is M-dimensional, and the *sensor output vector,* $\mathbf{y}_p(t)$, is P-dimensional. The *disturbance input vector,* $\mathbf{u}_D(t)$, is M_D-dimensional and will be thought to come from the *Disturbance Generator*:

$$\begin{cases} \mathbf{u}_D = \mathbf{\Theta}\mathbf{z}_D \\ \dot{\mathbf{z}}_D = \mathbf{F}\mathbf{z}_D;\ \mathbf{z}_D(0) = \mathbf{z}_0 \end{cases} \tag{2}$$

where the *disturbance state,* $\mathbf{z}_D(t)$, is N_D-dimensional. All matrices in (1)-(2) will have the appropriate compatible dimensions. Such descriptions of persistent disturbances were first used in [5] to describe signals of known form but unknown amplitude. Equation (2) can be rewritten as in [3] in a form that is not a dynamical system, which is sometimes easier to use:

$$\begin{cases} \mathbf{u}_D = \boldsymbol{\Theta}\,\mathbf{z}_D \\ \mathbf{z}_D = \mathbf{L}\phi_D \end{cases} \tag{3}$$

where ϕ_D is a vector composed of the known basis functions for the solution of $\mathbf{u}_D = \boldsymbol{\Theta}\,\mathbf{z}_D$, i.e., ϕ_D are the basis functions which make up the known form of the disturbance, and $\mathbf{L}$ is a matrix of dimension N_D x dim(ϕ_D). For the analysis performed in this paper, the amplitude of the disturbance does not need to be known, so $(\mathbf{L},\boldsymbol{\Theta})$ can be unknown. For a better understanding of the disturbance generator, consider the example of a disturbance generator for a *step disturbance;* in the form of equation (2), a step disturbance would have $\Theta = 1$ and $\mathbf{F} = 0$, in the form of equation (3), a step disturbance would have $\phi_D \equiv 1$.

In [5]-[6], as with much of the control literature, it is assumed that the plant and disturbance generator parameter matrices, $(\mathbf{A},\mathbf{B},\mathbf{C},\boldsymbol{\Gamma},\boldsymbol{\Theta},\mathbf{F})$, are known. This knowledge of the plant and its disturbance generator allows the Separation Principle of Linear Control Theory to be invoked to arrive at a State-Estimator based, linear controller which can suppress the persistent disturbances via feedback. In this paper, we will *not* assume that the plant and disturbance generator parameter matrices, $(\mathbf{A},\mathbf{B},\mathbf{C},\boldsymbol{\Gamma},\boldsymbol{\Theta})$, are known. But, we will assume that the disturbance generator parameter from (2), $\mathbf{F}$, is known, i.e., the form of the disturbance functions is known. In many cases, knowledge of $\mathbf{F}$ is not a severe restriction, since the disturbance function is often of known form but unknown amplitude.

Our control objective will be to cause the output of the plant, $\mathbf{y}_p(t)$, to asymptotically track the output of a known reference model, $\mathbf{y}_m(t)$. The Reference Model is given by

$$\begin{cases} \dot{\mathbf{x}}_m = \mathbf{A}_m\mathbf{x}_m + \mathbf{B}_m\mathbf{u}_m;\ \mathbf{x}_m(\mathbf{0}) = \mathbf{x}_0^m \\ \mathbf{y}_m = \mathbf{C}_m\mathbf{x}_m \end{cases} \tag{4}$$

where the *reference model state,* $\mathbf{x}_m(t)$, is an N_m-dimensional vector. The *reference model output,* $\mathbf{y}_m(t)$, must have the *same* dimension as the plant output, $\mathbf{y}_p(t)$. The excitation of the reference model is accomplished via the vector, $\mathbf{u}_m(t)$, which is generated by

$$\dot{\mathbf{u}}_m = \mathbf{F}_m\mathbf{u}_m;\ \mathbf{u}_m(0) = \mathbf{u}_0^m \tag{5}$$

It is assumed that the reference model is stable and the model parameters, $(\mathbf{A}_m, \mathbf{B}_m, \mathbf{C}_m, \mathbf{F}_m)$, *are known.*

As in [5]-[6], we define the *Ideal Trajectories* for the plant given by (1) as linear combinations of the plant states, the control inputs, and the disturbance inputs:

$$\begin{cases} \mathbf{x}_* = \mathbf{S}_{11}^* \mathbf{x}_m + \mathbf{S}_{12}^* \mathbf{u}_m + \mathbf{S}_{13}^* \mathbf{z}_D \\ \mathbf{u}_* = \mathbf{S}_{21}^* \mathbf{x}_m + \mathbf{S}_{22}^* \mathbf{u}_m + \mathbf{S}_{23}^* \mathbf{z}_D \end{cases} \tag{6}$$

where $\mathbf{x}_*(t)$ is the *ideal trajectory*, $\mathbf{u}_*(t)$ is the *ideal control*, $\mathbf{u}_*(t)$ and

$$\begin{cases} \dot{\mathbf{x}}_* = \mathbf{A}\mathbf{x}_* + \mathbf{B}\mathbf{u}_* + \mathbf{\Gamma}\mathbf{u}_D; \ \mathbf{x}_*(0) = \mathbf{x}_0 \\ \mathbf{y}_* = \mathbf{C}\mathbf{x}_* = \mathbf{y}_m \end{cases} \tag{7}$$

Note that the *ideal output*, $\mathbf{y}_*(t)$, matches the reference model output, $\mathbf{y}_m(t)$. If such ideal trajectories exist, they will produce exact output tracking.

By substituting the ideal trajectories given in (6) into (7) and by using the disturbance generator given by (2), the ideal trajectories can be made to match the reference model (4)-(5) with the following *Model Matching Conditions*:

$$\begin{cases} \mathbf{A}\mathbf{S}_{11}^* + \mathbf{B}\mathbf{S}_{21}^* = \mathbf{S}_{11}^* \mathbf{A}_m \\ \mathbf{A}\mathbf{S}_{12}^* + \mathbf{B}\mathbf{S}_{22}^* = \mathbf{S}_{11}^* \mathbf{B}_m + \mathbf{S}_{12}^* \mathbf{F}_m \\ \mathbf{A}\mathbf{S}_{13}^* + \mathbf{B}\mathbf{S}_{23}^* + \mathbf{\Gamma}\mathbf{\Theta} = \mathbf{S}_{13}^* \mathbf{F} \\ \mathbf{C}\mathbf{S}_{11}^* = \mathbf{C}_m \\ \mathbf{C}\mathbf{S}_{12}^* = \mathbf{0} \\ \mathbf{C}\mathbf{S}_{13}^* = \mathbf{0} \end{cases} \tag{8}$$

The model matching conditions given in (8) are necessary and sufficient conditions for the existence of ideal trajectories. Solutions to these matching conditions must exist for later analysis, but explicit solutions need never be known for the adaptive controller design. Necessary and sufficient conditions for the existence and uniqueness of solutions to (8) are given in [9]. We repeat this result here for completeness and the proof is given in the Appendix found in Part II.

Lemma 1: *If CB is nonsingular, there exist unique solutions to the Linear Matching Conditions (8) when* $T(s) \equiv C(sI - A)^{-1}B$ *shares no transmission zeros with the eigenvalues of* A_m, F_m, or F.

The desired control objective is for the output of the plant to asymptotically track the output of the reference model. We define the *output error vector* as:

$$\mathbf{e}_y \equiv \mathbf{y}_p - \mathbf{y}_m \tag{9}$$

To achieve the desired control objective, we want $\mathbf{e}_y \xrightarrow[t\to\infty]{} 0$. We define the *state tracking error* as follows:

$$\mathbf{e}_* \equiv \mathbf{x}_p - \mathbf{x}_* \tag{10}$$

Using (7) and (10), we can write the output error vector as:

$$\mathbf{e}_y \equiv \mathbf{y}_p - \mathbf{y}_m = \mathbf{y}_p - \mathbf{y}_* = \mathbf{C}\mathbf{x}_p - \mathbf{C}\mathbf{x}_* = \mathbf{C}\mathbf{e}_* \tag{11}$$

Furthermore, if we let $\Delta\mathbf{u} \equiv \mathbf{u}_p - \mathbf{u}_*$, from (1) and (7) we have

$$\dot{\mathbf{e}}_* = \mathbf{A}\mathbf{e}_* + \mathbf{B}\Delta\mathbf{u} \tag{12}$$

For analysis purposes, we define a *Fixed Gain Controller*

$$\mathbf{u}_p = \mathbf{u}_* + \mathbf{G}_e^*\mathbf{e}_y \tag{13}$$

If we use the fixed gain control law (13) in the plant given by (1), combined with the definition of $\dot{\mathbf{x}}_*$ from (7) and the output error vector in the form of equation (11), we obtain:

$$\dot{\mathbf{e}}_* = \left(\mathbf{A} + \mathbf{B}\mathbf{G}_e^*\mathbf{C}\right)\mathbf{e}_* \tag{14}$$

We can summarize the above by the following:

Theorem 1. *If* $(\mathbf{A},\mathbf{B},\mathbf{C})$ *is output feedback stabilizable with a gain* $\mathbf{G}_e^*$, *i.e., the eigenvalues of* $\mathbf{A}_C \equiv \mathbf{A} + \mathbf{B}\mathbf{G}_e^*\mathbf{C}$ *are all to the left of the jω-axis, then the fixed gain controller, (13), will produce asymptotic output tracking, i.e.,* $e_y \xrightarrow[t\to\infty]{} 0$.

If all the plant parameters, $(\mathbf{A},\mathbf{B},\mathbf{C},\mathbf{\Gamma},\mathbf{\Theta},\mathbf{F})$, are known, then the fixed gain controller given by (13) with a state estimator for $\mathbf{z}_D$ would be adequate for asymptotic tracking. Note that output feedback stabilization of $(\mathbf{A},\mathbf{B},\mathbf{C})$ can be accomplished when

$$M + P + N_D > N_p \tag{15}$$

and $(\mathbf{A},\mathbf{B},\mathbf{C})$ is controllable and observable [9]. In (13), detailed knowledge of the parameter matrices is not required, suggesting that an adaptive control scheme might be possible under our original assumptions that $(\mathbf{A},\mathbf{B},\mathbf{C},\mathbf{\Gamma},\mathbf{\Theta})$ are unknown and **F** from (2) is known.

Consider the plant given by (1) with the disturbance generator given by (3). Our control objective for this system will be accomplished by an *Adaptive Control Law* of the form:

$$\mathbf{u}_p = \mathbf{G}_m\mathbf{x}_m + \mathbf{G}_u\mathbf{u}_m + \mathbf{G}_e\mathbf{e}_y + \mathbf{G}_D\phi_D \tag{16}$$

where $\mathbf{G}_m, \mathbf{G}_u, \mathbf{G}_e$, and $\mathbf{G}_D$ are matrices of the appropriate compatible dimensions, whose definitions will be given later. We develop the gain adaptation laws to make asymptotic output tracking possible by first forming the following which are intended to simplify our notation:

$$\begin{cases} \Delta \mathbf{G}_u \equiv \mathbf{G}_u - \mathbf{S}_{22}^* \\ \Delta \mathbf{G}_m \equiv \mathbf{G}_m - \mathbf{S}_{21}^* \\ \Delta \mathbf{G}_e \equiv \mathbf{G}_e - \mathbf{G}_e^* \\ \Delta \mathbf{G}_D \equiv \mathbf{G}_D - \mathbf{S}_{23}^* \mathbf{L} \end{cases} \tag{17}$$

The starred gains in (17) are for analysis and come from the ideal trajectory, $\mathbf{x}_*$, of equation (6) with $\mathbf{z}_D$ in the form given in (3), which is then substituted into the fixed gain controller (13). Using (6), (7), and the adaptive control law (16), we can define:

$$\begin{aligned} \Delta \mathbf{u} &= \mathbf{u}_p - \mathbf{u}_* \\ &= \Delta \mathbf{G}_u \mathbf{u}_m + \Delta \mathbf{G}_m \mathbf{x}_m + \left(\Delta \mathbf{G}_e + \mathbf{G}_e^* \right) \mathbf{e}_y + \Delta \mathbf{G}_D \phi_D \end{aligned} \tag{18}$$

Then, via (11), (12), and (18), with appropriate definitions, we have:

$$\begin{aligned} \dot{\mathbf{e}}_* &= \mathbf{A}\mathbf{e}_* + \mathbf{B}\Delta \mathbf{u} \\ &= \left(\mathbf{A} + \mathbf{B}\mathbf{G}_e^* \mathbf{C} \right) e_* + \mathbf{B} \left[\Delta \mathbf{G}_u \quad \Delta \mathbf{G}_m \quad \Delta \mathbf{G}_e \quad \Delta \mathbf{G}_D \right] \eta \\ &= \mathbf{A}_C \mathbf{e}_* + \mathbf{B} \Delta \mathbf{G} \eta \end{aligned} \tag{19}$$

where $\eta \equiv [\mathbf{u}_m^T \ \mathbf{x}_m^T \ \mathbf{e}_y^T \ \phi_D^T]^T$ is the *vector of available information*. We combine (12) and (19) to obtain the *Tracking Error System*:

$$\begin{cases} \dot{\mathbf{e}}_* = \mathbf{A}_C \mathbf{e}_* + \mathbf{B} \Delta \mathbf{G} \eta \\ \mathbf{e}_y = \mathbf{C}\mathbf{e}_* \end{cases} \tag{20}$$

Now we specify the *Adaptive Gain Laws*:

$$\dot{\mathbf{G}} = -\mathbf{e}_y \eta^T \mathbf{H} \tag{21}$$

where $\mathbf{H} = [\mathbf{h}_{ii}]$, $i=1,2,\ldots,4$ is an arbitrary, positive definite matrix (i.e., $\mathbf{H} > 0$). This yields

$$\begin{cases} \dot{\mathbf{G}}_u = -\mathbf{e}_y \mathbf{u}_m^T \mathbf{h}_{11} \\ \dot{\mathbf{G}}_m = -\mathbf{e}_y \mathbf{x}_m^T \mathbf{h}_{22} \\ \dot{\mathbf{G}}_e = -\mathbf{e}_y \mathbf{e}_y^T \mathbf{h}_{33} \\ \dot{\mathbf{G}}_D = -\mathbf{e}_y \phi_D^T \mathbf{h}_{44} \end{cases} \tag{22}$$

Our Adaptive Controller is specified by (16) with the above adaptive gain laws (22). Note that none of the starred gains used in the earlier analysis appear in the realizable control law, (16) and (22). Next we will analyze the stability of this controller.

The *closed-loop adaptive system* consists of (1)-(5), (9), (16), and (22). Using (20) and (21), we have

$$\begin{cases} \dot{\mathbf{e}}_* = \mathbf{A}_C\mathbf{e}_* + \mathbf{B}\Delta\mathbf{G}\eta \\ \Delta\dot{\mathbf{G}} = \dot{\mathbf{G}} = -\mathbf{e}_y\eta^T\mathbf{H} \\ \mathbf{e}_y = \mathbf{C}\mathbf{e}_* \end{cases} \tag{23}$$

where $\mathbf{A}_C \equiv \mathbf{A} + \mathbf{B}\mathbf{G}_e^*\mathbf{C}$. We are able to obtain (23) from (21) because $\Delta\mathbf{G} \equiv \mathbf{G} - \mathbf{G}_*$ where $\mathbf{G}_* \equiv [\mathbf{S}_{22}^* \ \mathbf{S}_{21}^* \ \mathbf{G}_e^* \ \mathbf{S}_{23}^*\mathbf{L}]$ is constant (although generally unknown). The stability of the nonlinear system (23) can be analyzed using Lyapunov Theory. We form the positive definite functions:

$$\begin{cases} V_1(\mathbf{e}_*) \equiv \frac{1}{2}\mathbf{e}_*^T\mathbf{P}\mathbf{e}_* \\ V_2(\Delta\mathbf{G}) \equiv \frac{1}{2}trace(\Delta\mathbf{G}\mathbf{H}^{-1}\Delta\mathbf{G}^T) \end{cases} \tag{24}$$

where $\mathbf{P} > 0$ is the solution of the following pair of equations:

$$\begin{cases} \mathbf{A}_C^T\mathbf{P} + \mathbf{P}\mathbf{A}_C = -\mathbf{Q}; \ \mathbf{Q} > 0 \\ \mathbf{P}\mathbf{B} = \mathbf{C}^T \end{cases} \tag{25}$$

These equations are usually known as the *Kalman-Yacubovic Conditions*. The existence of a symmetric positive definite solution of (25) is known to be equivalent to the following condition:

$$\mathbf{T}_C(s) \equiv \mathbf{C}(s\mathbf{I} - \mathbf{A}_C)^{-1}\mathbf{B} \text{ is strictly positive real } (SPR) \tag{26}$$

For a proof of this equivalence, see [12] App. B. The *strict positive realness* of $\mathbf{T}_C(s)$ means that for some $\sigma > 0$ and for all ω real,

$$\operatorname{Re}\mathbf{T}_C(-\sigma + j\omega) \geq 0 \tag{27}$$

If we calculate the derivatives, $\dot{V}_i$, along the trajectories of (23), we have, using (25), that

$$\begin{cases} \dot{V}_1 = -\frac{1}{2}\mathbf{e}_*^T\mathbf{Q}\mathbf{e}_* + \mathbf{e}_*^T\mathbf{P}\mathbf{B}\Delta\mathbf{G}\eta \\ = -\frac{1}{2}\mathbf{e}_*^T\mathbf{Q}\mathbf{e}_* + \mathbf{e}_y^T v \\ v \equiv \Delta\mathbf{G}\eta \end{cases} \tag{28}$$

and

$$\begin{aligned}\dot{V}_2 &= trace(\Delta\dot{\mathbf{G}}\mathbf{H}^{-1}\Delta\mathbf{G}^T) \\ &= trace(-\mathbf{e}_y\eta^T\mathbf{H})(\mathbf{H}^{-1}\Delta\mathbf{G}^T) \\ &= -trace(\mathbf{e}_y\nu^T) = -trace(\mathbf{e}_y^T\nu) = -\mathbf{e}_y^T\nu\end{aligned} \tag{29}$$

We can form $V \equiv V_1 + V_2 \Rightarrow \dot{V} = -\frac{1}{2}\mathbf{e}_*^T\mathbf{Q}\mathbf{e}_*$ with $\dot{V} \le 0$. Consequently, Lyapunov theory guarantees stability of the zero equilibrium point of (23) and all trajectories of (24) will remain bounded. This guarantees that both $\mathbf{e}_*$ and $\Delta\mathbf{G}$ are bounded.

We can summarize the above by the following *Closed-Loop Stability Result:*

Theorem 2. Suppose the following are true:

(1) All $\mathbf{u}_m(t)$ are bounded (i.e., all eigenvalues of $\mathbf{F}_m$ are in the closed left-half plane);
(2) The reference model (4) is stable (i.e., all eigenvalues of $\mathbf{A}_m$ are in the open left-half plane);
(3) ϕ_D is bounded (i.e., all eigenvalues of $\mathbf{F}$ are in the closed left-half plane and any eigenvalues on the imaginary axis are simple);
(4) (**A**, **B**, **C**) is Almost Strict Positive Real (ASPR), i.e., $\mathbf{T}_C(s) \equiv \mathbf{C}(s\mathbf{I} - \mathbf{A}_C)^{-1}\mathbf{B}$ is strictly positive real.

Then $\mathbf{e}_*$ and $\Delta\mathbf{G}$ are bounded and $\mathbf{e}_* \xrightarrow[t\to\infty]{} 0$ and $\mathbf{e}_y \equiv \mathbf{y}_p - \mathbf{y}_m = \mathbf{C}\mathbf{e}_* \xrightarrow[t\to\infty]{} 0$.

See the Appendix in Part II for a proof of Theorem 2.

This stability analysis shows that asymptotic tracking occurs and the adaptive gains remain bounded. It does *not* prove that $\Delta\mathbf{G} \xrightarrow[t\to\infty]{} \mathbf{0}$. In fact, the gain adaptation laws (22) may not converge to the starred gains in (8); however, this is not required for the adaptive controller to achieve its goals.

3 Conclusions for Part I

We have reviewed our adaptive control theory here. This theory accounts for adaptive model tracking and for leakage of the disturbance term into the Q modes. However, the results require that the error system be minimum phase. In Part II, we will show how to modify the adaptive control with residual mode filters to deal with non-minimum phase systems.

References

1. Wen, J.T., Balas, M.J.: Robust adaptive control in Hilbert space. Journal of Mathematical Analysis and Application 143(1), 1–26 (1989)
2. Balas, M.J.: Finite-dimensional direct adaptive control for discrete-time infinite-dimensional linear systems. Journal of Mathematical Analysis and Applications 196(1), 153–171 (1995)
3. Fuentes, R.J., Balas, M.J.: Direct adaptive rejection of persistent disturbances. Journal of Mathematical Analysis and Applications 251(1), 28–39 (2000)
4. Frost, S.A., Balas, M.J., Wright, A.D.: Direct adaptive control of a utility-scale wind turbine for speed regulation. International Journal of Robust and Nonlinear Control 19(1), 59–71 (2009), doi:10.1002/rnc.1329
5. Johnson, C.D.: Theory of disturbance-accommodating controllers. Control & Dynamic Systems. In: Leondes, C.T. (ed.) Advances in Theory and Applications, vol. 12, pp. 387–489. Academic Press, New York (1976)
6. Balas, M.J.: Finite-dimensional controllers for linear distributed parameter systems: Exponential stability using Residual Mode Filters. J. Mathematical Analysis & Applications 133, 283–296 (1988)
7. Frost, S.A., Balas, M.J., Wright, A.D.: Modified adaptive control for region 3 operation in the presence of wind turbine structural modes. In: Proceedings 49th AIAA Aerospace Sciences Meeting, Orlando (2010)
8. Fuentes, R.J., Balas, M.J.: Robust Model Reference Adaptive Control with Disturbance Rejection. In: Proceedings of the American Control Conference (2002)
9. Kimura, H.: Pole assignment by gain output feedback. IEEE Trans. Automatic Control AC-20(4), 509–516 (1975)
10. Balas, M., Gajendar, S., Robertson, L.: Adaptive Tracking Control of Linear Systems with Unknown Delays and Persistent Disturbances (or Who You Callin' Retarded?). In: AIAA Guidance, Navigation and Control Conference, Chicago, IL (August 2009)
11. Balas, M., Fuentes, R.: A Non-orthogonal Projection Approach to Characterization of Almost Positive Real Systems with an Application to Adaptive Control. In: Proceedings of American Control Conference, Boston (2004)
12. Vidyasagar, M.: Nonlinear Systems Analysis, 2nd edn. Prentice-Hall, New Jersey (1993)
13. Balas, M.: Trends in Large Space Structure Control Theory: Fondest Hopes; Wildest Dreams. IEEE Trans. Automatic Control AC-27, 522–535 (1982)
14. Balas, M.: Control and Dynamic Systems: Advances in Theory and Application. In: Leondes, C.T. (ed.) Toward a More Practical Control Theory for Distributed Parameter Systems, vol. 18. Academic Press, New York (1982)
15. Balas, M.: Nonlinear Finite-Dimensional Control of a Class of Nonlinear Distributed Parameter Systems Using Residual Mode Filters: A Proof of Local Exponential Stability. J. Mathematical Analysis & Applications 162, 63–70 (1991)
16. Balas, M.: Finite-Dimensional Controllers for Linear Distributed Parameter Systems: Exponential Stability Using Residual Mode Filters. J. Mathematical Analysis & Applications 133, 283–296 (1988)
17. Bansenauer, B., Balas, M.: Reduced-Order Model Based Control of the Flexible Articulated-Truss Space Crane. AIAA Journal of Guidance Control and Dynamics 18, 135–142 (1995)

18. Wright, A., Balas, M.: Design of State-Space-Based Control Algorithms for Wind Turbine Speed Regulation. ASME Journal of Solar Energy Engineering 125, 386–395 (2003)
19. Frost, S.A., Balas, M.J., Wright, A.D.: Augmented Adaptive Control of a Wind Turbine in the Presence of Structural Modes, Mechatronics Special Issue on Wind Energy, IFAC (to appear, 2011)
20. Frost, S.A., Balas, M.J., Wright, A.D.: Modified adaptive control for Region 3 operation in the presence of turbine structural modes. In: Proceedings American Control Conference, Baltimore, MD (2010)
21. Frost, S.A., Balas, M.J., Wright, A.D.: Modified adaptive control for Region 3 operation in the presence of turbine structural modes. In: Proceedings 29th AIAA Aerospace Sciences Meeting and Exhibit Wind Energy Symposium (2010)

Adaptive Control of Non-minimum Phase Modal Systems Using Residual Mode Filters: Part II

Mark J. Balas and Susan A. Frost

Abstract. In Part II, we extend our adaptive control theory to accommodate problematic modal subsystems of a plant that inhibit the adaptive controller by causing the open-loop plant to be non-minimum phase. We will modify the adaptive controller with a Residual Mode Filter (RMF) to compensate for problematic modal subsystems, thereby allowing the system to satisfy the requirements for the adaptive controller to have guaranteed convergence and bounded gains. Also, we will apply the above theoretical results to a simple flexible structure example to illustrate the behavior with and without the residual mode filter.

1 Introduction

In Part II, we continue the development of the adaptive control approach. We will keep the consecutive equation numbering from Part I as well as the same reference list. We modify the adaptive control using the idea of Residual Mode Filters (RMF) introduced for fixed gain controllers in [6]. In this paper the RMF will be used to eliminate the effect of modes that prevent the almost strict positive realness (ASPR) of the overall system by being non-minimum phase. We have maintained the same reference list here as in the previous paper to make both papers more readable.

2 Residual Mode Filter Augmentation of Adaptive Controller

In some cases the plant in (1) does not satisfy the requirements of ASPR. Instead, there maybe be a modal subsystem that inhibits this property. This section will

Mark J. Balas
Department of Electrical and Computer Engineering, University of Wyoming, Laramie, WY USA

Susan A. Frost
Intelligent Systems Division, NASA Ames Research Center, Moffett Field, CA USA

present new results for our adaptive control theory. We will modify the adaptive controller with a Residual Mode Filter (RMF) to compensate for the troublesome modal subsystem, or the Q modes, as was done in [6] for fixed gain non-adaptive controllers. Here we present the theory for adaptive controllers modified by RMFs. In a previous paper, we examined the RMF with adaptive control, but assumed that there was no leakage of the disturbance into the Q modes [7]. Here we will deal with the issue of disturbances propagating through these modes.

Let us assume that (1) can be partitioned into the following modal form:

$$\begin{cases} \begin{bmatrix} \dot{x} \\ \dot{x}_Q \end{bmatrix} = \begin{bmatrix} A & 0 \\ 0 & A_Q \end{bmatrix} \begin{bmatrix} x \\ x_Q \end{bmatrix} + \begin{bmatrix} B \\ B_Q \end{bmatrix} u_p + \begin{bmatrix} \Gamma \\ \varepsilon \Gamma_Q \end{bmatrix} u_D \\ y_p = \begin{bmatrix} C & C_Q \end{bmatrix} \begin{bmatrix} x \\ x_Q \end{bmatrix}; \varepsilon \geq 0 \end{cases} \tag{1}$$

Define $x_p \equiv \begin{bmatrix} x \\ x_Q \end{bmatrix}$, $A_p = \begin{bmatrix} A & 0 \\ 0 & A_Q \end{bmatrix}$, $B_p = \begin{bmatrix} B \\ B_Q \end{bmatrix}$, $\Gamma_p = \begin{bmatrix} \Gamma \\ \varepsilon\Gamma_Q \end{bmatrix}$, $C_p^T = \begin{bmatrix} C \\ C_Q \end{bmatrix}$, and

Disturbance Generator $\begin{cases} \dot{z}_D = F z_D \\ u_D = \theta z_D \end{cases}$ or $z_D = L\phi_D$ as before in (2)-(3).

The Output Tracking Error and control objective remain as in (4)-(5), *i.e.* $e_y \equiv y_p - y_m \xrightarrow[t\to\infty]{} 0$.

However, now we will only assume that the subsystem (A, B, C) is Almost Strictly Positive Real, rather than the full un-partitioned plant (A_p, B_p, C_p), and the modal subsystem (A_Q, B_Q, C_Q) will be *known and open-loop stable, i.e.,* A_Q is stable. Also note that this subsystem is directly affected by the disturbance input. Recall that ASPR means $CB > 0$ and $P(s) = C(sI - A)^{-1}B$ is minimum phase. So, in summary, the actual plant has an ASPR subsystem and a known modal subsystem that is stable but inhibits the property of ASPR for the full plant. Hence, this modal subsystem must be compensated or filtered away.

We define the *Residual Mode Filter* (RMF):

$$\begin{cases} \dot{\hat{x}}_Q = A_Q \hat{x}_Q + B_Q u_p \\ \hat{y}_Q = C_Q \hat{x}_Q \end{cases} \tag{2}$$

And the *compensated tracking error*:

$$\tilde{e}_y \equiv e_y - \hat{y}_Q \tag{3}$$

Now we let $e_Q \equiv \hat{x}_Q - x_Q$ and obtain:

$$\dot{e}_Q = A_Q e_Q - \varepsilon \Gamma_Q u_D \tag{4}$$

Consequently,

$$\begin{aligned} \tilde{e}_y \equiv e_y - \hat{y}_Q &= C\Delta x + C_Q x_Q - [C_Q x_Q + C_Q e_Q] \\ &= C\Delta x - C_Q e_Q \end{aligned} \tag{5}$$

As in [1]-[2], we define the Ideal Trajectories, but only for the ASPR Subsystem:

$$\begin{cases} \dot{x}_* = Ax_* + Bu_* + \Gamma u_D \\ y_* = Cx_* = 0 \end{cases} \tag{6}$$

with $\begin{cases} x_* = S_1^* z_D \\ u_* = S_2^* z_D \end{cases}$. This is equivalent to the *Matching Conditions:*

$$\begin{cases} S_1^* F = AS_1^* + BS_2^* + \Gamma\theta \\ CS_1^* = 0 \end{cases} \tag{7}$$

which are known to be uniquely solvable when *CB* is nonsingular. However, we do *not* need to know the actual solutions for this adaptive control approach.

Let $\begin{cases} \Delta x \equiv x - x_* \\ \Delta u \equiv u_p - u_* \\ \Delta\tilde{y} \equiv \tilde{e}_y = C\Delta x - C_Q e_Q \end{cases}$. Then we have

$$\begin{cases} \Delta\dot{x} = A\Delta x + B\Delta u \\ \Delta\tilde{y} = C\Delta x - C_Q e_Q \end{cases} \tag{8}$$

because, from (33), $y_* = 0$. This system can be rewritten:

$$\begin{cases} \begin{bmatrix} \Delta\dot{x} \\ \dot{e}_Q \end{bmatrix} = \begin{bmatrix} A & 0 \\ 0 & A_Q \end{bmatrix} \begin{bmatrix} \Delta x \\ e_Q \end{bmatrix} + \begin{bmatrix} B \\ 0 \end{bmatrix} \Delta u = \overline{A} \begin{bmatrix} \Delta x \\ e_Q \end{bmatrix} + \overline{B}\Delta u + \varepsilon\overline{\Gamma}_Q u_D \\ \Delta\tilde{y} = \begin{bmatrix} C & -C_Q \end{bmatrix} \begin{bmatrix} \Delta x \\ e_Q \end{bmatrix} = \overline{C} \begin{bmatrix} \Delta x \\ e_Q \end{bmatrix} \end{cases} \tag{9}$$

Now we have the following:

Lemma. $\left(\overline{A} = \begin{bmatrix} A & 0 \\ 0 & A_Q \end{bmatrix}, \overline{B} = \begin{bmatrix} B \\ 0 \end{bmatrix}, \overline{C} = \begin{bmatrix} C & -C_Q \end{bmatrix} \right)$ is ASPR if and only if (A, B, C) is ASPR.

Proof: $\overline{CB} = \begin{bmatrix} C & -C_Q \end{bmatrix} \begin{bmatrix} B \\ 0 \end{bmatrix} = CB > 0$ and

$$
\begin{aligned}
\overline{P}(s) &\equiv \overline{C}(sI - \overline{A})^{-1}\overline{B} \\
&= \begin{bmatrix} C & -C_Q \end{bmatrix} \begin{bmatrix} (sI - A)^{-1} & 0 \\ 0 & (sI - A_Q)^{-1} \end{bmatrix} \begin{bmatrix} B \\ 0 \end{bmatrix} \\
&= C(sI - A)^{-1} B = P(s)
\end{aligned}
$$

is minimum phase. End of proof.

So there exists G_e^* such that $(\overline{A}_C \equiv \overline{A} + \overline{B}G_e^*\overline{C}, \overline{B}, \overline{C})$ is Strictly Positive Real (SPR) when (A, B, C) is ASPR. Consequently, as is well known from the Kalman-Yacubovic Theorem, there exists $\overline{P}, \overline{Q} > 0$ such that

$$
\begin{cases} \overline{A}_C^T \overline{P} + \overline{P}\overline{A}_C = -\overline{Q} \\ \overline{P}\overline{B} = \overline{C}^T \end{cases} \tag{10}
$$

We now write the *modified adaptive control law with RMF*:

$$
\begin{cases} u_p \equiv G_u u_m + G_m x_m + G_e \tilde{e}_y + G_D \phi_D \\ \tilde{e}_y \equiv y_p - \hat{y}_Q \\ \dot{\hat{x}}_Q = A_Q \hat{x}_Q + B_Q u_p \\ \hat{y}_Q = C_Q \hat{x}_Q \end{cases} \tag{11}
$$

with modified adaptive gains given by

$$
\begin{cases} \dot{G}_u = -\tilde{e}_y u_m h_u ; h_u > 0 \\ \dot{G}_m = -\tilde{e}_y x_m h_m ; h_m > 0 \\ \dot{G}_e = -\tilde{e}_y \tilde{e}_y^T h_e ; h_e > 0 \\ \dot{G}_D = -\tilde{e}_y \phi_D^T h_D ; h_D > 0 \end{cases} \tag{12}
$$

Finally, we have the following stability result:

Theorem 3. In (9), let (A, B, C) ASPR, A_Q stable, ϕ_D bounded. Then the Modified Adaptive Controller with RMF in (19)-(20) produces $e_y = y_p$ and e_Q ultimately bounded into a ball of radius $R_* \equiv \varepsilon \frac{\left(1 + \sqrt{p_{\max}}\right)}{a\sqrt{p_{\min}}} M_v$ with exponential rate and bounded adaptive gains (G_e, G_D).

Proof: From (19), we have $u_p \equiv G_e \tilde{e}_y + G_D \phi_D$, so we can write

$$\begin{aligned}\Delta u &\equiv u_p - u_* \\ &= [G_e \tilde{e}_y + G_D \phi_D] - [S_2^* L]\phi_D \\ &= G_e^* \tilde{e}_y + \Delta G \eta\end{aligned}$$

where $\begin{cases} \Delta G_e \equiv G_e - G_e^* \\ \Delta G_D \equiv G_D - (S_2^* L) \\ \Delta G \equiv G - G_* = [\Delta G_e \quad \Delta G_D] \\ \eta \equiv \begin{bmatrix} \tilde{e}_y \\ \phi_D \end{bmatrix} \end{cases}$. Then

$$\begin{cases} \dot{\zeta} = \bar{A}\zeta + \bar{B}\Delta u = \bar{A}_C \zeta + \bar{B} w + \varepsilon \bar{\Gamma}_Q u_D \\ \tilde{e}_y = \bar{C}\zeta \end{cases} \tag{13}$$

with $\zeta \equiv \begin{bmatrix} \Delta x \\ e_Q \end{bmatrix}$, $w \equiv \Delta G \eta, \bar{A}_C \equiv \bar{A} + \bar{B} G_e^* \bar{C}$.

From (20), we can see that

$$\dot{G} = \Delta \dot{G} = -\tilde{e}_y \eta^T h; h \equiv \begin{bmatrix} h_e & 0 \\ 0 & h_D \end{bmatrix} > 0 \tag{14}$$

Since (A, B, C) is ASPR, and by the lemma, so is $(\bar{A}, \bar{B}, \bar{C})$, we can we can use the following result from [8] where $v \equiv \bar{\Gamma}_Q u_D$ is bounded because the disturbance $u_D = L\phi_D$ is bounded.

Result. *Consider the nonlinear, coupled system of differential equations,*

$$\begin{cases} \dot{\varsigma} = \bar{A}_c \varsigma + \bar{B}\left(G(t) - G^*\right)\eta + \varepsilon v \\ \tilde{e}_y = \bar{C}\varsigma \\ \dot{G}(t) = -\tilde{e}_y \eta^{\mathrm{T}} h - aG(t) \end{cases} \tag{15}$$

where G^* is any constant matrix and h is any positive definite constant matrix, each of appropriate dimension. Assume the following:

i) the triple $(\bar{A}, \bar{B}, \bar{C})$ is SPR,

ii) there exists $M_K > 0$ such that $\left\| \left(G^*\right)^T G^* \right\| \le M_K$, using the trace norm,

iii) there exists $M_v > 0$ such that $\sup_{t \ge 0} \left\| v(t) \right\| \le M_v$,

iv) there exists a > 0 such that $a \le \dfrac{q_{\min}}{2p_{\max}}$, *and*

v) h *satisfies* $\left\| h^{-1} \right\|_2 \le \left(\dfrac{\varepsilon M_v}{a M_K} \right)^2$, *where* p_{min}, p_{max} *are the minimum and maximum eigenvalues of* $\overline{P}$ *and* q_{min} *is the minimum eigenvalue of* $\overline{Q}$ *in the system*

$$\begin{cases} \overline{A}_C^T \overline{P} + \overline{P}\overline{A}_C = -\overline{Q} \\ \overline{P}\overline{B} = \overline{C}^T \end{cases}.$$

Then the matrix G(t) is bounded and the state $\zeta(t)$ *exponentially approaches the ball of radius*

$$R_* \equiv \varepsilon \frac{\left(1+\sqrt{p_{\max}}\right)}{a\sqrt{p_{\min}}} M_v \text{ with } \varepsilon > 0.$$

From this result, we have ζ is ultimately bounded into the ball of radius R_*, which leads to $e_y \equiv y_p = y_p - y_* = C\Delta$ and e_Q ultimately bounded as well. Therefore $G = G_* + \Delta G$ is bounded, as desired. #

Consequently, the radius of the error ball $R_* \equiv \varepsilon \dfrac{\left(1+\sqrt{p_{\max}}\right)}{a\sqrt{p_{\min}}} M_v$ is determined by the size of ε, which is related to the amount of disturbance leakage into the Q modes. It can be seen that, when there is no leakage of the disturbance into the Q modes ($\varepsilon = 0$), the convergence is asymptotic to zero.

Also, when $\Gamma = B$ and $\Gamma_Q = B_Q$, it is possible to choose $S_1^* = 0$ and $S_2^* = -\theta$ in (34). Then, even if $\varepsilon = 1$, the tracking error will asymptotically go to zero.

3 Simulation Results with RMF

In this section we will apply the above theoretical results to a very simple flexible structure example to illustrate the behavior with and without the Residual Mode Filter. The structure has a rigid body mode and two flexible modes given by:

$$P(s) = \frac{1+s}{s^2} - \frac{3}{s^2+s+1} + \frac{1}{s^2+s+2} = \frac{s^5+s^4+3s^3+0s^2+3s+1}{s^6+2s^5+4s^4+3s^3+2s^2}.$$

This example can obviously be extended to have many more flexible modes. But we are only trying to illustrate the value of the RMF approach. More seriously complex flexible structures are being addressed but will have to await future papers.

This plant has two non-minimum phase zeros at 0.422±0.9543*j* and thus does not meet the ASPR condition. However, when the middle mode $P_Q(s) = -\dfrac{s}{s^2+s+1}$ is removed, the plant becomes:

$P(s)=\dfrac{1+s}{s^2}+\dfrac{1}{s^2+s+2}=\dfrac{s^3+3s^2+3s+2}{s^4+s^3+2s^2}$ which is minimum phase and has a state space realization given by:

$$\begin{cases}\dot{x}_p = Ax_p + B(u_p + u_D)\\ y_p = Cx_p\end{cases} \text{ with } A=\begin{bmatrix}0&1&0&0\\0&0&1&0\\0&0&0&1\\0&0&-2&-1\end{bmatrix},\ \Gamma = B = \begin{bmatrix}0\\0\\0\\1\end{bmatrix},\ C^T = \begin{bmatrix}2\\3\\3\\1\end{bmatrix}$$

with $CB=1$, so CB is nonsingular. Therefore, (A,B,C) is ASPR.

The reference model to be tracked is

$$\begin{cases}\dot{x}_m = (-1)x_m + (1)u_m\\ y_m = (1)x_m\end{cases}$$

which is excited by steps generated by $\dot{u}_m = (0)u_m$. The matching conditions are known to be solvable, but their solution is not needed to apply the theory.

The RMF generated by $P_Q(s) = -\dfrac{3}{s^2+s+1}$ is represented by $A_Q = \begin{bmatrix}0&1\\-1&-1\end{bmatrix}$,

$\Gamma_Q = B_Q = \begin{bmatrix}0\\1\end{bmatrix}$, $C_Q = \begin{bmatrix}-3 & 0\end{bmatrix}$. And we see that $C_Q B_Q = 0$.

The adaptive controller given by (38) - (39) is implemented with h_u=10, h_m=1, h_e=10, h_D=100, and a=0. The disturbance is a nondimensional step of size 10. Setting ε=1, we obtain figs. 1 and 2 from a MatLab/Simulink simulation. The

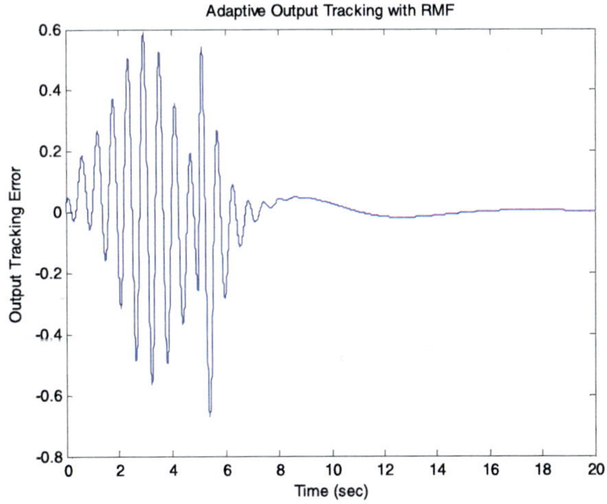

Fig. 1 Nondimensional output tracking response with adaptive controller augmented with RMF.

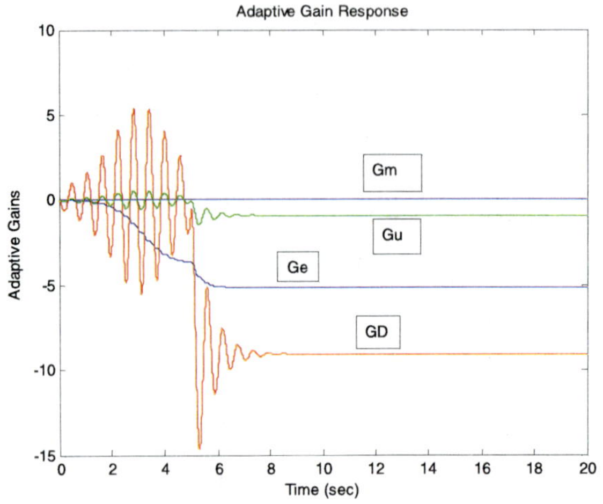

Fig. 2 Adaptive gains, Ge=error gain, Gd=disturbance gain.

output tracking error is shown to converge to zero in fig. 1. The adaptive gains also converge in fig. 2. This illustrates the behavior of the adaptive controller plus the second order RMF. Without the RMF, the plant and adaptive controller are immediately unstable in closed-loop.

4 Conclusions

We have proposed a modified adaptive controller with a residual mode filter. The RMF is used to accommodate problematic modes in the system that inhibit the adaptive controller, in particular the ASPR condition. This new theory accounts for adaptive model tracking and for leakage of the disturbance term into the Q modes. A simple three mode example shows that the RMF can restore stability to an otherwise unstable adaptively controlled system. This is done without seriously modifying the adaptive controller design.

References

1. Wen, J.T., Balas, M.J.: Robust adaptive control in Hilbert space. Journal of Mathematical Analysis and Application 143(1), 1–26 (1989)
2. Balas, M.J.: Finite-dimensional direct adaptive control for discrete-time infinite-dimensional linear systems. Journal of Mathematical Analysis and Applications 196(1), 153–171 (1995)
3. Fuentes, R.J., Balas, M.J.: Direct adaptive rejection of persistent disturbances. Journal of Mathematical Analysis and Applications 251(1), 28–39 (2000)
4. Frost, S.A., Balas, M.J., Wright, A.D.: Direct adaptive control of a utility-scale wind turbine for speed regulation. International Journal of Robust and Nonlinear Control 19(1), 59–71 (2009), doi:10.1002/rnc.1329

5. Johnson, C.D.: Theory of disturbance-accommodating controllers. Control & Dynamic Systems. In: Leondes, C.T. (ed.) Advances in Theory and Applications, vol. 12, pp. 387–489. Academic Press, New York (1976)
6. Balas, M.J.: Finite-dimensional controllers for linear distributed parameter systems: Exponential stability using Residual Mode Filters. J. Mathematical Analysis & Applications 133, 283–296 (1988)
7. Frost, S.A., Balas, M.J., Wright, A.D.: Modified adaptive control for region 3 operation in the presence of wind turbine structural modes. In: Proceedings 49th AIAA Aerospace Sciences Meeting, Orlando (2010)
8. Fuentes, R.J., Balas, M.J.: Robust Model Reference Adaptive Control with Disturbance Rejection. In: Proceedings of the American Control Conference (2002)
9. Kimura, H.: Pole assignment by gain output feedback. IEEE Trans. Automatic Control AC-20(4), 509–516 (1975)
10. Balas, M., Gajendar, S., Robertson, L.: "Adaptive Tracking Control of Linear Systems with Unknown Delays and Persistent Disturbances (or Who You Callin' Retarded?)". In: AIAA Guidance, Navigation and Control Conference, Chicago, IL (August 2009)
11. Balas, M., Fuentes, R.: A Non-orthogonal Projection Approach to Characterization of Almost Positive Real Systems with an Application to Adaptive Control. In: Proceedings of American Control Conference, Boston (2004)
12. Vidyasagar, M.: Nonlinear Systems Analysis, 2nd edn. Prentice-Hall, New Jersey (1993)
13. Balas, M.: Trends in Large Space Structure Control Theory: Fondest Hopes; Wildest Dreams. IEEE Trans. Automatic Control AC-27, 522–535 (1982)
14. Balas, M.: Control and Dynamic Systems: Advances in Theory and Application. In: Leondes, C.T. (ed.) Toward a More Practical Control Theory for Distributed Parameter Systems, vol. 18, Academic Press, New York (1982)
15. Balas, M.: Nonlinear Finite-Dimensional Control of a Class of Nonlinear Distributed Parameter Systems Using Residual Mode Filters: A Proof of Local Exponential Stability. J. Mathematical Analysis & Applications 162, 63–70 (1991)
16. Balas, M.: Finite-Dimensional Controllers for Linear Distributed Parameter Systems: Exponential Stability Using Residual Mode Filters. J. Mathematical Analysis & Applications 133, 283–296 (1988)
17. Bansenauer, B., Balas, M.: Reduced-Order Model Based Control of the Flexible Articulated-Truss Space Crane. AIAA Journal of Guidance Control and Dynamics 18, 135–142 (1995)
18. Wright, A., Balas, M.: Design of State-Space-Based Control Algorithms for Wind Turbine Speed Regulation. ASME Journal of Solar Energy Engineering 125, 386–395 (2003)
19. Frost, S.A., Balas, M.J., Wright, A.D.: Augmented Adaptive Control of a Wind Turbine in the Presence of Structural Modes, Mechatronics Special Issue on Wind Energy, IFAC (2011) (to appear)
20. Frost, S.A., Balas, M.J., Wright, A.D.: Modified adaptive control for Region 3 operation in the presence of turbine structural modes. In: Proceedings American Control Conference, Baltimore, MD (2010)
21. Frost, S.A., Balas, M.J., Wright, A.D.: Modified adaptive control for Region 3 operation in the presence of turbine structural modes. In: Proceedings 29th AIAA Aerospace Sciences Meeting and Exhibit Wind Energy Symposium (2010)

Appendix

Proof of **Lemma 1.** The Linear Matching Conditions (8) can be rewritten:

$$\begin{cases} AS_1 + BS_2 = S_1 L_m + H_1 \\ CS_1 = H_2 \end{cases}$$

where $S_1 \equiv \lfloor S_{11}^* \quad S_{12}^* \quad S_{13}^* \rfloor$, $S_2 \equiv \lfloor S_{21}^* \quad S_{22}^* \quad S_{23}^* \rfloor$, $L_m \equiv \begin{bmatrix} A_m & B_m & 0 \\ 0 & F_m & 0 \\ 0 & 0 & F \end{bmatrix}$,

and $\begin{cases} H_1 \equiv [0 \quad 0 \quad -\Gamma\theta] \\ H_2 \equiv [C_m \quad 0 \quad 0] \end{cases}$.

Suppose CB is nonsingular. Use the coordinate transformation W from Lemma 2 in [11] to put (A, B, C) into *normal form:*

$$\begin{cases} \dot{y} = \bar{A}_{11} y + \bar{A}_{12} z_2 + CBu \\ \dot{z}_2 = \bar{A}_{21} y + \bar{A}_{22} z_2 \end{cases}$$

i.e., there exists $W \equiv \begin{bmatrix} C \\ W_2^T P_2 \end{bmatrix}$ such that $WAW^{-1} \equiv \bar{A} = \begin{bmatrix} \bar{A}_{11} & \bar{A}_{12} \\ \bar{A}_{21} & \bar{A}_{22} \end{bmatrix}$,

$WB = \begin{bmatrix} CB \\ 0 \end{bmatrix} \equiv \bar{B}$, and $CW^{-1} = [I_m \quad 0] \equiv \bar{C}$ which implies that

$$\begin{aligned} \bar{S}_1 L_m &= WS_1 L_m = WAW^{-1} WS_1 + WBS_2 - WH_1 = \bar{A}\bar{S}_1 + \bar{B} S_2 - \bar{H}_1 \\ &= \bar{A}\bar{S}_1 + \begin{bmatrix} CB \\ 0 \end{bmatrix} S_2 - \bar{H}_1 \end{aligned}$$

and $H_2 = CW^{-1}WS_1 = \bar{C}\bar{S}_1 = [I \quad 0]\bar{S}_1 = \bar{S}_a$ where $\bar{S}_1 \equiv WS_1 = \begin{bmatrix} \bar{S}_a \\ \bar{S}_b \end{bmatrix}$. From this we have that $\begin{bmatrix} H_2 \\ \bar{S}_b \end{bmatrix} L_m = \bar{A} \begin{bmatrix} H_2 \\ \bar{S}_b \end{bmatrix} + \begin{bmatrix} CB \\ 0 \end{bmatrix} S_2 - \begin{bmatrix} \bar{H}_a \\ \bar{H}_b \end{bmatrix}$ which implies that

$$\begin{cases} S_2 = (CB)^{-1}[H_2 L_m + \bar{H}_a - (\bar{A}_{11} H_2 + \bar{A}_{12} \bar{S}_b)] \\ \bar{S}_b L_m = \bar{A}_{22} \bar{S}_b + (\bar{A}_{21} H_2 - \bar{H}_b) \end{cases}.$$

Now, if $(\bar{A}_{22}, L_m)$ share no eigenvalues, it is well known [5] that we can solve the above for a unique $\bar{S}_b$ and conversely, then $\bar{S}_1 = \begin{bmatrix} H_2 \\ \bar{S}_b \end{bmatrix}$, $S_2 = (CB)^{-1}[H_2 L_m + \bar{H}_a - (\bar{A}_{11} H_2 + \bar{A}_{12} \bar{S}_b)]$ and $\bar{A}\bar{S}_b - \bar{S}_b L_m = \bar{H}_b$. Since $(\bar{A}_{22}, L_m)$ share no eigenvalues, this is the same as $\bar{A}_{22}$ sharing no eigenvalues with A_m, F_m or F. But the eigenvalues of $\overline{A}_{22}$ from its normal form are known to be the transmission zeros of the open-loop system (A,B,C); see e.g. [13]. Thus, we have proved the result. End of Proof.

Proof of **Theorem 2**

It was already shown that e_* and ΔG are bounded. To prove that $e_* \xrightarrow[t\to\infty]{} 0$, we must use the following version of Barbalat's lemma; see [19] pp. 210-211:

Lemma: If *f(t)* is a real, differentiable function on (0,∞) with $\lim_{t\to\infty} f(t)$ finite and $\frac{df}{dt}$ uniformly continuous, then $\lim_{t\to\infty} \frac{df}{dt} = 0$.

We have already seen that $\dot{V}(t) \le 0$; therefore $V(t) - V(0) = \int_0^t \dot{V}(\tau)d\tau \le 0$ or $0 \le V(t) \le V(0)$ where $V(0) < \infty$. Hence $\lim_{t\to\infty} V(t)$ is finite. Also, $\dot{V}(t)$ is bounded because

$$\begin{aligned}\ddot{V}(t) &= -\left(e_*^T Q \dot{e}_*\right) \le \|e_*\|\|Q\|\|\dot{e}_*\| \\ &= \|e_*\|\|Q\|\|A_C e_* + B\Delta G \eta\| \\ &\le \|e_*\|\|Q\|\left(\|A_C\|\|e_*\| + \|B\|\|\Delta G\|\|\eta\|\right)\end{aligned}$$

and e_* and ΔG are bounded by the previous argument via Lyapunov theory. Also η is bounded since u_m is bounded, A_m is stable, $e_y = Ce_*$ is bounded, and ϕ_D is bounded. Thus $\dot{V}(t) = \int_0^t \ddot{V}(\tau)d\tau$ is uniformly continuous and Barbalat's Lemma may be applied to yield:

$0 = \lim_{t\to\infty} \dot{V}(t) = -\lim_{t\to\infty}\left(e_*^T Q e_*\right)$. Since $Q > 0$, we have $e_* \xrightarrow[t\to\infty]{} 0$, as desired. End of Proof.

Global Tracking Control Structures for Nonlinear Singularly Perturbed Aircraft Systems

Anshu Siddarth and John Valasek

Abstract. The problem of simultaneous tracking of both fast and slow states for a general class of nonlinear singularly perturbed systems is addressed. A motivating example is an aircraft tracking a prescribed fast moving target, while simultaneously regulating speed and/or one or more kinematic angles. Previous results in the literature have focused on tracking outputs that are a function of the slow states alone. Moreover, global asymptotic tracking has been demonstrated only for a class of nonlinear systems that possess a unique real root for the fast states. For a more general class of nonlinear systems only local tracking results have been proven. In this paper, control laws that accomplish global tracking of both fast and slow states is developed using geometric singular perturbation methods. Global exponential stability is proven using the composite Lyapunov function approach and an upper bound necessary condition for the scalar perturbation parameter is derived. Controller performance is demonstrated through simulation of a combined longitudinal lateral/directional maneuver for a nonlinear, coupled, six degree-of-freedom model of the F/A-18A Hornet. Results presented in the paper show that the controller accomplishes global asymptotic tracking even though the desired reference trajectory requires the aircraft to switch between linear and nonlinear regimes. Asymptotic tracking while keeping all the closed-loop signals bounded and well behaved is also demonstrated. Additionally the controller is independent of the scalar perturbation parameter nor does it require knowledge of it.

1 Introduction

This paper addresses systems that possess both slow and fast dynamics. This multiple time-scale behaviour is either a system characteristic (for example, aircraft and flexible beam structures) or arises due to implementation of a fast controller

Anshu Siddarth · John Valasek
Vehicle Systems & Control Laboratory, Department of Aerospace Engineering, Texas A&M University, College Station, TX 77843-3141
e-mail: anshun1@tamu.edu, valasek@tamu.edu

(for example, systems with fast actuators and/or fast sensors). The control objective is to develop a stable tracker for these two time-scale systems that would regulate both slow and fast states simultaneously. The singular perturbation approach[13] has been the foremost technique employed in the literature to examine the behaviour of these two time-scale systems. In this approach, the system dynamics are approximated by two lower-order subsystems. The slow subsystem captures the dominant phenomena assuming that the fast variables evolve infinitely many times faster, and have settled down onto a manifold. The fast subsystem addresses the neglected phenomena, and assumes that the slow variables remain constant. It has been shown that the complete system behaviour can be approximated by the dynamics of the slow subsystem provided the fast subsystem is uniformly asymptotically stable about the manifold [6, 10]. These results of singular perturbation methods have made it the most favourable model-reduction technique in the control literature[14].

The design of nonlinear tracking control laws for the slow variables using singular perturbation methods has received a lot of attention in the past. The typical methodology is to design two separate controllers for each of the two subsystems, and then apply their composite or sum to the full-order system. A tracking control law is designed for the slow subsystem whereas a stabilizing controller is designed for the fast subsystem. This is done to restrict the fast variables onto a manifold. Global asymptotic tracking of the composite control structure is guaranteed only if the manifold is unique. This manifold is the set of fixed points of the fast dynamics expressed as a smooth function of the slow variables and the control inputs; hence it is not always unique. To enforce the uniqueness condition, previous studies in the literature have:

1. Assumed that the system has a unique manifold[4, 8]
2. Considered nonlinear systems that have a unique manifold. This is satisfied by two time-scale systems that are nonlinear in the slow states and linear in the fast states[11]

For a general class of nonlinear systems such as aircraft, approximate approaches that guarantee local stability have been proposed. One approach is to consider the fast variables as control inputs for the slow subsystem. Reference[12] used this approach to design nonlinear flight test trajectories for velocity, angle-of-attack, sideslip angle and altitude by using the fast angular rates as the control variables. This control was augmented with an outer-loop controller that determines the control surface deflections needed to ensure that the angular rates track the output of the inner-loop. More recently the same concept has been employed for the control of generic reentry vehicles[7]. Another approach proposed in Reference[16] considered the general class of nonlinear singularly perturbed systems and computed local approximations of the manifold that helped conclude local stability for the complete system.

All of the approaches discussed above demonstrate slow state tracking either locally or globally by restricting the fast states, and, they address the output tracking problem for two time-scale systems with fast actuators. But for systems whose dynamics inherently possess different time-scales, both the slow and the fast states constitute the output vector. For example, during air combat maneuvering an aircraft is typically required to track a fast moving target while regulating speed (slow variable) and/or one or more kinematic and aerodynamic angles. In this case the fast states cannot be restricted to simply stabilize onto a manifold. The reduced-order approach therefore appears to be inadequate for a general class of output tracking problem. Reference[1] formulated optimal control laws to accomplish fast state tracking using invariant measures for systems with oscillatory fast dynamics.

In this paper, state feedback control laws are developed for a general class of nonlinear singularly perturbed systems to accomplish slow and fast state tracking simultaneously. The paper makes two major contributions. First, the approach developed here employs the reduced-order technique without imposing any assumptions about the fast manifold. This is done by extending the previous work of the authors[16] so as to not require computation of the manifold. Second, global exponential tracking is proved using the composite Lyapunov approach[10]. From the stability analysis it is shown that this approach applies to all classes of singularly perturbed systems, with the global exponential stabilization results of a class of singularly perturbed systems being a special case[3]. Additionally, the technique is independent of the scalar perturbation parameter and an upper bound on this parameter is derived as a necessary condition for stability results to hold. These results are demonstrated by simulation for a nonlinear, coupled, six degree-of-freedom model of the F/A-18A Hornet.

The paper is organized as follows. Section 2 mathematically formulates the control problem and briefly reviews the necessary concepts for model reduction from geometric singular perturbation theory. Control laws and the main results of the paper are detailed in Section 3. Section 4 presents simulation results, and conclusions are presented in Section 5.

2 Problem Formulation and Model Reduction

The following nonlinear singularly perturbed model represents the class of two time-scale dynamical systems addressed in this paper

$$\dot{\mathbf{x}} = \mathbf{f}(\mathbf{x},\mathbf{z}) + \mathbf{g}(\mathbf{x},\mathbf{z})\mathbf{u} \tag{1}$$

$$\varepsilon\dot{\mathbf{z}} = \mathbf{l}(\mathbf{x},\mathbf{z}) + \mathbf{k}(\mathbf{x},\mathbf{z})\mathbf{u} \tag{2}$$

$$\mathbf{y} = \begin{bmatrix} \mathbf{x} \\ \mathbf{z} \end{bmatrix} \tag{3}$$

where $\mathbf{x} \in \mathbb{R}^m$ is the vector of slow variables, $\mathbf{z} \in \mathbb{R}^n$ is the vector of fast variables, $\mathbf{u} \in \mathbb{R}^p$ is the input vector and $\mathbf{y} \in \mathbb{R}^{m+n}$ is the output vector. $\varepsilon \in \mathbb{R}^+$ is the singular perturbation parameter that satisfies $0 < \varepsilon << 1$. The vector fields $\mathbf{f}(.), \mathbf{g}(.), \mathbf{l}(.)$ and $\mathbf{k}(.)$ are assumed to be sufficiently smooth and $p \geq (m+n)$. The control objective is to drive the output so as to track sufficiently smooth, bounded, time-varying trajectories, such that $\mathbf{x}(t) \rightarrow \mathbf{x}_r(t)$ and $\mathbf{z}(t) \rightarrow \mathbf{z}_r(t)$ as $t \rightarrow \infty$.

2.1 Reduced-Order Modeling

The singularly perturbed model considered in Eqs.1,2 is expressed in the *slow time scale* t. In this time-scale the slow states evolve at an ordinary rate whereas the fast states move at a rate of $O\left(\frac{1}{\varepsilon}\right)$. This system can be equivalently expressed in the *fast time-scale* τ such that the fast states evolve at an ordinary rate and the slow variables move slowly at a rate of $O(\varepsilon)$

$$\mathbf{x}' = \varepsilon\left[\mathbf{f}(\mathbf{x},\mathbf{z}) + \mathbf{g}(\mathbf{x},\mathbf{z})\mathbf{u}\right] \tag{4}$$

$$\mathbf{z}' = \mathbf{l}(\mathbf{x},\mathbf{z}) + \mathbf{k}(\mathbf{x},\mathbf{z})\mathbf{u} \tag{5}$$

where $'$ represents a derivative with respect to $\tau = \frac{t-t_0}{\varepsilon}$ and t_0 is the initial time. Geometric singular perturbation theory[6] examines the behaviour of these singularly perturbed systems by studying the geometric constructs of reduced-order models obtained by substituting $\varepsilon = 0$ in Eqs.1,2 and Eqs.4,5. This results in the *Reduced Slow Subsystem*

$$\dot{\mathbf{x}} = \mathbf{f}(\mathbf{x},\mathbf{z}) + \mathbf{g}(\mathbf{x},\mathbf{z})\mathbf{u} \tag{6}$$

$$\mathbf{0} = \mathbf{l}(\mathbf{x},\mathbf{z}) + \mathbf{k}(\mathbf{x},\mathbf{z})\mathbf{u} \tag{7}$$

and the *Reduced Fast Subsystem*

$$\mathbf{x}' = \mathbf{0} \tag{8}$$

$$\mathbf{z}' = \mathbf{l}(\mathbf{x},\mathbf{z}) + \mathbf{k}(\mathbf{x},\mathbf{z})\mathbf{u} \tag{9}$$

The reduced slow subsystem is a locally flattened space of the complete system (Eqs.1,2). Since the vector fields were assumed to be sufficiently smooth there exists a smooth diffeomorphism that maps the complete system onto this locally flattened space. The set of points $(\mathbf{x},\mathbf{z},\mathbf{u}) \in \mathbb{R}^m \times \mathbb{R}^n \times \mathbb{R}^p$ is a smooth manifold $\mathscr{M}_0$ of dimension $m+p$ that satisfies the algebraic Eq.7:

$$\mathscr{M}_0 : \mathbf{z} = \mathbf{Z}_0(\mathbf{x},\mathbf{u}) \tag{10}$$

This set of points is identically the fixed points of the reduced fast subsystem (Eq.9). Thus the manifold $\mathscr{M}_0$ is invariant. The flow on this manifold is described by the differential equation

$$\dot{\mathbf{x}} = \mathbf{f}\left(\mathbf{x},\mathbf{Z}_0(\mathbf{x},\mathbf{u})\right) + \mathbf{g}\left(\mathbf{x},\mathbf{Z}_0(\mathbf{x},\mathbf{u})\right)\mathbf{u} \tag{11}$$

Fenichel[6] proved that the dynamics of a singularly perturbed system of the form represented in Eqs.1,2 is constrained within $O(\varepsilon)$ of Eq.11 if the reduced fast subsystem is stable about $\mathscr{M}_0$. If the dynamics of Eq.11 are locally asymptotically stable about the manifold, then it can be concluded that the complete system is also locally asymptotically stable. Global asymptotic stability conclusions about the complete system can only be made if the manifold is unique, which is a result from differential topology and center manifold theory [6].

3 Control Formulation and Stability Analysis

The central idea in the development is the following. If the manifold is unique and an asymptotically stable fixed point of the reduced fast subsystem, the complete system follows the dynamics of the reduced slow subsystem globally. Therefore, for a tracking problem addressed in this paper it is desired that this manifold lie exactly on the desired fast state reference for all time. *This condition can be enforced if the nonlinear algebraic set of equations is augmented with a controller that enforces the reference to be the unique manifold.* Additionally, this controller should also be capable of simultaneously driving the slow states to their specified reference. These ideas are mathematically formulated and analyzed in the following subsections.

3.1 Control Law Development

The objective is to augment the two time-scale system with controllers such that the system follows smooth, bounded, time-varying trajectories $[\mathbf{x}_r(t), \mathbf{z}_r(t)]^T$. The first step is to transform the problem into a non-autonomous stabilization control problem. Define the tracking error signals as

$$\mathbf{e}(t) = \mathbf{x}(t) - \mathbf{x}_r(t) \quad (12)$$

$$\xi(t) = \mathbf{z}(t) - \mathbf{z}_r(t) \quad (13)$$

Substituting Eqs.1,2, the tracking error dynamics are expressed as

$$\dot{\mathbf{e}} = \mathbf{f}(\mathbf{x},\mathbf{z}) + \mathbf{g}(\mathbf{x},\mathbf{z})\mathbf{u} - \dot{\mathbf{x}}_r \triangleq \mathbf{F}(\mathbf{e},\xi,\mathbf{x}_r,\mathbf{z}_r,\dot{\mathbf{x}}_r) + \mathbf{G}(\mathbf{e},\xi,\mathbf{x}_r,\mathbf{z}_r)\mathbf{u} \quad (14)$$

$$\varepsilon\dot{\xi} = \mathbf{l}(\mathbf{x},\mathbf{z}) + \mathbf{k}(\mathbf{x},\mathbf{z})\mathbf{u} - \varepsilon\dot{\mathbf{z}}_r \triangleq \mathbf{L}(\mathbf{e},\xi,\mathbf{x}_r,\mathbf{z}_r,\varepsilon\dot{\mathbf{z}}_r) + \mathbf{K}(\mathbf{e},\xi,\mathbf{x}_r,\mathbf{z}_r)\mathbf{u} \quad (15)$$

The control law is formulated using the reduced-order models for the complete stabilization problem, which is obtained using the procedure developed in Section 2.
Reduced Slow Subsystem

$$\dot{\mathbf{e}} = \mathbf{F}(\mathbf{e},\xi,\mathbf{x}_r,\mathbf{z}_r,\dot{\mathbf{x}}_r) + \mathbf{G}(\mathbf{e},\xi,\mathbf{x}_r,\mathbf{z}_r)\mathbf{u}_0 \quad (16)$$

$$\mathbf{0} = \mathbf{L}(\mathbf{e},\xi,\mathbf{x}_r,\mathbf{z}_r,\mathbf{0}) + \mathbf{K}(\mathbf{e},\xi,\mathbf{x}_r,\mathbf{z}_r)\mathbf{u}_0 \quad (17)$$

Reduced Fast Subsystem

$$\mathbf{e}' = \mathbf{0} \quad (18)$$

$$\xi' = \mathbf{L}(\mathbf{e},\zeta,\mathbf{x}_r,\mathbf{z}_r,\mathbf{z}_r') + \mathbf{K}(\mathbf{e},\xi,\mathbf{x}_r,\mathbf{z}_r)(\mathbf{u}_0 + \mathbf{u}_f) \quad (19)$$

It is known that the fast tracking error ξ will settle onto the manifold that is a function of the error **e** and control input **u**, which may not necessarily be the origin. To steer both errors to the origin, the control input must be designed such that the origin becomes the unique manifold of the reduced slow system (Eqs.16,17). Therefore, the slow controller $\mathbf{u}_0$ is designed to take the form

$$\begin{bmatrix} \mathbf{G}(\mathbf{e},\xi,\mathbf{x}_r,\mathbf{z}_r) \\ \mathbf{K}(\mathbf{e},\xi,\mathbf{x}_r,\mathbf{z}_r) \end{bmatrix} \mathbf{u}_0 = -\begin{bmatrix} \mathbf{F}(\mathbf{e},\xi,\mathbf{x}_r,\mathbf{z}_r,\dot{\mathbf{x}}_r) \\ \mathbf{L}(\mathbf{e},\xi,\mathbf{x}_r,\mathbf{z}_r,\mathbf{0}) \end{bmatrix} + \begin{bmatrix} A_e\mathbf{e} \\ A_\xi\xi \end{bmatrix} \tag{20}$$

where $A_\mathbf{e}$ and A_ξ specify the desired closed-loop characteristics. With this choice of slow control, the reduced fast subsystem becomes

$$\mathbf{e}' = \mathbf{0} \tag{21}$$
$$\xi' = \mathbf{L}(\mathbf{e},\xi,\mathbf{x}_r,\mathbf{z}_r,\mathbf{z}_r') - \mathbf{L}(\mathbf{e},\xi,\mathbf{x}_r,\mathbf{z}_r,\mathbf{0}) + A_\xi\xi + \mathbf{K}(\mathbf{e},\xi,\mathbf{x}_r,\mathbf{z}_r)\mathbf{u}_f \tag{22}$$

To stabilize the fast subsystem, the fast control $\mathbf{u}_f$ is designed as

$$\begin{bmatrix} \mathbf{G}(\mathbf{e},\xi,\mathbf{x}_r,\mathbf{z}_r) \\ \mathbf{K}(\mathbf{e},\xi,\mathbf{x}_r,\mathbf{z}_r) \end{bmatrix} \mathbf{u}_f = \begin{bmatrix} \mathbf{0} \\ \mathbf{L}(\mathbf{e},\xi,\mathbf{x}_r,\mathbf{z}_r,\mathbf{0}) - \mathbf{L}(\mathbf{e},\xi,\mathbf{x}_r,\mathbf{z}_r,\mathbf{z}_r') \end{bmatrix} \tag{23}$$

Thus, the composite control $\mathbf{u} = \mathbf{u}_0 + \mathbf{u}_f$ satisfies

$$\begin{bmatrix} \mathbf{G}(\mathbf{e},\xi,\mathbf{x}_r,\mathbf{z}_r) \\ \mathbf{K}(\mathbf{e},\xi,\mathbf{x}_r,\mathbf{z}_r) \end{bmatrix} \mathbf{u} = -\begin{bmatrix} \mathbf{F}(\mathbf{e},\xi,\mathbf{x}_r,\mathbf{z}_r,\dot{\mathbf{x}}_r) \\ \mathbf{L}(\mathbf{e},\xi,\mathbf{x}_r,\mathbf{z}_r,\mathbf{z}_r') \end{bmatrix} + \begin{bmatrix} A_e\mathbf{e} \\ A_\xi\xi \end{bmatrix} \tag{24}$$

assuming that the rank of $\begin{bmatrix} \mathbf{G}(.) \\ \mathbf{K}(.) \end{bmatrix} \geq (m+n)$.

The complete closed-loop and reduced slow subsystem for this control law are given as

$$\dot{\mathbf{e}} = A_e\mathbf{e} \tag{25}$$
$$\varepsilon\dot{\xi} = A_\xi\xi. \tag{26}$$

and

$$\dot{\mathbf{e}} = A_e\mathbf{e} \tag{27}$$
$$\mathbf{0} = A_\xi\xi. \tag{28}$$

respectively. Observe that with the proposed control law the nonlinear algebraic set of equations (Eq.17) have been transformed to a linear set of equations (Eq.28). With the proper choice of A_ξ, it is guaranteed that $\xi = \mathbf{0}$ is the unique manifold for both the complete and the reduced slow subsystems. Furthermore, this manifold is exponentially stable as can be deduced from the reduced fast subsystem

$$\mathbf{e}' = \mathbf{0} \tag{29}$$
$$\xi' = A_\xi\xi \tag{30}$$

Remark 1. The control law proposed in Eq.24 is independent of the perturbation parameter ε. Furthermore it is a function of $\mathbf{z}_r'$ that implies that the reference trajectory chosen for the fast states must be faster when compared to the reference of the slow states. Additionally, as for all singular perturbation techniques to work the closed-loop eigenvalues A_e and A_ξ must be chosen so as to maintain the time-scale separation.

3.2 *Stability Analysis*

Complete system stability is analyzed using the composite Lyapunov function approach[10]. Suppose that there exist positive definite Lyapunov functions $V(t,\mathbf{e}) = \mathbf{e}^T\mathbf{e}$ and $W(t,\xi) = \xi^T\xi$ for the reduced subsystems, with continuous first-order derivatives satisfying the following properties:

1. $V(t,\mathbf{0}) = 0$ and $\gamma_1||\mathbf{e}||^2 \le V(t,\mathbf{e}) \le \gamma_2||\mathbf{e}||^2 \; \forall t \in \mathbb{R}^+, \mathbf{e} \in \mathbb{R}^m, \gamma_1 = \gamma_2 = 1$,
2. $(\nabla_{\mathbf{e}}V(t,\mathbf{e}))^T A_{\mathbf{e}}\mathbf{e} \le -\alpha_1\mathbf{e}^T\mathbf{e}, \quad \alpha_1 = 2|\lambda_{\min}(A_{\mathbf{e}})|$,
3. $W(t,\mathbf{0}) = 0$ and $\gamma_3||\xi||^2 \le W(t,\xi) \le \gamma_4||\xi||^2 \; \forall t \in \mathbb{R}^+, \xi \in \mathbb{R}^n, \gamma_3 = \gamma_4 = 1$,
4. $(\nabla_\xi W(t,\xi))^T A_\xi\xi \le -\alpha_2\xi^T\xi, \quad \alpha_2 = 2|\lambda_{\min}(A_\xi)|$.

Next, consider the composite Lyapunov function $\nu(t,\mathbf{e},\xi) : \mathbb{R}^+ \times \mathbb{R}^m \times \mathbb{R}^n \to \mathbb{R}^+$ defined by the weighted sum of $V(t,\mathbf{e})$ and $W(t,\xi)$ for the complete closed-loop system

$$\nu(t,\mathbf{e},\xi) = (1-d)V(t,\mathbf{e}) + dW(t,\xi), \quad 0 < d < 1 \tag{31}$$

The derivative of $\nu(t,\mathbf{e},\xi)$ along the closed-loop trajectories Eqs.25,26 is given by

$$\dot{\nu} = (1-d)(\nabla_{\mathbf{e}}V)^T\dot{\mathbf{e}} + d(\nabla_\xi W)^T\dot{\xi} \tag{32}$$

$$\dot{\nu} = (1-d)(\nabla_{\mathbf{e}}V)^T A_{\mathbf{e}}\mathbf{e} + \frac{d}{\varepsilon}(\nabla_\xi W)^T A_\xi\xi \tag{33}$$

$$\dot{\nu} \le -(1-d)\alpha_1\mathbf{e}^T\mathbf{e} - \frac{d}{\varepsilon}\alpha_2\xi^T\xi \tag{34}$$

$$\dot{\nu} \le -\begin{bmatrix}\mathbf{e}\\ \xi\end{bmatrix}^T \begin{bmatrix}(1-d)\alpha_1 & 0\\ 0 & \frac{d}{\varepsilon}\alpha_2\end{bmatrix} \begin{bmatrix}\mathbf{e}\\ \xi\end{bmatrix} \tag{35}$$

Following the approach proposed in Reference[3], add and subtract $2\alpha\nu(t,\mathbf{e},\xi)$ to Eq.35 to get

$$\dot{\nu} \le -\begin{bmatrix}\mathbf{e}\\ \xi\end{bmatrix}^T \begin{bmatrix}(1-d)\alpha_1 & 0\\ 0 & \frac{d}{\varepsilon}\alpha_2\end{bmatrix} \begin{bmatrix}\mathbf{e}\\ \xi\end{bmatrix} + 2\alpha(1-d)V + 2\alpha dW - 2\alpha\nu \tag{36}$$

where $\alpha > 0$. Substitute in Eq.36 for the Lyapunov functions $V(t,\mathbf{e})$ and $W(t,\xi)$ to get

$$\dot{\nu} \le -\begin{bmatrix}\mathbf{e}\\ \xi\end{bmatrix}^T \begin{bmatrix}(1-d)\alpha_1 - 2\alpha(1-d) & 0\\ 0 & \frac{d}{\varepsilon}\alpha_2 - 2\alpha d\end{bmatrix} \begin{bmatrix}\mathbf{e}\\ \xi\end{bmatrix} - 2\alpha\nu \tag{37}$$

If ε satisfies

$$\varepsilon < \varepsilon^* = \frac{\alpha_2}{2\alpha} \tag{38}$$

and provided $\alpha_1 > 2\alpha$, then from the definitions of α_2, α, and d it can be concluded that the matrix in Eq.37 is positive definite. Then the derivative of the Lyapunov function is lower-bounded by

$$\dot{v} \leq -2\alpha v \tag{39}$$

Since the composite Lyapunov function lies within the following bounds

$$(1-d)\gamma_1||\mathbf{e}||^2 + d\gamma_3||\xi||^2 \leq v(t,\mathbf{e},\xi) \leq (1-d)\gamma_2||\mathbf{e}||^2 + d\gamma_4||\xi||^2 \tag{40}$$

or,

$$\gamma_{11}\left|\left|\begin{bmatrix}\mathbf{e}\\ \xi\end{bmatrix}\right|\right|^2 \leq v(t,\mathbf{e},\xi) \leq \gamma_{22}\left|\left|\begin{bmatrix}\mathbf{e}\\ \xi\end{bmatrix}\right|\right|^2 \tag{41}$$

where $\gamma_{11} = \min((1-d)\gamma_1, d\gamma_3)$ and $\gamma_{22} = \min((1-d)\gamma_2, d\gamma_4)$, the derivative of the Lyapunov function can be expressed as

$$\dot{v} \leq -2\alpha\gamma_{11}\left|\left|\begin{bmatrix}\mathbf{e}\\ \xi\end{bmatrix}\right|\right|^2 \tag{42}$$

From the definition of the constants γ_{11}, γ_{22}, and α, and invoking Lyapunov's Direct Method[9], *uniform exponential stability in the large of* $(\mathbf{e} = \mathbf{0}, \xi = \mathbf{0})$ *can be concluded.* Furthermore, since the reference trajectory $\mathbf{x}_r(t)$ and $\mathbf{z}_r(t)$ is bounded, it is concluded that the states $\mathbf{x}(t) \rightarrow \mathbf{x}_r(t)$ and $\mathbf{z}(t) \rightarrow \mathbf{z}_r(t)$ as $t \rightarrow \infty$. Since the matrix $\begin{bmatrix}\mathbf{G}(.)\\ \mathbf{K}(.)\end{bmatrix}$ is restricted to be full rank, examining the expression for $\mathbf{u}$ in Eq.24 it is concluded that $\mathbf{u} \in \mathcal{L}_\infty$.

Remark 2. Recall that for the special case of state regulation the system dynamics in Eqs.14,15 become autonomous. In such a case, the result of global exponential stability is obtained with less-restrictive conditions on the Lyapunov functions $V(\mathbf{e})$, $W(\xi)$, and consequently $v(\mathbf{e},\xi)$. Similar conclusions were made in Reference[3] for the stabilization problem of a special class of singularly perturbed systems where the control affects only the fast states. Note that for the special class of systems considered in Reference[3], the non-diagonal elements of the matrix in Eq.37 are nonzero, and the bound on the parameter ε is slightly different.

Remark 3. From Eq.37, a conservative upper bound for α is $\alpha < \frac{\alpha_1}{2}$, and consequently $\varepsilon^* \approx \frac{\alpha_2}{\alpha_1}$. Therefore, qualitatively this upper bound is indirectly dependent upon the choice of the closed-loop eigenvalues.

4 Numerical Simulation

The purpose of the example is to demonstrate the methodology and controller performance for an under-actuated, nonlinear, singularly perturbed system. The system studied is a nonlinear, coupled, six degree-of-freedom F/A-18A Hornet aircraft[5]. The combined longitudinal-lateral/directional maneuver requires tracking of the fast variables, in this case body-axis pitch and roll rates, while maintaining zero sideslip angle. Closed-loop characteristics such as stability, accuracy, speed of response and robustness are qualitatively analyzed. The maneuver consists of an aggressive vertical climb with a pitch rate of 25 deg/sec, followed by a roll at a rate of 50 deg/sec, while maintaining zero sideslip angle. The Mach number and angle-of-attack are assumed to be input-to-state stable. The initial conditions are a Mach number of 0.4 at 15,000 feet, an angle-of-attack of 10 deg, and elevon angle of -11.85 deg. All other states are zero. The F/A-18A Hornet model is expressed in stability axes. Since it is difficult to cast the nonlinear aircraft model into the singular perturbation form of Eq.1-2 , the perturbation parameter ε is introduced in front of those state variables that have the fastest dynamics. This is done so that the results obtained for $\varepsilon = 0$ will closely approximate the complete system behaviour (with $\varepsilon = 1$). This is called the forced perturbation technique, and is commonly used in the aircraft literature [2, 12]. Motivated by experience and previous results, the six slow states are Mach number M, angle-of-attack α, sideslip angle β and the three kinematic states: bank angle ϕ, pitch-attitude angle θ, and heading angle ψ. The three body-axis angular rates (p,q,r) constitute the fast states. The control variables for this model are elevon δ_e, aileron δ_a, and rudder δ_r and are assumed to have sufficiently fast enough actuator dynamics. The convention used is that a positive deflection generates a negative moment. The throttle η is maintained constant at 80%, because slow engine dynamics require introduction of an additional time-scale in the analysis; this is a consideration which is beyond the scope of this paper. The aerodynamic stability and control derivatives are represented as nonlinear analytical functions of aerodynamic angles and control surface deflections. Quaternions are used to represent the kinematic relationships from which the Euler angles are extracted. The details of these relationships are discussed in Reference[15].

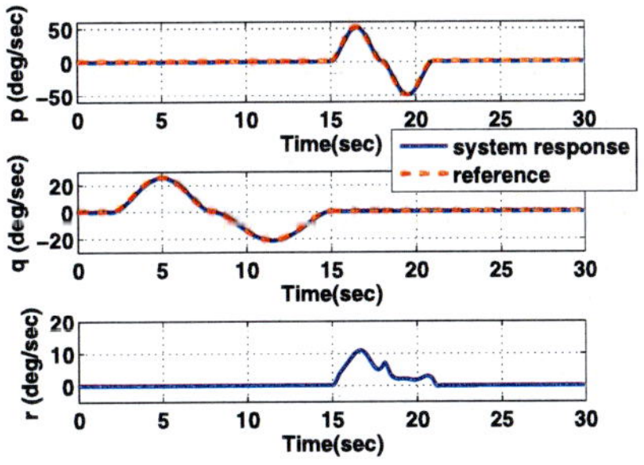

Fig. 1 Body-Axis Angular Rates

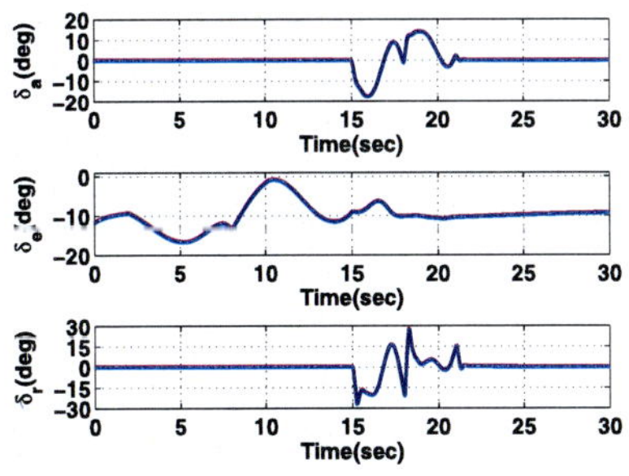

Fig. 2 Control Surface Deflections

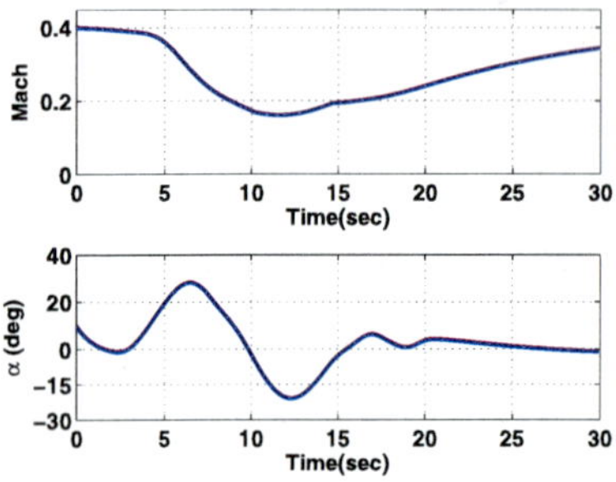

Fig. 3 Mach Number and Angle-of-Attack

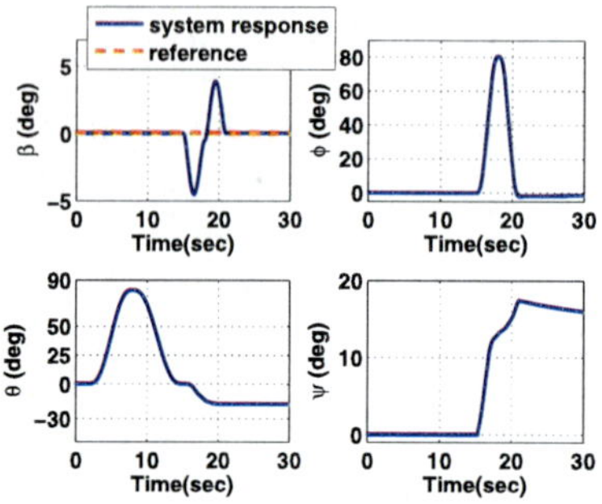

Fig. 4 Sideslip Angle and Kinematic Angles

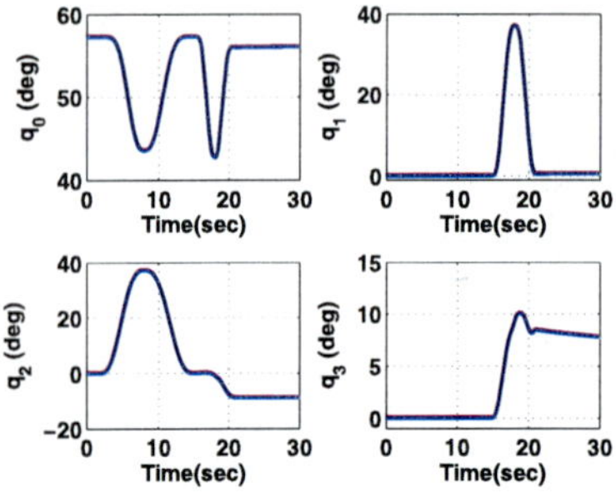

Fig. 5 Quaternion Parameters

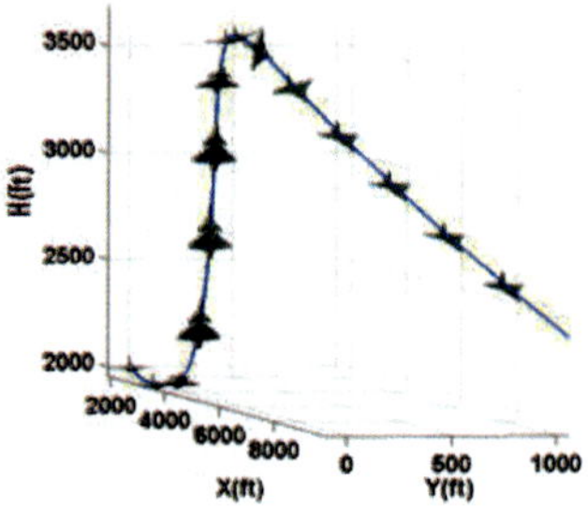

Fig. 6 Three-Dimensional Trajectory

Results and Discussion

Simulation results in Figures 1-6 show that all controlled states closely track their references. At two seconds the aircraft is commanded to perform a vertical climb, and after eight seconds the pitch rate command changes direction and Mach number drops. The lateral/directional states and controls are identically zero until the roll command is introduced at time equals 15 seconds. Observe that all of the states asymptotically track the reference. Figure 2 shows that the elevon deflection remains within specified limits[5] throughout the vertical climb, and the commanded roll produces a sideslip angle which is negated by application of the rudder. The aileron and the rudder deflections remain within bounds while the aircraft rolls and comes back to level flight. The maximum pitch-attitude angle is 81 deg, maximum bank angle is 81 deg (Figure 4), and the maximum sideslip error is ± 4deg. The quaternions and the complete trajectory are shown in Figures 5 and 6 respectively. From Figure 6, note that after completing the combined climb and roll maneuver, the aircraft is commanded to remain at zero sideslip angle, roll rate, and pitch rate. It then enters a steady dive with all other aircraft states bounded. The controller response is judged to be essentially independent of the reference trajectory designed. The robustness properties of the controller are quantified by the upper bound ε^*. For this example, the design variables are $d = 0.5$, $\alpha_1 = 10$, $\alpha = 2$, and $\alpha_2 = 15$, so the upper bound becomes $\varepsilon^* = 7.5$. Therefore for all $\varepsilon < \varepsilon^*$ global asymptotic tracking is guaranteed and in this case $\varepsilon = 1$.

5 Conclusions

A control law for global asymptotic tracking of both the slow and the fast states for a general class of nonlinear singularly perturbed systems was developed. A composite control approach was adopted to satisfy two objectives. First, it enforces the specified reference for the fast states to be 'the unique manifold' of the fast dynamics for all time. Second, it ensures that the slow states are tracked simultaneously as desired. Stability of the closed-loop signals was analyzed using the composite Lyapunov approach, and controller performance was demonstrated through numerical simulation of a nonlinear, coupled, six degree-of-freedom model of an F/A-18A Hornet. The control laws were implemented without making any assumptions about the nonlinearity of the six degree-of-freedom aircraft model. Based on the results presented in the paper, the following conclusions are drawn. First, both positive and negative angular rate commands were seen to be perfectly tracked by the controller and consistent tracking was guaranteed independent of the desired reference trajectory. Second, throughout the maneuver the controller demonstrated global asymptotic tracking even though the desired reference trajectory requires the aircraft to switch between linear and nonlinear regimes. This robust performance of the controller was shown to hold for all $\varepsilon < \varepsilon^* = 7.5$. Third, all closed-loop signals were bounded and the control surface deflections computed were smooth and within specified limits. Fourth, this technique does not require the knowledge of the perturbation parameter ε. This is an important consideration for systems such as aircraft, where quantifying this parameter can be difficult.

Acknowledgements. This material is based upon work supported in part by the U.S. Air Force Office of Scientific Research under contract FA9550-08-1-0038, with technical monitor Dr.Fariba Fahroo. This support is gratefully acknowledged by the authors. Any opinions, findings, conclusions or recommendations expressed in this material are those of the authors and do not necessarily reflect the views of the U.S. Air Force.

References

1. Arstein, Z., Gaitsgory, V.: Tracking fast trajectories along a slow dynamics: a singular perturbation approach. SIAM Journal of Control and Optimization 35(4), 1487–1507 (1997)
2. Calise, A.J.: Singular perturbation methods for variational problems in aircraft flight. IEEE Transactions on Automatic Control 21, 345–353 (1976)
3. Chen, C.C.: Global exponential stabilization for nonlinear singularly perturbed systems. In: IEEE Proceedings of Control Theory and Applications, vol. 145, pp. 377–382 (1998)
4. Choi, H.L., Lim, J.T.: Gain scheduling control of nonlinear singularly perturbed time-varying systems with derivative information. International Journal of Systems Science 36(6), 357–364 (2005)

5. Fan, Y., Lutze, F.H., MCliff, E.: Time-optimal lateral maneuvers of an aircraft. Journal of Guidance, Control and Dynamics 18, 1106–1112 (1995)
6. Fenichel, N.: Geometric singular perturbation theory for ordinary differential equations. Journal of Differential Equations 31, 53–98 (1979)
7. Georgie, J., Valasek, J.: Evaluation of longitudinal desired dynamics for dynamic-inversion controlled generic reentry vehicles. Journal of Guidance Control and Dynamics 26, 811–819 (2003)
8. Grujic, L.T.: On the theory and synthesis of nonlinear non-stationary tracking singularly perturbed systems. Control Theory and Advanced Technology 4(4), 395–409 (1988)
9. Ioannou, P., Sun, J.: Robust Adaptive Control. Prentice Hall Inc., Englewood Cliffs (2003)
10. Kokotovic, P., Khalil, H.K., Reilly, J.O.: Singular Perturbation Methods in Control: Analysis and Design. Academic Press, London (1986)
11. Li, L., Sun, F.C.: An adaptive tracking controller design for nonlinear singularly perturbed systems using fuzzy singularly perturbed model. IMA Journal of Mathematical Control and Information 26, 395–415 (2009)
12. Menon, P., Badgett, M.E., Walker, R.: Nonlinear flight test trajectory controllers for aircraft. Journal of Guidance 10, 67–72 (1987)
13. Naidu, D.S.: Singular Perturbation Methodology in Control Systems. IEEE Control Engineering Series, vol. 34 (1988)
14. Naidu, D.S., JCalise, A.: Singular perturbations and time scales in guidance and control of aerospace systems: A survey. Journal of Guidance, Control and Dynamics 24(6), 1057–1078 (2001)
15. Schaub, H., Junkins, J.L.: Analytical Mechanics of Space Systems. AIAA Education Series (2003)
16. Siddarth, A., Valasek, J.: Kinetic state tracking of a class of singularly perturbed systems. AIAA Journal of Guidance, Navigation and Control Accepted (to appear, 2011)

Motion Planning for a Fixed-Wing MAV Incorporating Closed-Loop Dynamics Motion Primitives and Safety Maneuvers

Michael Gros, Moritz Niendorf, Alfred Schöttl, and Walter Fichter

Abstract. In the following a new two-staged motion planning algorithm with aircraft safety guarantees for obstacle cluttered environments is presented. The first planning stage consists of a probabilistic roadmap global planner that implicitly accounts for kinematic constraints of the plant and generates waypoints. These are used as an orientation marker for the on-line sampling-based second planning stage incorporating motion primitives based on the closed-loop dynamics of a nonlinear 6 degrees of freedom model of a fixed-wing mini aerial vehicle (MAV). Limitations in turnaround time during on-line execution are accounted for by limiting the planning depth of the tree. Safe trajectory traversing is assured with the introduction of safety maneuvers in the horizontal and vertical plane that have to be feasible at every node. Results of simulation runs are presented for a scenario including a narrow passage and an unknown obstacle.

Michael Gros
Institute of Flight Mechanics and Control,
Pfaffenwaldring 7a, 70569 Stuttgart
e-mail: `michael.gros@ifr.uni-stuttgart.de`

Moritz Niendorf
Institute of Flight Mechanics and Control,
Pfaffenwaldring 7a, 70569 Stuttgart
e-mail: `moritz.niendorf@googlemail.com`

Alfred Schöttl
Chief Engineer NGC, MBDA Deutschland, Landshuter Straße 26,
85716 Unterschleißheim
e-mail: `alfred.schoettl@mbda-systems.de`

Walter Fichter
Professor, Institute of Flight Mechanics and Control,
Pfaffenwaldring 7a, 70569 Stuttgart
e-mail: `fichter@ifr.uni-stuttgart.de`

1 Introduction

Motion planning for fixed-wing MAVs in obstacle cluttered environments is a challenging task, especially when aircraft safety has to be assured continuously and in case of a high density of obstacles and narrow passages. Moreover, this becomes even more severe coping with unknown static and/or dynamic obstacles during runtime. The main purpose of this work is to combine the advantages of recently introduced planning algorithms to a realistic real-time planning framework. To this end, improvements are introduced in the fields of planning safety due to safety maneuvers and the generation of motion primitives (MPs) with a closed-loop model. Further, a tree-based local planner that combines advantages of depth-first and breadth-first search techniques as well as the incorporation of local sensor information for obstacle avoidance is presented.

A popular strategy to solve the planning task in obstacle cluttered environments is to decompose the motion planning problem into a coarse, discretized global planner, and a finer local planner that takes into account dynamic constraints [1]. Global planners employ roadmap-based representations whose query phase uses graph search algorithms, e.g. the A* algorithm [1], to produce a waypoint path connected with straight line segments. As representation of the dynamics of vehicles planners often sample motion primitives in order to reduce computational effort [1,2,3]. Frazzoli [2] introduced the so called maneuver automaton, a finite state machine that interconnects two classes of motion primitives, namely trim trajectories and maneuver trajectories. Hwangbo et al. [1] combined a grid-based global planner with a simple tree-based local planning framework and generated motion primitives for a fixed-wing UAV in 3D-slalom scenarios.

Obstacle avoidance during runtime is crucial when unmanned aerial vehicles (UAVs) have to fly through dynamic environments. Thus it makes sense to directly embed range information such as from a laser range finder (LRF) or a radar distance sensor into the motion planner. Such an approach is reported in [4] and [5] in the framework of rapidly exploring random trees (RRTs) and potential functions, respectively. Most safety measures considered in planning algorithms are applied to rotary wing MAVs [2,6]. The problem is more challenging in case of the nonholonomic constraints of a fixed-wing MAV. An incorporation of safety maneuvers for fixed-wing aircraft into the planning process cannot be found in current planning algorithms.

In order to design an algorithm that is tailored to real-time application, the following measures were carried out: A modification of the global planner so that it can cope with restrictions on the flight path angle and heading difference angle; the capability of avoiding unknown static and dynamic obstacles is enhanced by supporting the local planner with available range sensor information. To provide a fixed turnaround time during a planning step of the local planner, the planning horizon and the number of nodes stored to be considered for the next planning step, are limited; safety maneuvers at any node of the planning tree must be considered. These points are functionalities that are added with the respect to the original work of Hwangbo et al. [1].

The paper is outlined as follows. Section 2 gives a detailed description of the nonlinear closed-loop dynamics of the fixed-wing aircraft and the generation of motion primitives. The probabilistic roadmap based global planner is presented in section 3, the tree based local planner, including safety maneuvers and the simulated LRF, is introduced in section 4. A simulation scenario with an unknown obstacle and a narrow passage is discussed in section 5. Summary and outlook are given in section 6.

2 Simulation Model and Motion Primitives Generation

2.1 Problem Formulation

For a nonlinear system, $\dot{\mathbf{x}} = f(\mathbf{x}(t),\mathbf{u}(t))$, $\mathbf{x}(0) = \mathbf{x}_0$, where $\mathbf{x} \in R^n$ represents a state in the state space X, $\mathbf{x}_0$ being the initial state at time t=0 and $\mathbf{u} \in R^m$ being the input to the system, motion planning can be stated as path planning in a state space with first-order differential constraints as opposed to the classical path planning problem that is formulated in the configuration space; a big advantage is the incorporation of kinematic and dynamic constraints of a system, i.e. the incorporation of finite accelerations in the simulation of a system in an obstacle cluttered environment can be crucial for collision avoidance due to the system's inertia.

The dimension of the configuration space C is the number of the degrees of freedom of a body, thus a configuration $\mathbf{q} \in C$ determines a rigid body's position and attitude; the configuration $\mathbf{q}$ and its first order derivative are included in the state $\mathbf{x}$. Constraints imposed on $\mathbf{x}$ by kinematic and dynamic bounds on the system and by static and moving obstacles can be expressed by defining a subspace $X_{\text{free}} \subseteq R^n$ that contains all feasible states; moving obstacles are represented as static obstacles at time t, it is assumed that their trajectory is fully known. Constraints on the input $\mathbf{u}$ are covered by the subset $U \subseteq R^m$, leading to $\mathbf{x}(t) \in X_{\text{free}}(t)$; $\mathbf{u}(t) \in U$ are selected by a local planner. The system is run from the initial state $\mathbf{x}_0$ until the state reaches a terminal set X_g. Associated to X_g is the stopping time $t_g := inf(t\!: \mathbf{x}_g \in X_g)$. The purpose of this framework is to plan a trajectory across an obstacle cluttered environment, thus for X_g only position information is specified.

2.2 Simulation Model with Motion Primitives

A nonlinear 6 degrees of freedom (6-DoF) model of a fixed-wing MAV is used for this work. Aerodynamic data for the model was generated using Digital DATCOM [7] and validated with flight data. The state vector x consists of the position in an Earth-fixed reference frame (x,y,z), Euler angles (ϕ,θ,ψ), body-fixed velocities (u,v,w) and the aircraft angular velocity vector (p,q,r). The open-loop control input u to the system is represented by u = ($\delta_a, \delta_e, \delta_r, \delta_t$), the subscripts indicate aileron, elevator and rudder as well as throttle. Instead of the application of the open-loop input u over a specified time interval during the generation of Motion Primitives (MPs), a simulation with a closed-loop system is preferred due to its well-known stability advantages compared to an open-loop representation. As can be depicted from figure 1, flight path variables are chosen as reference signals $\mathbf{r}(t) \in U_r, U_r \subseteq R^{m_r}$; closing the loop, the dimension of the input vector can be reduced by one to $m_r=3$, assuming that zero body-fixed lateral acceleration a_y is desired. This is achieved via feedback of a_y to the rudder input δ_r. The closed-loop system is now defined as

$$\dot{\mathbf{x}} = f(\mathbf{x}(t), \mathbf{r}(t)),\ \mathbf{x}(0) = \mathbf{x}_0,\ \mathbf{r} = (\phi_{com}, \gamma_{com}, V_{com}), \tag{1}$$

where ϕ_{com} is the commanded roll angle. For a small angle of attack α and sideslip angle β, ϕ matches the bank angle μ. Thus, assuming $\phi \approx \mu$, ϕ serves as the reference signal of the lateral motion. The commanded flight path angle γ_{com} and the commanded absolute speed V_{com} serve as reference signals of the longitudinal motion. The full feedback policy, including turn coordination and turn compensation, can be deducted from figure 1. For motion primitive generation it is defined that the MAV should be capable to execute straight flights as well as turns at different turn rates, both with different flight path angles. Thus, the elements of $\mathbf{r}$ are limited to a finite number of values. The commanded absolute speed V_{com} is held constant while providing n_ϕ different ϕ_{com} and n_γ different γ_{com}, resulting in $n_r = n_\phi \cdot n_\gamma$ different sets of $\mathbf{r}$, namely r_i, with $i=1 \ldots n_r$.

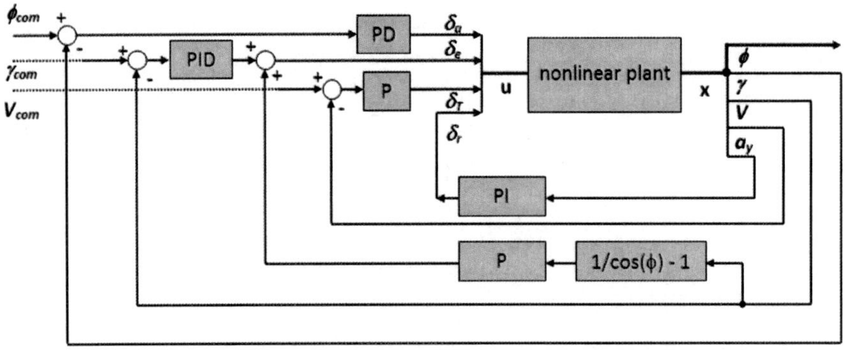

Fig. 1 Structure of the nonlinear closed-loop dynamics of our fixed-wing MAV.

2.3 *Motion Primitives Generation*

The sampling of motion primitives (MPs) decomposes the input of a given dynamic system into a finite number of control actions. The consecutive execution of MPs is determined by an optimization process in the local planner. Several ways to generate motion primitives exist including the recording of open-loop actions by a pilot during a maneuver or the application of an optimal control policy as defined in [2].

This work presents a straight forward way of MP sampling from the closed-loop system (1) in the interval $t \in (t_s, t_f]$ with the initial state $\mathbf{x}(t_s)=\mathbf{x}_s$ represented by $\mathbf{r}_s=\mathbf{r}_i(t_s)$. In contrast to the work of Frazzoli [2], this work introduces two different classes of MPs, namely steady-state MPs and non steady-state MPs which apply for small changes in the heading angle χ. Steady-state MPs begin and end in steady states, whereas in non steady-state MPs at least one boundary state is not a steady-state[1]. For steady-state MPs a constant reference signal $\mathbf{r}(t)= \mathbf{r}_i=const.$ is applied during the interval $t \in (t_s, t_f]$. The maximum of the settling time t_{set} of the underlying controllers for ϕ, γ assures that the commanded reference almost has been reached, thus leading to

$$\lim_{t \to t_{s\phi}} (\phi_{com} - \phi) < e_\phi, \lim_{t \to t_{s\gamma}} (\gamma_{com} - \gamma) < e_\gamma, t_{set} = \max(t_{s\phi}, t_{s\gamma}), \tag{2}$$

where e defines the maximum allowed controller error. As a consequence, every motion primitive ends in a quasi steady state $\mathbf{x}_f \mid \dot{\mathbf{x}}_f \approx 0$, excluding position; unsteady transitions between two consecutive MPs are prevented. The final state $\mathbf{x}_f$ can be associated with vector $\mathbf{r}_f = \mathbf{r}_i(t_f)$ and for the initial state of a MP, $\mathbf{x}_s$ is associated with $\mathbf{r}_s=\mathbf{r}_i(t_s)$. With a limitation time of the upper boundary t_{max} and lower boundary t_{min} to generate motion primitives of roughly the same length, t_f is constrained to $t_{max} \geq t_f \geq t_{min}$.

With the steady-state MP class small heading changes cannot be realized; to enable those maneuvers in the horizontal plane ($\gamma_s=\gamma_f=0°$), non steady-state MPs are introduced. Figure 2 depicts non-steady state MP generation. From an initial roll angle of $\phi_s(t_s)=0°$, a reference roll angle $\phi_{sub}(t_1) \neq 0°$ is commanded in the interval $t_s < t_1 \leq t_{node}$, followed by the application of $\phi_f(t_2)=0°$ with $t_{node} < t_2 \leq t_{set}(\phi=0°)$. The members of ϕ_{sub} are chosen from ϕ_{com} in such a way that only reference signals that command a small ϕ are taken in order to produce a small $\Delta\chi$. A number of n_{sub} steady-state MPs with $\phi_s(t_s)=0°$ and $\phi_f(t_f)=\phi_{sub}$ are used for the generation of non-steady-state MPs. Those MPs are then divided into n_{node} different nodes at time intervals of $t_{node}=t_{set}(\phi_{sub})/n_{node}$ leading to n_{node}-1 non steady-state intermediate nodes per MP. Subsequently n_{node}-1 simulations are conducted with $\phi_{com}(t_1) = \phi_{sub}$ to each intermediate node, which are placed in intervals of $t_{node} \times j$, $j=1 \ldots n_{node}-1$; from each intermediate node, the horizontal roll-angle reference signal $\phi_{com}(t_2)=0°$ is applied until eq. (2) is satisfied.

---

[1] MP naming was slightly changed compared to [2] due to differences in the classification. The set of steady-state MPs contains both trim and maneuver MPs, whereas non steady-state MPs represent pieces for maneuvers to refine the maneuver library. MPs based on constant, respectively piecewise constant controller references were used rather than optimal control laws as in [2].

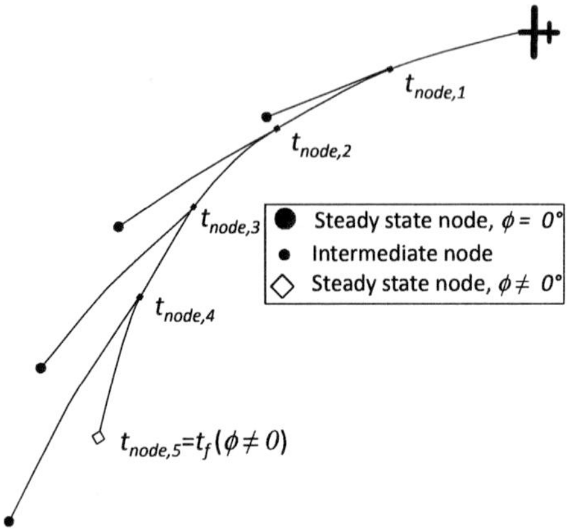

Fig. 2 Non steady-state motion primitive generation for the MAV with n_{node}=5. $n_{unsteady}$=9 non-steady-state MPs result from 1 steady-state MP.

Each node to node connection in this class is considered as a single non steady-state MP in the tree based planner, resulting in $n_{unsteady} = 2\cdot n_{node} - 1$ MPs with a non steady initial and/or final state for each member of ϕ_{sub}. Assuming that for the initial states $\mathbf{x}_s$ and end states $\mathbf{x}_f$ of steady-state MPs any combination of reference vectors r_i with i=1...n_r can be applied, the total number n_{total} of steady-state and non-steady state MPs to generate is defined by $n_{total}=n_r^2 + n_{sub}\cdot(n_{unsteady}-1)$.

3 Probabilistic Roadmap Global Planner

A common strategy to solve the motion planning problem in obstacle cluttered environments is the decomposition into a coarsely discretized global and a finer discretized local planning problem. The geometric representation of obstacles in both the global and the local planner is provided by axis aligned bounding boxes in the Earth-fixed reference frame, straight line segments in the global planner are surrounded by oriented bounding boxes (OBB). In this setup collision detection can be implemented efficiently using the method provided in Ref. [8].

Algorithm 1. PRM Roadmap Construction *(nNodes)*

1: *nodes* ← sample *nNodes* random configurations
2: **for** all *nodes*
3: **find** *k_nearest* nearest neighbors
4: **if** collision check and $\gamma \leq \gamma_{max}$ **then** *roadmap* ← *edge;* **end;**

A probabilistic roadmap (PRM) algorithm is used as global planning method to produce waypoints that are connected with straight segments for the local planner a priori. Reference [9] delivers an introduction to the basic PRM algorithm. As we see the PRM algorithm can be extended to geometric constraints for nodes and edges to implicitly satisfy the kinematic constraints of the aircraft.

The PRM algorithm is divided into a construction and a query phase. As can be seen from algorithm 1, roadmap construction begins with the sampling of *nNodes* different random configurations, in our case only the position (x,y,z) of a node with Euclidean distance as metric is considered, since the system representation can be seen as a mass point representation for that purpose. For every node a nearest neighbor search is conducted. The node structure is realized by a kd-tree representation [10]. Several constraints have to be satisfied during the construction phase before an edge connection between two nodes is possible:

- Edges included in the surrounding OBB are checked for collision to provide a collision free volume for the execution of the local planner.
- A constrained flight path angle γ_{max} of the MAV requires a constraint on the slope of the edge namely $\gamma \leq \gamma_{max}$.

Algorithm 2. PRM Roadmap Query

1: PRM Roadmap Construction $(\mathbf{q}_0,\mathbf{q}_g)$
2: *shortest path*$(\mathbf{q}_0,\mathbf{q}_g) \leftarrow$ modified A* between $\mathbf{q}_0,\mathbf{q}_g$ with Dubins metric

Algorithm 2 depicts the necessary steps in the query phase. Initial configuration $\mathbf{q}_0= (x_0,y_0,z_0,\chi_0,\gamma_0,\phi_0)$, i.e. position, heading, flight path angle, and the goal configuration $\mathbf{q}_g= (x_g,y_g,z_g,\chi_g,\gamma_g,\phi_g)$, with a given position and arbitrary χ_g, γ_g have to be connected to their k nearest neighbors in the roadmap; the roll angle ϕ is left arbitrary in the entire global planning process. As a second step a modified A* graph search algorithm is applied to find the shortest path from $\mathbf{q}_0$ to $\mathbf{q}_g$ in the roadmap. The basic A* algorithm [11] is an informed search algorithm that is proven to be complete. The modification made for this algorithm further adds flight path constraints limiting

- the maximum allowed heading difference $\Delta\chi_{max}$ due to a limited turning radius and
- the minimum length of a straight line segment L_{min} by approximating it as the length of an arc piece by $L_{min} = 2 \cdot \Delta\chi_{max} \cdot R_{min} \cdot c_{safety}$, where c_{safety} is a safety factor and R_{min} is the turning radius determined by the motion primitive with the smallest turning radius.

Only if these constraints are satisfied the specific node is included into the A* graph search and the heuristic in the form of a modified Dubins heuristic, as will be described below, is evaluated.

For a vehicle with minimum turning radius in the plane it can be proven that the shortest connection of a start and a goal configuration is always a part of a set with 6 path types, with mostly 3 segments consisting of arcs and straight segments. Those paths are referred to as Dubins paths. Here an analytical solution for Dubins paths from an initial configuration with position and heading to a goal position with the final heading left arbitrary is used [12]. An extension to the Dubins heuristic for an approximation of the trace in a 3-dimensional workspace is presented in Ref. [1] and is used in this work in a modified form, as can be seen in algorithm 3. Additionally, after the Dubins heuristic value for the horizontal plane $L_{h,Dubins}$ is obtained, a check of the flight path angle constraint γ_{max} is conducted in step 3 of the algorithm. As long as this constraint is not satisfied, full turns with the minimum turning radius R_{min} of the MAV are added to the Dubins heuristic value. As modification to the algorithm of Ref. [1], more accurate trace lengths for a fixed-wing aircraft are obtained by adding the height difference $\Delta z = z_g - z_i$ to the Dubins heuristic for the vertical plane $L_{v,Dubins}$ in step 4.

Algorithm 3. 3D Dubins Metric ($\mathbf{q}_1$,$\mathbf{q}_2$)

1: $L_{h,Dubins} \leftarrow$ Compute Dubins Heuristic in Horizontal Plane ($\mathbf{q}_1$,$\mathbf{q}_2$)

2: h $\leftarrow$ $L_{h,Dubins}$

3: **while** $\Delta z/h \geq sin(\gamma_{max})$ ***do*** $h \leftarrow h + 2\pi \times R_{min}$; **end;**

4: $L_{v,Dubins} \leftarrow h + \Delta z$

5: $L_{3D\,Dubins} \leftarrow L_{h,Dubins} + L_{v,Dubins}$

It is possible that the modified Dubins heuristic overestimates the distance with the consequence that the A* triangle equation does not hold; thus an optimal solution is not guaranteed. However, in interaction with the probabilistic roadmap algorithm the modified Dubins heuristic seems to produce a better path with respect to the kinematic constraints of a fixed-wing aircraft than the admissible Euclidean distance heuristic; a detailed study evaluating this issue is still due.

Once A* returns a shortest path according to the 3D Dubins heuristic as a sequence of nodes from $\mathbf{q}_0$ to $\mathbf{q}_g$, an additional post processing step is executed that attempts to connect non-neighbored nodes directly to each other. This automatically leads to an optimization of the path length, i.e. a shortened path, if a connection is collision free and feasible with respect to the above mentioned constraints.

4 Tree-Based Local Planner with Motion Primitives

This new local planner builds a tree from an initial state $\mathbf{x}_0$ to a goal region around the goal state $\mathbf{x}_g$, with only position specified a priori, by the interconnection of MPs. PRM waypoints serve as intermediate goal regions which are represented by a sphere with a certain radius around $\mathbf{x}_g$. Due to the discrete nature of MPs it is unlikely that a goal state is reached exactly; however, there is a trade-off. A too narrow goal region decreases the chance that a waypoint is hit; a too wide goal region decreases the guidance "property" of waypoints for the local planner in an obstacle cluttered environment. A goal region of the size less than the minimal turning radius has been found acceptable through simulation of the model considered here.

The planner performs an informed search based on a partially greedy cost functional J as opposed to Ref. [1] where a planner based on a greedy search is used. It is desired to obtain depth-first behavior in the free space that results in fast search-space exploration and a breadth-first exploration when close to obstacles or in the proximity of a goal region to minimize the probability to end up in a local minimum. The structure of the local planner is given by algorithm 4.

From chapter 3 it is clear that every node in the local planner can be sufficiently represented in the configuration space with $\mathbf{q}=(x,y,z,\chi,\gamma,\phi)$, even though the system of eq. (1) is defined in the state-space. For a MP that starts at $\mathbf{q}_s$, ends at $\mathbf{q}_f$ and an intermediate goal node $\mathbf{q}_{int}$ from the PRM path, the cost functional J is defined as

$$J(\mathbf{q}_f) = f(\mathbf{q}_s, \mathbf{q}_f) + g(\mathbf{q}_f, \mathbf{q}_{int}) + h(\mathbf{q}_f, \mathbf{q}_{obs}), \tag{3}$$

where $f(\mathbf{q}_s,\mathbf{q}_f)$ is the trace length of the MP and $g(\mathbf{q}_f,\mathbf{q}_{int})$ is the greedy cost-to-go with the 3D Dubins heuristic between $\mathbf{q}_f$ and $\mathbf{q}_{int}$. The term $h(\mathbf{q}_f,\mathbf{q}_{obs})$ represents the value of a potential function in the line of sight of the MAV between the position coordinates of $\mathbf{q}_f$ and an obstacle position $\mathbf{q}_{obs}$. In the potential function

$$h(\mathbf{q}_f, \mathbf{q}_{obs}) = K_{LRF} \cdot \max_{LRF,j}(1/(|\mathbf{q}_f - \mathbf{q}_{obs}| - 1)), \tag{4}$$

$\mathbf{q}_{obs}$ is determined by the simulation with a laser range finder (LRF) similar to Ref. [4], equipped with a beam of limited range d_{LRF}, variable pitch angle $\Delta\theta$ and azimuth angle $\Delta\psi$. The LRF simulation produces a grid of size j at each end configuration $\mathbf{q}_f$ of a MP with distance information to the next obstacles that intersect with the beam due to its pivoting; for simplicity reasons an ideal sensor was assumed without false hits or latency. The potential function always accounts for the worst case, thus the obstacle position $\mathbf{q}_{obs}$ with smallest distance to $\mathbf{q}_f$ was chosen from the grid.

In the tree building process, every MP is seen as a new branch of the tree originating from a node and will be incorporated into the planning process only if one of four subsequent safety maneuvers of the MAV in the horizontal or vertical plane is possible and if no collision occurs. Safety maneuvers were not simulated explicitly but represented by OBB placeholders and collision checked for every MP. In total there are two safety maneuvers for each the horizontal respectively the vertical plane. It is assumed that a safety maneuver is conducted at the edge of the flight envelope with a minimum turning radius that is by far lower than the minimum radius of steady-state MPs.

Similar to the A* algorithm this planner stores a list of nodes, named *open_list*, sorted in the order of their cost. Conversely to A*, this list is finite to limit the computational effort at each cycle. Only nodes that are members of the *open_list* are candidates for a further expansion of the tree. The node with the lowest cost is labeled *current_node* and thus is added to the tree. In the next step, the cost-functional of the nodes at the end of the branches of the *current_node* is evaluated. Subsequently, all collision-free nodes are added to the *open_list* which is then sorted again (line 9-10). If the size of *open_list* is bigger than allowed, nodes with the worst cost functional are removed. Therefore the size of the *open_list* is always a trade-off between computational effort and a convergence to the optimal solution since it is proven that the basic A* algorithm is complete [10].

Algorithm 4. Tree of Motion Primitives (simplified algorithm)

1: *current_node* ← *open_list*(1) ← $node_{MAV}$ ← initial node at state $\mathbf{x}_0$; *age* = 0;
2: **while** not in goal area around $\mathbf{x}_g$;
3: **if** in local goal area around $\mathbf{q}_{\text{int}}$
4: $\mathbf{q}_{\text{int}}$, local goal ← next waypoint from PRM;
5: *open_list*: **for** all nodes: calculate cost *J* **end;** sort; limit size;
6: *current_node* ← *open_list(1);*
7: **else**
8: *age* ← *age* +1;
9: *current_node:* **for** all branches: calculate cost *J;* collision/safety maneuver check; **end;**
10: *open_list*: save all feasible nodes of branches (from line 9); sort; limit size;
11: *current_node* ← *open_list*(1);
12: check if local or global goal area is reached;
13: **if** $age\text{-}age_{old} > planning\_depth$
14: *path*←recursively find path from *current_node* to current MAV configuration $node_{MAV}$;
15: $node_{MAV}$ ← move MAV to the next steady-state node;
16: **if** $node_{MAV}$ = *current_node* **then** execute safety_maneuver; **end;**
17: *open_list:* delete branches whose origin nodes possess timestamp ≤ *age*;
18: age_{old}← age;
19: **end**

A limitation of the computation time is achieved by a limitation of the number of newly added motion primitives in each planning step. This is realized by the variable *planning_depth*. After reaching the *planning_depth*, the MAV trajectory is recalculated recursively in the tree from the node with the best value, i.e. *current_node*, to the actual MAV node $node_{MAV}$ and the MAV is propagated to the next steady-state node (line 15). In case of a planning failure (line 16) a safety maneuver is executed. Parts of the tree that cannot be reached anymore by the MAV are pruned away.

The use of a modified Dubins heuristic for the greedy part of the cost functional gives a more accurate estimation of the vehicle trajectory compared to the Euclidean distance. This maximizes the influence of distance deviation and minimizes the deviation in vehicle orientation when farther away from the goal region to achieve a depth-first exploration in the free space; if the tree advances to a goal region in a narrow passage, the greedy cost-functional value of branches that possess a small deviation in orientation to the goal region will be lower and the limited length of the *open_list* causes a widening of the tree at the same hierarchy level which resembles breadth first behavior. The LRF provides local obstacle information that can be used in the repulsive potential to push the vehicle trajectory towards the free space X_{free} when in the proximity of obstacles; MP that point to close obstacles are penalized heavily by the potential function and vice versa.

5 Simulation Test Runs

For the motion primitive generation of the MAV a set of n_r= 35 different reference vectors **r** were chosen as shown in table 1. To enable the aircraft to execute small heading changes in the range of $1° \leq \Delta\chi \leq 15°$, the number of intermediate nodes for non steady-state MPs was chosen to n_{node}=5, resulting in $n_{unsteady}$=9 non steady-state MPs per member of ϕ_{sub}=[-20°,-10°,10°,20°] with n_{sub}=4 members.

This led to a total number of n_{total}=1257 applicable MPs; 170 MPs are excluded due to a violation of the time constraint t_{max}, leading to 1087 MPs that have to be administered and applied by the tree-based local planner.

Symmetry properties of the aircraft dynamics in the *x-z* plane could be used to reduce the MP set; a saving in cost functional evaluations due to the x-z plane symmetry could only be achieved for symmetric flight initial conditions, i.e. $\phi(t_s)=0°$, and only for the greedy part of *J*. Due to the minimal achievable savings the MP set is not using symmetry properties.

Constraints to the global planner were chosen to γ_{max}=10°,$\Delta\chi_{max}$=60°, R_{min}=38m. LRF parameters were set to d_{LRF}=80m, $\Delta\theta$ = [-20°, 20°], $\Delta\ \psi$= [-45°,45°].

Table 1 Reference Signal Discretization

ϕ_{com}	-45°	-20°	-10°	0°	10°	20°	45°
γ_{com}		-15°	-10°	0°	10°	15°	
V_{com}				20m/s			

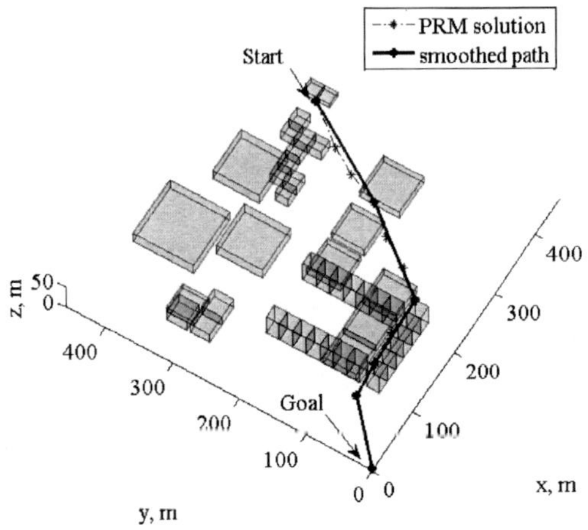

Fig. 3 The PRM solution of an exemplary scenario with a narrow passage in the order of R_{min}. Due to the narrow passage, PRM parameters had to be chosen to *nNodes*=10000, *nNeighbour*=500 to provide a dense sampling of the map.

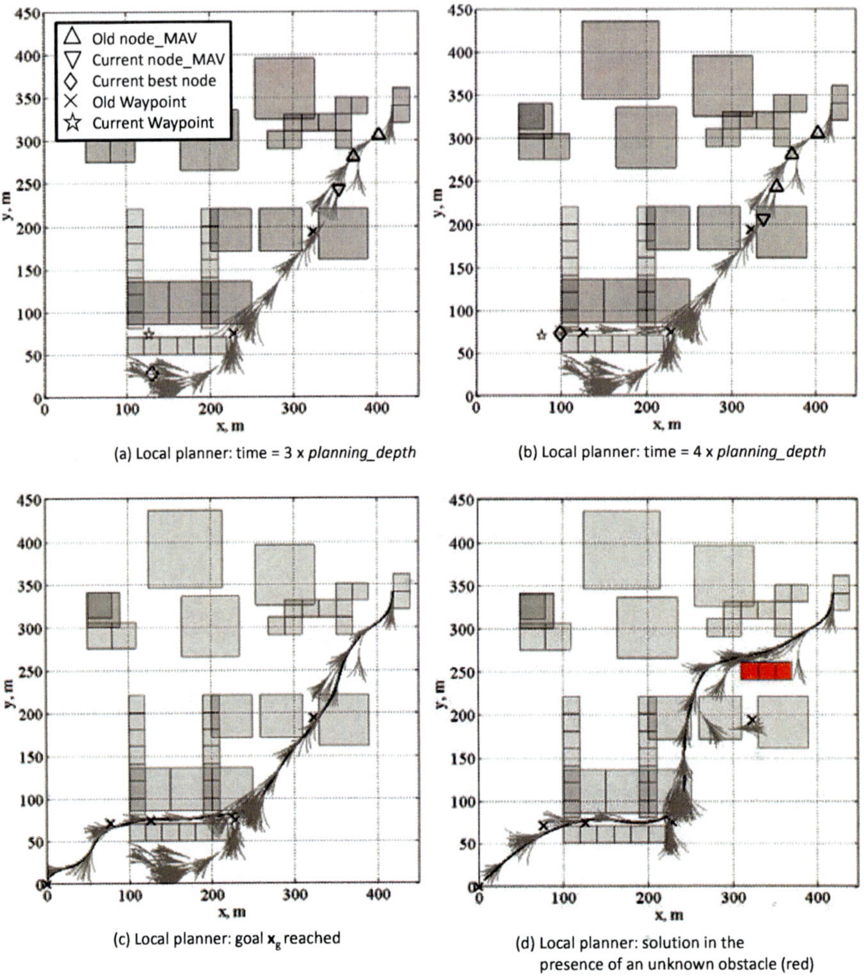

Fig. 4 Every figure shows the whole tree including any former members of *open_list*. Free space is traversed fast by the planner. The tree grows in breadth in front of the narrow passage, as shown in figures (a)-(b). Firstly, in (a) the best node lays outside of the narrow passage, but in (b) a path inside the narrow passage is explored that has the best value for the cost functional J. Afterwards it is switched to the next waypoint. Figure (c) depicts the finished motion plan from $\mathbf{x}_0$ to $\mathbf{x}_g$ with *planning_depth*=10, max(*open_list*)=20. A solution of the local planner in the presence of an unknown obstacle (red) is shown in figure (d). The path generated by the PRM planner is the same as in figure 3.

A simulation in an urban environment with focus on narrow passages and unknown obstacles was chosen to demonstrate the capabilities of the introduced planning framework.

Figure 3 shows the PRM solution and the smoothed path in an environment containing a narrow passage between buildings with a passageway width less than

R_{min}. The number of samples *nNodes* and the nearest neighbor parameter *k_nearest* have to be set to relatively high values compared to wider passages to find a way through the narrow passage.

Planning steps of the local planner are illustrated in figure 4a-c. It can be seen that depth first behavior is achieved in the free space, in contrary to the entrance of the narrow passage where it takes several steps and the tree widens until the next waypoint inside the narrow passage is reached. The safe traversing through the narrow passage is only possible since the upward vertical safety maneuver is always feasible at that time; it guarantees a recovery of the MAV even if the narrow passage is blocked by an unknown obstacle during runtime. The complete trajectory for the scenario is shown in figure 4c.

Figure 4d depicts the same scenario with an additional unknown static obstacle blocking the path to the second PRM waypoint. It can be seen that the obstacle is avoided and as soon as the *current_node* has reached the goal area of the second PRM waypoint, the cost is calculated for the next waypoint and thus, the motion plan is optimized ignoring the blocked waypoint. An advantage over pure depth-first planners is the usability of the node information stored in the *open_list*. In the example, the obstacle is passed on the right side even though a passageway to the left was proposed first due to a better value of the partial greedy cost functional J.

6 Summary and Outlook

In this work a two-staged motion planner with safety maneuvers based on motion primitives has been demonstrated including a straight forward method of motion primitive generation for closed-loop fixed-wing MAVs. The introduced framework is a step towards a computationally inexpensive on-line planner for small UAV that is able to function safely in a varying urban environment. A modification of a PRM based planner and a new tree based planner for safe trajectory traversing of fixed-wing aircraft were designed and demonstrated in an environment that included a narrow passage and the avoidance of an unknown obstacle.

More detailed parameter studies are being performed for the local planner and the interaction of the global planner with the local planner. So far, results show good performance, including acceptable runtime behavior.

References

1. Hwangbo, M., Kuffner, J., Kanade, T.: Efficient Two-Phase 3D Motion Planning for Small Fixed-Wing UAVs. In: Proceedings of the 2007 IEEE Int. Conf. on Robotics & Automation. IEEE, Los Alamitos (2007)
2. Frazzoli, E., Dahleh, M.A., Feron, E.: Real-Time Motion Planning for Agile Autonomous Vehicles. Journal of Guidance, Control and Dynamics 25(1), 116–129 (2002)
3. Kuwata, Y., Fiore, G., Frazzoli, E.: Real-time Motion Planning with Applications to Autonomous Urban Driving. IEEE Transactions on Control Systems Technology 17(5) (September 2009)

4. Saunders, J., Call, B., Curtis, A., et al.: Static and Dynamic Obstacle Avoidance in Miniature Air Vehicles. In: AIAA Infotech@Aerospace, Arlington VA, pp. 2005–6950 (2005)
5. Scherer, S., Singh, S., Chamberlain, L., et al.: Flying Fast and Low Among Obstacles. In: International Conference on Robotics and Automation (ICRA), pp. 2023–2029 (2007)
6. Wzorek, M., Doherty, P.: Reconfigurable Path Planning for an Autonomous Unmanned Aerial Vehicle. In: ICHIT 2006, vol. 2, pp. 242–249 (November 2006)
7. McDonnell Douglas Astronautics Company, The USAF Stability and Control Digital DATCOM, Vol. I, Users Manual, St. Louis Division (April 1979)
8. Gottschalk, S., Lin, M., Manocha, D.: OBBTree: A Hierarchical Structure for Rapid Inteference Detection. In: Proceedings of the 23rd Annual Conference on Computer Graphics and Interactive Techniques, pp. 171–180. ACM, New York (1996)
9. Kavraki, L.E., Svestka, P., Latombe, J.C., et al.: Probabilistic Roadmaps for Path Planning in High-Dimensional Configuration Spaces. IEEE Transactions on Robotics and Automation 12(4) (August 1996)
10. Atramentov, A., LaValle, S.M.: Efficient Nearest Neighbor Searching for Motion Planning. In: Proceedings of the IEEE International Conference on Robotics and Automation, vol. 1, pp. 632–637 (2002)
11. Hart, P., Nilsson, N.J., Raphael, B.: A Formal Basis for the Heuristic Determination of Minimum Cost Paths. IEEE Transactions on Systems Science and Cybernetics 4(2), 100–107 (1968)
12. Enright, J., Frazzoli, E., Savla, K., Bullo, F.: On Multiple UAV Routing with Stochastic Targets: Performance Bounds and Algorithms. In: Proceedings of the AIAA Conference on Guidance, Navigation and Control (2005)

Novel Dynamic Inversion Architecture Design for Quadrocopter Control

Jian Wang, Thomas Bierling, Leonhard Höcht, Florian Holzapfel,
Sebastian Klose, and Alois Knoll

Abstract. This paper presents a novel controller architecture for a quadrocopter. A two-loop controller using dynamic inversion is designed that allows direct commands for position and heading angle. The inner loop controls the body-fixed angular rates. And the outer loop achieves the position control. With this structure, the position dynamic equation appears in an elegant form. The derived controller is capable of decoupling the strongly coupled dynamics of the quadrocopter, maximizing the transmission bandwidth of the position control, as well as eliminating the singularity caused by the attitude control (i.e. pitch angle at 90 degree). Pseudo-control hedging is applied in the position loop to account for limitations, saturations, actuator dynamics and delay in the inner loop. The effectiveness of the designed controller is demonstrated by an implementation on a quadrocopter equipped with an ARM7 onboard processor.

Nomenclature

B	Body-fixed frame
W	World frame, deduced from NED frame with user-defined x-axis
M_{BW}, M_{WB}	Transformation matrices between B frame and W frame
L, M, N	The moments around x, y and z axis of B frame, respectively
p, q, r	Angular rates around x, y and z axis of B frame, respectively
$(\vec{r})_W$	Position vector denoted in W frame
$(\vec{V})_W^W$	Velocity vector defined w.r.t. W frame denoted in W frame

Jian Wang · Thomas Bierling · Leonhard Höcht · Florian Holzapfel
Institute of Flight System Dynamics, TU München, Germany, 85748

Sebastian Klose · Alois Knoll
Institute of Robotics and Embedded Systems, TU München, Germany, 85748

$\left(\vec{a}\right)_W^{WW}$	Acceleration vector defined w.r.t. W frame denoted in W frame
$\left(\dot{\vec{a}}\right)_W^{WWW}$	Acceleration vector differentiated w.r.t. W frame
$\vec{\gamma}_W$	Gravitational vector in W frame
$\vec{f}_B$	Specific force vector in B frame, accelerometer output.
T	Total thrust of the quadrocopter

1 Introduction

Recent technological progress in low-cost MEMS-based sensors, actuators and energy storage devices enables the development of miniature vertical take-off and landing (VTOL) systems. The quadrocopter is one of the most preferred types for many civil applications as well as research platforms. There are many advantages, like easy construction and steering principle, VTOL and hovering ability. However, because of the nonlinear dynamic behavior, the control and guidance of these vehicles is a subject of research, especially in applications such as search and rescue, surveillance, inspection, and so on. For these applications, high stability, high precision hovering ability, high bandwidth, and high maneuverability are important.

Previous effort on nonlinear dynamic inversion control for Micro Aerial Vehicles (MAV) include three-loop design corresponding to inversion of rotational, attitude and path dynamics in separated cascaded loops[1]. A more common controller architecture is two-loop design [2]-[4], where the outer loop is the position control and the inner loop provides attitude control, as illustrated in Fig. 1. Both control loops have relative degree 2 dynamics. But they have a limited bandwidth and a singularity occurs at a Pitch angle of 90° when Euler angles are used [2].

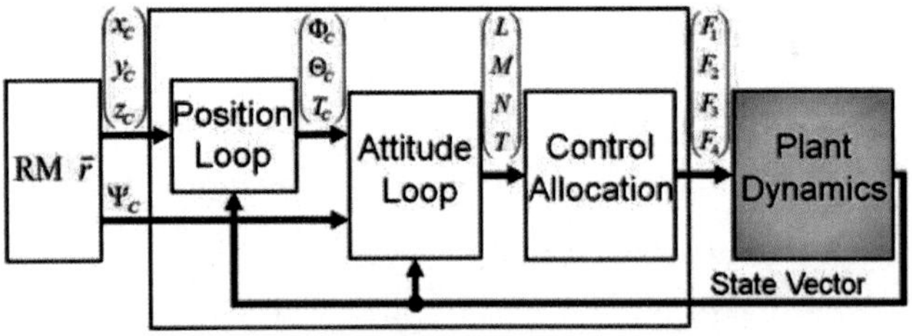

Fig. 1 Conventional two loops control architecture

The focus of this work is on the design of a baseline controller using the nonlinear dynamic inversion method. The system is capable to utilize the high bandwidth of the system and full actuator range without instability caused by actuator saturation. It can thus be used as a solid basis to apply augmented control philosophies, e.g. adaptive control. In addition to the functional requirement, the control algorithm has to be implemented in the embedded hardware and has to fulfill real-time requirements while limited memory and onboard processing capacity have to be considered. Detailed derivation of the control law design is illustrated, followed by the flight test data to verify the design. For the flight test, vision sensors are used to aid the inertial sensors embedded onboard. Details about the visual processing and the experimental setup can be found in [2]. Specific sensor data fusion, trajectory planning and protection mechanisms for takeoff and landing are integrated for the flight test.

2 Dynamics of the Quadrocopter

One well-known inherent quadrocopter characteristics is the strong coupling in pitch-yaw-roll. A tradeoff often has to be made between maximization of the system bandwidth and dynamic decoupling. Both problems can be confronted in an elegant way if we look at the reaction of the quadrocopter on changes of rotational rates of the propellers as follows:

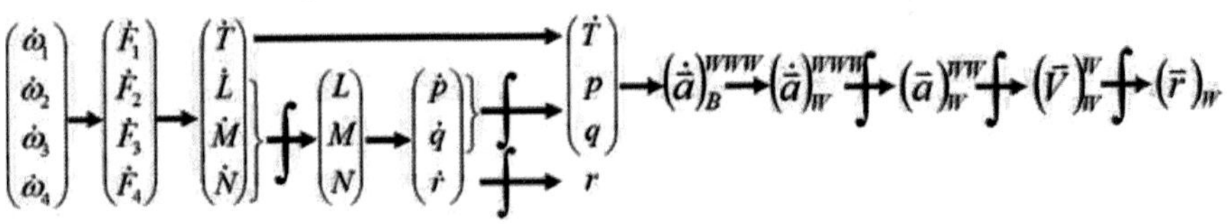

Fig. 2 Signal flow diagram of the quadrocopter dynamics

As shown in Fig. 2, a change in the rotational rate of each propeller results in changes in the thrust of each motor, which gives the change of total thrust and the body-fixed rotational moments. These moments result in body fixed angular accelerations, and consequently the angular rates. Then the angular rates p, q and the change in the total thrust yield a change in the acceleration with respect to the world frame. The change in acceleration in the end results in change of the position by three integrations.

The resulting change in acceleration in the body frame consists of 2 integrations in the x_B-y_B-plane, whereas in the direction of the z_B-axis acceleration directly results from the change in thrust. Different axes of the position dynamics are coupled with different dynamic orders. It is difficult to perform exact input-output feedback from position to the motor thrusts. A dynamic inversion of the position dynamics with relative degree 3 however is possible.

As can be seen in the signal flow diagram in Fig. 2, attitude angles are not included, but only implicitly appear in the transformation matrix. The inputs for the translation dynamics extend to change of acceleration, i.e. angular rate and change of thrust. In other words, position can be controlled by a more direct input, which certainly increases the bandwidth of the position control. Thus the dynamic inversion of the position dynamics with relative degree 3 is dynamically correct, and the complexity of the dynamic inversion is manageable with potential increase in the bandwidth.

For the yaw control, there will be coupling from position dynamics if the azimuth angle is used as the control variable. From Fig. 2, the yaw dynamics, however, is not inherently coupled with the translational dynamics. Hence a simple way to solve the problem is to control the yaw rate directly, but with an integral part in the error controller. Considering the drift rate of the gyro (200º /hour), the heading can be well controlled, without coupling with the translational dynamics.

In summary, the new control structure is an outer loop position control of relative degree 3 and an inner loop rate control of relative degree 1. In the next chapter, the mathematic derivation for the dynamic inversion is explained in detail.

3 Dynamic Inversion

Dynamic inversion is an approach where feedback linearization is applied to the outputs of interests. It addresses the problem of controller design by transforming a nonlinear system to a linear one by feedback. The transformed plant, as an equivalent linear system, may be analyzed by all means of linear system and control theory. [5] [6]

3.1 Inner Loop – Rotational Dynamics

The inner loop commands angular rates and generates moment commands $\left(\vec{M}_{des}^{G}\right)_B$ for the control allocation. The rotational dynamics are well known. By neglecting the aerodynamic moment in the moment dynamics, the desired moment command is directly obtained by,

$$\left(\vec{M}_{des}^{G}\right)_B = I_{BB}^{G} \cdot \left(\dot{\vec{\omega}}^{OB}\right)_B + \left(\vec{\omega}^{OB}\right)_B \times I_{BB}^{G} \cdot \left(\vec{\omega}^{OB}\right)_B \tag{1}$$

3.2 Outer Loop-Translation Dynamics

The command to the outer loop is the desired position in the W-frame, from which it generates an angular rate command that is issued to the inner loop. Starting from Newton's 2nd law, we can derive the translational equation of motion by assuming that the W-frame can be used as inertial frame,

$$m \cdot (\vec{a})_W^{WW} = \left(\vec{F}^{G}\right)_W + \left(\vec{F}_{Grav}^{G}\right)_W + \left(\vec{F}_{Aero}^{G}\right)_W \tag{2}$$

For a quadrocopter the aerodynamic lift and side force are negligible and the aerodynamic drag coefficient can be assumed to be constant for simplicity. With these assumptions, equation (2) can be rewritten as,

$$(\vec{a})_W^{WW} = M_{WB}\vec{f}_B + \vec{\gamma}_W - d_W = M_{WB}\begin{bmatrix} 0 \\ 0 \\ -T/m \end{bmatrix} + \begin{bmatrix} 0 \\ 0 \\ g \end{bmatrix} - d_W \tag{3}$$

Where $d_W = \frac{1}{2m} C_D \rho S \left\| (\vec{V})_W^W \right\| \cdot (\vec{V})_W^W$. As the inner loop inputs have not appeared, Eqn. (3) is differentiated

$$(\dot{\vec{a}})_W^{WWW} = \dot{M}_{WB}\vec{f}_B + M_{WB}\dot{\vec{f}}_B - \dot{d}_W \tag{4}$$

Where $\dot{d}_W = \frac{1}{2m} C_D \rho S \left[\frac{\left[(\vec{V})_W^W\right]^T \cdot (\vec{a})_W^{WW}}{\left\|(\vec{V})_W^W\right\|} \cdot (\vec{V})_W^W + \left\|(\vec{V})_W^W\right\| \cdot (\vec{a})_W^{WW} \right]$ and the Euler differentiation rule can be used,

$$\dot{M}_{WB} = M_{WB} \cdot \Omega^{WB} \text{, and } \Omega^{WB} = skew\left(\vec{\omega}^{WB}\right)_B \tag{5}$$

Hence Eqn. (4) becomes

$$(\dot{\vec{a}})_W^{WWW} = M_{WB}\begin{bmatrix} -T/m \cdot q \\ T/m \cdot p \\ -\dot{T}/m \end{bmatrix} - \dot{d}_W \tag{6}$$

Now the angular rates explicitly appear in the equation. The first order time derivative of acceleration is the third order time derivative of the position. So the relative degree for position dynamics is three as expected. By the zeros in the first two rows of the specific force vector $\vec{f}_B$, the yaw rate is algebraically cancelled out.

The coupled dynamics can be inverted analytically using Eqn. (6). Here we can replace the 3rd order time derivative of the position with the pseudo control $\nu = (\dot{\vec{a}})_W^{WWW}$

$$\begin{bmatrix} q \\ p \\ \dot{T} \end{bmatrix} = m\begin{bmatrix} -1/T & 0 & 0 \\ 0 & 1/T & 0 \\ 0 & 0 & -1 \end{bmatrix} M_{BW} \cdot \left(\nu + \dot{d}_W\right) \tag{7}$$

A distinguished advantage here is that, the attitude of the quadrocopter (be it Euler angle or Quaternion) is not a controlled state. This leads to the dynamic inversion equation of such a simple form and does not result in singularities usually caused by the attitude. The only singularity in the above inversion is thrust = 0, which can be easily remedied in the implementation. Theoretically this control law is capable of flying loops, inverted fast descending, etc.

The overview of the outer loop design is illustrated in Fig. 3,

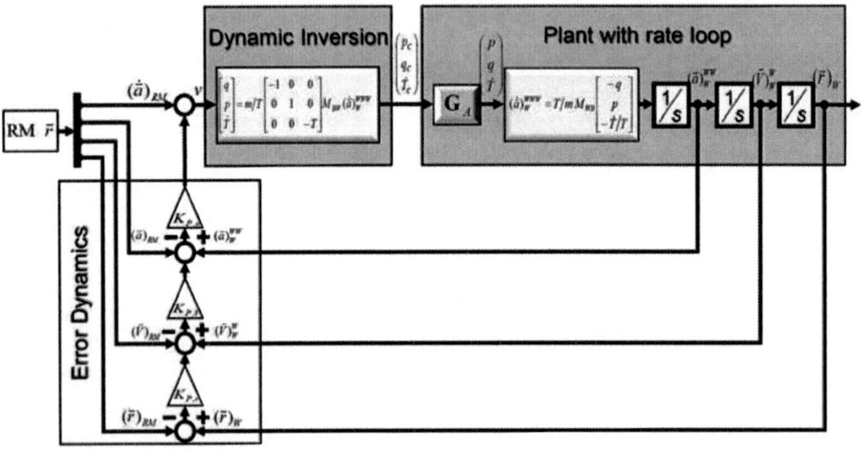

Fig. 3 Overview of the position controller

4 Implementation of the Two-Loop Design

The controller was implemented on an ARM 7 processor, using the Simulink Quadrocopter Framework [7]. Critical parts of the control system like the inner loop, control allocation and data fusion are running with an update rate of 1 kHz [2]. With such a fast update rate, control deviation due to the dynamic inversion error can be quickly compensated.

4.1 Quadrocopter Specifics

The structure of the nonlinear dynamics of the quadrocopter is well known but some of the parameters have to be measured, which are listed in Table 1 below.

'l' is the arm length between the motor and center of mass. k_n and k_m are motor specific parameters and M_{prop} is the yaw moment generated by the propellers , $F_{prop} = k_n \cdot n^2$, $M_{prop} = k_m \cdot F_{prop}$ [8].

Table 1 Quadrocopter Specific Parameters

Mass (kg)	0.68	$I_{xx} (= I_{yy})$	0.007	$k_n (10^{-8}$ N/rpm$^2)$	5.7
l(m)	0.17	I_{zz}	0.012	k_m(m)	0.016

4.2 Actuator Saturations and Sensor Limitation

The thrust per motor is in the range of 0.05 N to 3.5 N [8]. Thus the moment generated can be calculated easily. To control the moment while keeping the total thrust unchanged, the thrust difference per motor have to be symmetric. Thus the control range of each motor around the hover point is about ±1.6N. The maximum moment and angular acceleration for each axis are calculated and listed in Table 2.

Table 2 Actuator Saturations

$M_x(=M_y)$(Nm)	±0.544	M_z(Nm)	±0.102
$\dot{p}(=\dot{q})$(rad/s^2)	±77.7	$\dot{r}$(rad/s^2)	±8.5

For sensors, the Gyro is limited to a maximum value of 300°/s for the angular rates. The accelerometer measures accelerations with a full-scale range of ±1.5g, while the operation range is ±3g. The visual tracking system [2] is using a model based tracking algorithm with a stereo camera setup consisting of two standard webcams. The tracking system has an update rate of 25Hz and a speed limitation of about 1m /s due to motion blur in the images. The accuracy level varies with the lighting situation, the quadrocopter attitude and speed.

4.3 Reference Model

The reference model, or command filter is designed to generate smooth trajectories which is physically possible for the vehicle to fly. System relative degree, actuator dynamics, as well as actuator saturations and sensor limitations are taken into account.

4.3.1 Inner Loop Reference Model

The moment dynamics have angular rates as outputs and moments as inputs (see Eqn. 1). These dynamics have only relative degree one. However, in order to explicitly account for motor dynamics, a second order reference model is used instead of a first order reference model. Hence the second order time derivative can be limited. The additional pole can be placed with a small time constant: 0.002s.

In the experiments [8], the time constants of the propulsion dynamics are found to be 1/80s for increasing thrust and 1/40s for decreasing thrust. A simulation model has been constructed to assess and maximize the bandwidth of the inner loop, which determines the time constant of the slower pole.

4.3.2 Outer Loop Reference Model

For the outer loop, there are three integrations between the inputs (angular rates and the change of Thrust) and outputs (position in W frame). Hence a third order reference model is used. The time scale separation has to be considered in the outer loop. Compared with the attitude control inner loop shown in Fig. 1, the designed rate control inner loop allows smaller time scale separation, i.e. higher bandwidth, for the position control outer loop. In flight test, the increase in the bandwidth is not so distinct, as the major limitation comes from the vision system

[2]. Nevertheless, the optimal Eigenvalues of the outer loop designed in [2], where attitude control was used as inner loop, were at '-4', while the optimal Eigenvalues of the outer loop are now assigned at '-5'. The differential equation of the 3rd order reference model is shown below, with ω_0=5.

$$\dddot{y}_R + 3\omega_0 \ddot{y}_R + 3\omega_0^2 \dot{y}_R + \omega_0^3 \cdot y_R = \omega_0^3 \cdot y_c \tag{8}$$

4.4 Error Controller

With exact dynamic inversion the outputs would exactly follow the reference trajectories. Due to the model uncertainties such as parameter error and sensor error, the 3rd order time derivatives of the vehicle position differ from the dedicated pseudo control. The effect is propagated through the integration and leads to a difference between plant outputs and reference signals. The error controller uses feedback of the state errors, augmented by an integral error control to achieve steady state accuracy. [9].

For the outer loop, the error dynamics is expressed by Eqn. (9), where $e = y_{ref} - y$

$$v = \dddot{y}_{ref} + K_{d2} \cdot \ddot{e} + K_{d2} \cdot K_d \cdot \dot{e} + K_{d2} \cdot K_d \cdot K_p \cdot e + K_i \cdot \int e \cdot dt \tag{9}$$

The error dynamics can be adjusted by the coefficients K, to follow the reference dynamics. The integral gain can be determined by pole placement and a small value is chosen to ensure steady state accuracy while not affecting the performance.

$$K_{d2} = 3\omega_0,\ K_d = \omega_0,\ K_p = \frac{\omega_0}{3} \tag{10}$$

4.5 Control Allocation

For the quadrocopter, the relationship between the forces & moments and the propulsion controls are invertible. The desired force and moment commands are denoted with the subscript 'des'.

$$\begin{pmatrix} L_{des} \\ M_{des} \\ N_{des} \\ T \end{pmatrix} = \begin{bmatrix} 0 & -l & 0 & l \\ l & 0 & -l & 0 \\ k_m & -k_m & k_m & -k_m \\ 1 & 1 & 1 & 1 \end{bmatrix} \cdot \begin{pmatrix} F_1 \\ F_2 \\ F_3 \\ F_4 \end{pmatrix} \tag{11}$$

4.6 Pseudo Control Hedging

Pseudo Control Hedging (PCH) is implemented to 'hide' the actuator dynamics from the error dynamics [10]. The expected reaction of the plant $\hat{v}$ is calculated using Eqn. (6), in which the angular rates are taken from the Gyro measurements, the thrust and change of thrust are estimated by means of the modeled actuator dynamics $\hat{G}_A\left(\dot{T},T\right)$, as they are not measureable.

$$\hat{\nu} = \hat{F}\left(p, q, \hat{G}_A\left(\dot{T},T\right)\right) \quad (12)$$

Then the hedging signal or the expected reaction deficit can be calculated by $\nu_h = \nu - \hat{\nu}$.

The dynamics of the reference model is decelerated by the expected reaction deficit. Another function of PCH is to prevent the integrator to wind up in the error dynamics [10]. This can be demonstrated in experiment (Fig. 4(e)) by introducing an external displacement disturbance to the hovering quadrocopter.

4.7 Sensor Data Fusion

To achieve high bandwidth control, fast data fusion is important. There is not enough computation power on the ARM 7 processor for a standard full state Kalman Filter with fast update rate. Instead, a model based Kalman state estimator is implemented to fuse the sensor data. The dynamic nature and system orders are taken into account in the filter, i.e. third order dynamics in x- and y-axes and second order dynamic in z- axis. Hence the inputs are time derivatives of the accelerations for x- and y- axes calculated using Eqn. (6) and acceleration in z axis obtained from the accelerometer. In total there are 9 states: 3 axes position, 3 axes velocities, and accelerations in x- and y- axes and acceleration bias in z axis. Acceleration bias in x- and y- axes is unobservable as an AHRS (Attitude Heading Reference System) filter is used to estimate the quaternions. There are five measurements, three positions obtained from the visual tracking system and two accelerations in x-y axes from the accelerometer. The constant Kalman gain matrix (9x5) is calculated as the optimal Kalman gain for the system offline, given the process noise and measurement covariance [11].

The vision system uses two standard web cameras to track the quadrocopter, based on edge matching algorithms [2]. The measured output is the position vector in the World frame at approximately 25Hz with variable accuracy (2cm -10cm). The latency of the vision system is about 100ms and it is taken into account by adjusting the position measurement with velocity compensation.

The most important advantage of the state estimator is that it is running onboard with an update rate of 1 kHz. Therefore, disturbances detected by the IMU can be compensated within milliseconds. The filter also significantly reduces the noise originated from the thermal accelerometer measurements [13]. Overall, the filter shows very fast reaction and little noise.

5 Experimental Results

In order to show the performance of the new controller, different trajectories like circle, infinity sign, and step commands have been tested with data recorded

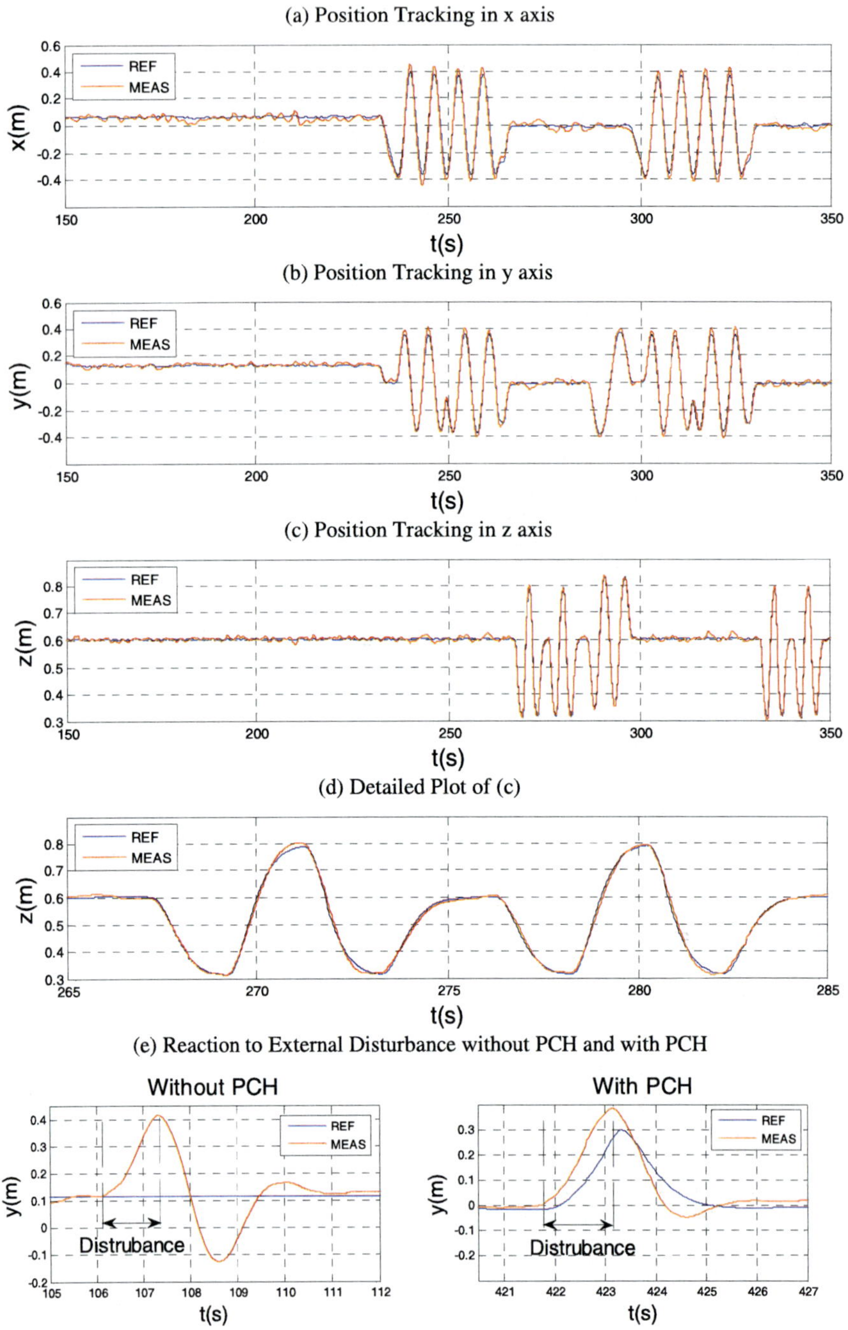

Fig. 4 Experimental Results of Trajectory Flights

online. To show the accuracy of the system, the reference commands and the vision sensor measurement are plotted. It can be seen in Fig. 4 (a-d) that the position tracking is nearly perfect: the control errors are relatively small (within 5cm) esp. in the z axis. The inter-axis coupling is negligible. The robustness and performance of the system has been demonstrated in two international trade fairs, 'Embedded World 2010' and 'ELECTRONICA 2010' held in Germany, where it was flying approximately 8 hours every day during the fairs.

The effect of PCH can be seen in (e) and (f) of Fig. 4. Big overshot caused by the integrator wind-up can be compensated by the hedging signals that change the reference signals in case of actuator saturations.

6 Conclusion

With the current position controller and the vision sensor, the quadrocopter is able to fly with good accuracy and a comparatively higher bandwidth. Based on the available structure, new control theory and application, like adaptive control and advanced data fusion are of interest to the author and to be developed.

Acknowledgments

The authors gratefully acknowledge the support by the International Graduate School of Science and Engineering (IGSSE), TU-München, project 4.03 Image Aided Flight Control.

References

1. Holzapfel, F., Sachs, G.: Dynamic Inversion Based Control Concept with Application to an Unmanned Aerial Vehicle. In: AIAA Guidance, Navigation and Control Conference and Exhibit: AIAA-2004-4907 (2004)
2. Klose, S., Wang, J., et al.: Markerless Vision Assisted Flight Control of a Quadrocopter. In: IEEE 2010 RSJ International Conference on Intelligent Robots and Systems (2010)
3. Voos, H.: Nonlinear Control of a Quadrotor Micro-UAV using Feedback-Linearization. In: Proc. of IEEE International Conference on Mechatronics (2009)
4. Bouabdallah, S., Siegwart, R.: Backstepping and Sliding-mode Techniques Applied to an Indoor Micro Quadrotor. In: Proc. of the IEEE International Conference on Robotics and Automation (2005)
5. Marquez, H.: Feedback linearization. In: Nonlinear Control Systems – Analysis and Design. Wiley, Canada (2003)
6. Khalil, H.: Feedback linearization. In: Nonlinear Systems, 3rd edn. Prentice Hall, Englewood Cliffs (2002)
7. Achtelik, M.: Simulink Quadrocopter Framework. Semesterarbeit, Technische Universität München (2009)
8. Achtelik, M.: Nonlinear and Adaptive Control of a Quadcopter, Diplomarbeit, Technische Universität München (2010)

9. Holzapfel, F.: Nichtlineare adaptive Regelung eines unbemannten Fluggerätes. PhD thesis, Technische Universität München (2004)
10. Johnson, E.: Limited Authority Adaptive Flight Control. PhD thesis, Georgia Institute of Technology (2000)
11. Franklin, F., Powell, D., et al.: Digital Control of Dynamic System, 2nd edn. Addison-Wesley, Reading (1990)
12. Ascending technologies GmbH, Hummingbird Autopilot (2010), `http://www.asctec.de` (accessed July 27, 2010)
13. MEMSIC, Inc., Thermal Accelerometer (August 27, 2010), `http://www.memsic.com`

Parallel Implementation of Constrained Nonlinear Model Predictive Controller for an FPGA-Based Onboard Flight Computer

Alexander Joos and Walter Fichter

Abstract. Model Predictive Control (MPC) is an established control method in various application areas. Its ability of taking constraints into account makes it interesting also for automatic flight control. However, the computational complexity of MPC schemes usually limits its application. This paper describes a simple formulation of a constrained nonlinear MPC (NMPC) approach that can be realized on small onboard computers based on Field Programmable Gate Arrays (FPGAs). In contrary to classical implementations of MPCs a computationally expensive optimization problem can be avoided while even nonlinear prediction models and constraints can be considered. This is accomplished through parallel time-domain simulations. To this end, the parallel implementation properties of FPGAs are exploited. The 3d-kinematics is proposed as prediction model for the NMPC to plan the aircraft state trajectory (position and attitude) taking constraints and obstacles into account. Simulation results with a nonlinear 6 degree of freedom simulation model verify the functionality. Feasibility of hardware synthesis of parallel predicted models for the NMPC approach on an FPGA is shown by analysis.

1 Introduction

Flight safety is an important topic for manned aircraft as well as for unmanned aerial vehicles (UAVs). Thus there is a need for automatic aircraft control algorithms that are able to cope with constraints in the aircraft states, inputs and with obstacles incorporating nonlinear dynamics of an aircraft. One possible approach to solve such problems is Nonlinear Model Predictive Control (NMPC)

Alexander Joos
Ph.D. Candidate, Institute of Flight Mechanics and Control, Pfaffenwaldring 7a, 70569 Stuttgart
e-mail: alexander.joos@ifr.uni-stuttgart.de

Walter Fichter
Professor, Institute of Flight Mechanics and Control, Pfaffenwaldring 7a, 70569 Stuttgart
e-mail: fichter@ifr.uni-stuttgart.de

with constraints. NMPCs have the ability to generate optimal inputs for nonlinear plants that can not sufficiently be controlled by MPCs with linear prediction models under the consideration of constraints.

The basic idea behind NMPC is to solve repeatedly an optimal control problem with finite horizon. This means that an optimal control input is to be found that reduces a cost function in which deviations from a commanded state are weighted while considering constraints of the inputs and of the plant. This powerful control algorithm is therefore a good candidate to control unmanned aerial vehicles (UAVs) while considering nonlinear aspects of the plant and constraints on inputs and the states. On the other hand NMPC requires strong computing power to solve nonlinear constrained optimization problems in real-time, which makes this approach hard to implement on small computers as they are usually used as onboard flight control computers for small UAVs.

In literature, some references can be found that cope with the implementation of MPCs on small computers with moderate computing power. Reference [1] demonstrates the implementation of a constrained MPC on an FPGA that uses a linear prediction model. For optimization the interior point method with dense matrix formulation is used to solve the quadratic programming problem. A linear aircraft model with 4 states and 1 input is used to demonstrate the control of the altitude by the use of the elevator. The ability of the FPGA to compute code in parallel is not used. Parallel implementation on an FPGA is used in Ref. [2], but it is restricted to linear systems with constraints. In Ref. [3] an implementation of an explicit MPC is described in which the optimization problem is precomputed and the real-time control problem is simply reduced to the evaluation of a piecewise linear function. The limitation of this approach is that the memory requirements increase rapidly with an increase of the problems dimensions. An algorithm specific processor for embedded unconstrained MPC is proposed in Ref. [4]. The prediction model in the demonstrated application is a linear state space model. For solving the optimization problem Newton's Method is used. Matrix operations are computed in an auxiliary unit that acts as a matrix coprocessor. Results are presented from controlling the linear model of a rotating antenna with 2 states and 1 input, constraints are not considered. In Ref. [5] the application to a nonlinear glucose regulation control problem with constraints on the input is demonstrated by HIL simulations. The nonlinear model is linearized for several intervals to generate multiple linear state space matrices for use in the MPC algorithm and discretized with a sample rate of 5 min, which is far too large for flight applications. The application of a nonlinear MPC for high level control for a fixed wing UAV is described in Ref. [6]. The control problem uses an error dynamics model with 2 states and 1 input that bases on the 2d-kinematics in horizontal plane. The input is constrained. The high level control is implemented in a PC104 onboard flight computer and tested in HIL simulations as well as with MATLAB simulations. Although the PC104 is a relatively powerful onboard computer (comparable to a desktop PC) and the prediction model is of small dimension, 4 Hz update rate is the limit that is reported in this reference.

In summary, online computed MPCs on small computers are restricted to low-dimensional problems with linear prediction models. Then, the optimization problem for linear MPC is the solution of a quadratic program that can be solved efficiently even online [7, 8]. In case of NMPC applications [6] powerful and relatively large and heavy computers are used together with models of low complexity (2d-kinematics). Thus, it is not applicable for small fixed wing aircraft.

Reference [9] describes an optimization method based on brute force method for NMPC that does not suffer from complex optimization. In general this optimization method is also computationally very expensive since numerous prediction models representing the plant have to be propagated with numerous input candidates in each control step. No implementations on small flight control computers for UAV application are reported in literature.

In this paper, the design of a constrained NMPC based on optimization with time-domain simulations is presented. It is shown that a UAV state trajectory control problem can be solved even with a low number of models that have to be predicted. Furthermore a partially parallel implementation of the prediction models is proposed to achieve feasibility.

The paper is structured as follows. After a review of classical MPC formulations in chapter 2, a parallel implementable approach is introduced. As prediction model for high level state trajectory planning for the aircraft the 3d-kinematics model is proposed. A method to refine the commands of the NMPC without additional computational effort follows. Simulation results of a nonlinear 6 degrees of freedom (6-DoF) simulation model controlled by a constrained NMPC with the nonlinear 3d-kinematics prediction model are demonstrated in applications with collision avoidance and a kind of unconventional landing scenario. Both scenarios require a state trajectory planning by the NMPC. The last chapter presents results from hardware synthesis of parallel prediction models for an FPGA.

The first contribution of this paper is a feasibility proof of a NMPC based on time-domain simulation optimization, applied to a UAV state trajectory planning problem even with a low number of prediction models. The computationally most expensive part of this kind of constrained NMPC formulation is the real-time model prediction. Therefore, a second contribution is a parallel implementation scheme of model prediction for a constrained NMPC on a FPGA-based onboard flight control computer as it is presented e.g. in Ref. [10]. This enables the application of constrained NMPC to small flight vehicles.

2 Classic Nonlinear Model Predictive Control Formulation

The optimization problem for a NMPC can be described by

$$\min_{\overline{\mathbf{u}}} J(\overline{\mathbf{u}}) = \min_{\overline{\mathbf{u}}} \int_{t_0}^{t_0+T_p} F(\overline{\mathbf{x}}(\tau), \overline{\mathbf{u}}(\tau)) d\tau \qquad (1)$$

subject to

$$\dot{\overline{\mathbf{x}}} = \mathbf{f}(\overline{\mathbf{x}}(\tau), \overline{\mathbf{u}}(\tau)), \quad \overline{\mathbf{x}}_0 = \mathbf{x}(t_0) \tag{2}$$

and with equality and inequality constraints $\mathbf{g}_1$ and $\mathbf{g}_2$:

$$\mathbf{g}_1(\overline{\mathbf{x}}(\tau), \overline{\mathbf{u}}(\tau)) = 0, \quad \mathbf{g}_2(\overline{\mathbf{x}}(\tau), \overline{\mathbf{u}}(\tau)) \geq 0 \qquad \forall \tau \in [t_0, t_0 + T_p] \tag{3}$$

$\overline{\mathbf{x}}$ and $\overline{\mathbf{u}}$ are predicted states and inputs respectively, $J(\overline{\mathbf{u}})$ is the cost function that shall be minimized with respect to $\overline{\mathbf{u}}$, and T_p is the prediction horizon.

NMPC requires a computationally expensive online solution for a nonlinear optimization problem [7]. Reference [7] names three approaches from which only direct solutions using a finite parameterization of the controls and/or constraints are normally applicable for on-line application. A direct solution for this problem can be achieved by parameterization of the controls at every predicted sampling interval and holding them constant for that interval [7]. To calculate the cost function the systems dynamics described by equation 2 are solved by numerical integration. Equality and inequality constraints are evaluated with the predicted inputs and states as well. The resulting optimization problem can then be solved with sequential quadratic programming [11]. The matrices for this approach are often dense, which makes the solution with this approach computationally expensive. The "simultaneous approach" in Ref. [7] can fail when the optimization cannot be completed in time. In this case, feasibility is not guaranteed with this approach.

All of the above methods have the disadvantages of being computationally extensive or even unsolvable on computers with low computational power. Therefore in the following section, a simpler approach to the solution and on-line implementation of NMPC problems is presented.

3 NMPC Approach with Parallel Implementation

3.1 General Description of the Proposed NMPC Approach

The idea is to propagate a nonlinear simulation model with a certain set of combinations of control input candidates [9] that are quantized in amplitude and discretized in time and to evaluate the resulting state vectors and inputs in a cost function. In this paper, to save computational cost, the prediction of the models is proposed to be done partially in parallel on an FPGA which is especially suited for this task. The optimization problem is therefore reduced to finding the cost function with the smallest value and taking the corresponding input of the first propagation step as the optimal control input.

The parameters of two NMPC setups demonstrated in this paper are listed in Table 1. To introduce the idea, NMPC 1 will be considered in the following. Later in this document results will be shown also with NMPC 2.

Table 1 Parameters of the proposed NMPC used for simulation.

Parameters of the proposed approach	NMPC 1	NMPC 2
Prediction Horizon	$T_p = 7$s	$T_p = 7$s
Number of variable control inputs	$n = 2\ (p_c, q_c)$	$n = 2\ (p_c, q_c)$
Time intervals with constant control input	$m = 2$ (3s, 4s)	$m = 2$ (3s, 4s)
Number of discrete candidates per control input	$d = 3$	$d = 9$
Discrete candidates for each control input in the first time interval with constant control input (discrete candidates in the second time interval are chosen ≈ 15% smaller)	$p_c = (j_p\text{-}1)\cdot 0.1$ rad/s; $j_p = 0, 1, 2$ $q_c = (j_q\text{-}0.8)\cdot 0.07$ rad/s; $j_q = 0, 1, 2$	$p_c = (j_p\text{-}4)\cdot 0.03$ rad/s; $j_p = 0,\ldots,8$ $q_c = (j_q\text{-}3.9)\cdot 0.02$ rad/s; $j_q = 0,\ldots,8$
Prediction steps in each of the m time intervals	$h_1 = 3, h_2 = 4$	$h_1 = 3, h_2 = 4$
Number of predicted models	d^n^1+d^n^m = 90	d^n^1+d^n^m = 6642
Number of resulting predicted state vectors	d^n^m = 81	d^n^m = 6561
Number of cost functions to evaluate	d^n^m = 81	d^n^m = 6561
Number of model prediction steps in total	$h_1\cdot$ d^n^1+ $h_2\cdot$ d^n^m = 351	$h_1\cdot$d^n^1+ h_2d^n^m = 26487
Constraints in attitude	$\lvert\phi\rvert < 30°, \lvert\theta\rvert < 15°$	$\lvert\phi\rvert < 30°, \lvert\theta\rvert < 15°$
Sample rate of the NMPC controller	10 Hz	10 Hz

The prediction model of the NMPC is the nonlinear 3 dimensional kinematics model with prediction starting condition $\bar{\mathbf{x}}_0 = \mathbf{x}(t_0)$ where $\mathbf{x}(t_0)$ is the actual state (position and attitude) of the aircraft ($s = \sin$, $c = \cos$):

$$\begin{pmatrix} \dot{x} \\ \dot{y} \\ \dot{z} \\ \dot{\phi} \\ \dot{\theta} \\ \dot{\psi} \end{pmatrix} = \begin{bmatrix} c\theta c\psi & -c\phi s\psi + s\phi s\theta c\psi & s\phi s\psi + c\phi s\theta c\psi & 0 & 0 & 0 \\ c\theta s\psi & c\phi c\psi + s\phi s\theta s\psi & -s\phi c\psi + c\phi s\theta s\psi & 0 & 0 & 0 \\ -s\theta & s\phi c\theta & c\phi c\theta & 0 & 0 & 0 \\ 0 & 0 & 0 & 1 & \dfrac{s\phi s\theta}{c\theta} & \dfrac{c\phi s\theta}{c\theta} \\ 0 & 0 & 0 & 0 & c\phi & -s\phi \\ 0 & 0 & 0 & 0 & \dfrac{s\phi}{c\theta} & \dfrac{c\phi}{c\theta} \end{bmatrix} \begin{pmatrix} u \\ v \\ w \\ p_c \\ q_c \\ r \end{pmatrix} \quad (4)$$

Inputs to the prediction model $\bar{\mathbf{u}}$ are bodyfixed velocities $(u,v,w)^T$ and turn rates $(p,q,r)^T$, the predicted states $\bar{\mathbf{x}}$ are euler angles $(\phi,\theta,\psi)^T$ to describe the attitude of the aircraft, and earthfixed "north-east-down" position $(x,y,z)^T$. Inputs u, v, w, r are chosen as constants in the demonstrated results in this paper. Each of the $n = 2$ variable inputs (roll and pitch rate) can take $d = 3$ discrete candidate values $(p_c, q_c)^T$ per time interval with constant discrete value (NMPC 1).

The number of predicted models over a first time interval is $d^n = 9$, each with one of nine constant combinations of input values p_c and q_c. The resulting nine state vectors of these nine predicted models for the first time interval are used as start conditions for another time period in which again the input values can take $d^n = 9$ combinations of discrete values. In the second time period all nine combinations of input values are used with all nine resulting predicted state vectors of the first time period. This leads to $d^{n \cdot m} = 81$ predicted state vectors after the second time interval. Figure 1a) shows the predicted positions of the 81 predicted state vectors as they are calculated in one single NMPC update. Figure 1b) shows the NMPC update with 1 Hz NMPC update rate. In the simulation results in this paper the NMPC update rate is 10 Hz, so in the results shown later in this paper there are 10 times more predicted state vectors than displayed in Fig. 1b).

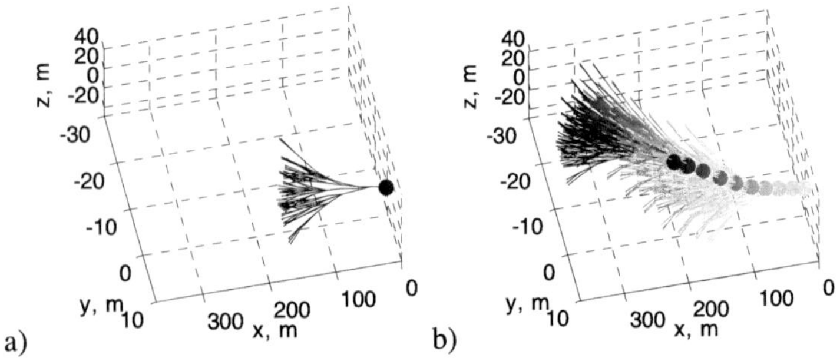

Fig. 1 a) Predicted positions in an example with 81 predicted state vectors (position and attitude) according to Table 1, NMPC 1. b) Prediction trees calculated with an NMPC update rate of 1 Hz.

A cost function J is evaluated with all $d^{n \cdot m}$ resulting state vectors and corresponding inputs:

$$J_b = \sum_{i=t_0}^{i=t_0+T_p} F(\bar{\mathbf{x}}_b(i), \bar{\mathbf{u}}_b(i)) + c_{\phi b}(i) + c_{\theta b}(i) + c_{Pb}(i) + \frac{k(i)}{dist_b(i)} \tag{5}$$

$$b \in \mathrm{N} \wedge 1 \leq b \leq d^{n \cdot m}$$

where $i = t_0, t_0+\delta t, \ldots, t_0+T_p$ and T_p is the prediction horizon, δt is the length of a time interval with constant value of control input in the numerical integration of the dynamic prediction model. $\mathbf{x}_{\text{ref}}$ is a commanded final state for the aircraft (position and attitude). To reach a commanded point with a commanded attitude, deviations of the predicted state (position and attitude) from $\mathbf{x}_{\text{ref}}$ and predicted inputs are weighted in F. Approaching the reference point from a certain direction is fulfilled by appropriate weights for x- and y- direction and a weight for a deviation between the actual heading of the aircraft and the direction to the commanded final position. Constraints c_ϕ, c_θ, c_P for roll angle, pitch angle and position are considered as barrier functions:

$$c_j = \begin{cases} k_j > 0 & \text{if constraint violated} \\ 0 & \text{else} \end{cases} \qquad j \in \{\phi, \theta, P\} \tag{6}$$

Constraint c_P is used e.g. for obstacle avoidance and positions below ground level. Avoiding the obstacle is supported by a cost that increases when the distance *dist* between predicted position and the obstacle decreases. The optimization problem in this case is reduced to finding the one cost function $J_{\min}$ with the lowest value out of the 81 resulting cost functions (Eq. 7). The optimal input is the predicted control input of the first time interval that belongs to cost function $J_{\min}$.

$$\min\left\{J_b / b \in \mathrm{N} \wedge 1 \le b \le d^{n \cdot m}\right\} = J_{\min} \tag{7}$$

This input is applied to the plant for one control cycle and the procedure is repeated in every control cycle as it is usually done in MPC approaches (Fig. 1b).

3.2 *Refinement of NMPC Commands*

In this chapter an idea is presented how to generate finer and therefore more precise NMPC commands without increase of computational effort. One possibility is to use more and finer discretized candidates as in NMPC 2 (Table 1) what leads to an increase in computational cost. Another idea is to use the same number of candidates, but finer candidates for p and q in the first of the m time intervals in the NMPC and add these to the actual roll rate $p(t_0)$ and pitch rate $q(t_0)$ of the plant. The model prediction tree (Fig. 1) is therefore being computed starting from the actual position, attitude, roll and pitch rate of the aircraft.

$$\begin{aligned} p_c &= (j_p - 1)\cdot 0.03 + p(t_0) \qquad j_p = 0,1,2 \\ q_c &= (j_q - 0.8)\cdot 0.021 + q(t_0) \quad j_q = 0,1,2 \end{aligned} \tag{8}$$

The presented approach introduced in Chapter 3 reduces the computational effort significantly by parallel implementation of prediction models (this can be done in parallel in each of the m time intervals) and if a low number of discrete input candidates are used. Of course, the choice of number of discrete values each control input can take is a trade-off between computational effort and optimality of the solution. Simulation results will show that even low numbers of input candidates can be used due to the repeated computation of optimal controls within the NMPC scheme with high update rate.

4 Simulations Results for UAV State Trajectory Planning

4.1 *Scenarios and Control Structure*

At first a scenario is presented in which the aircraft starts from an arbitrary position and attitude and shall reach a given setpoint of position and attitude while avoiding an obstacle without violating constraints. This requires some kind of state trajectory planning which is solved in this paper by NMPC. For this scenario, results with high and low numbers of discrete input candidates in the prediction model are demonstrated. In a second scenario a kind of unconventional landing is demonstrated that also requires state trajectory planning due to the initial starting state relative to the commanded final state. In both scenarios no steady state conditions have to be fulfilled and no planned state trajectory is required a priori. The NMPC controlled aircraft has to find a feasible state trajectory to a commanded final state (position and attitude) without violating constraints and in scenario 1 while avoiding an obstacle automatically.

The control structure consists of two loops. An inner Linear Quadratic Regulator (LQR) control loop (low level controller) commands aileron, elevator, rudder and thrust. It is designed as a proportional/integral LQR controller. Since the aircraft has only 4 control inputs, only the 4 aircraft states: bodyfixed rates p, q and the bodyfixed velocities u and v can be controlled exactly by the low level controller. Bodyfixed down-velocity w follows from the pitch stability of the aircraft and varies only slightly around a constant value. Yaw rate r follows from the dynamics and the kinematics of the aircraft, e.g. $r = g/V \sin(\phi)$ in case of a steady state coordinated turn, with g = gravitational acceleration and V = speed of the aircraft [12].

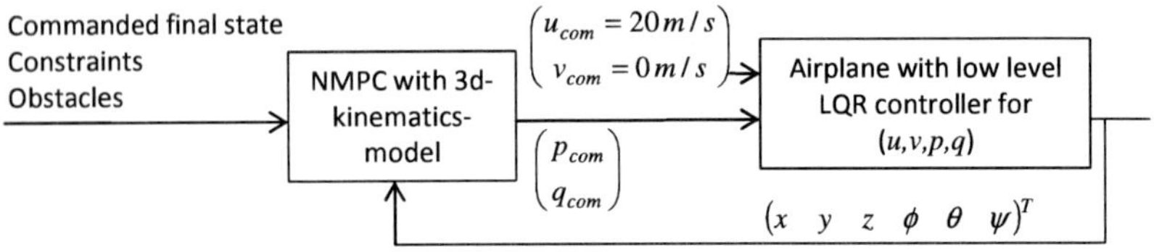

Fig. 2 Control structure of the demonstrated example.

The outer loop (high level controller) is a NMPC controller that generates commands for pitch and roll rate $(p_{com}, q_{com})^T$ according to the methodology introduced in the previous chapter.

Simulations in which the 3d-kinematics in the NMPC was predicted with $r = 0$ rad/s and with $r = g/V \sin(\phi)$ have shown both comparable control results. In the following, results will be shown with prediction input $r = 0$ rad/s. Although the prediction error is higher than with $r = g/V \sin(\phi)$, the control problem can be solved due to the high update rate (10 Hz) of the high level NMPC controller that generates changing commands $(p_{com}, q_{com})^T$ every 0.1s.

With this control structure the 3d-kinematics model can be used as adequate prediction model for the nonlinear MPC as will be demonstrated in the following.

4.2 Simulation Results

In this chapter simulation results are demonstrated with the NMPC approach as proposed in chapter 3.1. For simulation purposes a nonlinear 6-DoF aircraft simulation model is used. The states of the 6-DoF simulation model are bodyfixed velocities $(u,v,w)^T$ and turn rates $(p,q,r)^T$, Euler angles $(\phi,\theta,\psi)^T$ to describe the attitude of the aircraft, and earthfixed "north-east-down" position $(x,y,z)^T$. The aircraft dynamics in the simulation are modeled with aerodynamic force and moment coefficients gathered from USAF Digital DATCOM, gravity forces, control forces due to control surface deflection that are also generated with DATCOM, and propeller forces.

Table 1 lists the parameters of two NMPCs used for the following simulation results. In the first scenario (Fig. 3) the aircraft starts from an initial position $(x = 0\text{m}, y = 0\text{m}, z = -100\text{m})^T$ in earthfixed "north-east-down" coordinates with $\phi = \psi = 0°$, $\theta = 3°$. The commanded final state is $\mathbf{x}_{\text{ref}} = (x = -1000\text{m}, y = -1000\text{m}, z = 0\text{m}, \phi = 0°, \theta = 3°, \psi = \pm 90°)^T$ where $\psi = \pm 90°$ is e.g. the heading of a runway. Figure 3 shows 3d-plots of NMPC 1 and NMPC 2 (Table 1) applied to a collision avoidance problem.

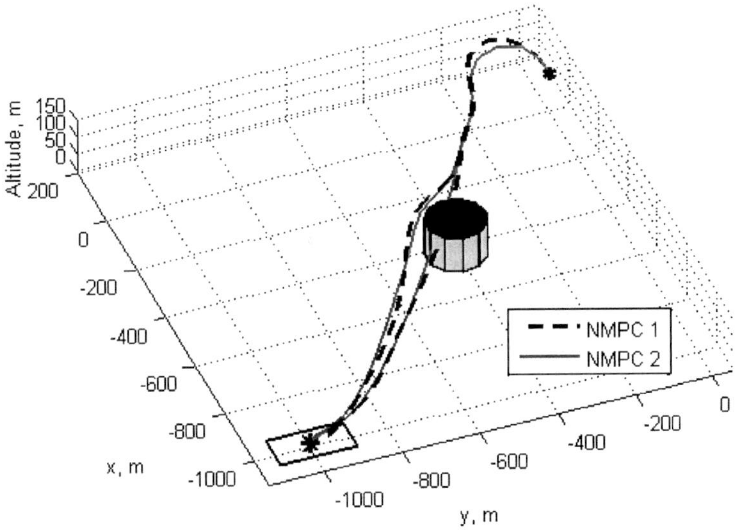

Fig. 3 Scenario 1: 3d-plots of NMPC 1 (81 predicted state vectors/controller update) and NMPC 2 (6561 predicted state vectors/controller update) both with and without collision avoidance.

Throughout the flight, constraints in roll and pitch angle shall not be violated which is fulfilled as can be seen in Fig. 4 for NMPC 1. During the first turn the NMPC controlled aircraft uses the maximal allowed angles in order to change the direction as fast as possible. The obstacle has been placed right in the obstacle-free path (curves that go through the cylinder in Fig. 3). With Fig. 3 it can be stated, that the results of NMPC 1 and NMPC 2 are comparable. The planning and obstacle avoidance problem is solved in either case.

Both controllers prevent a violation of the constraints (results for NMPC 1 in Fig. 4) and both manage to avoid the obstacle (Fig. 3). Remember that NMPC 2 uses a higher number of discrete values for each control input (factor 3, Table 1) and a far higher number of predicted models per controller update than NMPC 1 (factor ≈ 74, Table 1).

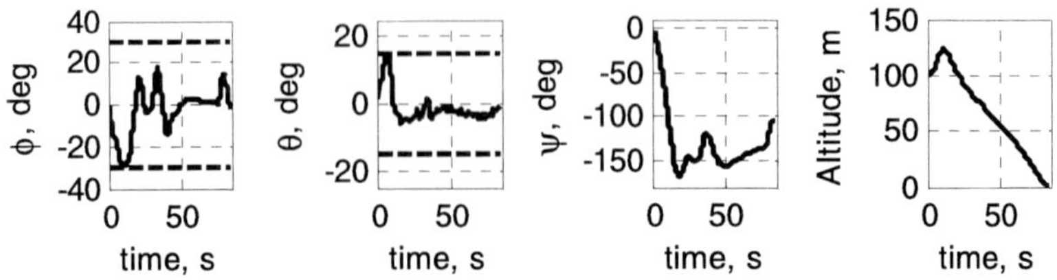

Fig. 4 States of the aircraft during flight according to Fig. 3 with NMPC 1 applied. Constraints are plotted as dashed lines.

The trade-off between optimality and computational expense is won by NMPC 1. This has been observed in numerous simulations in which NMPC 1 has shown that it can solve control problems even with this low number of discrete candidates for each control inputs. The reason for this is the repeated optimization scheme of the NMPC at an update rate of 10 Hz. The prediction tree with the 81 predicted state vectors is calculated every 0.1s and the commands for the turn rates are updated with the same rate. The control commands of the NMPC are displayed in Fig. 5. It can be seen that the commands for roll and pitch rate are well controlled by the low level controller even if the commanded rate by the NMPC changes quickly (e.g. q at about 2.6 s). Due to this fast update in commands from the NMPC it is possible to command even little changes in the attitude.

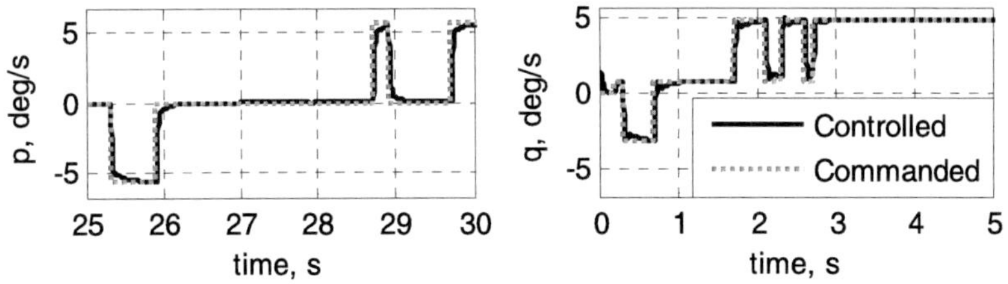

Fig. 5 Roll rate and pitch rate commanded by NMPC 1 and controlled by the LQR low level controller according to the scenario 1 in Fig. 3.

In a second scenario the aircraft starts from an initial position $(x = 0\text{m}, y = 0\text{m}, z = -150\text{m})^T$ with $\phi = \psi = 0°$, $\theta = 3°$. The task is to reach a final state $\mathbf{x}_{\text{ref}} = (x = 10\text{m}, y = 30\text{m}, z = 0\text{m}, \phi = 0°, \theta = 3°, \psi = 0/180°)^T$ which lies below the initial position, while taking constraints $|\phi| < 30°$, $|\theta| < 20°$ into account. With these constraints there is no straight path the aircraft could fly to reach the commanded final state. The NMPC finds a feasible state trajectory to accomplish the reference state $\mathbf{x}_{\text{ref}}$. Of course, since this is an optimization problem with numerous reference states (final position, final attitude) and constraints, not all states can always be controlled exactly in this kind of landing scenarios.

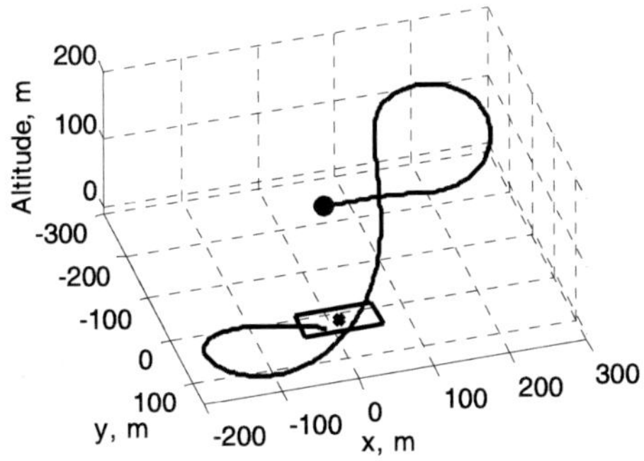

Fig. 6 Scenario 2 with final position right below the initial position.

4.3 *Results for Refined NMPC Commands*

Figure 7 shows the commands of the same scenario as in Fig. 3 but with far smoother commands for p and q than in Fig. 5 (method as described in chapter 3.2, Eq. 8). As in Fig. 1 (NMPC 1, Table 1), 81 resulting state vectors are predicted and evaluated in cost functions every 0.1s. Simulation results are comparable to the results shown in Fig. 3 with less gain in altitude during the first turn and smoother behavior in pitch θ.

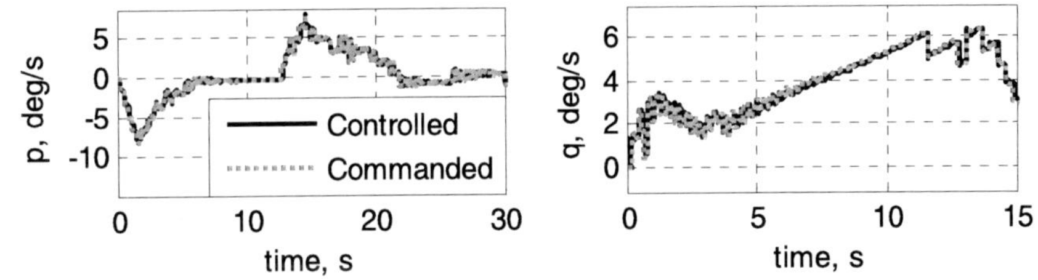

Fig. 7 Refined roll rate and pitch rate commanded by the NMPC and controlled by the LQR low level controller according to a scenario as in Fig. 3.

5 Hardware Synthesis of Parallel Model Prediction for an FPGA

Model prediction is the computationally most expensive part of the presented constrained NMPC approach. To this end, in this section feasibility of parallel model prediction on an FPGA for the constrained NMPC problem is demonstrated with the nonlinear 3d-kinematics prediction model as it is used in the previous chapter. Trigonometric functions are calculated through interpolation of look-up-tables for cosine and its inverse to enable parallelization.

In the following, required resources of an FPGA are reported for several configurations of parallel computed nonlinear 3d-kinematics models (Table 2). The code has been implemented in the DK Design Suite of Mentor Graphics that allows the programming of an FPGA in a C-like language called Handel-C. The presented computational effort results are generated in the DK Design Suite. The FPGA for which the code is built is a Xilinx Spartan 3A DSP.

FPGAs contain configurable logic blocks (CLBs) to synthesize the program code in hardware. Each CLB contains slices and each slice is a group of logic blocks such as look-up-tables (LUTs) and flip-flops (FFs) [13]. To save logic blocks, Arithmetic Logical Units (ALUs) can be used by the compiler e.g. for multiplication. In the following the number of occupied slices will be regarded as indicator for hardware effort of the program code.

It can be stated, that with an increasing number of parallel implemented prediction models, the number of occupied slices on the FPGA increases approximately linearly (compare cases 1, 5, 7 in Table 2). If more than one prediction step per parallel implemented prediction model is calculated (this is done in series on the FPGA) more slices are required (compare case 2 against case 1) and the total computing time increases due to serial computations.

Table 2 Comparison of required resources of the FPGA with different implementations of model prediction.

Case	LUTs	FFs	ALUs	Occupied Slices	Time delay of longest path	Serial model prediction steps of each parallel prediction model	Parallel implemented prediction models	Time intervals with constant control input	Start values for all predicted models
1	6% (2890)	2% (1381)	7% (9)	7% (1878)	23.01 ns	1	9	1	Identical
2	9%	5%	7%	14%	23.01 ns	5	9	1	Identical
3	13%	13%	7%	30%	23.01 ns	2	9	2	Identical
4	28%	21%	7%	52%	23.01 ns	10	9	2	Identical
5	29%	12%	39%	39%	23.01 ns	1	49	1	Identical
6	36%	16%	39%	51%	23.01 ns	1	49	1	Different
7	52%	21%	64%	70%	25.44 ns	1	81	1	Identical

For a simple estimation of maximal number of model prediction steps that can be computed on the FPGA in one second it is assumed that no other code is running on the FPGA, so case 7 can be considered in which 70% of the FPGA-slices are used. Further assumptions are, that each parallel prediction model consists of approximately 100 lines of code and that 81 model prediction steps can be calculated in parallel (case 7 in Table 2). The clock-rate of the FPGA for this estimation is calculated as 1/(time delay of the longest path).

Thus the number of model prediction steps that can be calculated in one second on the FPGA is:

$$\frac{\text{clock rate} \cdot \text{parallel model prediction steps}}{\text{lines of code per model prediction step}} = \frac{\text{model prediction steps}}{s} \tag{9}$$
$$\frac{1}{25.44 \cdot 10^{-9} s} \cdot \frac{1}{100} \cdot 81 \approx 3.2 \cdot 10^{7} \frac{1}{s}$$

Computation time T for 81 parallel model prediction steps with each 100 lines of code is: $T = 100 \cdot 25.44 \cdot 10^{-9}\text{s}/81 = 3.1 \cdot 10^{-8}\text{s}$. Since communication aspects between the FPGA and a serial computing unit on the onboard computer and the calculation and evaluation of a cost function are not regarded in this simple estimation the calculated number of maximal model prediction steps/s is just an upper bound that cannot be reached in a full NMPC implementation. It is expected that the maximal number of computable model prediction steps will be reduced by a factor of 0.1-0.01 in relation to the value in equation 9. This is still significantly more than the number of model prediction steps used in the simulations above that verified feasibility of the proposed approach (Table 1, Fig. 3-9).

6 Conclusion

This paper demonstrates the feasibility of a proposed parallel implementable constrained NMPC approach to a UAV state trajectory planning problem with and without collision avoidance. The presented approach is based on time-domain optimization which requires the prediction of time-domain models. A method to refine the NMPC commands without additional computational effort has been introduced. The functional feasibility of the constrained NMPC with 3d-kinematics as prediction model even with a low number of control candidates has been shown in simulations. To save computational cost, parallel implementation of prediction models on an FPGA has been proposed and implementation feasibility has been shown through hardware synthesis of different configurations of parallel prediction models for an FPGA.

The presented constrained NMPC approach enables the use of constrained NMPC with dimensions that were hardly implementable on small UAVs before.

References

1. He, M., Ling, K.V.: Model Predictive Control on a Chip. In: International Conference on Control and Automation (ICCA 2005), Budapest, Hungary, June 27-29 (2005)
2. Ling, K.V., Wu, B.F., Maciejowski, J.M.: Embedded Model Predictive Control (MPC) using FPGA. In: The International Federation of Automatic Control, Seoul, Korea, July 6-11 (2008)
3. Johansen, T.A., Jackson, W., Schreiber, R., et al.: Hardware Synthesis of Explicit Model Predictive Controllers. IEEE Transactions on Control Systems Technology 15(1), 191–197 (2007)
4. Vouzis, P.D., Bleris, L.G., Arnold, M.G., et al.: A Custom-made Algorithm-Specific Processor for Model Predictive Control. In: IEEE ISIE, Montreal, Quebec, Canada, July 9-12 (2006)

5. Vouzis, P.D., Bleris, L.G., Arnold, M.G., et al.: A System-on-a-Chip Implementation for Embedded Real-Time Model Predictive Control. IEEE Transactions On Control Systems Technology 17(5), 1006–1017 (2009)
6. Kang, Y., Hedrick, J.K.: Linear Tracking for a Fixed-Wing UAV Using Nonlinear Model Predictive Control. IEEE Transactions on Control Systems Technology 17(5), 1202–1210 (2009)
7. Findeisen, R., Allgöwer, F.: An Introduction to Nonlinear Predictive Control. In: 21st Benelux Meeting on Systems and Control, Veldhoven (2002)
8. Maciejowski, J.M.: Predictive Control with Constraints. Prentice Hall, Englewood Cliffs (2002)
9. Passino, K.M.: Biomimicry for Optimization, Control and Automation. Springer, Heidelberg (2005)
10. Weimer, F., Trittler, M., Joos, A., et al.: FPGA-Based Onboard Computer System for Mini Aerial Vehicles. In: International Micro Air Vehicle Conference and Flight Competition, Braunschweig, Germany, July 6-9 (2010)
11. Grimm, W., Well, K.H.: Nichtlineare Optimierung. Institute of Flight Mechanics and Control. Universität Stuttgart, Germany (2001)
12. Butter, U.: Flugregelung. Institute of Flight Mechanics and Control. Universität Stuttgart, Germany (2009)
13. Xilinx, Spartan-3 Generation FPGA User Guide. Version 1.5 (2010), http://www.xilinx.com/support/documentation/user_guides/ug331.pdf (accessed August 24, 2010)

Robust Linear-Parameter Varying Autopilot Design for a Tail/Thrust Vector Controlled Missile

Berno J.E. Misgeld, Marco Darcis, and Thomas Kuhn

Abstract. A robust autopilot design methodology using linear parameter varying transformations is presented and applied to a high-agile surface launched air defence missile, which is currently developed by Diehl-BGT-Defence. The lateral dynamics of the tail/thrust vector controlled missile are modelled as a second-order quasi-linear parameter varying (LPV) system. The incidence angle is used as exogenous variable, which is assumed to be estimated during missile flight. Decoupled lateral dynamics are assumed because of the application of a bank-to-turn manoeuvre plane angle control approach. Lateral single channel flight controllers are designed via H_∞-optimal control and μ-synthesis with the LPV lateral dynamics, which are extended by uncertain models of control actuating system, time-delay and body bending model. The flight controllers for lateral dynamics are designed at a number of operating points described by the LPV model over the MACH flight envelope. The controllers are implemented using a gain scheduling approach, where an altitude dependent gain loss in the control loop is compensated with the inverse normalised air density. The flight controllers were implemented in the nonlinear simulation environment and tested in extreme flight manoeuvres. All flight controllers showed good damping and acceleration tracking performance and were stable during nonlinear simulations.

1 Introduction

Flight control systems for modern missiles have to face greater challenges in terms of control system design. The missile autopilot has to be designed for a large flight envelope of the missile. For high-agile missiles, this means high incidence manoeuvres, mainly at the beginning and during the endgame of flight. In addition to demanded large lateral acceleration capacity, the flight control system has to be designed to engage manoeuvring targets over an extended velocity and altitude range. The resulting multivariable system to be controlled is highly nonlinear and

Berno J.E. Misgeld · Marco Darcis · Thomas Kuhn
Diehl BGT Defence, Alte Nußdorfer Straße 13, 88662 Überlingen, Germany

time-varying. The autopilot has to guarantee feedback control properties like fast and accurate acceleration reference tracking, good rate damping and robust stability and performance.

Traditionally, missile autopilot design is done using gain scheduling control [1], [2]. The nonlinear, time-varying system is linearised using a first-order Taylor approximation at a so called trim condition, which is a quasi-equilibrium, obtained at a certain time of flight. Classical linear controller design techniques can then be used for the resulting linear time-invariant system. This procedure is usually repeated for a family of operating points, for example Mach number, dynamic pressure, incidence angle, motor burn time and manoeuvre plane angle. The resulting controller parameters are interpolated depending on the current operating conditions during missile flight. This approach worked well in most applications, but problems with a rapidly changing incidence angle, which leads to a fast parameter scheduling variable in the frequency range of the incidence angle mode, may occur [3]. A further disadvantage of this approach is the rising number of controller design points resulting from the extended flight envelope of modern missiles. This leads to a higher development cost, or demands the introduction of automated tuning methods [4].

With the introduction of robust linear control design techniques, new possibilities for missile autopilot design were given. The definition of certain requirements, as well as uncertainties to the design process is possible with methods like H_∞-optimal control and μ-synthesis. As a result of the optimisation process, a controller is given, which guarantees robustness and performance properties. The application of robust control techniques to missile autopilot design showed promising results [5]-[9], although more than one robust controller is needed to cover the flight envelope of the missile. This problem can be overcome with scheduling/blending techniques [10] and was successfully demonstrated for the industrial missile autopilot example in [9]. The autopilot of a high-agile tail/thrust vector control air-to-air missile of this industrial application was designed for thrust/ burnout phase, varying altitude and incidence angles at a given dynamic pressure operating point. In addition to the successfully demonstrated controller blending, flight controller anti-windup was proven to work reliably during real world flight tests [11]. The approach of [9] was further developed in a missile control approach for a high-agile tail/thrust vector control ground-to-air missile, with industrial application [12]. In this approach, the robust controllers are designed at certain Mach operating points and gain scheduled over altitude. Uncertainties included in the approach are parametric airframe uncertainties resulting from linearisation over the incidence angle range and at different motor burn times. This approach worked well for control performance tests using a flight test validated nonlinear six-degrees of freedom model. Since the nonlinear lateral dynamics, depending on the incidence angle are modelled as uncertainties, the control design approach is considered suboptimal in performance terms.

To include the nonlinear incidence angle dependent lateral dynamics to a linear time-invariant controller design, linear parameter varying (LPV) control methods can be used [3]. LPV-based methods consist of reformulating the nonlinear plant to a linear time-varying model for which then a linear controller design technique can be applied [3], [13]. In the approach presented in [3], the nonlinear lateral dynamics are brought to a quasi-LPV form by a state transformation. The approach of [3] is adopted in this paper and revised. The nonlinear lateral dynamics are reformulated for quasi-LPV transformation and augmented by uncertainties in Section 2. In Section 3 the controller design is presented via μ-synthesis. Different to [3], the controller is designed as a (1, 2) controller for acceleration tracking and yaw/pitch rate damping. Implementation aspects are presented in Section 4. The controller is tested in simulations with the nonlinear lateral missile model and finally in a nonlinear validated 6DOF model in Section 5. In addition to the two lateral LPV channel controllers a roll controller is used to damp the roll rate and to control the manoeuvre plane. Finally, this paper ends with results and a discussion in Section 6.

2 System Model

2.1 Lateral Dynamics

The lateral yaw dynamics at a certain operating point can be described by the second order nonlinear state-space model of the form

$$\frac{d}{dt}\begin{bmatrix}\beta\\ r\end{bmatrix}=\begin{bmatrix}\frac{1}{mV_0}\cos(\beta)Y(\beta)\\ \frac{1}{I_{yy}}N(\beta)\end{bmatrix}+\begin{bmatrix}0 & -1\\ 0 & N_r\end{bmatrix}\begin{bmatrix}\beta\\ r\end{bmatrix}+\begin{bmatrix}\frac{1}{mV_0}b_y\\ \frac{1}{I_{yy}}b_n\end{bmatrix}\zeta$$
$$\begin{bmatrix}\beta\\ r\\ a_{y,cg}\end{bmatrix}=\begin{bmatrix}1 & 0\\ 0 & 1\\ \frac{Y(\beta)}{V_0} & 0\end{bmatrix}\begin{bmatrix}\beta\\ r\end{bmatrix}+\begin{bmatrix}0\\ 0\\ \frac{b_y}{m}\end{bmatrix}\zeta \quad , \tag{1}$$

where β is the sideslip-angle, r is the yaw rate, m is the missile mass, I_{yy} is the missile inertia, ζ the rudder position given by the control actuating system (CAS) and V_0 is the absolute velocity. Note that the output equation for $a_{y,cg}$ is nonlinearly dependent on β. This term is linearised and regarded as uncertainty in the control design approach. The force and moment terms b_y, b_n in the control input matrix are linearised at low sideslip-angles

$$b_y=\bar{q}S_{ref}\frac{\partial c_y}{\partial\beta}$$

$$b_n=\bar{q}S_{ref}D_{ref}\frac{\partial c_n}{\partial\beta}$$

and contain aerodynamic and thrust-vector control (TVC) derivative values, with $\bar{q}$, the dynamic pressure, S_{ref}, D_{ref}, the aerodynamic reference area and diameter, respectively. Similar to b_y, b_n, the term N_r is the aerodynamic yaw moment derivative with respect to the yaw rate $N_r = \bar{q}\frac{S_{ref}D_{ref}}{I_{yy}}\frac{D_{ref}}{V_0}\frac{\partial c_n}{\partial r}$. Note that b_y, b_n are nonlinear with respect to the sideslip-angle and time varying with respect to TVC. This effect is neglected in the quasi-LPV transformation but specified as parametric uncertainty in the controller design. The nonlinear aerodynamic force and moment entries in (1) are generated from aerodynamic data at each Mach design point as table interpolation data

$$\begin{aligned} Y(\beta) &= \bar{q}S_{ref}c_y(\beta) \\ N(\beta) &= \bar{q}S_{ref}D_{ref}c_n(\beta) \end{aligned}.$$

The model of the lateral dynamics (1) is given at each Mach number of the flight envelope of the missile for the controller design grid. All model data is generated at sea level altitude and stored for controller design. The equation of (1) is extended by a linear CAS model, described by a second order differential equation with uncertain damping D_{CAS} and Eigenfrequency ω_{CAS}

$$\begin{aligned} \frac{d}{dt}\begin{bmatrix} x_1 \\ x_2 \end{bmatrix} &= \begin{bmatrix} 0 & 1 \\ -\omega_{CAS}^2 & -2D_{CAS}\omega_{CAS} \end{bmatrix}\begin{bmatrix} x_1 \\ x_2 \end{bmatrix} + \begin{bmatrix} 0 \\ \omega_{CAS}^2 \end{bmatrix}\zeta_C \\ \zeta &= \begin{bmatrix} 1 & 0 \end{bmatrix}\begin{bmatrix} x_1 \\ x_2 \end{bmatrix} \end{aligned}. \tag{3}$$

For controller design and test, the model of (1) is extended by a model of the structural vibrations (body bending). Only the first bending mode is modelled using a low damped oscillator state space model of second order

$$\begin{aligned} \begin{bmatrix} \dot{x}_{bb,1} \\ \dot{x}_{bb,2} \end{bmatrix} &= \begin{bmatrix} 0 & 1 \\ -\omega_{bb}^2 & -2\zeta_{bb}\omega_{bb} \end{bmatrix} + \begin{bmatrix} 0 \\ 1 \end{bmatrix}\zeta \\ \begin{bmatrix} \Delta r \\ \Delta a_{y,CG} \\ \Delta a_{y,IMU} \end{bmatrix} &= \begin{bmatrix} 0 & k_{r,bb} \\ -k_{a,bb}\omega_{bb}^2 & -2k_{a,bb}\zeta_{bb}\omega_{bb} \\ -k_{a,bb}\omega_{bb}^2 & -2k_{a,bb}\omega_{bb}\zeta_{bb} \end{bmatrix}\begin{bmatrix} x_{bb,1} \\ x_{bb,2} \end{bmatrix} + \begin{bmatrix} 0 \\ k_{a,bb} \\ k_{a,bb} \end{bmatrix}\zeta \end{aligned} \tag{4}$$

in which ω_{bb} and ζ_{bb} are the body bending frequency and the damping. The input ζ is the rudder command, calculated from fin deflections and the outputs are the delta rates and accelerations, applied by addition to equation (1). $k_{a,bb}$ and $k_{q,bb}$ are the rate and acceleration gains, calculated from constants, derived by a mechanical finite element model and validated by body bending tests [9]. Finally, the time-delay of the flight computer was modelled as two sample time instances, with a continuous first order Padé approximation to

$$G_{r,a}(s)=\frac{-\frac{n_d s}{T_s}+2}{\frac{n_d s}{T_s}+2}, \tag{5}$$

where T_s is the system sampling time, n_d =2 is the assumed time delay. Equation (5) is applied to the outputs of (1). The flight controller model, consisting of equations (1) to (5) is given in Fig. 1.

2.2 LPV System

The nonlinear system of lateral dynamics that is considered here shall be transformed to a linear system, whose parameter matrices **A**, **B** and **C** depend on an exogenous time-varying variable θ

$$\begin{aligned}\dot{\mathbf{x}}&=\mathbf{A}(\theta)\mathbf{x}+\mathbf{B}(\theta)\mathbf{u}\\ \mathbf{y}&=\mathbf{C}(\theta)\mathbf{x}\end{aligned}, \tag{6}$$

with input vector **u**, state vector **x** and output vector **y**. Note that the exogenous time-varying parameter of the quasi-LPV system form (6) is actually not exogenous, but the internal angle-of-attack state and is assumed to be estimated in the flight control system. The nonlinear system considered here is an output-nonlinear square system with state dependent nonlinear matrix **f**(**z**), linear matrix **A**(**z**), **B**(**z**)

$$\begin{aligned}\frac{d}{dt}\begin{bmatrix}\mathbf{z}\\ \mathbf{w}\end{bmatrix}&=\mathbf{f}(\mathbf{z})+\mathbf{A}(\mathbf{z})\begin{bmatrix}\mathbf{z}\\ \mathbf{w}\end{bmatrix}+\mathbf{B}(\mathbf{z})\mathbf{u}\\ \mathbf{y}&=\mathbf{C}\mathbf{z}\end{aligned}, \tag{7}$$

which is to be transformed to the quasi-LPV system

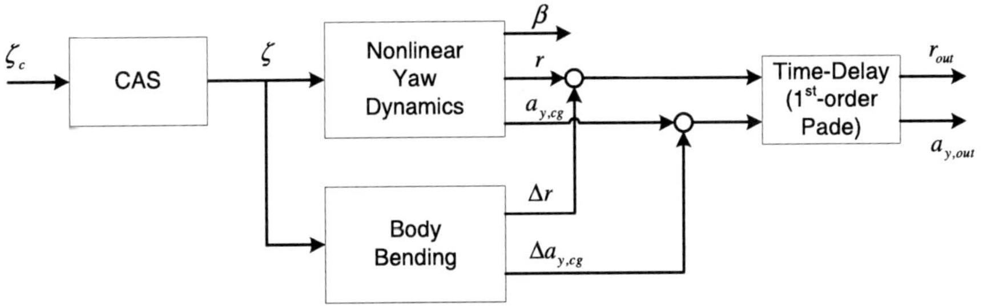

Fig. 1 Model of the lateral dynamics, including control actuating system (CAS), structural dynamics (body bending) and time-delay of the flight computer

$$\frac{d}{dt}\begin{bmatrix}\mathbf{z}\\ \mathbf{w}^*\end{bmatrix} = \mathbf{A}^*(\mathbf{z})\begin{bmatrix}\mathbf{z}\\ \mathbf{w}^*\end{bmatrix} + \mathbf{B}^*(\mathbf{z})\mathbf{u}^*, \qquad (8)$$
$$\mathbf{y} = \mathbf{C}\mathbf{z}$$

with the transformed input vector $\mathbf{u}^*$ and state vector $\mathbf{w}^*$. The requirement for the transformation from (7) to (8) is $\mathbf{f}(\mathbf{0}) = \mathbf{0}$ and that there exist nonlinear continuous, differentiable functions $\mathbf{w}_{eq}(\mathbf{z}), \mathbf{u}_{eq}(\mathbf{z})$ for which holds

$$\mathbf{0} = \mathbf{f}(\mathbf{z}) + \mathbf{A}(\mathbf{z})\begin{bmatrix}\mathbf{z}\\ \mathbf{w}_{eq}(\mathbf{z})\end{bmatrix} + \mathbf{B}(\mathbf{z})\mathbf{u}_{eq}(\mathbf{z}). \qquad (9)$$

Using (9) and the introduction of a new state $\mathbf{w}^* = \mathbf{w} - \mathbf{w}_{eq}(z)$ and an input transformation $\mathbf{u}^* = \mathbf{u} - \mathbf{u}_{eq}(\mathbf{z})$, equation (7) can be transformed to the quasi-LPV system. For the transformation the matrices $\mathbf{A}(\mathbf{z}), \mathbf{B}(\mathbf{z})$ are rewritten to

$$\mathbf{A}(\mathbf{z}) = \begin{bmatrix}\mathbf{A}_{zz}(\mathbf{z}) & \mathbf{A}_{zw}(\mathbf{z})\\ \mathbf{A}_{wz}(\mathbf{z}) & \mathbf{A}_{ww}(\mathbf{z})\end{bmatrix}; \qquad \mathbf{B}(\mathbf{z}) = \begin{bmatrix}\mathbf{B}_z(\mathbf{z})\\ \mathbf{B}_w(\mathbf{z})\end{bmatrix}. \qquad (10)$$

After some algebraic reformulation, the quasi-LPV system is obtained as

$$\frac{d}{dt}\begin{bmatrix}\mathbf{z}\\ \mathbf{w} - \mathbf{w}_{eq}(\mathbf{z})\end{bmatrix} = \begin{bmatrix}\mathbf{A}_{zz}(\mathbf{z}) & \mathbf{A}_{zw}(\mathbf{z})\\ \mathbf{A}_{wz}(\mathbf{z}) - \frac{\partial \mathbf{w}_{eq}(\mathbf{z})}{\partial \mathbf{z}}\mathbf{A}_{zz}(\mathbf{z}) & \mathbf{A}_{ww}(\mathbf{z}) - \frac{\partial \mathbf{w}_{eq}(\mathbf{z})}{\partial \mathbf{z}}\mathbf{A}_{zw}(\mathbf{z})\end{bmatrix}\begin{bmatrix}\mathbf{z}\\ \mathbf{w} - \mathbf{w}_{eq}(\mathbf{z})\end{bmatrix}$$
$$+ \begin{bmatrix}\mathbf{B}_z(\mathbf{z})\\ \mathbf{B}_w(\mathbf{z}) - \frac{\partial \mathbf{w}_{eq}(\mathbf{z})}{\partial \mathbf{z}}\mathbf{B}_z(\mathbf{z})\end{bmatrix}\left[\mathbf{u} - \mathbf{u}_{eq}(\mathbf{z})\right] \qquad (11)$$

Finally, for the nonlinear yaw dynamics of (1) the quasi-LPV model is given by

$$\frac{d}{dt}\begin{bmatrix}\beta\\ r - r_{eq}\end{bmatrix} = \begin{bmatrix}0 & -1\\ 0 & N_r + \frac{\partial r_{eq}(\beta)}{\partial \beta}\end{bmatrix}\begin{bmatrix}\beta\\ r - r_{eq}\end{bmatrix} + \begin{bmatrix}\frac{b_y}{mV_0}\\ \frac{b_n}{I_{yy}} - \frac{\partial r_{eq}(\beta)}{\partial \beta}\frac{b_y}{mV_0}\end{bmatrix}\left[\zeta - \zeta_{eq}\right] \qquad (12)$$

with the partial derivative function

$$\frac{\partial r_{eq}(\beta)}{\partial \beta} = \frac{\cos\beta \frac{b_n}{b_y}\frac{\partial Y(\beta)}{\partial \beta} - \sin\beta \frac{b_n}{b_y}Y(\beta) - \frac{\partial N(\beta)}{\partial \beta}}{mV_0\frac{b_n}{b_y} + I_{yy}N_r}, \qquad (13)$$

and the state equilibrium equations

$$\zeta_{eq}(\beta) = \frac{-\frac{N_r \cos\beta}{mV_0} Y(\beta) - \frac{N(\beta)}{I_{yy}}}{\frac{b_n}{I_{yy}} + \frac{N_r b_y}{mV_0}}$$

$$r_{eq}(\beta) = \frac{\cos\beta \frac{b_n}{b_y} Y(\beta) - N(\beta)}{mV_0 \frac{b_n}{b_y} + I_{yy} N_r} . \tag{14}$$

3 Controller Design

To conduct a controller design the quasi-LPV model (12) has to be modelled as a linear time-invariant (LTI) system. Therefore, the sideslip-angle β is assumed to be constant. The linear model is obtained at a certain sideslip-angle and can be used for controller design. To include the uncertainties of the varying sideslip-angle into the controller design, the parameters for varying incidence angle are determined over the incidence angle range and modelled as parametric uncertainty. As a nominal value the zero sideslip-angle parameters are used. In addition to the uncertainty, modelled to sideslip-angle dependent parameters, parametric uncertainty was assumed to the control input matrix parameters b_y, b_n. As mentioned before, this is done to include incidence angle dependent variation in the control efficiency. Note that this effect is reduced with an active TVC, but has to be taken into account. The plant of lateral dynamics is augmented with equations (2)-(5).

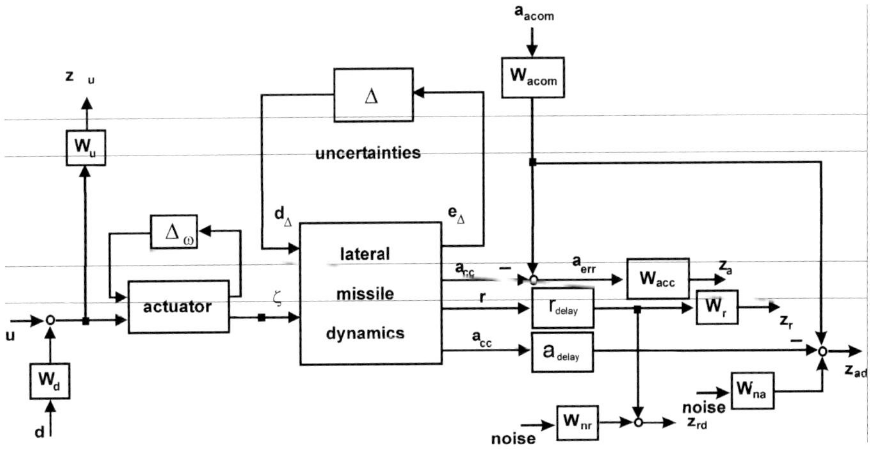

Fig. 2 Plant interconnection structure with linear time-invariant lateral missile dynamics for controller synthesis

No body bending model was included in the controller design. The augmented model with weighting function for controller synthesis is shown in Fig. 2. From Fig. 2 it can be seen that the controller design is conducted as a (1, 2) controller, consisting of a rate feedback and an acceleration tracking loop.

The controller output u (input to the actuator) is fitted with a complex weighting to limit the actuation bandwidth

$$W_u(s) = \frac{k_u(z_u s + 1)}{p_u s + 1} \tag{15}$$

where k_u is the static gain, z_u is zero and p_u is the pole of the transfer function. Parameters for (15) are determined for each Mach controller design point from crossover frequency and gain at high/low frequency. W_{acom} is a constant weight which specifies a ratio of acceleration tracking to rate damping loop in the μ-synthesis. Sensitivity weightings for yaw rate and acceleration loop are used to specify the desired damping and acceleration tracking properties of the loop. The weight W_r for the damping loop is modelled as constant (gain) and the weight $W_{acc}(s)$ is a transfer function

$$W_{acc}(s) = \frac{k_{acc}(z_{acc} s + 1)}{p_{acc} s + 1}, \tag{16}$$

with parameters k_{acc}, z_{acc} and p_{acc} are again derived from gain at low frequencies (steady state error for the acceleration loop), crossover frequency and high frequency gain. The weighting for the acceleration loop $W_{acc}(s)$ does not have integral action, since the linearised dynamics of the quasi-LPV system have integral action in the acceleration loop. Finally, constant weightings W_{na} and W_{nr} are introduced which cover the effects of measurement noise that corrupt the feedback channels. The actuator and the lateral dynamics are extended by parametric uncertainties, which are represented by the structured (real) block uncertainty

$$\Delta = \begin{bmatrix} \delta_{b_y} & & & \\ & \delta_{b_n} & & \\ & & \ddots & \\ & & & \delta_{D_{CAS}} \end{bmatrix}, \tag{17}$$

where $\delta_{b_y}, \delta_{b_y}, \dots \delta_{D_{CAS}}$ are real scalar perturbations $-1 \le \delta_i \le 1$ and correspond to the uncertainties in the model equations. The plant is brought to the general control configuration [14], [16] and the μ-synthesis is conducted with DK-iteration algorithm using the Robust Control Toolbox [15]. After controller design and test of robust stability, step response tests were conducted with the linear design model, containing model uncertainty. A test result for a higher super sonic Mach number is given in Fig. 3 as an example for the results achieved in linear simulations, where the upper part of the figure shows the damping loop and the lower part of the figure shows the acceleration tracking loop.

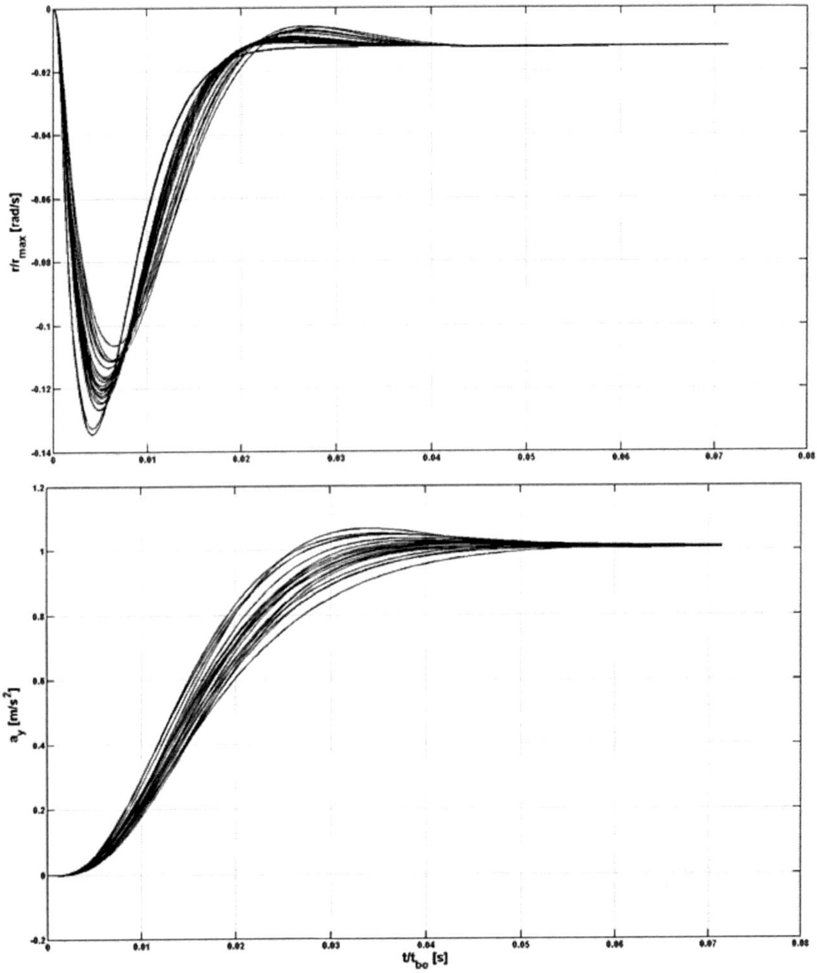

Fig. 3 Step response tests with the design model including uncertainty at a higher super-sonic Mach number with damping (upper) and acceleration tracking (lower) part of the figure.

4 Flight Controller Implementation

Similar to the previous implementation of the robust controller [12] five lateral flight controllers were designed for the complete Mach envelope. Before implementation, the controllers were reduced to 7th-order with additional check of robust stability. In addition to the implementation of the robust controllers, tuned with the equilibrium linearised quasi-LPV system at a certain sideslip-angle, each controller consists of a state transformation for the specific Mach number, see

Fig. 4. Note that the centre-of-gravity acceleration can be used in the nonlinear model of yaw motion as a feedback value (Fig.4). Sign changes have to be regarded when implementing the LPV transformations for the pitch channel. The controllers were gain scheduled depending on the normalised air density as a function of altitude

$$k_{lat} = \frac{\rho(h)}{\rho_0}. \tag{18}$$

with $\rho(h)$, the air density corresponding to the current altitude and ρ_0, the sea level air density. Acceleration error and rate input to the controller were divided by the factor k_{lat}. Note that the dynamic pressure for the state transformations has also to be adopted since the transformation is dependent on angle-of-attack or sideslip angle. The variation of thrust vector effects over altitude were not considered in the scheduling approach and are assumed to be covered by the control effectiveness uncertainties used in the controller design. During simulations over varying altitude a slight degradation of control performance towards slower response times could be observed, which is addressed to the TVC altitude dependent effects.

Finally, signal filtering and processing were applied to the acceleration input of the controller. The filters were implemented in the nonlinear 6-degrees of freedom model only and are needed for two reasons. Firstly, the body bending effects that were not included in the controller design can reduce control performance in the acceleration path due to the high gain of the body bending Eigenfrequency peak. Therefore, a notch filter was implemented to filter the inertial measurement unit (IMU) measured acceleration. The notch filter frequency was scheduled depending on motor burn phase. Secondly, the robust flight controller was designed with centre-of-gravity (COG) acceleration. To avoid lever arm effects in the acceleration signal due to IMU displacement from COG a lever arm correction filter was implemented as a second order filter with time variable lever arm l_a to

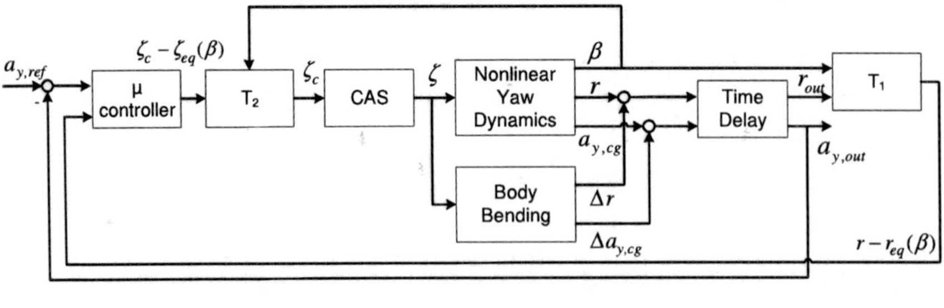

Fig. 4 Flight controller implementation with sideslip-angle dependent state and input transformations T_1 and T_2, respectively.

$$G_{acc}(s) = l_a \frac{\omega_a^2 s}{s^2 + 2\zeta_a \omega_a + \omega_a^2}, \tag{19}$$

which is used to filter mainly low frequency effects of the lever arm rate bias on acceleration. Another reason for the band limitation is the body bending effects on the rate signal which should not be amplified. Input to (19) is the notch filtered yaw rate. The output of (19) is subtracted from the notch filtered measured acceleration signal.

5 Nonlinear Simulation

Nonlinear simulations were conducted with two models. In a first step, a controller designed for a certain Mach number is implemented and tested in the nonlinear model of lateral yaw motion (1), which is extended by the CAS model. The controller response to acceleration reference steps is tested over the whole manoeuvre range at that Mach condition. In a second step, the controller is implemented in the nonlinear 6-degrees-of-freedom simulation as a three axis (roll, pitch and yaw) autopilot. The roll controller already in use for the ground-to-air missile was used to complete the three axis autopilot. The 6DOF model contains full nonlinear wind tunnel measured aerodynamics, a detailed model of the control actuation system, the body bending model, a motor model and a model of the thrust vector control system. Fig. 5 shows an acceleration reference step response series with the nonlinear model of yaw motion at a medium super-sonic Mach number. The acceleration is scaled on the maximum allowable acceleration over all flight conditions, which is determined by the structural limit. A slight tendency to overshoots can be seen towards higher acceleration reference values. This effect is due to the controller design at lower incidence angles with the fixed quasi-LPV model. However, since the overshoot is less than 10%, a gain scheduling as a function of incidence angle is not required. Although additional performance may be gained with the introduction of a gain scheduling of inner- and outer-control loop. Note that there is no steady-state error in Fig. 5. This is because of the integral action of the quasi-LPV dynamics. Integral action and the resulting zero steady-state error are the result of an accurate equilibrium transformation. Steady-state errors can be observed if uncertainties are introduced to the model, for example to $Y(\beta)$ or $N(\beta)$. The lower part of Fig. 5 shows the resulting rudder position and the side-slip-angle, which are scaled on the maximum allowable rudder position and the global maximum sideslip-angle, respectively.

An example of a simulation with the nonlinear 6DOF model is given in Fig. 6. The multivariable loop response is given for an acceleration step on the yaw channel of 50% maximum possible acceleration value. Note that the missile motor is in burnout condition leading to a drop in missile velocity below Mach 1.0 and a resulting increase in the incidence angle towards the absolute incidence angle limit for the missile. After a relatively fast initial response to the step command, a small

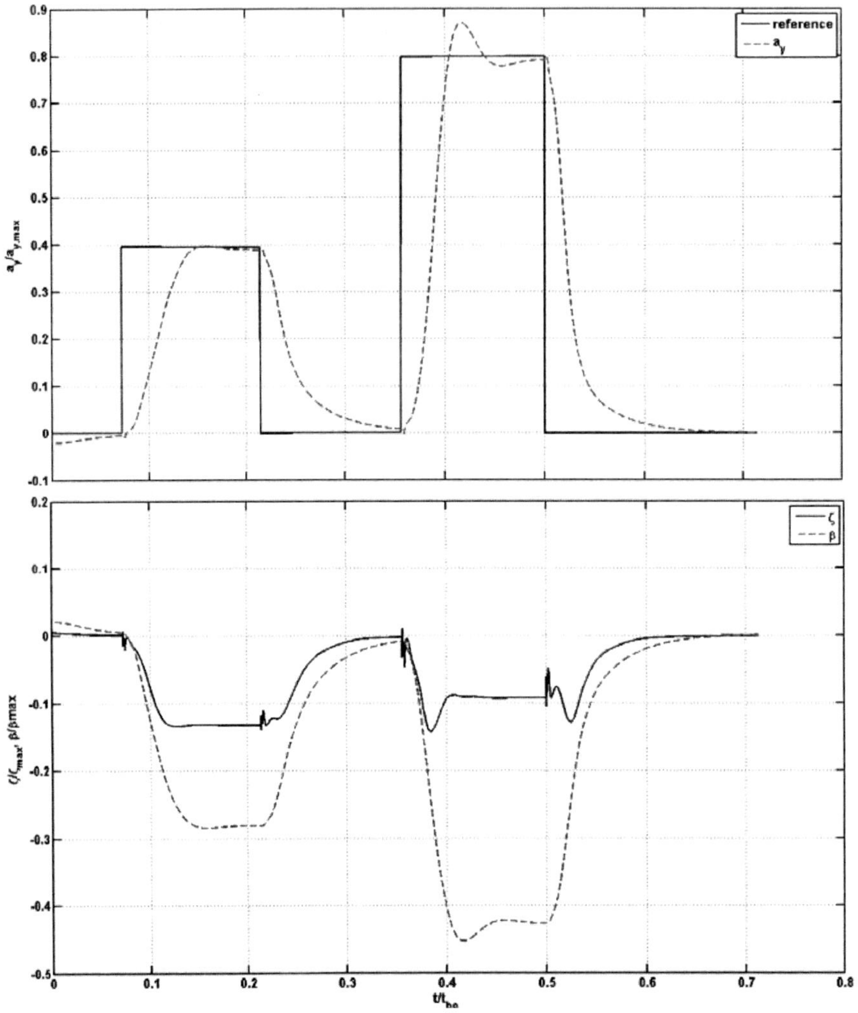

Fig. 5 Response to acceleration steps with nonlinear yaw model at medium super-sonic Mach.

steady-state error can be seen, which is suggested to derive from the drop of missile velocity. The multivariable loop coupling is small compared to the step in acceleration (see z-axis acceleration). This is also true for the nick/yaw rate and the roll rate, which is not shown in the Fig. 6. Similar results to Fig. 6 were obtained during step response tests with the nonlinear 6DOF model. A slight tendency to overshoots could be observed with reference steps leading to high incidence angles. Small steady-state errors were observed for rapidly changing velocity, that is on large manoeuvres without burning motor, or small manoeuvres with burning motor.

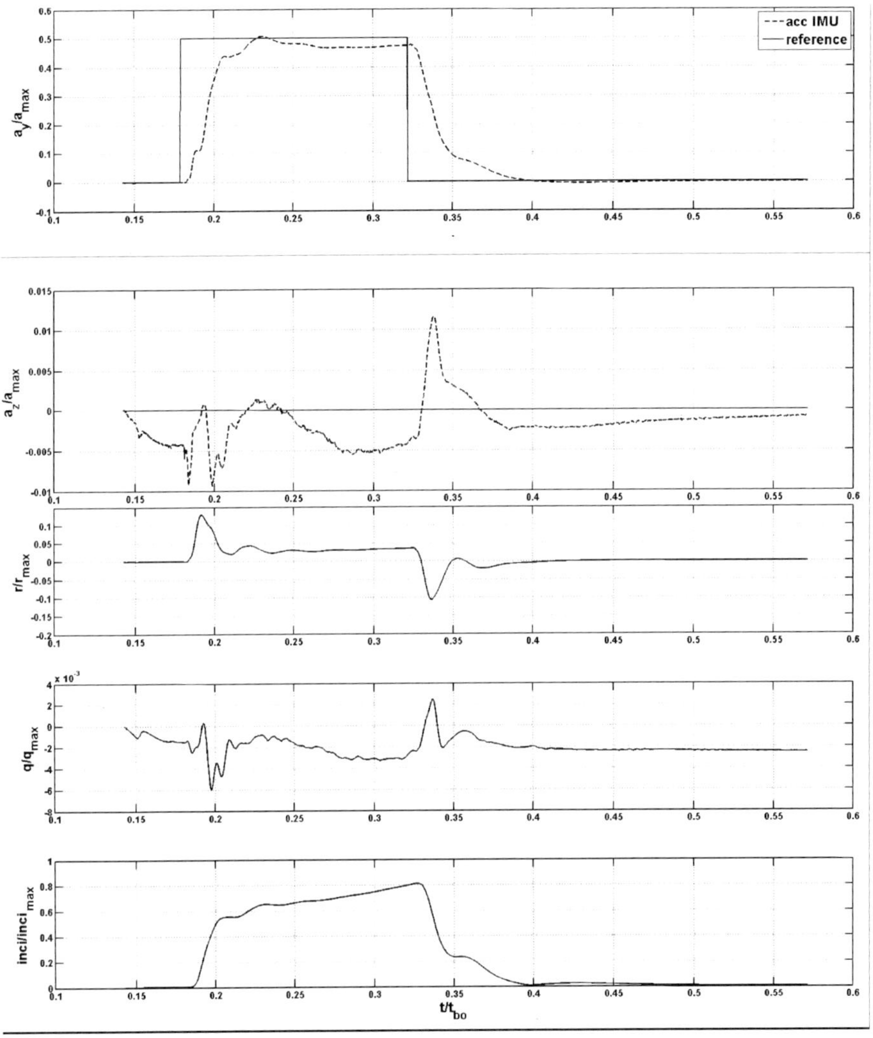

Fig. 6 Nonlinear 6DOF model response to an acceleration step command in the yaw channel around Mach 1 during motor burnout.

6 Results and Discussion

A robust autopilot design using a transformation of the nonlinear yaw dynamics to a quasi-linear parameter varying system was presented. The robust autopilot was tuned with the quasi-LPV model at a low sideslip-angle for the yaw dynamics of a high-agile ground-to-air missile. The resulting autopilot was first tested in a nonlinear model of yaw motion and in a second phase tested in a detailed and validated nonlinear 6-degrees-of-freedom model. The controller was designed for different Mach numbers, used as operating conditions and scheduled over the air density. Uncertainties of the quasi-LPV system dependent variable and control surface effectiveness were included to the design. Good results were achieved in simulations with a tendency to small overshoots (below 10%) and small steady-state error when leaving the operating conditions of a single controller. As a result, five controllers were designed to cover the full flight envelope of the missile. However, the controllers have to be combined using a blending methodology to obtain a reasonable flight controller, which will be the subject of future work. Furthermore, the quasi-LPV transformation functions contained parameters which were time dependent, like mass and inertia but were defined as constant parameters in the approach. It is suggested that the introduction of a time-dependency as a function of burn-time will lead to further performance gains of this approach. It should be noted, that the control system reaction to rudder/elevator saturation was not investigated in this study as there were no means of anti-windup measures included to this approach. In addition to that, the angle-of-attack and the sideslip-angle were assumed to be known, instead of using estimators for both values. The investigation of these effects will be subject to future work.

References

1. Nesline, F.W., Zarchan, P.: Robust Instrumentation Configurations for Homing Missile Flight Control. In: Proceedings of the AIAA Guidance, Navigation and Control Conference, Danvers, MA, August 1980, pp. 209–219 (1980)
2. Rugh, W.J., Shamma, J.S.: Research on gain scheduling. Automatica 36, 1401–1425 (2000)
3. Shamma, J.S., Cloutier, J.R.: Gain-Scheduled Missile Autopilot Design Using Linear Parameter Varying Transformations. Journal of Guidance, Control and Dynamics 16(2), 256–263 (1993)
4. Kleinwächter, F., Kuhn, T.: Robuste Auslegung strukturfester Flugregler mittels Polvorgabe und Parameteroptimierung. DGLR-Jahrestagung, Friedrichshafen (2005)
5. Reichert, R.T.: Application of $H\infty$ Control to Missile Autopilot Design. In: AIAA Guidance, Navigation and Control Conference, Boston, MA, pp. 1065–1072 (1989)
6. Wise, K.A., Mears, B.C., Poolla, K.: Missile Autopilot Design Using $H\ \infty$ Optimal Control With μ-Synthesis. In: Proceedings of the American Control Conference, San Diego, CA, pp. 2363–2367 (1990)
7. Yang, S.M., Huang, N.H.: Application of $H\infty$ Control to Pitch Autopilot of Missiles. IEEE Transactions on Aerospace and Electronic Systems 32(1), 426–433 (1996)
8. Jackson, P.: Applying μ-Synthesis to Missile Autopilot Design. In: Proceedings of the 29th IEEE Conference on Decision and Control, Honolulu, HI, pp. 2993–2998 (1990)
9. Buschek, H.: Full Envelope Missile Autopilot Design Using Gain Scheduled Robust Control. Journal of Guidance, Control and Dynamics 22(1), 115–122 (1999)
10. Hyde, R.A.: H_∞ Aerospace Control Design. Springer, Heidelberg (1995)
11. Dold, R., Buschek, H.: Flight Test of a Scheduled μ-Synthesis Autopilot for an Air-To-Air Missile. In: AIAA Guidance, Navigation and Control Conference, Montreal, Canada (2001)
12. Misgeld, B.J.E., Dold, R., Kuhn, T., Buschek, H.: Robust Autopilot Design for a High-Agile Ground-to-Air Missile. DGLR-Jahrestagung, Hamburg (2010)

13. Tsourdos, A., Zbikowski, R., White, B.: Robust Autopilot for a Quasi-Linear Parameter-Varying Missile Model. Journal of Guidance, Control and Dynamics 24(2), 287–295 (2001)
14. Skogestad, S., Postlethwaite, I.: Multivariable Feedback Control. John Wiley and Sons, Chichester (2005)
15. Chiang, R.Y., Safonov, M.G.: Robust Control Toolbox, User's Guide. The MathWorks Inc., Natick, MA (1992)
16. Zhou, K., Doyle, J., Glover, K.: Robust and Optimal Control. Prentice Hall, New Jersey (1996)

Part III
Sensors, Data Fusion and Navigation

A Single Frequency Strapdown Algorithm for Integrating IMUs in ECEF-Frame

Johann Dambeck and Benjamin Braun

Abstract. This contribution will introduce a strapdown navigation algorithm for integrating inertial measurement units (IMU), which – due to its inherent ability to propagate navigation states at the same frequency as the integrating IMU is providing measurements – is called single frequency strapdown algorithm. The algorithm will be derived w.r.t. Earth-centred Earth-fixed reference frame. Numerical results for a simulated kinematic trajectory are presented.

1 Introduction

The integration of measurements of an inertial measurement unit (IMU) for propagation of (kinematic) navigation states like position, velocity and orientation is an integral part in many air, land and sea navigation applications. The derivation from Newtonian physics results in a system of first order ordinary differential equations for the navigation states given measurements of the specific force in the body-fixed frame (shorthand *b*-frame) centred at and along/about the orthogonal (calibrated) IMU sensing axes w.r.t. inertial space. These inertial measurements are provided by so-called non-integrating IMUs. Since most precise IMUs are time-integrating, i.e. provide short equidistant time-integrals of specific force and angular rate, the commonly derived inertial navigation differential equations cannot be directly used. Common engineering practice is to calculate the required non-integrating IMU measurements by dividing the integrating IMU measurements by the integration time interval, but this corresponds to first order error terms and thus is only useful for low grade IMUs or applications with a high vibration level which would counteract the benefit of higher order terms in strapdown navigation algorithms. In contrast to this simple averaging approach, inertial

Johann Dambeck
MBDA, Germany
e-mail: dambeck@tum.de

Benjamin Braun
Technische Universität München, Germany
e-mail: benjamin.braun@tum.de

navigation experts have derived so-called dual frequency strapdown algorithms [9, 6, 8, 4, 11] based on Laning-Bortz orientation parameterization [7, 1] i.e. a rotation matrix is parameterized by the exponential of a skew matrix containing the three Laning-Bortz orientation parameters. These algorithms are "dual frequency" because integrating IMU measurements at a higher frequency are used to propagate navigation states at a lower frequency (relations of 3:1 and 4:1 have been published). This contribution will introduce a strapdown navigation algorithm for integrating IMUs, which – due to its inherent ability to propagate navigation states at the same frequency as the integrating IMU is providing measurements – is called *single frequency strapdown algorithm.* The algorithm will be derived w.r.t. Earth-centred Earth-fixed reference frame (shorthand *e*-frame). The paper is organized as follows: First, the navigation algorithm for non-integrating IMUs in *e*-frame is introduced in section 2. Then, the orientation solution using integrating gyroscope measurements is derived in section 3. In section 4 the transport rate is calculated as intermediate result for the position and velocity solution of section 5 using integrating accelerometer measurements. The resulting single frequency strapdown algorithm is presented as a template in section 6. Finally, numerical results for a simulated kinematic trajectory are shown in section 7.

2 Navigation Algorithm for Non-integrating IMUs in *e*-Frame

Before addressing the single frequency strapdown algorithm for integrating IMUs the commonly known strapdown navigation differential equations expecting non-integrating IMU inputs are briefly revisited. These will serve as basis for the new algorithm in the sequel. Independent of the chosen navigation state parameterization the system of ordinary differential equations for propagating the navigation states using non-integrating IMU measurements is of the form

$$\dot{\mathbf{z}}(t)=\mathbf{f}\left(\mathbf{z}(t),\mathbf{s}(t)\right), \tag{1}$$

where $\mathbf{z}(t)$ is the navigation state vector (containing position, velocity and orientation) and $\mathbf{s}(t)$ is the sensor measurement vector (containing IMU measured specific linear force and angular rate). In the most general case the navigation state vector $\mathbf{z}(t)$ consists of the Cartesian position vector $\mathbf{x}_e(t)$, the Cartesian velocity vector $\mathbf{v}_e(t)$ and the rotation matrix $\mathbf{R}_{eb}(t)$ where e represents the Earth-centred Earth-fixed reference frame and b the body-fixed frame. The sensor measurement vector $\mathbf{s}(t)$ is constituted of the vector of specific force in b-frame $\mathbf{f}_b(t)$ measured by the accelerometer triad and the vector of angular rate in b-frame w.r.t. inertial (i-) frame $\omega_{ib}(t)$ measured by the gyro triad:

$$\mathbf{z}(t):=\begin{bmatrix}\mathbf{x}_e(t)\\ \mathbf{v}_e(t)\\ \mathbf{R}_{eb}(t)\end{bmatrix} \quad \text{and} \quad \mathbf{s}(t):=\begin{bmatrix}\mathbf{f}_b(t)\\ \boldsymbol{\omega}_{ib}(t)\end{bmatrix}.$$

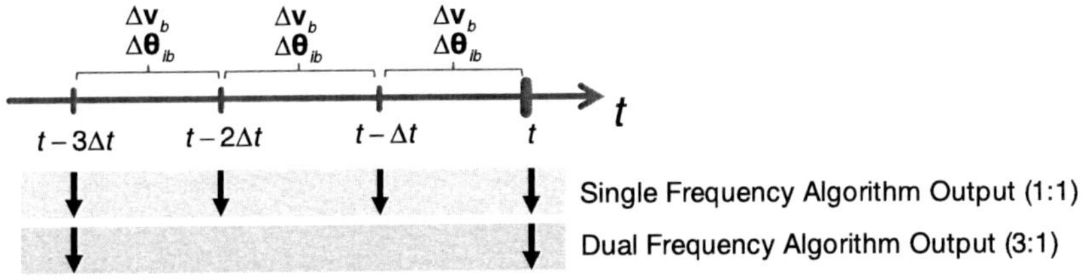

Fig. 1 Input / Output ratio of single and dual frequency strapdown algorithms

With this choice of navigation state parameters the nonlinear 1st order ordinary navigation differential equations becomes [2]:

$$\dot{\mathbf{x}}_e(t) = \mathbf{v}_e(t) \tag{2}$$

$$\dot{\mathbf{v}}_e(t) = \mathbf{R}_{eb}(t)\cdot\mathbf{f}_b(t) + \boldsymbol{\gamma}_e\left(\mathbf{x}_e(t),t\right) - 2\boldsymbol{\omega}_{ie}\times\mathbf{v}_e(t) \tag{3}$$

$$\dot{\mathbf{R}}_{eb}(t) = \mathbf{R}_{eb}(t)\boldsymbol{\Omega}_{ib}(t) - \boldsymbol{\Omega}_{ie}\mathbf{R}_{eb}(t) \tag{4}$$

where

$\mathbf{x}_e(t)$	(Cartesian) position in e-frame
$\mathbf{v}_e(t)$	(Cartesian) velocity in e-frame
$\mathbf{R}_{eb}(t)$	Rotation from b- into e-frame, $\mathbf{R}_{eb}(t)\mathbf{R}_{eb}^{\mathrm{T}}(t) = \mathbf{I} \wedge \det(\mathbf{R}_{eb}(t)) = +1$
$\mathbf{f}_b(t)$	Specific force in b-frame
$\omega_{ib}(t)$	Angular rate of b-frame w.r.t. i-frame, $\Omega_{ib}(t) := [\omega_{ib}(t)\times]$
ω_{ie}	Angular rate of e-frame w.r.t. i-frame, $\Omega_{ie} := [\omega_{ie}\times]$
$\gamma_e(\mathbf{x}_e(t), t)$	Gravity vector incl. tidal acceleration in e-frame
t	Time

This system of navigation differential equations in e-frame for a non-integrating IMU is nonlinear only due to the gravity field of the Earth (and tidal accelerations) and has time varying coefficients, i.e. the linear acceleration $\mathbf{f}_b(t)$ and angular rate $\omega_{ib}(t)$ measured by a non-integrating IMU. So far, a numerical integration routine of sufficient order like Runge-Kutta [5] could be used to numerically solve this system of first order differential equations since no closed-form solution is known. Integrating IMUs however measure time integrals of linear accelerations $\mathbf{f}_b(t)$ and angular rates $\omega_{ib}(t)$, i.e. velocity increments $\Delta\mathbf{v}_b(t)$ and angle increments $\Delta\theta_{ib}(t)$ defined by

$$\Delta\mathbf{v}_b\left(t-\Delta t,t\right) := \int_{t-\Delta t}^{t}\mathbf{f}_b\left(\tau\right)\mathrm{d}\tau, \qquad \Delta\boldsymbol{\theta}_{ib}\left(t-\Delta t,t\right) := \int_{t-\Delta t}^{t}\boldsymbol{\omega}_{ib}\left(\tau\right)\mathrm{d}\tau .$$

It will be shown how to solve the system of navigation differential equations in e-frame (2), (3) and (4) for an integrating IMU with sufficiently high error order by keeping the integrating IMU input rate as navigation state output rate. Figure 1 illustrates the input/output ratios of the single and dual frequency algorithms.

3 Solution of the Rotational Initial Value Problem

The orientation differential equation in e-frame (4) describes the propagation of the orientation of b-frame w.r.t. e-frame knowing the IMU measured rates $\Omega_{ib}(t)$ and the Earth's rate of rotation Ω_{ie}. Since the Earth's rate of rotation is very accurately known [10] it is independent of position and velocity and has no explicit time dependency. In this section the corresponding orientation difference equation incorporating integrating gyro measurements is derived.

3.1 Decomposition and Discretization of the Rotational Initial Value Problem

For simplicity the initial value problem (4) can be decomposed into two initial value problems

$$\dot{\mathbf{R}}_{ie}(t) = \mathbf{R}_{ie}(t)\mathbf{\Omega}_{ie}, \qquad \mathbf{R}_{ie}(t_0) = \mathbf{R}_{ie_0} \text{ and} \tag{5}$$

$$\dot{\mathbf{R}}_{ib}(t) = \mathbf{R}_{ib}(t)\mathbf{\Omega}_{ib}(t), \quad \mathbf{R}_{ib}(t_0) = \mathbf{R}_{ib_0}. \tag{6}$$

The desired solution $\mathbf{R}_{eb}(t)$ is obtained by multiplying the solutions of both initial value problems, i.e.

$$\mathbf{R}_{eb}(t) = \mathbf{R}_{ie}^{\mathrm{T}}(t)\cdot\mathbf{R}_{ib}(t).$$

The corresponding orientation difference equations can be written as

$$\mathbf{R}_{ie}(t+\Delta t) = \mathbf{R}_{ie}(t)\cdot\mathbf{T}_{e}(t,t+\Delta t), \quad \mathbf{R}_{ie}(t_0) = \mathbf{R}_{ie_0} \text{ and} \tag{7}$$

$$\mathbf{R}_{ib}(t+\Delta t) = \mathbf{R}_{ib}(t)\cdot\mathbf{T}_{b}(t,t+\Delta t), \quad \mathbf{R}_{ib}(t_0) = \mathbf{R}_{ib_0} \tag{8}$$

where $\mathbf{T}_e(t, t+\Delta t)$ and $\mathbf{T}_b(t, t+\Delta t)$ are yet unknown "transition" matrices. If these recursive equations would be known the solution of our initial value problem (4) would be

$$\mathbf{R}_{eb}(t+\Delta t) = \mathbf{R}_{ie}^{\mathrm{T}}(t+\Delta t)\cdot\mathbf{R}_{ib}(t+\Delta t) = \mathbf{T}_{e}^{\mathrm{T}}(t,t+\Delta t)\cdot\underbrace{\mathbf{R}_{ie}^{\mathrm{T}}(t)\cdot\mathbf{R}_{ib}(t)}_{=\mathbf{R}_{eb}(t)}\cdot\mathbf{T}_{b}(t,t+\Delta t)$$

$$\Leftrightarrow \mathbf{R}_{eb}(t+\Delta t) = \mathbf{T}_{e}^{\mathrm{T}}(t,t+\Delta t)\cdot\mathbf{R}_{eb}(t)\cdot\mathbf{T}_{b}(t,t+\Delta t). \tag{9}$$

3.2 Determination of the Transition Matrix of the e-Frame

The solution of the initial value problem (5) with constant coefficients due to a constant Earth's rate of rotation is

$$\mathbf{R}_{ie}(t) = \mathbf{R}_{ie_0}\cdot\mathrm{e}^{\mathbf{\Omega}_{ie}\cdot(t-t_0)}.$$

By definition $\mathbf{R}_{ie}(t)$ changes by a simple rotation about the Earth's axis of rotation, i.e. z-axis of e-frame, according to the Earth's rate of rotation ($\boldsymbol{\omega}_{ie} := (0,0,\omega_{ie})^{\mathrm{T}}$)

$$\mathbf{R}_{ie}(t+\Delta t) = \mathbf{R}_{ie}(t)\cdot \mathrm{e}^{\boldsymbol{\Omega}_{ie}\cdot\Delta t} = \mathbf{R}_{ie}(t)\cdot \mathbf{R}_3(\omega_{ie}\Delta t)$$

$$\Rightarrow \mathbf{T}_e(t,t+\Delta t) = \mathrm{e}^{\boldsymbol{\Omega}_{ie}\cdot\Delta t} = \mathbf{R}_3(\omega_{ie}\Delta t) = \begin{bmatrix} \cos(\omega_{ie}\Delta t) & -\sin(\omega_{ie}\Delta t) & 0 \\ \sin(\omega_{ie}\Delta t) & \cos(\omega_{ie}\Delta t) & 0 \\ 0 & 0 & 1 \end{bmatrix}. \quad (10)$$

3.3 Determination of the Transition Matrix of the b-Frame

Between two arbitrary coordinate frames $\dot{\mathbf{R}}(t) = \mathbf{R}(t)\cdot\boldsymbol{\Omega}(t)$ and $\mathbf{R}(t,\, t+\Delta t) = \mathbf{R}(t)\mathbf{T}(t,\, t+\Delta t)$. The general transition matrix $\mathbf{T}(t,\, t+\Delta t)$ is derived by expanding the general rotation matrix $\mathbf{R}(t,\, t+\Delta t)$ in a Taylor series

$$\mathbf{R}(t+\Delta t) = \mathbf{R}(t) + \dot{\mathbf{R}}(t)\cdot\Delta t + \ddot{\mathbf{R}}(t)\cdot\frac{\Delta t^2}{2!} + \dddot{\mathbf{R}}(t)\cdot\frac{\Delta t^3}{3!} + O(\Delta t^4) \quad (11)$$

and subsequently expressing the time derivatives of the rotation matrix $\mathbf{R}(t)$ by the orientation differential equation $\dot{\mathbf{R}}(t) = \mathbf{R}(t)\cdot\boldsymbol{\Omega}(t)$ and the corresponding time derivatives

$$\dot{\mathbf{R}} = \mathbf{R}\cdot\boldsymbol{\Omega},$$
$$\ddot{\mathbf{R}} = \dot{\mathbf{R}}\cdot\boldsymbol{\Omega} + \mathbf{R}\cdot\dot{\boldsymbol{\Omega}} = \mathbf{R}\cdot\left[\boldsymbol{\Omega}^2 + \dot{\boldsymbol{\Omega}}\right],$$
$$\dddot{\mathbf{R}} = \dot{\mathbf{R}}\cdot\left[\boldsymbol{\Omega}^2 + \dot{\boldsymbol{\Omega}}\right] + \mathbf{R}\cdot\left[\dot{\boldsymbol{\Omega}}\boldsymbol{\Omega} + \boldsymbol{\Omega}\dot{\boldsymbol{\Omega}} + \ddot{\boldsymbol{\Omega}}\right] = \mathbf{R}\cdot\left[\boldsymbol{\Omega}^3 + 2\boldsymbol{\Omega}\dot{\boldsymbol{\Omega}} + \dot{\boldsymbol{\Omega}}\boldsymbol{\Omega} + \ddot{\boldsymbol{\Omega}}\right].$$

Substitution of these time derivatives of $\mathbf{R}$ into (11) gives

$$\mathbf{R}(t+\Delta t) = \mathbf{R}(t)\left[\mathbf{I}_3 + \boldsymbol{\Omega}\Delta t + \left(\boldsymbol{\Omega}^2 + \dot{\boldsymbol{\Omega}}\right)\frac{\Delta t^2}{2!} + \left(\boldsymbol{\Omega}^3 + 2\boldsymbol{\Omega}\dot{\boldsymbol{\Omega}} + \dot{\boldsymbol{\Omega}}\boldsymbol{\Omega} + \ddot{\boldsymbol{\Omega}}\right)\frac{\Delta t^3}{3!} + O(\Delta t^4)\right].$$

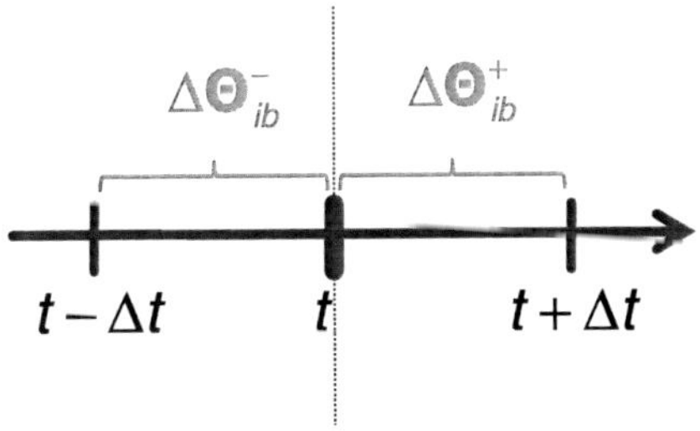

Fig. 2 Subsequent integrating IMU measurements.

One can easily identify the term in square brackets as transition matrix $\mathbf{T}(t, t + \Delta t)$. It just depends on the skew rate matrix Ω and its time derivatives

$$\mathbf{T}(t,t+\Delta t)=\mathbf{I}_3+\mathbf{\Omega}\cdot\Delta t+\left(\mathbf{\Omega}^2+\dot{\mathbf{\Omega}}\right)\cdot\frac{\Delta t^2}{2!}+\left(\mathbf{\Omega}^3+2\mathbf{\Omega}\dot{\mathbf{\Omega}}+\dot{\mathbf{\Omega}}\mathbf{\Omega}+\ddot{\mathbf{\Omega}}\right)\cdot\frac{\Delta t^3}{3!}+O\left(\Delta t^4\right). \tag{12}$$

For the particular recursive form (8) of the initial value problem (6) the skew rate matrix $\mathbf{\Omega}_{ib}$ and its time derivatives are required to calculate the transition matrix $\mathbf{T}_b(t, t + \Delta t)$ which have to be extracted from integrating IMU measurements. For this purpose, two subsequent integrating gyro triad measurements $\Delta\Theta_{ib}{}^+$ and $\Delta\Theta_{ib}{}^-$ (cf. figure 2) are expanded in Taylor series

$$\begin{aligned}
\Delta\mathbf{\Theta}_{ib}^+ &:= \int_t^{t+\Delta t}\mathbf{\Omega}_{ib}(\tau)\mathrm{d}\tau=\mathbf{\Omega}_{ib}(t)\cdot\Delta t+\dot{\mathbf{\Omega}}_{ib}(t)\cdot\frac{\Delta t^2}{2!}+\ddot{\mathbf{\Omega}}_{ib}(t)\cdot\frac{\Delta t^3}{3!}+O\left(\Delta t^4\right),\\
\Delta\mathbf{\Theta}_{ib}^- &:= \int_{t-\Delta t}^{t}\mathbf{\Omega}_{ib}(\tau)\mathrm{d}\tau=\mathbf{\Omega}_{ib}(t)\cdot\Delta t-\dot{\mathbf{\Omega}}_{ib}(t)\cdot\frac{\Delta t^2}{2!}+\ddot{\mathbf{\Omega}}_{ib}(t)\cdot\frac{\Delta t^3}{3!}+O\left(\Delta t^4\right).
\end{aligned} \tag{13}$$

These expansions represent relations between subsequent integrating gyro measurements $\Delta\Theta_{ib}$ and the non-integrating gyro measurement $\Omega_{ib}(t)$ and its time derivatives at time t. Inverting these Taylor series expansions and substituting the result into the expression for the transition matrix $\mathbf{T}_b(t, t + \Delta t)$ would give the desired result. Alternatively, we can express the transition matrix $\mathbf{T}_b(t, t + \Delta t)$ by products of two subsequent integrating gyro measurements and identify the coefficients by direct comparison

$$\begin{aligned}
\mathbf{T}_b(t,t+\Delta t) &= \mathbf{I}_3+a_1\Delta\mathbf{\Theta}_{ib}^+ +a_2\Delta\mathbf{\Theta}_{ib}^- +a_3\Delta\mathbf{\Theta}_{ib}^{+\,2}+a_4\Delta\mathbf{\Theta}_{ib}^+\Delta\mathbf{\Theta}_{ib}^- +a_5\Delta\mathbf{\Theta}_{ib}^-\Delta\mathbf{\Theta}_{ib}^+\\
&\quad +a_6\Delta\mathbf{\Theta}_{ib}^{-\,2}+a_7\Delta\mathbf{\Theta}_{ib}^{+\,3}+h.o.t.
\end{aligned} \tag{14}$$

With the angular increments (13) the required double products in (14) are as in (13) up to 3rd order terms in the time increment Δt

$$\begin{aligned}
\Delta\mathbf{\Theta}_{ib}^+\cdot\Delta\mathbf{\Theta}_{ib}^+ &= \mathbf{\Omega}_{ib}^2(t)\cdot\Delta t^2+\left(\mathbf{\Omega}_{ib}(t)\dot{\mathbf{\Omega}}_{ib}(t)+\dot{\mathbf{\Omega}}_{ib}(t)\mathbf{\Omega}_{ib}(t)\right)\cdot\tfrac{\Delta t^3}{2}+O\left(\Delta t^4\right)\\
\Delta\mathbf{\Theta}_{ib}^+\cdot\Delta\mathbf{\Theta}_{ib}^- &= \mathbf{\Omega}_{ib}^2(t)\cdot\Delta t^2-\left(\mathbf{\Omega}_{ib}(t)\dot{\mathbf{\Omega}}_{ib}(t)-\dot{\mathbf{\Omega}}_{ib}(t)\mathbf{\Omega}_{ib}(t)\right)\cdot\tfrac{\Delta t^3}{2}+O\left(\Delta t^4\right)\\
\Delta\mathbf{\Theta}_{ib}^-\cdot\Delta\mathbf{\Theta}_{ib}^+ &= \mathbf{\Omega}_{ib}^2(t)\cdot\Delta t^2+\left(\mathbf{\Omega}_{ib}(t)\dot{\mathbf{\Omega}}_{ib}(t)-\dot{\mathbf{\Omega}}_{ib}(t)\mathbf{\Omega}_{ib}(t)\right)\cdot\tfrac{\Delta t^3}{2}+O\left(\Delta t^4\right)\\
\Delta\mathbf{\Theta}_{ib}^-\cdot\Delta\mathbf{\Theta}_{ib}^- &= \mathbf{\Omega}_{ib}^2(t)\cdot\Delta t^2-\left(\mathbf{\Omega}_{ib}(t)\dot{\mathbf{\Omega}}_{ib}(t)+\dot{\mathbf{\Omega}}_{ib}(t)\mathbf{\Omega}_{ib}(t)\right)\cdot\tfrac{\Delta t^3}{2}+O\left(\Delta t^4\right)
\end{aligned} \tag{15}$$

and the triple product of the most actual integrating gyro measurement is correspondingly

$$\Delta\mathbf{\Theta}_{ib}^+\cdot\Delta\mathbf{\Theta}_{ib}^+\cdot\Delta\mathbf{\Theta}_{ib}^+=\mathbf{\Omega}_{ib}^3(t)\cdot\Delta t^3+O\left(\Delta t^4\right). \tag{16}$$

(13), (15) and (16) are substituted into (14) and the resulting equation is termwise compared with (12) to determine the coefficients of (14).

Comparison of coefficients linear in $\boldsymbol{\Omega}_{ib}(t)$ and its time derivatives gives three linear equations (of which two are linearly dependent) for two unknown coefficients

$$\begin{aligned}
\boldsymbol{\Omega}_{ib}(t)\cdot\Delta t: &\quad a_1 + a_2 = 1\\
\dot{\boldsymbol{\Omega}}_{ib}(t)\cdot\Delta t^2: &\quad \tfrac{1}{2}a_1 - \tfrac{1}{2}a_2 = \tfrac{1}{2}\\
\ddot{\boldsymbol{\Omega}}_{ib}(t)\cdot\Delta t^3: &\quad \tfrac{1}{6}a_1 + \tfrac{1}{6}a_2 = \tfrac{1}{6}
\end{aligned}$$

which are solved for a_1 and a_2

$$\begin{bmatrix} a_1 \\ a_2 \end{bmatrix} = -\frac{1}{2}\begin{bmatrix} -1 & -1 \\ -1 & 1 \end{bmatrix}\begin{bmatrix} 1 \\ 1 \end{bmatrix} = \begin{bmatrix} 1 \\ 0 \end{bmatrix}.$$

Comparison of coefficients bilinear/quadratic in $\boldsymbol{\Omega}_{ib}(t)$ and its time derivatives gives three linear equations for four unknown coefficients (underdetermined, thus we use the degree of freedom to preserve symmetry and preferably eliminate terms using oldest measurements by setting $a_6 := 0$)

$$\begin{aligned}
\boldsymbol{\Omega}_{ib}^2(t)\cdot\Delta t^2: &\quad a_3 + a_4 + a_5 + a_6 = \tfrac{1}{2}\\
\boldsymbol{\Omega}_{ib}(t)\dot{\boldsymbol{\Omega}}_{ib}(t)\cdot\Delta t^3: &\quad \tfrac{1}{2}a_3 - \tfrac{1}{2}a_4 + \tfrac{1}{2}a_5 - \tfrac{1}{2}a_6 = \tfrac{1}{3}\\
\dot{\boldsymbol{\Omega}}_{ib}(t)\boldsymbol{\Omega}_{ib}(t)\cdot\Delta t^3: &\quad \tfrac{1}{2}a_3 + \tfrac{1}{2}a_4 - \tfrac{1}{2}a_5 - \tfrac{1}{2}a_6 = \tfrac{1}{6}
\end{aligned}$$

which are solved for a_3, a_4 and a_5

$$\begin{bmatrix} a_3 \\ a_4 \\ a_5 \end{bmatrix} = \begin{bmatrix} 1 & 1 & 1 \\ 1 & -1 & 1 \\ 1 & 1 & -1 \end{bmatrix}^{-1}\begin{bmatrix} \frac{1}{2} \\ \frac{2}{3} \\ \frac{1}{3} \end{bmatrix} = \begin{bmatrix} \frac{1}{2} \\ -\frac{1}{12} \\ \frac{1}{12} \end{bmatrix}.$$

Comparison of coefficients cubic in $\boldsymbol{\Omega}_{ib}(t)$ simply gives $\boldsymbol{\Omega}_{ib}^3(t)\cdot\Delta t^3:\ a_7 = \tfrac{1}{6}$.

Substituting the derived coefficients a_1 ... a_7 in formula (14) for the transition matrix $\mathbf{T}_b(t, t+\Delta t)$ results in

$$\mathbf{T}_b(t,t+\Delta t) = \mathbf{I}_3 + \Delta\boldsymbol{\Theta}_{ib}^{+} + \tfrac{1}{2}\Delta\boldsymbol{\Theta}_{ib}^{+\,2} - \tfrac{1}{12}\left(\Delta\boldsymbol{\Theta}_{ib}^{+}\Delta\boldsymbol{\Theta}_{ib}^{-} - \Delta\boldsymbol{\Theta}_{ib}^{-}\Delta\boldsymbol{\Theta}_{ib}^{+}\right) + \tfrac{1}{6}\Delta\boldsymbol{\Theta}_{ib}^{+\,3} + O\left(\Delta t^4\right) \tag{17}$$

With the determined transition matrices (10) and (17) the recursive propagation equation for the rotation matrix describing the orientation of the b-frame relative to the e-frame in 4th error order is completely determined. Herein, beside constant coefficients, just two successive integrating gyro measurements are required. Using a longer history of integrating gyro measurements would mathematically give a more accurate result, but due to sensor noise and platform vibrations one can in general not expect higher orientation accuracy when incorporating higher order terms. To conclude, it is recommended to orthonormalize $\mathbf{R}_{eb}(t+\Delta t)$ from time to time in order to reduce numerical inaccuracies [3]:

$$\mathbf{R}_{eb}(t) = \mathbf{R}_{eb}(t)\left[\mathbf{R}_{eb}^{\mathrm{T}}(t)\mathbf{R}_{eb}(t)\right]^{-1/2}. \tag{18}$$

4 Angular Rate between e- and b-Frame from Integrating Gyro Measurements

The rate matrix $\boldsymbol{\Omega}_{eb}(t)$ will be required for the solution of the translational initial value problem. Hence a formula for the computation from integrating gyro measurements will be derived in this section. For the calculation of $\boldsymbol{\Omega}_{eb}(t)$ from $\boldsymbol{\Omega}_{ib}(t)$ the Earth's rate of rotation has to be transformed into b-frame and then compensated, i.e.

$$\boldsymbol{\Omega}_{eb}(t) = \boldsymbol{\Omega}_{ib}(t) - \mathbf{R}_{eb}^{\mathrm{T}}(t)\boldsymbol{\Omega}_{ie}\mathbf{R}_{eb}(t).$$

$\boldsymbol{\Omega}_{ib}(t)$ is calculated from gyro measurements by again making use of the Taylor expansions (13). The most trivial inversion of the Taylor series gives the central difference quotient

$$\boldsymbol{\Omega}_{ib}(t) = \tfrac{1}{2\Delta t}\left(\Delta\boldsymbol{\Theta}_{ib}^{+} + \Delta\boldsymbol{\Theta}_{ib}^{-}\right) + O\left(\Delta t^{2}\right)$$

and therefore the rate matrix $\boldsymbol{\Omega}_{eb}(t)$ finally gets

$$\boldsymbol{\Omega}_{eb}(t) = \tfrac{1}{2\Delta t}\left(\Delta\boldsymbol{\Theta}_{ib}^{+} + \Delta\boldsymbol{\Theta}_{ib}^{-}\right) - \mathbf{R}_{eb}^{\mathrm{T}}(t)\boldsymbol{\Omega}_{ie}\mathbf{R}_{eb}(t) + O\left(\Delta t^{2}\right). \tag{19}$$

Later, when collecting all partial results in the single frequency strapdown algorithm template in section 6, it will become clear that the 2nd error order of the central difference quotient is sufficient to use.

5 Solution of the Translational Initial Value Problem

The translational differential equations in e-frame (2) and (3) describe the propagation of position and velocity w.r.t. e-frame given the IMU measured specific forces $\mathbf{f}_b(t)$, the local gravity vector $\gamma_e(\mathbf{x}_e(t))$, the Earth's rate of rotation ω_{ie} and the solution of the rotational initial value problem $\mathbf{R}_{eb}(t)$. In this section difference equations for position and velocity incorporating integrating accelerometer measurements are derived.

5.1 Non-integrating from Integrating Accelerometer Measurements

Just like the angular rate $\boldsymbol{\Omega}_{ib}(t)$ in the orientation differential equation had to be replaced by integrating gyroscope triad increments $\Delta\Theta_{ib}(t)$, the specific force $\mathbf{f}_b(t)$ in the velocity differential equation has to be expressed by integrating accelerometer triad increments $\Delta\mathbf{v}_b(t)$. Thus, $\Delta\mathbf{v}_b(t)$ is analogously to (13) expanded in a Taylor series

$$\Delta \mathbf{v}_b^+ := \int_t^{t+\Delta t} \mathbf{f}_b(\tau)\mathrm{d}\tau = \mathbf{f}_b(t)\cdot\Delta t + \dot{\mathbf{f}}_b(t)\cdot\frac{\Delta t^2}{2!} + O(\Delta t^3),$$
$$\Delta \mathbf{v}_b^- := \int_{t-\Delta t}^{t} \mathbf{f}_b(\tau)\mathrm{d}\tau = \mathbf{f}_b(t)\cdot\Delta t - \dot{\mathbf{f}}_b(t)\cdot\frac{\Delta t^2}{2!} + O(\Delta t^3)$$

and due to the lower error order required again solved by trivial series inversion

$$\begin{aligned} \Delta\mathbf{v}_b^+ + \Delta\mathbf{v}_b^- &= 2\mathbf{f}_b(t)\cdot\Delta t + O(\Delta t^3) \\ \Delta\mathbf{v}_b^+ - \Delta\mathbf{v}_b^- &= \dot{\mathbf{f}}_b(t)\cdot\Delta t^2 + O(\Delta t^4) \end{aligned} \;\Rightarrow\; \begin{aligned} \mathbf{f}_b(t) &= \frac{\Delta\mathbf{v}_b^+ + \Delta\mathbf{v}_b^-}{2\Delta t} + O(\Delta t^2) \\ \dot{\mathbf{f}}_b(t) &= \frac{\Delta\mathbf{v}_b^+ - \Delta\mathbf{v}_b^-}{\Delta t^2} + O(\Delta t^2). \end{aligned} \tag{20}$$

5.2 *Velocity from Integrating Accelerometer Measurements*

The velocity difference equation is derived by Taylor series expansion

$$\mathbf{v}_e(t+\Delta t) = \mathbf{v}_e(t) + \dot{\mathbf{v}}_e(t)\cdot\Delta t + \ddot{\mathbf{v}}_e(t)\cdot\frac{\Delta t^2}{2!} + O(\Delta t^3). \tag{21}$$

$\dot{\mathbf{v}}_e(t)$ is given by the velocity differential equation (3)

$$\dot{\mathbf{v}}_e(t) = \mathbf{R}_{eb}(t)\mathbf{f}_b(t) + \gamma_e(\mathbf{x}_e(t)) - 2\boldsymbol{\omega}_{ie}\times\mathbf{v}_e(t)$$

and its time derivative by

$$\ddot{\mathbf{v}}_e(t) = \dot{\mathbf{R}}_{eb}(t)\mathbf{f}_b(t) + \mathbf{R}_{eb}(t)\dot{\mathbf{f}}_b(t) + \frac{\partial\gamma_e(\mathbf{x}_e(t))}{\partial\mathbf{x}_e^{\mathrm{T}}}\mathbf{v}_e(t) - 2\boldsymbol{\omega}_{ie}\times\dot{\mathbf{v}}_e(t).$$

Thus, using (20) and $\dot{\mathbf{R}}_{eb}(t) = \mathbf{R}_{eb}(t)\boldsymbol{\Omega}_{eb}(t)$, the velocity time derivatives are

$$\begin{aligned} \dot{\mathbf{v}}_e(t) &= \mathbf{R}_{eb}(t)\frac{\Delta\mathbf{v}_b^+ + \Delta\mathbf{v}_b^-}{2\Delta t} + \gamma_e(\mathbf{x}_e(t)) - 2\boldsymbol{\omega}_{ie}\times\mathbf{v}_e(t) + O(\Delta t^2) \\ \ddot{\mathbf{v}}_e(t) &= \mathbf{R}_{eb}(t)\left[\boldsymbol{\Omega}_{eb}(t)\frac{\Delta\mathbf{v}_b^+ + \Delta\mathbf{v}_b^-}{2\Delta t} + \frac{\Delta\mathbf{v}_b^+ - \Delta\mathbf{v}_b^-}{\Delta t^2}\right] + \frac{\partial\gamma_e(\mathbf{x}_e(t))}{\partial\mathbf{x}_e^{\mathrm{T}}}\mathbf{v}_e(t) \\ &\quad -2\boldsymbol{\omega}_{ie}\times\dot{\mathbf{v}}_e(t) + O(\Delta t^2). \end{aligned} \tag{22}$$

The Cartesian gravity vector $\gamma_e(\mathbf{x}_e(t))$ and the gravity vector gradient $\partial\gamma_e(\mathbf{x}_e(t))/\partial\mathbf{x}_e$ can be approximated by the gravity field of an equipotential ellipsoid of revolution [2]. For tactical grade IMUs this model is sufficient which furthermore can be truncated after the J_2 parameter as defined in [10]. For navigation and strategic grade IMUs more accurate models are required.

5.3 Position from Integrating Accelerometer Measurements

Analogously to (21), the position difference equation is obtained by expansion

$$\mathbf{x}_e(t+\Delta t)=\mathbf{x}_e(t)+\underbrace{\dot{\mathbf{x}}_e(t)}_{=\mathbf{v}_e(t)}\cdot\Delta t+\underbrace{\ddot{\mathbf{x}}_e(t)}_{=\dot{\mathbf{v}}_e(t)}\cdot\frac{\Delta t^2}{2!}+\underbrace{\dddot{\mathbf{x}}_e(t)}_{=\ddot{\mathbf{v}}_e(t)}\cdot\frac{\Delta t^3}{3!}+O\left(\Delta t^4\right). \quad (23)$$

The time derivatives of $\mathbf{x}_e(t)$ can be substituted by the known velocity $\mathbf{v}_e(t)$ and the velocity time derivatives (22).

6 Single Frequency Strapdown Algorithm in e-Frame

In this section, equations (9), (10), (17), (18), (19), (21), (22) and (23) required for propagation of position, velocity and orientation are combined in a template, constituting the *single frequency strapdown algorithm.*

Input:

- Integrating IMU measurements: $(\Delta\mathbf{v}_b^-, \Delta\mathbf{\Theta}_{ib}^-)$, $(\Delta\mathbf{v}_b^+, \Delta\mathbf{\Theta}_{ib}^+)$
- Previous navigation states: $\mathbf{x}_e(t), \mathbf{v}_e(t), \mathbf{R}_{eb}(t)$

Single Frequency Algorithm:

- Auxiliary equations

$$\mathbf{R}_{eb}(t)=\mathbf{R}_{eb}(t)\left[\mathbf{R}_{eb}^{\mathrm{T}}(t)\mathbf{R}_{eb}(t)\right]^{-1/2}$$

$$\mathbf{\Omega}_{eb}(t)=\frac{1}{2\Delta t}\left(\Delta\mathbf{\Theta}_{ib}^{+}+\Delta\mathbf{\Theta}_{ib}^{-}\right)-\mathbf{R}_{eb}^{\mathrm{T}}(t)\mathbf{\Omega}_{ie}\mathbf{R}_{eb}(t) \qquad +O\left(\Delta t^2\right)$$

$$\mathbf{T}_b(t)=\mathbf{I}_3+\Delta\mathbf{\Theta}_{ib}^{+}+\tfrac{1}{2}\Delta\mathbf{\Theta}_{ib}^{+2}-\tfrac{1}{12}\left(\Delta\mathbf{\Theta}_{ib}^{+}\Delta\mathbf{\Theta}_{ib}^{-}-\Delta\mathbf{\Theta}_{ib}^{-}\Delta\mathbf{\Theta}_{ib}^{+}\right)+\tfrac{1}{6}\Delta\mathbf{\Theta}_{ib}^{+3} \qquad +O\left(\Delta t^4\right)$$

$$\dot{\mathbf{v}}_e(t)=\mathbf{R}_{eb}(t)\frac{\Delta\mathbf{v}_b^{+}+\Delta\mathbf{v}_b^{-}}{2\Delta t}+\boldsymbol{\gamma}_e\left(\mathbf{x}_e(t)\right)-2\boldsymbol{\omega}_{ie}\times\mathbf{v}_e(t) \qquad +O\left(\Delta t^2\right)$$

$$\ddot{\mathbf{v}}_e(t)=\mathbf{R}_{eb}(t)\left[\mathbf{\Omega}_{eb}(t)\frac{\Delta\mathbf{v}_b^{+}+\Delta\mathbf{v}_b^{-}}{2\Delta t}+\frac{\Delta\mathbf{v}_b^{+}-\Delta\mathbf{v}_b^{-}}{\Delta t^2}\right]+\frac{\partial\boldsymbol{\gamma}_e\left(\mathbf{x}_e(t)\right)}{\partial\mathbf{x}_e^{\mathrm{T}}}\mathbf{v}_e(t)-2\boldsymbol{\omega}_{ie}\times\dot{\mathbf{v}}_e(t) \qquad +O\left(\Delta t^2\right)$$

- State propagation

$$\mathbf{x}_e(t+\Delta t)=\mathbf{x}_e(t)+\mathbf{v}_e(t)\Delta t+\dot{\mathbf{v}}_e(t)\frac{\Delta t^2}{2!}+\ddot{\mathbf{v}}_e(t)\frac{\Delta t^3}{3!} \qquad +O\left(\Delta t^4\right)$$

$$\mathbf{v}_e(t+\Delta t)=\mathbf{v}_e(t)+\dot{\mathbf{v}}_e(t)\Delta t+\ddot{\mathbf{v}}_e(t)\frac{\Delta t^2}{2!} \qquad +O\left(\Delta t^3\right)$$

$$\mathbf{R}_{eb}(t+\Delta t)=\mathbf{R}_3^{\mathrm{T}}(\omega_{ie}\Delta t)\mathbf{R}_{eb}(t)\mathbf{T}_b(t)$$

Output:

- Current navigation states: $\mathbf{x}_e(t+\Delta t), \mathbf{v}_e(t+\Delta t), \mathbf{R}_{eb}(t+\Delta t)$

Gravity and gravity gradient in e-frame could be modelled by [2]

$$\gamma_{e,i} = GM\left(-\frac{x_{e,i}}{\|\mathbf{x}_e\|^3} - a^2 J_2\left(\frac{3}{2}\frac{x_e}{\|\mathbf{x}_e\|^5} + 3\frac{z_e}{\|\mathbf{x}_e\|^5}\delta_{i3} - \frac{15}{2}\frac{x_{e,i}z_e^2}{\|\mathbf{x}_e\|^7}\right)\right) + \omega_{ie}^2(1-\delta_{i3})x_{e_i}, \quad i = 1,2,3$$

$$\begin{aligned}\frac{\partial\gamma_{e,i}(\mathbf{x}_e)}{\partial x_{e,j}} &= GM\left(-\frac{1}{\|\mathbf{x}_e\|^3}\delta_{ij} + 3\frac{x_{e_i}x_{e_j}}{\|\mathbf{x}_e\|^5}\right) - \frac{3}{2}GM\,a^2 J_2\left(\frac{1}{\|\mathbf{x}_e\|^5}\delta_{ij} - 5\frac{x_{e_i}x_{e_j}}{\|\mathbf{x}_e\|^7}\right)\\ &+\frac{15}{2}GM\,a^2 J_2\left(\frac{z_e^2}{\|\mathbf{x}_e\|^7}\delta_{ij} + 2\frac{x_{e_i}z_e}{\|\mathbf{x}_e\|^7}\delta_{j3} - 7\frac{x_{e_i}x_{e_j}z_e^2}{\|\mathbf{x}_e\|^9}\right)\\ &-3GM\,a^2 J_2\left(\frac{1}{\|\mathbf{x}_e\|^5}\delta_{i3}\delta_{j3} - 5\frac{x_{e_j}z_e}{\|\mathbf{x}_e\|^7}\delta_{i3}\right) + GM\,\omega_{ie}^2\left(\delta_{ij} - \delta_{i3}\delta_{j3}\right), \quad i,j = 1,2,3\end{aligned}$$

with GM and J_2 specified in [10] and δ_{ij}=1 for i=j and δ_{ij}=0 for $i \neq j$.

7 Numerical Results

The performance of the single frequency strapdown algorithm was analyzed by means of numerical simulation. Especially the accuracy and the robustness against noise were considered. For the sake of analytically knowing the derivatives different sinusoidal functions in the components of the position vector were chosen as kinematic trajectory (cf. figure 3). 100 Hz error-free integrating IMU measurements were easily derived from the analytical trajectory. By numerically integrating these generic measurements over two consecutive sample time intervals error-free velocity and angle increments $\Delta\mathbf{v}_b$ and $\Delta\boldsymbol{\theta}_{ib}$ were generated. For the analysis of the accuracy, the error-free IMU measurements were processed. For the verification of the robustness against noise, white noise was added to the generated IMU data. The results are additionally compared to the often used averaging approach which converts the integrating IMU measurements to non-integrating ones before integrating them with the conventional strapdown equations and Runge-Kutta integration scheme. The non-integrating measurements are obtained by dividing the velocity and angle increments by the sample time, $\bar{\mathbf{f}}_b(t) = \Delta\mathbf{v}_b(t\text{-}\Delta t,t)\,/\,\Delta t$ and $\bar{\boldsymbol{\omega}}_{ib}(t) = \Delta\boldsymbol{\theta}_{ib}(t\text{-}\Delta t,t)\,/\,\Delta t$, which correspond to a 1st order Taylor series expansion of $\Delta\mathbf{v}_b(t\text{-}\Delta t,t) = \int_{t-\Delta t}^{t}\mathbf{f}_b(\tau)\,d\tau$ and $\Delta\boldsymbol{\theta}_{ib}(t\text{-}\Delta t,t) = \int_{t-\Delta t}^{t}\boldsymbol{\omega}_{ib}(\tau)\,d\tau$. $\bar{\mathbf{f}}_b(t)$ and $\bar{\boldsymbol{\omega}}_{ib}(t)$ thus can be interpreted as mean specific forces and angular rates over the last time step Δt, which explains the naming averaging approach. Figure 4 shows the position and orientation errors with error-free IMU measurements. The horizontal and vertical position errors remain below 1.5e-3m. The orientation error does not exceed 6e-11°. Figure 5 illustrates the performance of the single frequency strapdown algorithm with measurement noise ($\sigma_{\Delta\mathbf{v}}$ = m/s, $\sigma_{\Delta\theta}$ = 2e-8 rad). The position error is lower than 3m. The orientation error remains below 2e-5° in roll, pitch and yaw axis.

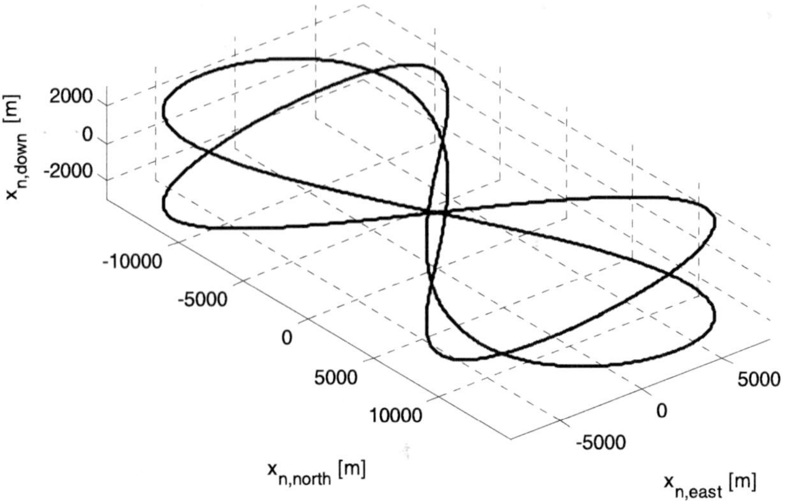

Fig. 3 True trajectory

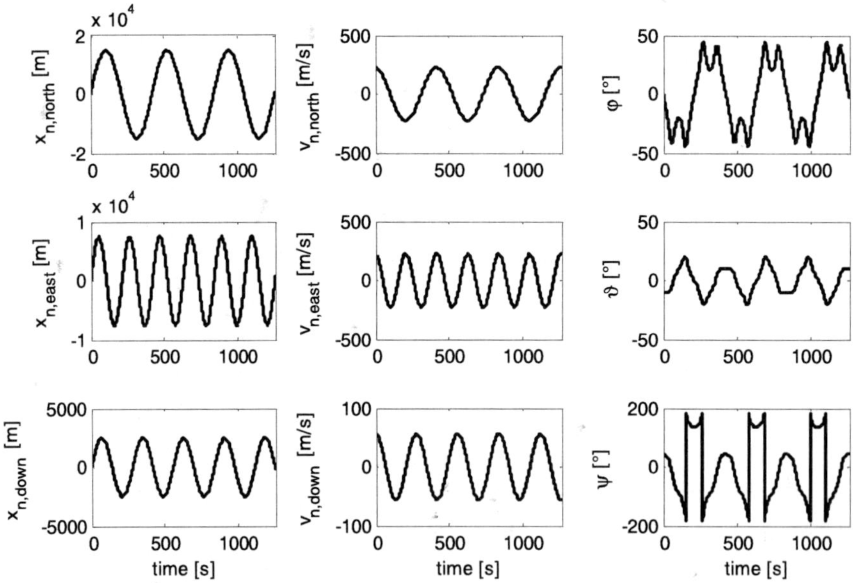

Fig. 4 True position, velocity and orientation along trajectory

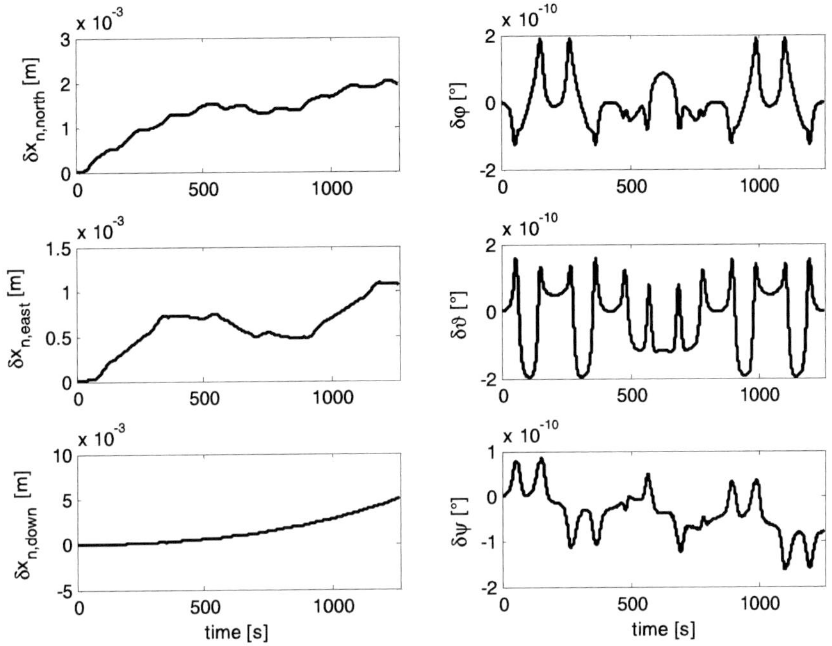

Fig. 5 Position and orientation error with Single Frequency Strapdown Algorithm and error-free IMU measurements

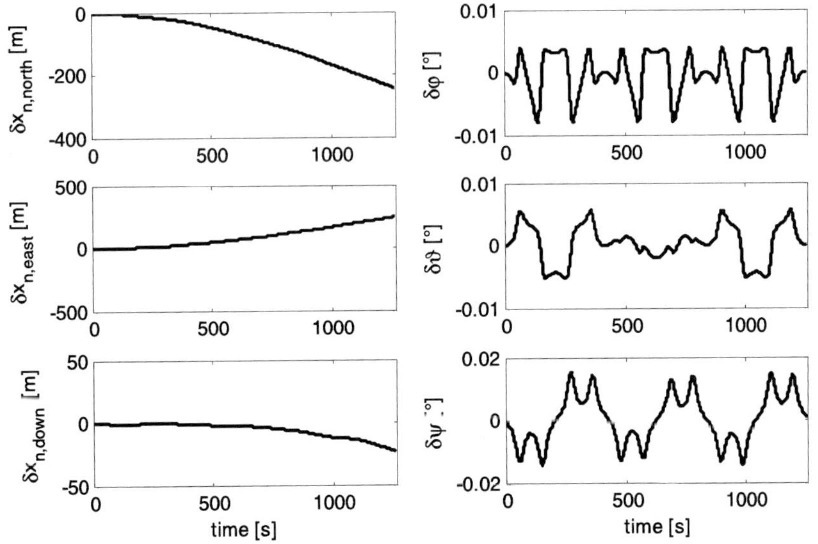

Fig. 6 Position and orientation error with Averaging Approach and error-free IMU measurements

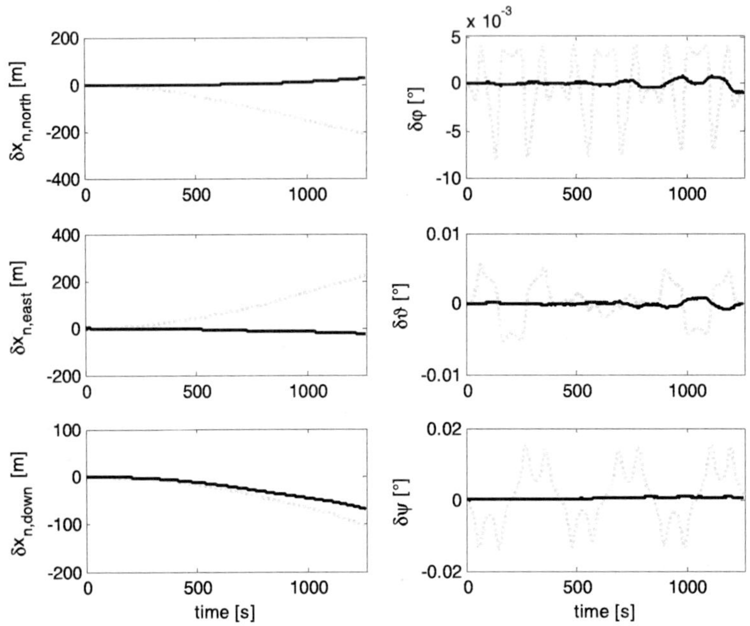

Fig. 7 Comparison of position, velocity and orientation error with noisy IMU measurements (σ_{Δ_v}=1e-4m/s, $\sigma_{\Delta\theta}$= 1e-6°) ▪▪▪▪ Averaging Approach, ▬▬ Single Frequency Strapdown Algorithm

8 Conclusion

The computational effort of the dual frequency algorithm is extremely low, but it has two drawbacks. First it is based on Laning-Bortz orientation parameters which lead to a quite nonlinear orientation differential equation in Laning-Bortz parameters, which is commonly approximated before it is solved. Second dual frequency strapdown algorithms require a higher integrating IMU input rate to provide the lower navigation state output rate.

The single frequency strapdown algorithm presented requires some more computations, but gives a result with a very transparent error order and allows to output the propagated navigation states with the same rate an integrating IMU is providing its measurements. This approach can easily be adapted to other parameterizations of the navigation state vector and the usage of quaternions as orientation parameters. The provided simulation results verify the accuracy of the single frequency strapdown algorithm and show its robustness for a suitable trajectory.

References

1. Bortz, J.E.: A New Mathematical Formulation for Strapdown Inertial Navigation. IEEE Transactions on Aerospace and Electronic Systems AES-7(1) (1971)
2. Dambeck, J.: Geodesy & Inertial Navigation Algorithms. lecture notes, Technical University of Munich (2010)

3. Farrell, J.L.: Integrated Aircraft Navigation. Academic Press, San Diego (1976)
4. Gusinsky, V.Z., Lesyuchevsky, V.M., Litmanovich, Y.A., Musoff, H., Schmidt, G.T.: New Procedure for Deriving optimized Strapdown Attitude Algorithms. Journal of Guidance, Control & Dynamics 20(4) (1997)
5. Hairer, E., Nørsett, S.P., Wanner, G.: Solving Ordinary Differential Equations I. Nonstiff Problems, 2nd edn. Springer, Heidelberg (2000)
6. Ignagni, M.B.: Optimal Strapdown Attitude Integration Algorithms. Journal of Guidance, Control & Dynamics 13(2) (1990)
7. Laning, J.H.: The vector analysis of finite rotations and angles. Massachusetts Institute of Technology, Cambridge, MIT/IL Special Rept. 6398-S-3 (1949)
8. Lee, J.G., Yong, J.Y., Mark, J.G., Tazartes, D.A.: Extension of Strapdown Attitude Algorithm for High-Frequency Base Motion. J. of Guidance, Control & Dynamics 13(4) (1990)
9. Miller, R.B.: A New Strapdown Attitude Algorithm. J. of Guidance, Control & Dynamics 6(4) (1983)
10. National Imaginary and Mapping Agency (NIMA): World Geodetic System 1984. Technical Report 8350.2, 3rd edn. (2000)
11. Savage, P.: Strapdown Inertial Navigation Integration Algorithm Design Part 1: Attitude Algorithms. Journal of Guidance, Control and Dynamics 21(1) (1998)
12. Savage, P.: Strapdown Inertial Navigation Integration Algorithm Design Part 2: Velocity and Position Algorithms. Journal of Guidance, Control and Dynamics 21(2) (1998)

Broadband Wind Estimation Algorithm for Gust Load Alleviation

Arndt Hoffmann, Kai Loftfield, and Robert Luckner

Abstract. Wind disturbances especially vertical gusts and turbulence can deteriorate the quality of measurements performed by the payload of utility aircraft. Gust load alleviation systems can reduce this effect significantly if the atmospheric disturbances are determined precisely and broadband up to the frequency range of the short period mode. A novel approach to determine vertical gusts and turbulence that can be used for future implementation in a gust load alleviation system based on feed forward control is presented. This approach uses a linear aircraft model of the longitudinal aircraft motion combined with the Dryden Turbulence Model within a Kalman filter framework. For performance analysis it is compared to an algorithm using simplified flight mechanical relations. For proof of the concept and its robustness, different kinds of disturbances such as discrete gusts and a bias in the angle of attack measurement are simulated and discussed.

1 Introduction

For a variety of missions, such as airborne gravity measurement, the flight of a utility aircraft has to be smooth even in disturbed atmosphere. In such a mission the aircraft is operated at low altitudes for optimal measurement quality. As a consequence the aircraft is often affected by gusts and turbulence producing additional aerodynamic forces and moments. The resulting aircraft accelerations disturb measurement quality and therefore reduce mission effectiveness and affect the pilot's

Dipl.-Ing. Arndt Hoffmann
Berlin Institute of Technology, Marchstr. 12, 10587 Berlin, Germany
e-mail: arndt.hoffmann@ilr.tu-berlin.de

Kai Loftfield
Berlin Institute of Technology, Marchstr. 12, 10587 Berlin, Germany
e-mail: loftfield@mailbox.tu-berlin.de

Prof. Dr.-Ing. Robert Luckner
Berlin Institute of Technology, Marchstr. 12, 10587 Berlin, Germany
e-mail: robert.luckner@tu-berlin.de

and the mission-engineer's workload. Gust load alleviation systems based on feed forward control can alleviate accelerations significantly if the atmospheric disturbances are determined precisely and broadband up to the frequency range of the short period mode. The predominating influences on the utility aircraft Stemme S15, a powered glider with a high aspect ratio and a single propeller engine, are vertical gusts and turbulence and therefore the wind-induced angle of attack (AOA). They cause additional vertical accelerations. For a broadband computation of the not directly measurable, but observable, wind-induced AOA, flight mechanical variables have to be measured accurate and broadband.

In Wise [1] an Extended Kalman filter (EKF) is used to determine the AOA and the angle of side slip (AOSS) based on the nonlinear six degree-of-freedom equations of motion and a detailed aircraft aerodynamic model. The EKF framework uses inertial measurements of Euler angles, body rates, body acceleration and measurement of dynamic pressure. This approach was successfully flight tested in the X-45 program. Heller [2] and Myschik [3] use a complementary filter approach based on linear equations to determine AOA and AOSS to extend the bandwidth of these signals. Braga [4] shows a low computational cost method dealing with time varying noise statistical properties using a new approach for an adaptive EKF. The method is validated in flight path reconstruction application, with simultaneous air data calibration for AOA, AOSS and static pressure sensors. Combining a linear model of the short period motion and the Dryden Turbulence Model, Hoffmann [5] has shown a Kalman-Bucy filter approach for precise and broadband wind-induced AOA determination in presence of atmospheric turbulence and observation noise. Myschik [6] discussed an integrated wind estimation/navigation system for use on general aviation aircraft. This integrated system enables on board determination of the actual wind condition and aerodynamic flow angles using inertial navigation data. A method for estimating wind fields for small and mini unmanned aerial vehicles is described by Langelaan [7]. The approach utilizes sensors that are already part of standard autopilot sensor suite. An approach how wind AOA can be computed using relations of flight mechanic variables that can be measured in sufficient quality is shown by Hahn [8].

In this paper a method using a linear aircraft model of the longitudinal aircraft motion combined with the Dryden Turbulence Model within a discrete Kalman filter framework is described. For proof of the concept and its robustness, different kinds of disturbances such as discrete gusts and a bias in the AOA measurement are simulated and discussed. The results of this approach are compared with results of the approach described in [8].

2 System Model

In this section the linear aircraft model and the Dryden Turbulence Model are described and it is shown how they are combined, for analysis purpose and Kalman filter design.

2.1 Notation

In the aircraft longitudinal motion that is considered here the flight path velocity $\underline{V}_K$ and the airspeed $\underline{V}_A$ are two dimensional vectors in the aircraft's plane of symmetry. The velocity components in x-direction are referred to with the letter u, the ones in z-direction with the letter w. The second index defines the reference frame (g = geodetic, b = body fixed). The velocity vectors are

$$\underline{V}_{Ab} = \begin{bmatrix} u_A \\ w_A \end{bmatrix}_b = \begin{bmatrix} \cos\alpha \\ \sin\alpha \end{bmatrix} V_A, \quad \underline{V}_{Kg} = \begin{bmatrix} u_K \\ w_K \end{bmatrix}_g = \begin{bmatrix} \cos\gamma \\ -\sin\gamma \end{bmatrix} V_K$$

$$\tan\alpha = \frac{w_{Ab}}{u_{Ab}}, \quad V_A = \sqrt{w_{Ab}^2 + u_{Ab}^2}, \quad \sin\gamma = -\frac{w_{Kg}}{V_K}, \quad V_K = \sqrt{w_{Kg}^2 + u_{Kg}^2} \ ,$$

where α is the angle of attack (AOA) and γ is the flight path angle.

2.2 Linear State Space Aircraft Model

The nonlinear equations of motion for an aircraft can be written in state-space form as follows:

$$\dot{\underline{x}}_{ac}(t) = a\left(\underline{x}_{ac}(t), \underline{u}_{ac}(t), \underline{z}(t), t\right) + b\left(\underline{x}_{ac}(t), \underline{u}_{ac}(t)\right) \quad (1)$$

$$\underline{y}_{ac}(t) = c\left(\underline{x}_{ac}(t), \underline{u}_{ac}(t), \underline{z}(t), t\right) \quad , \quad (2)$$

where $\underline{x}_{ac}$ is the state vector, $\underline{u}_{ac}$ is the control input vector, $\underline{y}_{ac}$ is the output vector and $\underline{z}$ is the disturbance vector. The functions $a()$, $b()$ and $c()$ model the nonlinear state dynamics and measurements. This 6 degree of freedom model is linearized for horizontal flight, with respect to the state vector $\underline{x}_{ac}$, the control input vector $\underline{u}_{ac}$, the output vector $\underline{y}_{ac}$ and the disturbance vector $\underline{z}$, resulting in the following linear state-space representation,

$$\dot{\underline{x}}_{ac} = \underline{\underline{A}}_{ac}\, \underline{x}_{ac} + \underline{\underline{B}}_{ac}\, \underline{u}_{ac} + \underline{\underline{E}}_{ac}\, \underline{z} + \underline{\xi}_{ac} \quad (3)$$

$$\underline{y}_{ac} = \underline{\underline{C}}_{ac}\, \underline{x}_{ac} + \underline{\underline{F}}_{ac}\, \underline{z} + \underline{\zeta}_{ac} \quad , \quad (4)$$

with process noise $\underline{\xi}_{ac}$ used to model system and observation noise $\underline{\zeta}_{ac}$ to model measurement uncertainties. The vector elements for the longitudinal motion are:

$$\underline{x}_{ac} = \begin{bmatrix} q_K \\ \alpha_K \\ V_K \\ \Theta \\ H \end{bmatrix}; \ \underline{u}_{ac} = \begin{bmatrix} \eta_F \\ \eta \end{bmatrix}; \ \underline{y}_{ac} = \begin{bmatrix} q_K \\ \alpha \\ V_A \\ V_K \\ \Theta \\ H \\ \dot{H} \end{bmatrix}; \ \underline{z} = \begin{bmatrix} \alpha_W \\ u_W \end{bmatrix}; \ \underline{\xi}_{ac} = \begin{bmatrix} \xi_{q_K} \\ \xi_{\alpha_K} \\ \xi_{V_K} \\ \xi_{\Theta} \\ \xi_H \end{bmatrix}; \ \underline{\zeta}_{ac} = \begin{bmatrix} \zeta_{q_K} \\ \zeta_{\alpha} \\ \zeta_{V_A} \\ \zeta_{V_K} \\ \zeta_{\Theta} \\ \zeta_H \\ \zeta_{\dot{H}} \end{bmatrix}$$

The aircraft states are body axis pitch rate q_K, kinematic AOA α_K, flight path velocity V_K, pitch angle Θ and altitude H. The output vector consists of the body axis pitch rate q_K, the AOA α, the airspeed V_A, the flight path velocity V_K as well as the pitch angle Θ, the altitude H and the vertical speed $\dot{H}$. Input is the elevator deflection η as well as the thrust power setting η_F. Wind-induced AOA α_W and the horizontal wind velocity u_W are the disturbances. Process noise $\underline{\xi}_{ac}$ is assumed to be zero mean Gaussian white noise with covariance $\underline{\underline{Q}}_{ac}$. Observation noise $\underline{\zeta}_{ac}$ is assumed to be zero mean Gaussian white noise with covariance $\underline{\underline{R}}_{ac}$.

$$\underline{\underline{R}}_{ac} = \text{diag}^2 \begin{bmatrix} \frac{0.166\times\pi}{180}\ \text{rad/s} \\ \frac{0.25\times\pi}{180}\ \text{rad} \\ 0.009\ \text{m/s} \\ 0.009\ \text{m/s} \\ \frac{0.25\times\pi}{180}\ \text{rad} \\ 0.0001\ \text{m} \\ 0.027\ \text{m/s} \end{bmatrix} \qquad \underline{\underline{Q}}_{ac} = \text{diag}^2 \begin{bmatrix} \frac{0.166\times\pi}{180}\ \text{rad/s}^2 \\ \frac{0.25\times\pi}{180}\ \text{rad/s} \\ 0.009\ \text{m/s}^2 \\ \frac{0.25\times\pi}{180}\ \text{rad/s} \\ 0.0001\ \text{m/s} \end{bmatrix} \tag{5}$$

The model is derived for horizontal flight for a reference state of $V_A = 50$ m/s, $H = 100$ m, no wind and a reference flight path angle $\gamma = 0°$.

2.3 Disturbance Model

The Dryden Turbulence Model is an approximation of real gust Power Spectral Density (PSD). To generate the PSD for simulation purpose, zero mean Gaussian white noise $\underline{r}(t)$ with covariance $\underline{\underline{I}}$ is shaped through a Dryden filter as shown in Fig. 1. For vertical and horizontal turbulence the transfer functions are given by (6) and (7), where σ is the standard deviation, L is the characteristic wave length, ω is the characteristic cut off frequency and T the characteristic time constant. The variable w_W specifies the vertical and u_W the horizontal wind component.

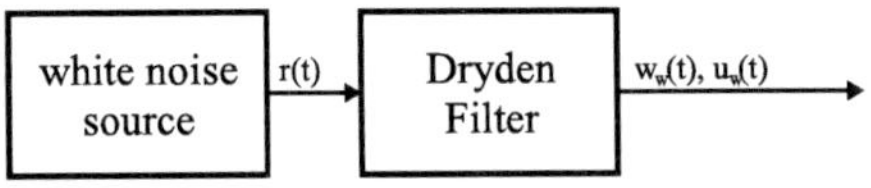

Fig. 1 Turbulence simulation

$$\hat{F}_{ww}(s) = \sqrt{\sigma_w^2 T_w}\frac{1+s\sqrt{3}T_w}{(1+sT_w)^2}\ ; \quad T_w = \frac{L_w}{V_K} = \frac{1}{\omega_w} \tag{6}$$

$$\hat{F}_{uw}(s) = \sqrt{2\sigma_u^2 T_u}\frac{1}{(1+sT_u)}\ ; \quad T_u = \frac{L_u}{V_K} = \frac{1}{\omega_u}\ . \tag{7}$$

The filter characteristics depend on height, terrain roughness and wind speed. The characteristic values of the Dryden filter for the reference state of $V_K = 50$ m/s, $H = 100$ m and moderate turbulence are, according to [9], as follows:

$$L_w = 100\ \text{m},\ L_u = 260\ \text{m},\ \sigma_w = 1\ \text{m/s},\ \sigma_u = 1.38\ \text{m/s}\ .$$

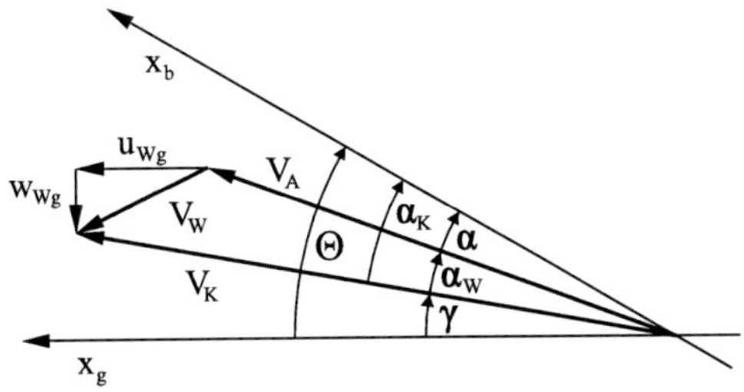

Fig. 2 Velocity vector diagram in the aircraft's plane of symmetry

The kinematical relation of aircraft and wind states for the longitudinal motion is given by the velocity vector equation $\underline{V}_K = \underline{V}_A + \underline{V}_W$ and is shown in Fig. 2. The two components of the wind velocity vector $\underline{V}_{Wg}$ in the geodetic reference frame are u_{Wg} and w_{Wg}. Using the relation shown in Fig. 2, the wind-induced AOA α_W can be calculated with (8)[1].

$$\sin\alpha_W = \frac{w_{Wg}}{V_A}\cos\gamma + \frac{u_{Wg}}{V_A}\sin\gamma \ . \tag{8}$$

For horizontal flight ($\gamma = 0°$) and small angles α_W is approximately

$$\alpha_W = \frac{w_{Wg}}{V_A} \ . \tag{9}$$

Using the linear relation (9) and the transfer functions (6) and (7), the Dryden Turbulence Model can be written in observer state space form as:

$$\underline{\dot{x}}_{Dry} = \underline{\underline{A}}_{Dry}\,\underline{x}_{Dry} + \underline{\underline{B}}_{Dry}\,\underline{r} + \underline{\xi}_{Dry} \tag{10}$$

$$\underline{z} = \underline{\underline{C}}_{Dry}\,\underline{x}_{Dry} \ . \tag{11}$$

And in detail:

$$\begin{bmatrix}\delta\dot{\alpha}_W^*\\ \delta\dot{\alpha}_W\\ \delta\dot{u}_W\end{bmatrix} = \begin{bmatrix}0 & -\omega_w^2 & 0\\ 1 & -2\omega_w & 0\\ 0 & 0 & -\omega_u\end{bmatrix}\begin{bmatrix}\delta\alpha_W^*\\ \delta\alpha_W\\ \delta u_W\end{bmatrix} + \begin{bmatrix}\frac{\sigma_w}{V_A}\sqrt{\omega_w^3} & 0\\ \frac{\sigma_w}{V_A}\sqrt{3\omega_w} & 0\\ 0 & \sigma_u\sqrt{2\omega_u}\end{bmatrix}\begin{bmatrix}r_1\\ r_2\end{bmatrix} + \begin{bmatrix}\xi_{\alpha_W^*}\\ \xi_{\alpha W}\\ \xi_{uW}\end{bmatrix}$$

$$\begin{bmatrix}\delta\alpha_W\\ \delta u_W\end{bmatrix} = \begin{bmatrix}0 & 1 & 0\\ 0 & 0 & 1\end{bmatrix}\begin{bmatrix}\delta\alpha_W^*\\ \delta\alpha_W\\ \delta u_W\end{bmatrix} \ . \tag{12}$$

The state α_W^* is an internal state without physical interpretation. Model uncertainties in the Dryden filter are regarded as process noise ξ_{Dry}. It is assumed to be zero mean Gaussian white noise with covariance $\underline{\underline{Q}}_{Dry}$.

[1] Sign convention as in [10].

$$\underline{\underline{Q}}_{Dry} = \mathrm{diag}^2 \begin{bmatrix} 5\times10^{-5}\times\frac{\pi}{180}\ \mathrm{rad/s^2} \\ 5\times10^{-4}\times\frac{\pi}{180}\ \mathrm{rad/s} \\ 3.5\times10^{-3}\ \mathrm{m/s^2} \end{bmatrix} \quad . \tag{13}$$

2.4 System Model

The aircraft and the disturbance model have to be coupled to become the system model that is used for filter design. The linear aircraft model (3) and (4) can be combined with the Dryden Turbulence Model (6) and (7). The system model state space equations can be written explicitly as follows:

$$\begin{bmatrix} \dot{\underline{x}}_{Dry} \\ \dot{\underline{x}}_{ac} \end{bmatrix} = \begin{bmatrix} \underline{\underline{A}}_{Dry} & \underline{\underline{0}} \\ \underline{\underline{E}}_{ac}\underline{\underline{C}}_{Dry} & \underline{\underline{A}}_{ac} \end{bmatrix} \begin{bmatrix} \underline{x}_{Dry} \\ \underline{x}_{ac} \end{bmatrix} + \begin{bmatrix} \underline{\underline{0}} \\ \underline{\underline{B}}_{ac} \end{bmatrix} \begin{bmatrix} \underline{u}_{ac} \end{bmatrix} + \begin{bmatrix} \underline{\underline{B}}_{Dry} \\ \underline{\underline{0}} \end{bmatrix} \begin{bmatrix} \underline{r} \end{bmatrix} + \begin{bmatrix} \underline{\xi}_{Dry} \\ \underline{\xi}_{ac} \end{bmatrix} \tag{14}$$

$$\begin{bmatrix} \underline{y}_{ac} \end{bmatrix} = \begin{bmatrix} \underline{\underline{F}}_{ac}\underline{\underline{C}}_{Dry} & \underline{\underline{C}}_{ac} \end{bmatrix} \begin{bmatrix} \underline{x}_{Dry} \\ \underline{x}_{ac} \end{bmatrix} + \begin{bmatrix} \underline{\zeta}_{ac} \end{bmatrix} \quad . \tag{15}$$

Since the input vector to the Dryden Turbulence model $\underline{r}$ is not measureable, the term $\begin{bmatrix} \underline{\underline{B}}_{Dry} & \underline{\underline{0}} \end{bmatrix}^{\top} \begin{bmatrix} \underline{r} \end{bmatrix}$ is regarded as a part of the process noise. In short the system model state space can be written as:

$$\dot{\underline{x}} = \underline{\underline{A}}\,\underline{x} + \underline{\underline{B}}\,\underline{u} + \underline{\xi} \tag{16}$$

$$\underline{y} = \underline{\underline{C}}\,\underline{x} + \underline{\zeta} \tag{17}$$

with:

$$\underline{\xi} = \begin{bmatrix} \underline{\xi}_{Dry} \\ \underline{\xi}_{ac} \end{bmatrix} + \begin{bmatrix} \underline{\underline{B}}_{Dry} \\ \underline{\underline{0}} \end{bmatrix} \begin{bmatrix} \underline{r} \end{bmatrix}, \qquad \underline{\zeta} = \underline{\zeta}_{ac} \tag{18}$$

$$\underline{\underline{Q}} = \begin{bmatrix} \underline{\underline{Q}}_{Dry} + \mathrm{diag}^2\left(\begin{bmatrix} \underline{\underline{B}}_{Dry} \end{bmatrix} \begin{bmatrix} 1 \\ 1 \end{bmatrix} \right) & \underline{\underline{0}} \\ \underline{\underline{0}} & \underline{\underline{Q}}_{ac} \end{bmatrix} , \tag{19}$$

where $\underline{\underline{Q}}$ is the covariance matrix used for filter design.

2.5 Frequency Range Analysis

To identify the frequency range where significant normal accelerations occur and to define the frequency range, for which the disturbance signal has to be reconstructed for feed forward gust alleviation, the amplitude response of the aircraft $\left|F_{n_z\,\alpha_W}\right|$, the Dryden filter $\left|F_{\alpha_W\,r_1}\right|$ and the combination of both $\left|F_{n_z\,r_1}\right|$ is plotted in Fig. 3. The characteristic high-pass behaviour of the aircraft can be clearly seen as well as the low-pass behavior of the Dryden filter. In combination two maxima result, one close two the Eigen-frequency of the phygoid mode ($\omega_{Ph} = 0.18$ rad/s) and the second close to the short period mode ($\omega_{SP} = 4.7$ rad/s). For gust load alleviation both

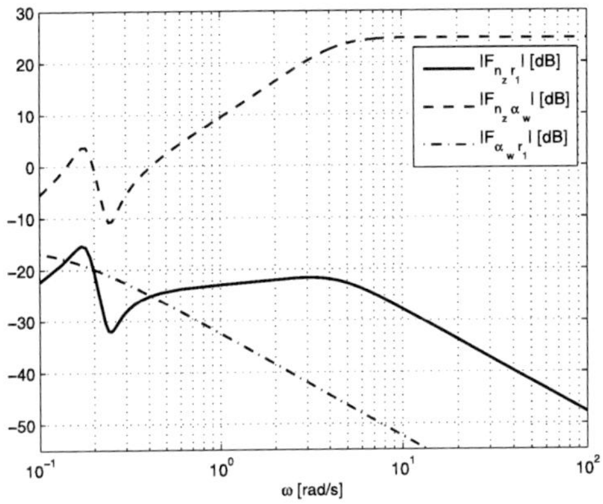

Fig. 3 Amplitude response

maxima are of interest and therefore the disturbance signal has to be reconstructed up to a frequency range of approximately 10 rad/s. For this analysis aeroelastic and instationary aerodynamic effects are not taken into account.

3 Kalman Filter

The presented approach uses a Kalman filter to estimate the states of the linear system model (16) and (17).

The Kalman filter estimates the state vector $\underline{x}$ by processing the output vector $\underline{y}$. This estimation is generated in two phases. The predict algorithm calculates a value according to the state equation and estimates the uncertainty of this value. Then the update algorithm computes a weighted average of the predicted value and the output value. To receive an optimum estimate, a Kalman gain $\underline{\underline{K}}$ is computed for which the trace of the estimate error covariance matrix $\underline{\underline{P}}$ is minimal. The estimate error covariance matrix is defined as $\underline{\underline{P}} = \mathrm{E}\left[(\underline{x}-\hat{\underline{x}})(\underline{x}-\hat{\underline{x}})^\top\right]$, where E denotes the expectation operator and $\hat{\underline{x}}$ the estimated state vector.

The notation used to illustrate the two different phases has two subscripts. The first subscript indicates the point in time, while the second indicates the processing of the k-th set of measurements. The predict algorithm propagates $x_{k-1|k-1}$ to $x_{k|k-1}$ and the update algorithm propagates $x_{k|k-1}$ to $x_{k|k}$. Where only one value is calculated for a time step, the second subscript is dropped.

Since it is fundamental to software processing and the nature of the Kalman filter, a discrete system model has to be used. Therefore the continuous linear system model (16) and (17) is discretized. This is done using a zero-order hold and results in the following equations.

$$\underline{x}_{k+1} = \underline{\underline{F}}\,\underline{x}_k + \underline{\underline{G}}\,\underline{u}_k + \underline{\xi}_k \tag{20}$$

$$\underline{y}_k = \underline{\underline{H}}\,\underline{x}_k + \underline{\zeta}_k \quad . \tag{21}$$

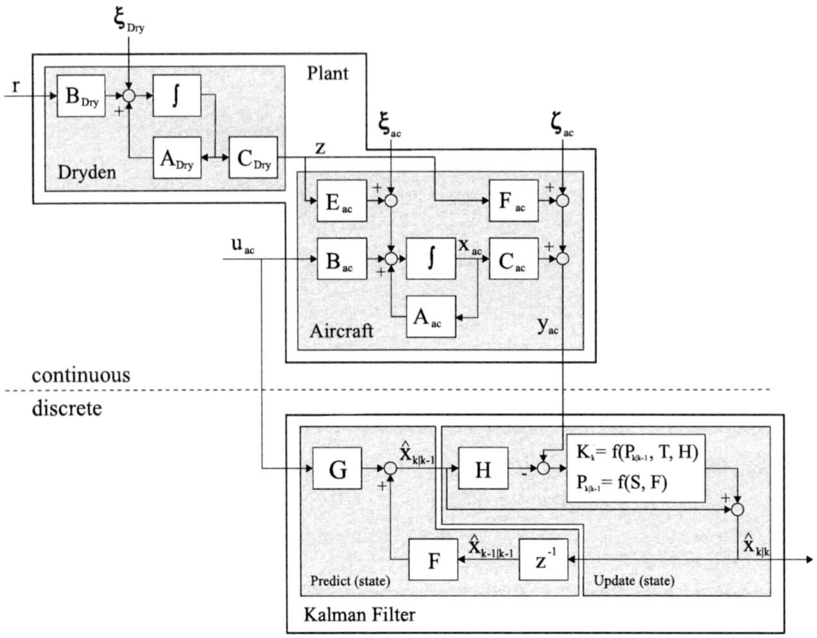

Fig. 4 Block diagram of plant and Kalman filter

The equations for the predict and update steps for the state vector and the estimate error covariance matrix are:

<u>Predict</u>:

$$\underline{\hat{x}}_{k|k-1} = \underline{\underline{F}}\,\underline{\hat{x}}_{k-1|k-1} + \underline{\underline{G}}\,\underline{u}_k \tag{22}$$

$$\underline{\underline{P}}_{k|k-1} = \underline{\underline{F}}\,\underline{\underline{P}}_{k-1|k-1}\underline{\underline{F}}^{\top} + \underline{\underline{S}} \tag{23}$$

<u>Update</u>:

$$\underline{\underline{K}}_k = \underline{\underline{P}}_{k|k-1}\underline{\underline{H}}^{\top}\left(\underline{\underline{T}} + \underline{\underline{H}}\,\underline{\underline{P}}_{k|k-1}\underline{\underline{H}}^{\top}\right)^{-1} \tag{24}$$

$$\underline{\hat{x}}_{k|k} = \underline{\hat{x}}_{k|k-1} + \underline{\underline{K}}_k\left(\underline{y}_k - \underline{\underline{H}}\,\underline{\hat{x}}_{k|k-1}\right) \tag{25}$$

$$\underline{\underline{P}}_{k|k} = \left(\underline{\underline{\mathrm{I}}} - \underline{\underline{K}}_k\,\underline{\underline{H}}\right)\underline{\underline{P}}_{k|k-1} \tag{26}$$

with $\underline{\underline{S}}$ and $\underline{\underline{T}}$ being the discrete process and observation noise derived from the process and observation noise described in Sect. 2.4. In addition, the Kalman filter can estimate constant errors (bias) in the measurement . The bias $\underline{b}_k$ can be interpreted as an additional input vector in the output equation (21) with the new feedthrough matrix $\underline{\underline{\tilde{D}}}$. Since $\underline{b}_k$ is a constant vector a new equation is added to the system model.

$$\underline{x}_{k+1} = \underline{\underline{F}}\,\underline{x}_k + \underline{\underline{G}}\,\underline{u}_k + \underline{\xi}_k \tag{27}$$

$$\underline{b}_{k+1} = \underline{b}_k \tag{28}$$

$$\underline{y}_k = \underline{\underline{H}} + \underline{\underline{\tilde{D}}}\,\underline{b}_k + \underline{\zeta}_k \ . \tag{29}$$

Equations (27) - (29) can be written in state space representation of the following form:

$$\begin{bmatrix} \underline{x}_{k+1} \\ \underline{b}_{k+1} \end{bmatrix} = \begin{bmatrix} \underline{\underline{F}} & \underline{\underline{0}} \\ \underline{\underline{0}} & \underline{\underline{I}} \end{bmatrix} \begin{bmatrix} \underline{x}_k \\ \underline{b}_k \end{bmatrix} + \begin{bmatrix} \underline{\underline{G}} \\ \underline{\underline{0}} \end{bmatrix} \begin{bmatrix} \underline{u}_k \end{bmatrix} + \begin{bmatrix} \underline{\underline{I}} \\ \underline{\underline{0}} \end{bmatrix} \begin{bmatrix} \underline{\xi}_k \end{bmatrix} \tag{30}$$

$$\begin{bmatrix} \underline{y}_k \end{bmatrix} = \begin{bmatrix} \underline{\underline{H}} & \underline{\underline{\tilde{D}}} \end{bmatrix} \begin{bmatrix} \underline{x}_k \\ \underline{b}_k \end{bmatrix} + \begin{bmatrix} \underline{\zeta}_k \end{bmatrix} \ . \tag{31}$$

Equations (30) and (31) are of a form on which the Kalman filter is applicable. The block diagram in Fig. 4 shows the structure of the system model and the Kalman filter. For simplification, the propagation of the estimate error covariance is not explicitly shown in the block diagram.

4 Algorithm Using Simplified Flight Mechanic Relations

For the longitudinal motion the following equation is exact as shown in Fig. 2

$$\alpha_W = \Theta - \gamma - \alpha \ . \tag{32}$$

Due to measurement quality the flight path angle is approximated by $\gamma \approx \arcsin\left(\frac{\dot{H}}{V_A}\right)$. With this approximation the estimation equation for α_W reads

$$\alpha_{W\,DLR} \approx \Theta - \arcsin\left(\frac{\dot{H}}{V_A}\right) - \alpha \ . \tag{33}$$

This estimation was successfully implemented in a gust load alleviation system by the German Aerospace Center (DLR) [8] and is here used for comparison to the presented algorithm.

5 Simulation Results

The system model and the Kalman filter, as shown in Fig. 4, were implemented in Simulink. The simulation sample time was 0.016 s. Fig. 5 - Fig. 7 show simulation results for the derived system model, the Kalman filter and the simplified flight mechanical approach. The aircraft is in trimmed condition, disturbed by Dryden turbulence as well as discrete $1-\cos$ gusts for α_W ($t = 10$ s and $t = 70$ s) and u_W ($t = 50$ s) and responding to elevator deflection. Simulated states of the real system are labeled with the index "real", estimated states with "est" and the simplified flight mechanical approach with "DLR". The strength of observation and process noise is as defined in Sects. 2.2 and 2.3. The simulations were performed on a standard PC.

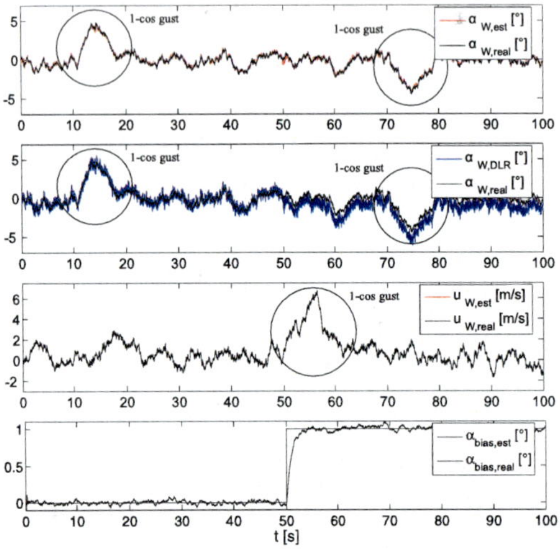

Fig. 5 Simulated time histories of the turbulent states and their estimates

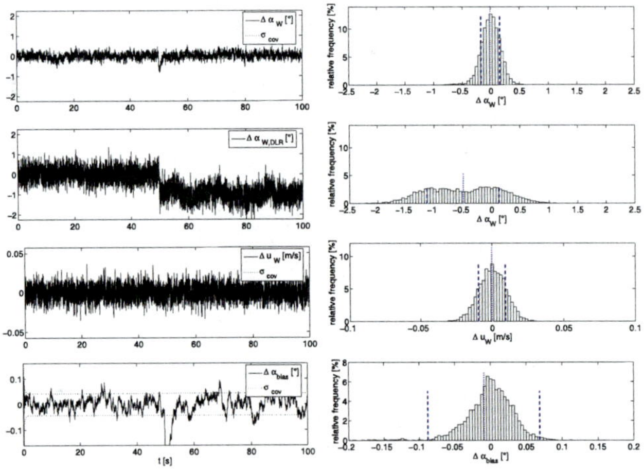

Fig. 6 Error analysis of the turbulent states and their estimates

The not directly measurable turbulence states and the bias state ($\alpha_{bias} = 1°; t = 50$ s) are compared with their estimates in Figs. 5 and 6.

Fig. 6 shows time histories of the error ($\Delta = x_{real} - x_{est}$) between the state variables and their estimates and relative frequency distributions of the error. The dotted lines indicate the mean value and the dashed lines the standard deviations of the error. The standard deviation, calculated from the trace of the estimation error covariance matrix σ_{cov} is shown for the estimations computed by the Kalman filter and is a measure for the quality of estimation. The correlation between the turbulence states, the bias and the estimated states computed by the Kalman filter show

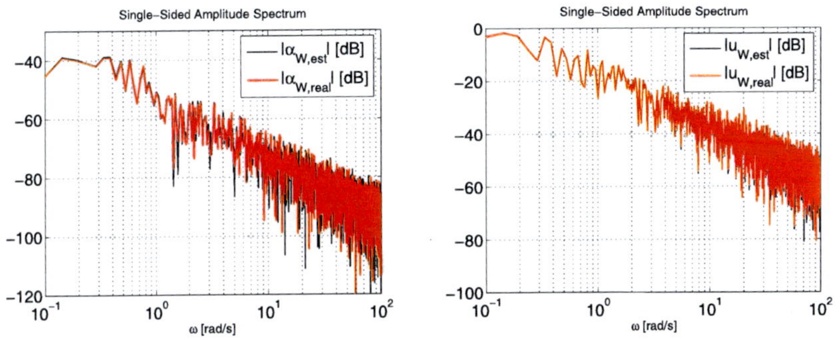

Fig. 7 Fourier Analysis of α_W (left) and u_W (right) and their estimation

to be sufficient for gust load alleviation. The simplified flight mechanical approach estimates the wind induced AOA well up to the occurrence of the bias in the AOA measurement. Due to this bias the relative frequency distribution for the estimated wind induced AOA error is broad and flat.

Fourier analysis in Fig. 7 shows a broadband correlation of the estimated turbulence states and the simulated states which emphasises the correlation seen in the time histories. The frequency range requirement derived for gust load alleviation, i.e. good agreement between real and estimated turbulence up to 10 rad/s, is easily met.

6 Conclusion

The application of a discrete Kalman filter using a combination of a linear aircraft model and the Dryden Turbulence Model to estimate the not directly measurable wind states has proven to be feasible and shows better results than an algorithm using simplified flight mechanical relations. Simulation results show a good and broadband correlation between the states and their estimates up to the short period mode which is sufficient for gust load alleviation. The estimations are robust against variations of the disturbance model and a bias in the AOA measurement.

For operational use in an aircraft this approach shall be developed further using an extended Kalman filter (EKF) based on the described concept and a non-linear aircraft model. The EKF has to be carefully tuned by observation and minimizing of the estimation error covariance matrix. The observability of each state under different conditions, e.g. no turbulence, has to be analyzed.

References

1. Wise, K.: Flight Testing of the X-45A J-UCAS Computational Alpha-Beta System. In: AIAA Guidance, Navigation and Control Conference and Exhibit, Keystone, Colorado, August 21-24 (2006)

2. Heller, M., Myschik, S., Holzapfel, F., Sachs, G.: Low-cost approach based on navigation data for determing angles of attack and sideslip for small aircraft. In: AIAA Guidance, Navigation and Control Conference and Exhibit, Austin, Texas, August 11 - 14 (2003)
3. Myschik, S., Heller, M., Holzapfel, F., Sachs, G.: Low-cost System zur Bestimmung von Anstell- und Schiebewinkel fuer Kleinflugzeuge. Deutscher Luft und Raumfahrt Kongress, Munich (2003)
4. Braga de Mendonca, C.: Adaptive stochastic Filtering for online aircraft flight path reconstruction. Journal of aircraft 44(5) (September-October 2007)
5. Hoffmann, A., Luckner, R.: Broadband prediction of wind-induced disturbances. Deutscher Luft und Raumfahrt Kongress, Hamburg (2010)
6. Myschik, S., Sachs, G.: Flight testing an integrated wind/airdata and navigation system for general aviation aircraft. In: AIAA Guidance Navigation and Control, Hilton Head, South Carolina (2007)
7. Langelaan, J.W.: Wind field estimation for small unmanned aerial vehicles. In: AIAA Guidance, Navigation and Control Conference, Toronto, Ontario, Canada, August 2 - 5 (2010)
8. Hahn, K.-U., Koenig, R.: Flight Test and Simulation Results of the Advanced Gust Management System LARS. In: Proc. of the AIAA Atmospheric Flight Mechanics Conference, Hilton Head, South Carolina, USA (1992)
9. U.S. Military Specification MIL-F-8785C (1980)
10. Brockhaus, R.: Flugregelung., 2. bearb. Auflage. Springer, Heidelberg (2001), ISBN 3-540-55416-5

Interval Analysis as a System Identification Tool

E. van Kampen, Q.P. Chu, and J.A. Mulder

Abstract. This paper shows how the theory of interval analysis can be utilized for system identification by applying it to two aerospace related problems. Interval numbers form an extension of the regular crisp numbers and they were first introduced in the 1960's to investigate how rounding errors propagate through systems implemented on digital computers. Since then interval analysis is mainly used as a tool for solving global nonlinear optimization problems where it is one of the few methods that can guarantee to find the global minimum of any nonlinear cost function. Here the properties of intervals are employed for system identification tasks. The first application is optimization of human perception modeling, i.e. identifying how a human pilot perceives visual cues and motion cues, based on the pilot input and output. In the second example trim points of a nonlinear aircraft model are identified simultaneously, with the guarantee that all trim points are found.

1 Introduction

System identification is a crucial part in many engineering problems. An example is in adaptive flight control, where an online aerodynamic model is required by the controller to adapt its control strategy after a structural failure or a control surface failure. System identification consists of defining a model structure (Fig. 1) and then optimizing the parameters in this model such that the model approximates the true system as closely as possible:

$$y = f(\mathbf{x}) \qquad \hat{y} = f(\mathbf{x};\mathbf{p}) \qquad \min_{\mathbf{p}\in\mathbf{P}} J(\mathbf{p}) = \|y - \hat{y}(\mathbf{p})\| \tag{1}$$

with $\mathbf{x}$ the input vector, $\mathbf{p}$ the parameter vector, y the true system output, and $\hat{y}$ the model output.

E. van Kampen · Q.P. Chu · J.A. Mulder
Faculty of Aerospace Engineering, Delft University of Technology, 2629 HS Delft, The Netherlands
e-mail: E.vanKampen@TUDelft.nl, Q.P.Chu@TUDelft.nl, J.A.Mulder@TUDelft.nl

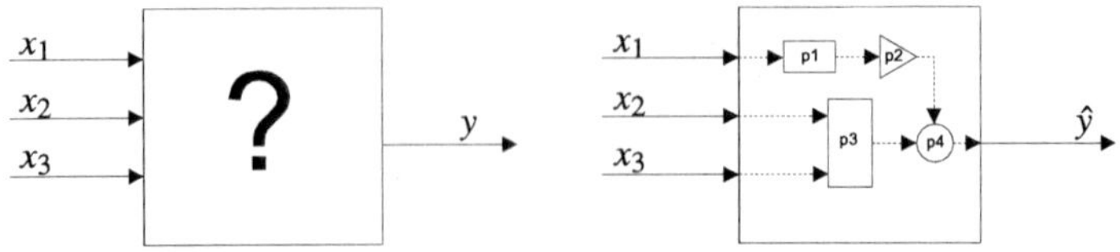

Fig. 1 First step of system identification: selecting the model structure.

This equation shows that system identification is a minimization problem that, depending on the model structure, can be nonlinear and have multiple local minima. A large number of methods is available to solve these types of problems, from which gradient-based methods and genetic algorithms are among the most frequently used.

With gradient based methods are meant any algorithm that directly or indirectly uses the slope of the cost function to find a (local) minimum. Some algorithms use this slope or gradient directly, such as the Newton method, while others still use slope information in a more subtle way. The Nelder-Mead algorithm[1] for example compares the function evaluations of the vertices of a given simplex and acts based on the ordering of these function evaluations, which can be seen as an indirect way of using slope information without explicit evaluation of the cost function derivative.

Genetic Algorithm[2] optimization methods are based on principles of biological evolution, such as natural selection and genetics. Both gradient-based optimization methods and genetic algorithms can get stuck in a local minimum of the cost function and their solution depends on the initialization of the algorithm. With a different starting point or population, a different solution can be found, and it is not possible to determine when the global minimum of the cost function has been reached.

In this paper interval methods are applied to the optimization problem. Interval methods are guaranteed to find the global minimum of the cost function, resulting in the optimal set of parameters.

1.1 Interval Analysis

Interval numbers are an extension of the regular numbers and consist of a lower bound, an upper bound, and all the real numbers in-between these bounds.

$$[x] = [a,b] = \{x^* \in \mathbb{R} | a \leqslant x^* \leqslant b\} \tag{2}$$

$$\inf([x]) = a \tag{3}$$

$$\sup([x]) = b \tag{4}$$

Interval analysis was introduced in the 1960's by Ramon Moore[3], who used it to analyze how rounding errors propagate on limited precision digital computers. Later, interval analysis was found to be an excellent tool for solving global nonlinear optimization problems[4].

The basic arithmetic operations on interval numbers are defined as:

$$[x] \Diamond [y] = \{x^* \Diamond y^* \,|x^* \in [x], y^* \in [y]\} \tag{5}$$

$$[a,b] + [c,d] = [a+c, b+d] \tag{6}$$
$$[a,b] - [c,d] = [a-d, b-c] \tag{7}$$
$$[a,b] \cdot [c,d] = [\min(ac,ad,bc,bd), \max(ac,ad,bc,bd)] \tag{8}$$
$$\frac{[a,b]}{[c,d]} = [a,b] \cdot [1/d, 1/c] \qquad if 0 \notin [c,d] \tag{9}$$

All interval arithmetic operations are based on the inclusion principle, sometimes referred to as the fundamental theorem of interval analysis, which states that the outcome of the operation on a subset of the input interval arguments is included in the outcome of the operation performed on the complete input intervals. The subsets can be smaller intervals or crisp numbers (thin intervals):

$$\begin{array}{l} x_1^* \in [x_1], x_2^* \in [x_2], \ldots, x_n^* \in [x_n] \quad \Rightarrow \\ f(x_1^*, x_2^*, \ldots, x_n^*) \subset [f([x_1], [x_2], \ldots, [x_n])] \end{array} \tag{10}$$

The next section explains how these interval numbers are used to solve global optimization problems.

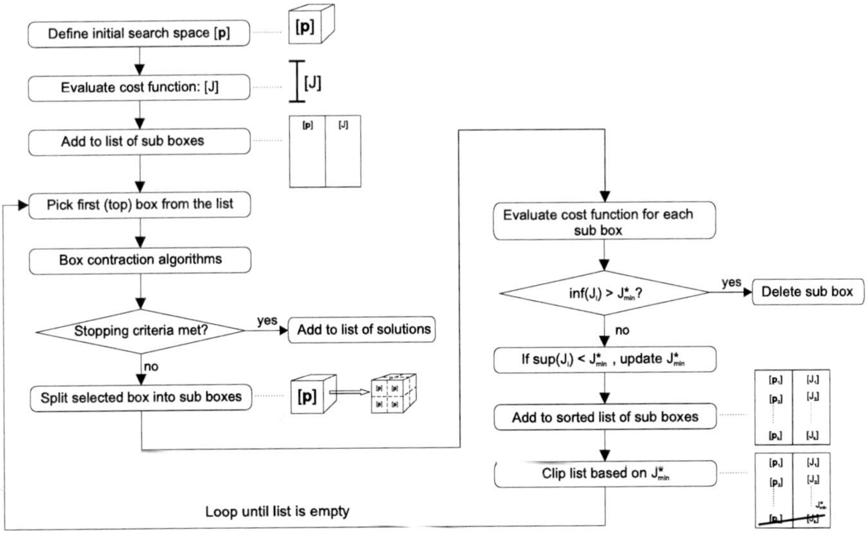

Fig. 2 Branch and bound algorithm for interval optimization.

1.2 Interval Optimization

Two key aspects of interval analysis play role in optimization: inclusion and subdivision. The interval inclusion theorem allows an infinite set of consecutive parameters to be evaluated as one bounded interval, ensuring that all possible outcomes are included in the interval outcome. Interval subdivision, often called box splitting, is the process of subdividing an interval into subintervals and thereby reducing the dependency effect, which is the overestimation that can occur when working with interval numbers. The combination of division(branching) and inclusion(bounding) is proceduralized in the so-called interval branch and bound methods. Branch and bound methods form the backbone of all interval optimization algorithms[5]. Figure 2 shows a schematic diagram of an interval branch and bound algorithm that is guaranteed to find the minimum of a nonlinear cost function.

Figure 3 shows an example of interval optimization using the branch and bound method. First the domain is subdivided and the cost function is evaluated, using interval arithmetic operations, over each sub-interval. The lowest cost function estimate ρ is set as the lowest upper bound of these function evaluations. Next, interval boxes in the domain that lead to an interval cost function evaluation with a lower bound higher than ρ can be removed from the search space. This process is repeated until the global minimum is found.

Some examples of applications of interval optimization are found in the field of neural network optimization[6] and integer ambiguity resolution[7, 8]

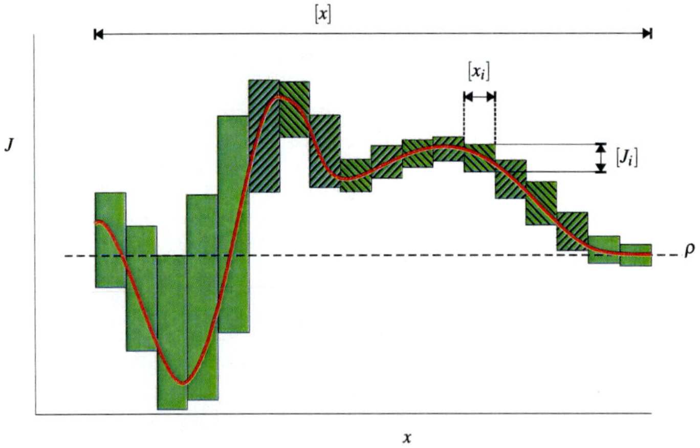

Fig. 3 Principles of applying the branch and bound algorithm (hatched boxes can be eliminated based on the minimum cost function value estimate ρ)

2 Human Perception Modeling

The first application is part of ongoing research on the cybernetic approach to assess simulator fidelity[9]. In order to make flight simulators more realistic, it is important to know what pilots see and feel in the simulator and how they react to these inputs. Therefore models of pilot perception are created using pilot inputs and responses from both the simulator and a real aircraft (see Fig. 4). With the information obtain by comparing these perception models, the motion cueing algorithms of the simulator can be adapted to increase simulator fidelity.

Fig. 4 SIMONA Research Simulator and Citation II laboratory aircraft of the Delft University of Technology.

Figure 5 gives the multi-loop closed-loop control task that is used to identify the human perception parameters[10].

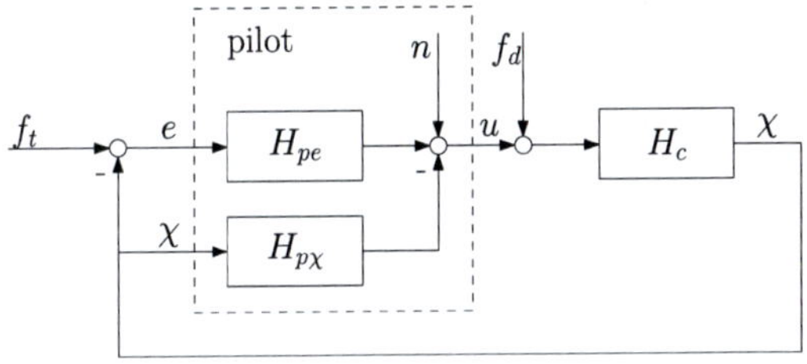

H_pe Visual perception channel
$H_p\chi$ Motion perception channel
H_c Controlled plant dynamics
f_t Target signal
f_d Disturbance signal
n Remnant
u Pilot output
e Error signal
χ Controlled state

Fig. 5 Multi-loop closed-loop control task.

The visual and motion channels are described by:

$$H_{pe}(j\omega) = K_v(1 + j\omega T_v)e^{-j\omega\tau_v}H_{nm}(j\omega) \tag{11}$$

$$H_{p\chi}(j\omega) = (j\omega)V H_{oto}(j\omega)K_m e^{-j\omega\tau_m}H_{nm}(j\omega). \tag{12}$$

In the visual perception channel, H_{pe}, K_v is the visual perception gain, T_v the visual lead time constant and τ_v the visual perception time delay. The control action of the pilot is limited by the neuromuscular dynamics H_{nm}. The physical motion perception channel, $H_{p\chi}$, includes the dynamics of the otoliths H_{oto}, the motion perception gain K_m and a motion perception time delay τ_m.

Some of these parameters can be obtained from previous research, leaving the set of parameters that needs to be identified as:

$$\mathbf{p} = \left[K_v \; T_v \; \tau_v \; K_m \; \tau_m \right]^T \tag{13}$$

The identification is performed in two steps. In the first step, non-parametric frequency response functions are estimated from the time domain data. The non-parametric frequency response functions can be estimated by using Fourier coefficients or linear time-invariant (LTI) models (e.g. autoregressive exogenous (ARX) models) and are used in the second step to fit a multimodal pilot model by adjusting the parameters. In the second step, a comparison is made between gradient-based optimization and interval optimization, see Fig. 6.

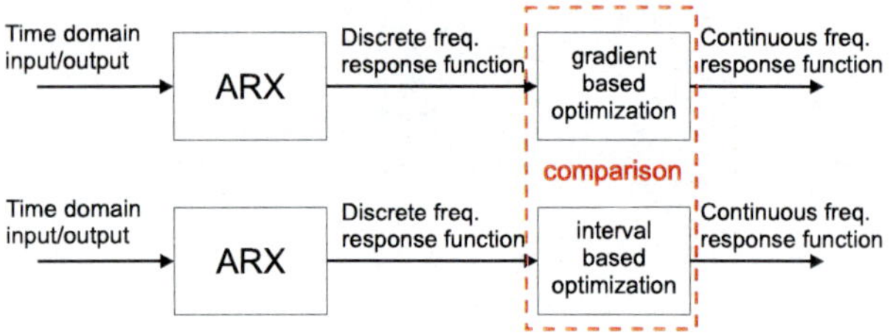

Fig. 6 Two-step identification.

The cost function for this problem is:

$$J = \sum_{\omega} \frac{\left|\hat{H}_{pe}(j\omega) - \tilde{H}_{pe}(j\omega, P)\right|^2}{\hat{\sigma}^2_{|\hat{H}_{pe}|}} + \frac{\left|\hat{H}_{p\chi}(j\omega) - \tilde{H}_{p\chi}(j\omega, P)\right|^2}{\hat{\sigma}^2_{|\hat{H}_{p\chi}|}}, \tag{14}$$

with:
$\hat{H}_{pe}(j\omega)$ Discrete frequency response for the visual channel
$\tilde{H}_{pe}(j\omega, P)$ Parameterized frequency response for the visual channel
$\hat{H}_{p\chi}(j\omega)$ Discrete frequency response for the motion channel
$\tilde{H}_{p\chi}(j\omega, P)$ Parameterized frequency response for the motion channel

Table 1 gives the results of the comparison between the interval branch and bound algorithm and the gradient-based optimization.

Table 1 Solution of the gradient-based optimization method for the baseline condition and several initial starting points.

	$x_0(0)$	$x_0(1/4)$	$x_0(1/2)$	$x_0(3/4)$	$x_0(1)$	Interval
K_v	0.6058	0.6058	0.6058	0.6058	0.6058	[0.6057, 0.6059]
T_v	0.0921	0.0921	0.0921	0.0921	0.0921	[0.1518, 0.1522]
τ_v	0.2621	0.2621	0.2621	0.2621	0.2621	[0.2620, 0.2622]
K_m	0.0001	0.0005	0.0001	0.0004	0.0004	[0.0011, 0.0012]
τ_m	0.0001	0.6706	0.0001	1.0000	1.0000	[0.3047, 0.3048]
J	24.2229	23.0905	24.2229	23.4161	23.4161	[18.3237, 18.3731]

As initial condition for the gradient-based optimization, 5 different points are selected, denoted by $x_0(0)$ to $x_0(1)$, which means 5 points are chosen that are equally distributed from one side of the search space $x_0(0)$ to the other side of

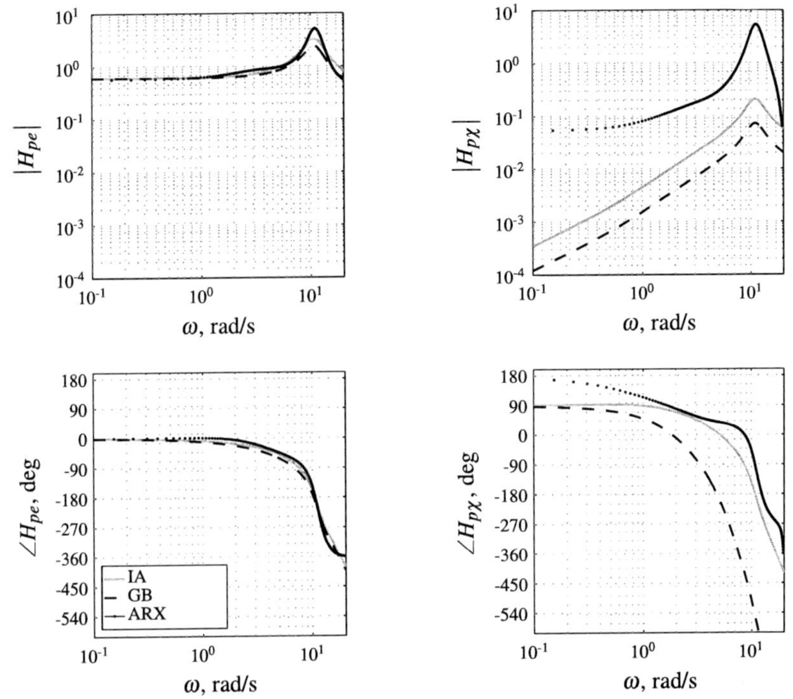

Fig. 7 Comparison between pilot visual(H_{pe}) channel and motion($H_{p\chi}$) channel frequency response functions for gradient-based optimization(GB) and interval optimization(IA).

the search space $x_0\,(1)$. The interval results are presented in the last column. Clearly the cost function J obtained with the interval methods is lower than those found by gradient-based optimization, which indicates that local minimums are present in the cost function and that the gradient method can get stuck here. Figure 7 confirms these results by showing that the frequency response functions found by the interval method(IA) are closer to the target(ARX) than those found by the gradient method(GB).

3 Aircraft Trim

The second application concerns identification of the trim points of aircraft. Trim points are defined as those combinations of states and controls for which the translational and rotational accelerations of the aircraft are zero. Analysis of the trimmed or steady state equations of motion of an aircraft is an important and essential part of the design phase, in particular for the control system. By examining the relation between the steady state control inputs and the flight conditions, the stability properties of the aircraft can be determined. Finding the solutions to the steady state equations of motion for nonlinear aircraft models is a global nonlinear optimization problem.

The equations of motion of the aircraft model that is examined here are [11]:

$$\begin{aligned}
\dot{u} &= rv - qw - g\sin\theta + \frac{q_d S}{m} C_X + \frac{T}{m} \\
\dot{v} &= pw - ru + g\cos\theta\sin\phi + \frac{q_d S}{m} C_Y \\
\dot{w} &= qu - pv + g\cos\theta\cos\phi + \frac{q_d S}{m} C_Z \\
\dot{p} I_X - \dot{r} I_{XZ} &= pq I_{XZ} - qr\left(I_Z - I_Y\right) + q_d S b C_l \\
\dot{q} I_Y &= pr\left(I_Z - I_X\right) - \left(p^2 - r^2\right) I_{XZ} + q_d S c C_m - r H_{eng} \\
\dot{r} I_Z - \dot{p} I_{XZ} &= pq\left(I_X - I_Y\right) - qr I_{XZ} + q_d S b C_n + q H_{eng}
\end{aligned}$$

where u,v,w are the translational velocities and p,q,r the rotational velocities. Converting this equation to interval notation and substituting the control and aerodynamic coefficients, in this case for an F-16 model, yields the following set of interval equations:

$$\begin{aligned}
[f_1] &= -g\sin([\theta]) + \tfrac{q_d S}{m}\left[C_{X_0}([\alpha],[\delta_e])\right] + \tfrac{1000}{m}[T] \\
[f_2] &= g\cos([\theta])\sin([\phi]) \\
&\quad + \tfrac{q_d S}{m}\left(-3.50\cdot 10^{-4}[\beta] + 1.83\cdot 10^{-5}[\delta_a] + 5.0\cdot 10^{-5}[\delta_r]\right) \\
[f_3] &= g\cos([\theta])\cos([\phi]) + \tfrac{q_d S}{m}\left(\left[C_{Z_0}([\alpha])\right]\left(1-[\beta]^2\right) - 1.33\cdot 10^{-4}[\delta_e]\right) \\
[f_4] &= \left[C_{l_0}([\alpha],[\beta])\right] + \left[\Delta C_{l,\delta_a}([\alpha],[\beta])\right][\delta_a] + \left[\Delta C_{l,\delta_r}([\alpha],[\beta])\right][\delta_r] \\
[f_5] &= \left[C_{m_0}([\alpha],[\delta_e])\right] + \left(x_{c.g.ref} - x_{c.g.}\right)\left[C_{Z_0}([\alpha])\right]\left(1-[\beta]^2\right) \\
&\quad - \left(x_{c.g.ref} - x_{c.g.}\right) 1.33\cdot 10^{-4}[\delta_e] \\
[f_6] &= \left[C_{n_0}([\alpha],[\beta])\right] + \left[\Delta C_{n,\delta_a}([\alpha],[\beta])\right][\delta_a] + \left[\Delta C_{n,\delta_r}([\alpha],[\beta])\right][\delta_r] \\
&\quad - \tfrac{c}{b}\left(x_{c.g.ref} - x_{c.g.}\right)\left(-3.5\cdot 10^{-4}[\beta] + 1.83\cdot 10^{-5}[\delta_a] + 5\cdot 10^{-5}[\delta_r]\right) \\
[f_7] &= \cos([\alpha])\cos([\beta])\sin([\theta]) - \sin([\alpha])\cos([\beta])\cos([\phi]) \\
&\quad - \sin([\beta])\sin([\phi])\cos([\theta])
\end{aligned} \tag{15}$$

where $[f_1]$ to $[f_6]$ are the accelerations that we try to get to zero and $[f_7]$ a kinematic constraint for horizontal flight. The aerodynamic and control coefficients C_{Z_0},C_{l_0},$\Delta C_{l,\delta_r}$, $\Delta C_{l,\delta_a}$,C_{m_0},$\Delta C_{n,\delta_r}$,$\Delta C_{n,\delta_a}$ are stored in data tables and an interval interpolation method (see fig. 8) is required to get these coefficients as intervals. Although table based data is used here, any type of aerodynamic model parameterization can be used, for example using multivariate splines [12], as long as the values can be bounded by interval functions.

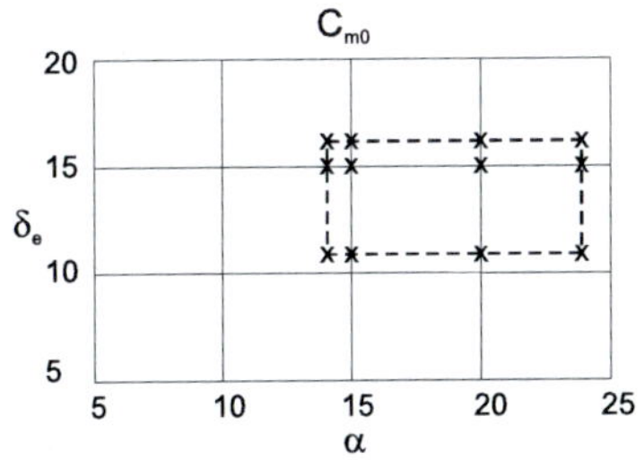

Fig. 8 Interval linear interpolation for C_{m_0} returns an interval constructed from the lowest and highest value of the evaluation of the marked points.

The parameter vector for this problem consists of the angle of attack, elevator deflection, thust level, sideslip angle, aileron deflection, rudder deflection, and pitch angle:

$$[P] = ([\alpha], [\delta_e], [T], [\beta], [\delta_a], [\delta_r], [\theta])$$

The goal is to find the values of these parameters such that the total acceleration of the aircraft is minimized:

$$J = \sum_{i=1}^{7} \|[f_i]\|$$

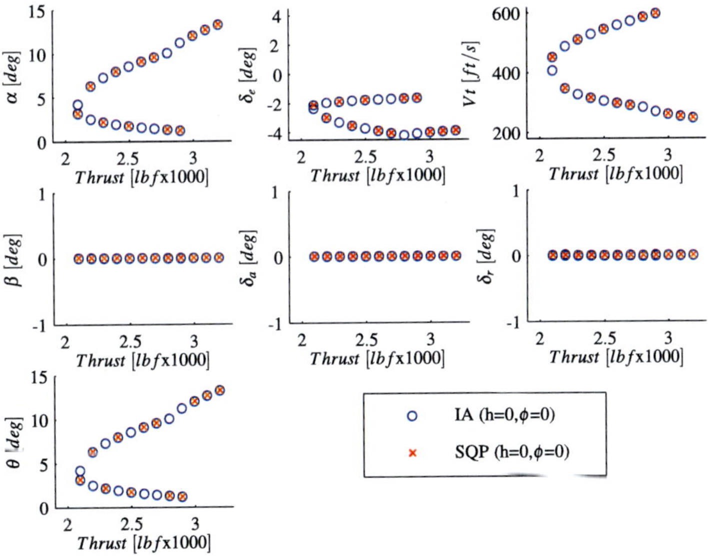

Fig. 9 Comparison between interval-trim (IA) and SQP-trim for the case of horizontal level flight with fixed thrust.

The interval branch and bound algorithm deletes parts of the search space that do not result in zero acceleration and therefore cannot contain a trim point. A comparison is made between the interval trim algorithm and the well known sequential quadratic programming (SQP) trim algorithm. Figure 9 gives the result of this comparison for the case of level horizontal flight with fixed thrust. It is well known that aircraft can have two distinct trim points for fixed thrust, one where the airspeed is high and angle of attack low, and one where the airspeed is low and the angle of attack is high. This latter trim point is in what is called the backside of the power curve. With SQP, only a single trim point is found for each thrust setting, while with the interval method, both trim points are found simultaneously. The SQP method relies on an initial guess, and by changing this guess the other trim point can be found. However, the problem with conventional trim methods is that it is unclear when all trim points are found. There might always be a different initial guess leading to another trim point. With interval trim this problem is resolved, since it can be guaranteed that all trim points are found simultaneously.

4 Discussion on the Implementation of Interval Methods

There are several software packages available that can perform interval operations. For the results in this paper the INTLAB[13] toolbox for MATLAB by S.H. Rump is used. From the definition of the basic interval operations (eq.5), it can be concluded that interval operations require more computation effort than crisp operations, ranging from double the number of computations for addition and subtraction, up to 8 times as many computations for multiplication and division. Additionally, the nature of the branch and bound algorithm can lead to an exponential growth in computation effort with an increase in the number of parameters, although this is not necessarily the case.

As a consequence, interval optimization algorithms will require more computational effort than their crisp counterparts. Whether to choose for an interval optimization method, depends very much on the application and the need for a guaranteed solution. If the latter is more important than the computation time, then interval methods should be considered. As an indication of the increase in computation time, the results in section 3 on aircraft trim take about 200 times as long to compute with interval methods. On the other hand, it will take more than one attempt with non-interval methods to get all the solutions, and even after 200 attempts there is no guarantee that all the solutions are found.

5 Conclusions

In this paper it is shown that the system identification problem often contains a nonlinear optimization problem. Conventional methods for solving this nonlinear optimization problem can get stuck in a local minimum of the cost function, resulting in a sub-optimal solution. Interval optimization is based on the inclusion property of interval operations and is an excellent tool for global nonlinear optimization. Interval optimization is guaranteed to find the global minimum of the cost function.

Two applications of interval optimization are presented. From the first application, human perception modeling, it is concluded that interval optimization estimates the parameters of a human perception model better than previously applied methods. For the second application, identifying aircraft trim points, interval analysis guarantees that all trim points are found.

Overall it can be concluded that interval optimization is a promising method for system identification problems that contains a nonlinear optimization. However, when dealing with systems that have a larger number of unknown parameters, or with cost functions whose evaluation requires many operations, a practical limit of computional effort will be reached faster by interval optimization methods than by crisp optimization methods.

References

1. Nelder, J.A., Mead, R.: A simplex method for function minimization. The Computer Journal 7, 308–313 (1965)
2. Goldberg, D.E.: Genetic Algorithms in Search, Optimization and Machine Learning. Addison-Wesley Publishing Company, Inc., Reading (1989)
3. Moore, R.E.: Interval Analysis. Prentice-Hall, Inc., Englewood Cliffs (1966)
4. Jaulin, L., Kieffer, M., Didrit, O., Walter, E.: Applied Interval Analysis. Springer, Heidelberg (2001)
5. Ichida, K., Fujii, Y.: An interval arithmetic method for global optimization. Computing 23(1), 85–97 (1979)
6. De Weerdt, E., Chu, Q.P., Mulder, J.A.: Neural network output optimization using interval analysis. Trans. Neur. Netw. 20(4), 638–653 (2009)
7. de Weerdt, E., van Kampen, E., Chu, Q.P., Mulder, J.A.: Integer ambiguity resolution using interval analysis. ION Journal of Navigation 55, 293–307 (2008)
8. van Kampen, E., de Weerdt, E., Chu, Q.P., Mulder, J.A.: Applied interval based integer ambiguity resolution. ION Journal of Navigation 56, 205–219 (2009)
9. Zaal, P.M.T., Pool, D.M., Chu, Q.P., van Paassen, M.M., Mulder, M., Mulder, J.A.: Modeling Human Multimodal Perception and Control Using Genetic Maximum Likelihood Estimation. Journal of Guidance, Control and Dynamics 32(4), 1089–1099 (2009)
10. Peter, M.T., Zaal, D.M., Pool, Q.P., Chu, M.M., Van Paassen, M., Mulder, M., Mulder, J.A.: Modeling human multimodal perception and control using genetic maximum likelihood estimation. Journal of Guidance, Control and Dynamics 32(4), 1089–1099 (2009)
11. Stevens, B., Lewis, F.: Aircraft Control and Simulation. John Wiley & Sons, Chichester (1992)
12. de Visser, C.C., Chu, Q.P., Mulder, J.A.: A new approach to linear regression with multivariate splines. Automatica 45(12), 2903–2909 (2009)
13. Rump, S.M.: INTLAB - INTerval LABoratory. In: Csendes, T. (ed.) Developments in Reliable Computing, pp. 77–104. Kluwer Academic Publishers, Dordrecht (1999), `http://www.ti3.tu-harburg.de/rump/`

Investigation of the Attitude Error Vector Reference Frame in the INS EKF

Stephen Steffes, Jan Philipp Steinbach, and Stephan Theil

Abstract. The Extended Kalman Filter is used extensively for inertial navigation. If initial attitude errors are small, many authors choose to represent the attitude states as a vector of small angles in the vehicle body frame. Some authors choose to represent this vector in the navigation frame instead, but the corresponding reduction of filter performance in the closed loop filter is not discussed. Performance is regained when switching to an open loop filter, but closed loop filters are widely desired. This paper investigates this performance reduction. To show the effect, Monte Carlo simulation results are shown for several cases with a simplified inertial navigation problem using a closed and open loop filter and attitude states in the body and inertial frames. A qualitative argument is given to explain the effects, which stem from a state propagation model that poorly reflects the true system model for this case. A method is proposed to regain performance by using an estimated inertial frame for the attitude states. This method is only beneficial when the attitude states are measured indirectly via the velocity state equation. Results with this new frame are shown and discussed.

1 Introduction

Attitude determination techniques continue to be widely researched [1]. For vehicle navigation, much of the literature defines a small angle attitude error vector (the

Stephen Steffes
DLR German Aerospace Center, Institute of Space Systems, Navigation and Control Systems Department, Robert-Hooke-Str. 7, 28359 Bremen, Germany
e-mail: stephen.steffes@dlr.de

Jan Philipp Steinbach
Helmut-Schmidt-University, University of the Federal Armed Forces Hamburg, Institute of Automation Technology, Holstenhofweg 85, 22043 Hamburg, Germany
e-mail: jan.steinbach@hsu-hh.de

Stephan Theil
DLR German Aerospace Center, Institute of Space Systems, Navigation and Control Systems Department, Robert-Hooke-Str. 7, 28359 Bremen, Germany
e-mail: stephan.theil@dlr.de

difference between the estimated and true attitude references) if the attitude error is expected to be relatively small. This vector is often represented in the body frame, but some authors choose to use the navigation frame instead. However, a discussion of the corresponding reduction of filter performance in the closed loop filter has not been seen by the present authors and is the motivation for this research.

The choice of frame is one factor which determines the complexity of the state transition and state update equations and the performance of the filter. Other than simpler state equations, a big advantage of using an inertial navigation frame (or a slowly changing navigation frame) is seen in real time applications where the filter corrections are calculated after the reference time of the update[2]. In this case, the attitude error states are not rotated in the filter propagation step, making the attitude updates insensitive to the computational and measurement delays. In contrast, if the body frame is used then the attitude corrections must be transformed to the current body frame when correcting the attitude states.

Attitude estimates are updated as part of the EKF update routine. In this work a small angle quaternion defines the error in the estimated attitude quaternion. A small angle attitude error vector is part of this quaternion and has additive errors to first order. Crassidis, Markley and Cheng[1], Crassidis[3], Gray[2] and Markley[4] use attitude errors in the B frame, Farrell[5] and Wendel[6] use attitude errors in the North-East-Down frame, and Gai[7], Gray[2] and Thompson and Quasius[8] use an inertial frame. All of these authors simply add the EKF update corrections to the attitude error estimates, which does not change the frame of the attitude error vector. If the update does change the frame of the attitude error vector then the attitude covariance states much be rotated to the new frame.

This paper investigates the use of a small angle attitude error vector in an inertial frame for an INS using the closed loop EKF. The closed and open loop EKFs are first presented as background for the discussion. A simple INS problem is then described where a small angle attitude error vector can be represented in either the inertial or body frames. System models, measurement models, state vectors and state transition equations are given for both attitude error representations. Simulation results are then given, which show the performance of using the closed and open loop EKFs in combination with both attitude error representations. The degraded performance of the closed loop EKF with attitude errors in the inertial frame is further analyzed. A new estimated inertial frame is proposed for the attitude error vector to improve performance of this case. Simulation results are given for this new case showing improved performance and are further discussed.

2 Extended Kalman Filter

The EKF is discussed in numerous references [9, 10, 11, 5, 12, 6] and is summarized here. The EKF can be used in a closed-loop or open-loop manner[12]. The equations for both methods are listed in Table 1. The notation t_{k^-} is used to denote the time immediately before the updates and t_{k^+} denotes the time immediately after. In the closed-loop method, corrections to the state estimates are fed back to the current

Table 1 Summary of continuous-discreet open-loop and closed-loop EKF equations [9, 12]. Equations on the left are only for the closed-loop filter, those on the right are for the open-loop filter, and the equations in the middle are for both types of filters.

	Closed-Loop	Open-Loop
System Model	$\dot{\mathbf{x}}(t) = f(\mathbf{x}(t),t) + \mathbf{w}(t);\ \mathbf{w}(t) \sim \mathcal{N}(\mathbf{0},Q(t))$	
Measurement Model	$\mathbf{z}_k = h_k(\mathbf{x}(t_{k^-})) + \mathbf{v}_k;\ k = 1,2,\ldots;\ \mathbf{v}_k \sim \mathcal{N}(\mathbf{0},R_k)$	
Initial Conditions	$\mathbf{x}(0) \sim \mathcal{N}(\hat{\mathbf{x}}(0),P(0))$	$\mathbf{x}(0) \sim \mathcal{N}(\hat{\mathbf{x}}(0),P(0));\ \delta\hat{\mathbf{x}}(0) = \mathbf{0}$
Other Assumptions	$E\left[\mathbf{w}(t)\mathbf{v}_k^T\right] = 0;\ \forall k, \forall t$	
State Propagation	$\dot{\hat{\mathbf{x}}}(t) = f(\hat{\mathbf{x}}(t),t)$	
Error State Propagation	—	$\delta\dot{\hat{\mathbf{x}}}(t) = f(\mathbf{x}(t),t) - f(\hat{\mathbf{x}}(t),t)$
Covariance Propagation	$\dot{P}(t) = F(\hat{\mathbf{x}}(t),t)P(t) + P(t)F(\hat{\mathbf{x}}(t),t)^T + Q(t)$	
Whole State Update	$\hat{\mathbf{x}}(t_{k^+}) = \hat{\mathbf{x}}(t_{k^-}) + K_k(\mathbf{z}_k - h_k(\hat{\mathbf{x}}(t_{k^-})))$	—
Error State Update	—	$\delta\hat{\mathbf{x}}(t_{k^+}) = \delta\hat{\mathbf{x}}(t_{k^-}) + K_k[\mathbf{z}_k - h_k(\hat{\mathbf{x}}(t_{k^-}) + \delta\hat{\mathbf{x}}(t_{k^-}))]$
Covariance Update	$P(t_{k^+}) = [I - K_k H_k]P(t_{k^-})$	
Kalman Gain Matrix	$K_k = P(t_{k^-})H_k^T\left[H_k P(t_{k^-})H_k^T + R_k\right]^{-1}$	
Definitions	$F(\hat{\mathbf{x}}(t),t) = [\partial f(\mathbf{x}(t),t)/\partial\mathbf{x}(t)]_{\mathbf{x}(t)=\hat{\mathbf{x}}(t)}$ $H_k \equiv H_k(\hat{\mathbf{x}}(t_{k^-})) = [\partial h_k(\mathbf{x}(t_k))/\partial\mathbf{x}(t_k)]_{\mathbf{x}(t_k)=\hat{\mathbf{x}}(t_{k^-})}$	

estimated state vector ($\hat{\mathbf{x}}$) during the update routine. The estimated state vector always represents the most accurate estimate of the states, which allows the most accurate calculation of the state estimate propagation. In the open-loop method, corrections to the state estimates are not fed back to the estimated state vector but are instead added to the current estimated error state vector ($\delta\hat{\mathbf{x}}$). The estimated state vector is corrected with the estimated error state vector to get the most accurate state estimates. This type of filter is commonly used when the state propagation calculations cannot be changed directly. For additive errors the error state vector is defined as:

$$\delta\mathbf{x}(t) \equiv \mathbf{x}(t) - \hat{\mathbf{x}}(t) \tag{1}$$

where $\mathbf{x}$ is the true (error free) state vector. Note that the Error State Propagation and Whole State Update equations in Table 1 all assume the errors are additive.

3 Simplified Navigation Problem

The following simplified navigation problem will be used in simulations to compare the closed and open EKF performance using an attitude error state vector in the body and an inertial frame. Most INS applications are much more complex than this example and the performance differences of the frame and filter type choices may be hidden by other effects. The presented navigation problem is as simple as possible to highlight the desired effects.

Consider a vehicle at an Earth fixed position with an IMU providing 100Hz measurements of vehicle specific force and angular velocity in the IMU body frame (B) with no errors. The B frame is the rectangular coordinate system of the IMU measurements. The navigator provides the position, velocity and attitude of the vehicle in an inertial navigation frame (I). The I frame is defined to be equal to the Earth Centered Earth Fixed (ECEF) frame at time $t = 0$. The ECEF frame is centered at the Earth's center of mass with the x-axis extending through the point (0° latitude, 0° longitude), the z-axis extending through the spin axis and the y-axis completing the right-handed coordinate system. The ECEF frame rotates about the z-axis relative to inertial at Earth rate ($\omega_{Earth} = 7.2921159e-5\ rad/s$). The Earth rotation vector in the I and ECEF frame is:

$$\Omega^I = \Omega^{ECEF} = [0, 0, \omega_{Earth}]^T \tag{2}$$

Note that in the following sections the position, velocity, attitude, acceleration and rotation variables are a function of time. However, the (t) time dependency notion is dropped to simplify the notation. This notion is only used to indicate discrete time points or for clarification.

The navigation algorithm calculates the vehicle state over time. The system model is given as [13, 14] (see Shuster[15] for quaternion algebra definitions):

$$\dot{\mathbf{r}}^I = \mathbf{v}^I \tag{3}$$

$$\dot{\mathbf{v}}^I = T(\mathbf{q}_B^I)\mathbf{a}^B + \mathbf{g}^I(\mathbf{r}^I) \tag{4}$$

$$\dot{\mathbf{q}}_B^I = \frac{1}{2}\mathbf{q}_B^I \otimes \begin{bmatrix} \omega_{IB}^B \\ 0 \end{bmatrix} \tag{5}$$

where $\mathbf{r}^I$ is inertial position, $\mathbf{v}^I$ is inertial velocity, $\mathbf{q}_B^I$ is the B to I frame quaternion and $T(\mathbf{q}_B^I)$ is the equivalent transformation matrix, $\mathbf{a}^B$ is the measured vehicle specific force, and ω_{IB}^B is the measured vehicle angular velocity vector with respect to I in the B frame. $\otimes$ is the quaternion multiplication operator and quaternions are represented as a column vector with the scalar element last. For this example the process noise $\mathbf{w}(t)$ is zero for all time. $\mathbf{g}^I$ is spherical gravity in the I frame and is calculated using:

$$\mathbf{g}^I(\mathbf{r}^I) = -\mu\mathbf{r}^I / \left\|\mathbf{r}^I\right\|^3 \tag{6}$$

where $\mu = 398600.4418 km^3/s^2$ is the standard gravitational parameter for Earth. The IMU measures the dynamics of the vehicle, which is at a fixed position relative to Earth. Therefore, the acceleration and rotation are constant in the B frame and:

$$\mathbf{a}^B = \mathbf{a}^B(0) = T(\mathbf{q}_I^B(0))(\Omega^I \times \Omega^I \times \mathbf{r}^I(0) - \mathbf{g}^I(\mathbf{r}^I(0))) \tag{7}$$

$$\omega_{IB}^B = \omega_{IB}^B(0) = T(\mathbf{q}_I^B(0))\Omega^I \tag{8}$$

The error states for $\mathbf{r}^I$ and $\mathbf{v}^I$ are additive and follow from eq. (1), but the error state for the attitude quaternion is multiplicative. To avoid the complications with using a quaternion in the state vector [1] a new quantity θ will be defined, which is a vector of small angles representing the attitude error and is approximately additive. θ will be represented later in both B and I frames, but until then the frame will be kept general. Consider two general reference frames $A1$ and $A2$. The error quaternion in the $A1$ frame can be defined as:

$$\mathbf{p}^{A1} \equiv \mathbf{q}_{A2}^{A1} \otimes \hat{\mathbf{q}}_{A1}^{A2} \tag{9}$$

For small angular errors:

$$\mathbf{p}^{A1} \approx [-\theta^{A1}/2, 1]^T \tag{10}$$

where θ^{A1} is a small angle rotation vector in the $A1$ frame. With this definition, the state vector is written as:

$$\mathbf{x} \equiv \left[\mathbf{r}^I, \mathbf{v}^I, \theta^{A1}\right]^T \tag{11}$$

and the entire error state vector $\delta\mathbf{x}$ is additive.

To derive the system model for θ^{A1}, take the derivative of eq. (9):

$$\dot{\mathbf{p}}^{A1} = \dot{\mathbf{q}}^{A1}_{A2} \otimes \hat{\mathbf{q}}^{A2}_{A1} + \mathbf{q}^{A1}_{A2} \otimes \dot{\hat{\mathbf{q}}}^{A2}_{A1} \tag{12}$$

From Savage [13]:

$$\dot{\mathbf{q}}^{A1}_{A2} = \frac{1}{2}\mathbf{q}^{A1}_{A2} \otimes \left[\omega^{A2}_{IA2}, 0\right]^T - \frac{1}{2}\left[\omega^{A1}_{IA1}, 0\right]^T \otimes \mathbf{q}^{A1}_{A2} \tag{13}$$

Substituting this into eq. (12) and using eq. (9) and eq. (10) gives:

$$\begin{aligned}
\begin{bmatrix} -\frac{1}{2}\dot{\theta}^{A1} \\ 0 \end{bmatrix} &\approx \frac{1}{2}\mathbf{q}^{A1}_{A2} \otimes \begin{bmatrix} \omega^{A2}_{IA2} \\ 0 \end{bmatrix} \otimes \hat{\mathbf{q}}^{A2}_{A1} - \frac{1}{2}\begin{bmatrix} \omega^{A1}_{IA1} \\ 0 \end{bmatrix} \otimes \mathbf{q}^{A1}_{A2} \otimes \hat{\mathbf{q}}^{A2}_{A1} \\
&\quad + \frac{1}{2}\mathbf{q}^{A1}_{A2} \otimes \hat{\mathbf{q}}^{A2}_{A1} \otimes \begin{bmatrix} \omega^{A1}_{IA1} \\ 0 \end{bmatrix} - \frac{1}{2}\mathbf{q}^{A1}_{A2} \otimes \begin{bmatrix} \omega^{A2}_{IA2} \\ 0 \end{bmatrix} \otimes \hat{\mathbf{q}}^{A2}_{A1} \\
\begin{bmatrix} \dot{\theta}^{A1} \\ 0 \end{bmatrix} &\approx \begin{bmatrix} \omega^{A1}_{IA1} \\ 0 \end{bmatrix} \otimes \mathbf{p}^{A1} - \mathbf{p}^{A1} \otimes \begin{bmatrix} \omega^{A1}_{IA1} \\ 0 \end{bmatrix} \\
&= \begin{bmatrix} \omega^{A1}_{IA1} - \frac{1}{2}\omega^{A1}_{IA1} \times \theta^{A1} - \omega^{A1}_{IA1} + \frac{1}{2}\theta^{A1} \times \omega^{A1}_{IA1} \\ \frac{1}{2}\omega^{A1}_{IA1} \cdot \theta^{A1} - \frac{1}{2}\theta^{A1} \cdot \omega^{A1}_{IA1} \end{bmatrix} \\
&= \begin{bmatrix} -\omega^{A1}_{IA1} \times \theta^{A1} \\ 0 \end{bmatrix}
\end{aligned} \tag{14}$$

Taking the vector part of this equation and substituting in the I and B frames for $A1$ yields the two system models:

$$A1 = I : \dot{\theta}^I \approx 0_{3\times 1} \tag{15}$$

$$A1 = B : \dot{\theta}^B \approx -\omega^B_{IB} \times \theta^B \tag{16}$$

To derive F the system model must first be explicitly stated in terms of the states. The equation for $\dot{\mathbf{v}}^I$ must be stated in terms of θ. If $A1 = I$ then, using the Taylor series expansion for a small angle rotation matrix[16] and eq. (9), eq. (4) becomes:

$$\begin{aligned}
A1 = I : \dot{\mathbf{v}}^I &= T(\mathbf{p}^I)\mathbf{a}^{\hat{I}} + \mathbf{g}^I\left(\mathbf{r}^I\right) \\
&\approx \left(I_{3x3} - (\theta^I \times) + \tfrac{1}{2}(\theta^I \times)(\theta^I \times) - \ldots\right)\mathbf{a}^{\hat{I}} + \mathbf{g}^I\left(\mathbf{r}^I\right)
\end{aligned} \tag{17}$$

where:

$$\mathbf{a}^{\hat{I}} \equiv T(\hat{\mathbf{q}}^I_B)\mathbf{a}^B \tag{18}$$

Note that only the frame of **a** is affected by the rotation, not the length of the vector. One can think of $\mathbf{a}^{\hat{I}}$ as the acceleration in an estimated I frame called $\hat{I}$. Finally, if $A1 = B$ then the velocity system equation instead becomes:

$$\begin{aligned} A1 = B: \dot{\mathbf{v}}^I &= T(\hat{\mathbf{q}}_B^I)T(\mathbf{p}^B)^T\mathbf{a}^B + \mathbf{g}^I\left(\mathbf{r}^I\right) \\ &\approx T(\hat{\mathbf{q}}_B^I)(I_{3x3} + (\theta^B\times) - \tfrac{1}{2}(\theta^B\times)(\theta^B\times) + \ldots)\mathbf{a}^B + \mathbf{g}^I\left(\mathbf{r}^I\right) \end{aligned} \tag{19}$$

If $A1 = I$ then the system model for the entire state vector $f(\mathbf{x}(t),t)$ consists of eq. (3), eq. (15) and eq. (17). If $A1 = B$ then it consists of eq. (3), eq. (16) and eq. (19). Calculating the Jacobian gives:

$$F(\hat{\mathbf{x}}(t),t) = \begin{bmatrix} 0_{3\times3} & I_{3\times3} & 0_{3\times3} \\ -\mu/\left\|\hat{\mathbf{r}}^I\right\|^3 I_{3\times3} & 0_{3\times3} & F_{v\theta} \\ 0_{3\times3} & 0_{3\times3} & F_{\theta\theta} \end{bmatrix} \tag{20}$$

where $0_{3\times3}$ is a 3×3 matrix of 0's, $I_{3\times3}$ is the 3×3 identity matrix, and the gravity term is a first order approximation[5]. For $A1 = I$ the values for $F_{v\theta}$ and $F_{\theta\theta}$ are:

$$A1 = I: F_{v\theta} \approx (I_{3\times3} - \frac{1}{2}(\hat{\theta}\times))(\mathbf{a}^{\hat{I}}\times) + \frac{1}{2}((\mathbf{a}^{\hat{I}}\times\hat{\theta})\times) + \ldots \tag{21}$$

$$= (\mathbf{a}^{\hat{I}}\times) = ((T(\hat{\mathbf{q}}_B^I)\mathbf{a}^B)\times) \tag{22}$$

$$A1 = I: F_{\theta\theta} = 0_{3\times3} \tag{23}$$

and for $A1 = B$ they are:

$$A1 = B: F_{v\theta} \approx -T(\hat{\mathbf{q}}_B^I)\left[(I_{3\times3} - \frac{1}{2}(\hat{\theta}\times))(\mathbf{a}^B\times) - \frac{1}{2}((\mathbf{a}^B\times\hat{\theta})\times) + \ldots\right] \tag{24}$$

$$= -T(\hat{\mathbf{q}}_B^I)(\mathbf{a}^B\times) \tag{25}$$

$$A1 = B: F_{\theta\theta} = -(\omega_{IB}^B\times) \tag{26}$$

where the fact that $\hat{\theta}^{A1} = 0_{3\times1}$ is used to reduce the $F_{v\theta}$ equations.

It is known that the vehicle position is fixed relative to Earth, which is used as a measurement to update the filter. The velocity in the ECEF frame is then zero, which can be calculated in terms of the inertial states by subtracting Earth rotation velocity from the inertial velocity in the ECEF frame [5]. The measurement model is then:

$$\mathbf{z}_k = T_I^{ECEF}(t_k)\mathbf{v}^I(t_{k^-}) - (T_I^{ECEF}(t_k)\mathbf{r}^I(t_{k^-}))\times\Omega^{ECEF} + \nu_k \tag{27}$$

where $\nu_k \sim \mathcal{N}(\mathbf{0},R_k)$ and $T_I^{ECEF_k}$ is the I to $ECEF$ transformation defined as:

$$T_I^{ECEF}(t_k) = \begin{bmatrix} cos(t_k * \omega_{Earth}) & sin(t_k * \omega_{Earth}) & 0 \\ -sin(t_k * \omega_{Earth}) & cos(t_k * \omega_{Earth}) & 0 \\ 0 & 0 & 1 \end{bmatrix} \tag{28}$$

Calculating the Jacobian gives:

$$H_k = \left[(\Omega^{ECEF} \times) T_I^{ECEF}(t_k) \; T_I^{ECEF}(t_k) \; 0_{3\times3} \right] \tag{29}$$

To keep the small angle approximation in eq. (10) as accurate as possible, the estimated value of θ^{A1} will be fed back to the whole state attitude estimate after each EKF update in the closed loop case. The notation t_{k+} will be used to denote the time after the update and before the feedback, and t_{k++} will denote the time immediately after the feedback. Therefore, immediately after the whole state update equation the following operations are done with the estimated states:

$$\hat{\mathbf{q}}_{A2}^{A1}(t_{k++}) = [-\hat{\theta}^{A1}(t_{k+})/2, 1]^T \otimes \hat{\mathbf{q}}_{A2}^{A1}(t_{k+}) \tag{30}$$

$$\hat{\theta}^{A1}(t_{k++}) = 0_{3\times1} \tag{31}$$

where $0_{3\times1}$ is a 3×1 vector of 0's, and $\hat{\mathbf{q}}_{A2}^{A1}(t_{k++})$ is renormalized to 1 after this operation. With this method, the estimated small angle vector $\hat{\theta}^{A1}$ is always $0_{3\times1}$ during propagation. A similar procedure is used by a number of authors[3, 4] and is often called a "reset" of the attitude states, however without the explicit notation used here. The effect this has on the true state θ^{A1} is found by starting with eq. (9) at t_{k++} and using eq. (10), eq. (30) and the fact that $\mathbf{q}_{A2}^{A1}(t_{k++}) = \mathbf{q}_{A2}^{A1}(t_{k+})$ (since the true attitude is not changed by the feedback) to get:

$$\begin{aligned} \mathbf{p}^{A1}(t_{k++}) &= \mathbf{q}_{A2}^{A1}(t_{k++}) \otimes \hat{\mathbf{q}}_{A1}^{A2}(t_{k++}) \\ &= \mathbf{q}_{A2}^{A1}(t_{k+}) \otimes \hat{\mathbf{q}}_{A1}^{A2}(t_{k+}) \otimes [\tfrac{1}{2}\hat{\theta}^{A1}(t_{k+}), 1]^T \\ &\approx [-\tfrac{1}{2}\theta^{A1}(t_{k+}), 1]^T \otimes [\tfrac{1}{2}\hat{\theta}^{A1}(t_{k+}), 1]^T \\ [-\tfrac{1}{2}\theta^{A1}(t_{k++}), 1]^T &\approx [-\tfrac{1}{2}\theta^{A1}(t_{k+}) + \tfrac{1}{2}\hat{\theta}^{A1}(t_{k+}), 1]^T \\ \theta^{A1}(t_{k++}) &\approx \theta^{A1}(t_{k+}) - \hat{\theta}^{A1}(t_{k+}) \end{aligned} \tag{32}$$

which agrees with earlier statements that θ is approximately additive. There is no change in the covariance matrix due to these operations since this merely moves information from one place to another and does not change the statistics associated with these states[4]. Which means:

$$P(t_{k++}) = P(t_{k+}) \tag{33}$$

4 Baseline Simulation Results

Four Monte Carlo simulations are used to give a quantitative measure of the performance reduction when using the closed loop EKF with an attitude error vector in the I frame. Using the closed and open loop EKF algorithms summarized in section 2, four separate simulations were run on the navigation problem discussed in section 3: closed loop with $A1 = B$ (CLB case), open loop with $A1 = B$ (OLB case), closed loop with $A1 = I$ (CLI case), and open loop with $A1 = I$ (OLI case).

For every case, the simulation was setup as follows. The filter starts at $t = 0$ and ends at $t = 10$. $\mathbf{r}^I(0) = [6378137m, 0, 0]^T$, which is at 0° latitude, 0° longitude on Earth's surface. $\mathbf{v}^I(0)$ is set to the Earth surface velocity, which is $\Omega^{ECEF} \times \mathbf{r}^I(0)$. $\mathbf{q}_B^I(0)$ is set to a random quaternion $[\hat{e}sin(\alpha/2), cos(\alpha/2)]^T$ with $\alpha \sim \mathscr{U}(0, 360°)$

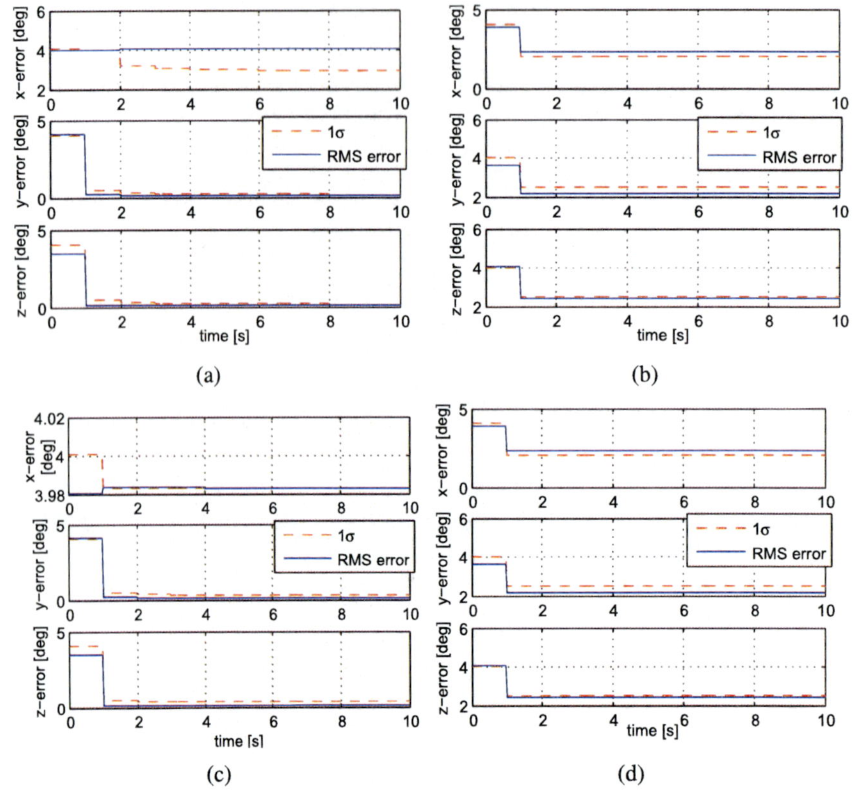

Fig. 1 RMS attitude error CLI (a), CLB (b), OLI (c) and OLB (d) Monte Carlo simulations.

and $\hat{e}$ uniformly random over the unit sphere. The estimated states are set to $\hat{\mathbf{r}}^I(0) = \mathbf{r}(0) + \nu_r$, $\hat{\mathbf{v}}^I(0) = \mathbf{v}(0) + \nu_v$, $\hat{\theta}^{A1}(0) = [0,0,0]^T$, and $\hat{\mathbf{q}}^I_B(0) = \mathbf{p}\prime \otimes \mathbf{q}^I_B$, where $\nu_r \sim \mathcal{N}(0, \sigma_r I_{3\times3})$, $\nu_v \sim \mathcal{N}(0, \sigma_v I_{3\times3})$, $\mathbf{p}\prime$ is the quaternion $[q_1, q_2, q_3, q_4]^T$ with $q_1, q_2, q_3 \sim \mathcal{N}(0, \sigma_\theta^2)$ and $q_4 = \sqrt{q_1^2 + q_2^2 + q_3^2}$. $P(0)$ is set to:

$$P(0) = \begin{bmatrix} \sigma_r^2 I_{3\times3} & 0_{3\times3} & 0_{3\times3} \\ 0_{3\times3} & \sigma_v^2 I_{3\times3} & 0_{3\times3} \\ 0_{3\times3} & 0_{3\times3} & \sigma_\theta^2 I_{3\times3} \end{bmatrix} \tag{34}$$

where $\sigma_r = 1m$, $\sigma_v = 0.1m/s$, and $\sigma_\theta = 0.07rad$ $(= 4°)$. As previously mentioned, there is no system noise, so $\mathbf{w}(t)$ is 0 at all time. There is also no measurement noise since it is known that the ECEF velocity is exactly zero, but using zero measurement noise can lead to division by zero in the Kalman gain matrix. Therefore, R_k is set to $(0.001m/s)^2 I_{3\times3}$. The state propagation equations are integrated at 100Hz using forward Euler for eq. (3), forward Euler with rotation correction[14] for eq. (4), a 3rd order quaternion integration method[16] for eq. (5) and the exact solutions for eq. (15) and eq. (16).

100 simulations were run for each Monte Carlo and the same set of initial conditions are used for each Monte Carlo. Figure 1 shows the results. For each case at each time point, the root mean square (RMS) of the attitude errors and the corresponding 1σ values for all 100 simulations are plotted. The position and velocity errors are all near or below their RMS 1σ values and are not shown.

5 Analysis of Simulation Results

The filter performances for the CLB, OLI, and OLB cases are all as expected; the RMS errors are all near or below the $1\sigma_{RMS}$ values. However, the performance of the CLI case is very poor. After the second measurement update (at 2sec) the error in attitude is well outside of the $1\sigma_{RMS}$ boundary. The fact that both open loop cases perform as expected suggests that the poor performance of the CLI case is related to feeding back the state corrections in the EKF update routine. The fact that both cases with $A1 = B$ perform as expected suggests that the CLI performance is also related to the state transition equations. Both the velocity and attitude propagation terms depend on the choice of $A1$, but the attitude terms (eq. (23) and eq. (26)) can be eliminated as suspect since they do not depend on the state corrections. The velocity terms (eq. (22) and eq. (25)) contain $\hat{\mathbf{q}}_B^I$, which is corrected after every EKF update with eq. (30). It seems that the velocity terms are causing the CLI performance problem since the attitude corrections come from updates with velocity measurements. The following qualitative approach explains why the CLI case inherently has poor performance in these simulations.

In eq. (22), $\mathbf{a}^B$ is first transformed with the attitude estimate into the $\hat{I}$ frame and this is used in a cross product matrix which is multiplied with the covariances of θ^I during the EKF propagation[1]. When the attitude estimate is corrected from the EKF update the direction of $\mathbf{a}^{\hat{I}}$ changes at this instant, even though the direction of $\mathbf{a}^I$ did not. In effect, the $\hat{I}$ frame changes even though the true attitude does not change at this instant. This is misinterpreted by the EKF as a change in the acceleration direction relative to θ^I, which causes poor filter performance since this does not reflect the system model.

In contrast, in eq. (25), $\mathbf{a}^B$ is first crossed with the attitude state covariances and then this cross product is transformed with the attitude estimate. Again the attitude estimate is changed by the update, but the relationship between $\mathbf{a}^B$ and the attitude covariance (i.e. the cross product matrix) is not affected. The EKF does not interpret the attitude update as a change in acceleration, but it does misinterpret it as a change in the entire cross product. Clearly, the EKF is not sensitive to this misinterpretation since the performance of the CLB case is OK.

Finally, in the OLI case the attitude corrections are accumulated in the error state vector instead of updating the attitude estimate. This eliminates the false acceleration direction changes in the CLI case, therefore providing better performance.

[1] Additionally, note that crossing two vectors in different frames yields a resultant vector in an undefined frame.

The Monte Carlo simulations used a relatively large initial attitude error of $\sigma_\theta = 4°$. As this error is reduced, the size of the attitude updates is reduced and the CLI case shows increasingly better performance. For the presented simulation setup, an initial attitude error of $\sigma_\theta \approx 0.05°$ was found to be the boundary between good and poor performance for the CLI case, where "good" performance means the RMS attitude and velocity errors are near or below their respective $1\sigma_{RMS}$ boundaries.

6 Closed Loop with $A1 = \hat{I}$ Case

The cross product in eq. (22) crosses a vector in the $\hat{I}$ frame with the covariances of θ in the I frame. Crossing vectors in two different frames in general does not make sense because the frame of the product is not defined. A better approach might be to use some $\theta^{\hat{I}}$ as the attitude state instead. To show how this simple, yet unconventional change affects the EKF performance, the system model and update equations will be derived and Monte Carlo results for the resulting system will be shown.

To start, $\theta^{\hat{I}}$ is first defined as:

$$\theta^{\hat{I}} \equiv T(\hat{\mathbf{q}}_B^I)T(\mathbf{q}_I^B)\theta^I \tag{35}$$

Using eq. (9) this becomes:

$$\theta^{\hat{I}} = T((\mathbf{p}^I)^{-1})\theta^I \tag{36}$$

This can be further reduced using a Taylor series expansion to:

$$\theta^{\hat{I}} \approx (I_{3x3} + (\theta^I\times) - \frac{1}{2}(\theta^I\times)(\theta^I\times) + \ldots)\theta^I \tag{37}$$

$$\theta^{\hat{I}} \approx \theta^I \tag{38}$$

It is already known that θ^I (and thus $\mathbf{p}^I$) are constant during the EKF propagation from eq. (15). Therefore, the system model for this state must be:

$$\dot{\theta}^{\hat{I}} = 0_{3\times 1} \tag{39}$$

To find the system model for velocity combine eq. (17) and eq. (38) to get:

$$\dot{\mathbf{v}}^I \approx (I_{3x3} - (\theta^{\hat{I}}\times) + \frac{1}{2}(\theta^{\hat{I}}\times)(\theta^{\hat{I}}\times) - \ldots)\mathbf{a}^{\hat{I}} + \mathbf{g}^I(\mathbf{r}^I) \tag{40}$$

Therefore, the F matrix for this case is the same as in the CLI case. Additionally, the measurement equations are unchanged since they are not affected by the choice of $A1$.

To keep small angle approximations as accurate as possible, the estimated value of $\theta^{\hat{I}}$ will be fed back to the whole state attitude estimate after each EKF update, analogous to the procedure in eq. (30) and eq. (31). Immediately after the whole state update equation the following operations are done:

$$\hat{\mathbf{q}}^I_B(t_{k++}) = [-\hat{\theta}^{\hat{I}}(t_{k+})/2, 1]^T \otimes \hat{\mathbf{q}}^I_B(t_{k+}) \tag{41}$$

$$\hat{\theta}^{\hat{I}}(t_{k++}) = 0_{3\times 1} \tag{42}$$

As a result, the $\hat{I}$ frame changes discretely at the update times. The resulting change in $\theta^{\hat{I}}$ is calculated by starting with eq. (36) at t_{k++} and using parts of eq. (32) and eq. (36) again to get:

$$\begin{aligned}
\theta^{\hat{I}}(t_{k++}) &= T((\mathbf{p}^I(t_{k++}))^{-1})\theta^I(t_{k++}) \\
&\approx T([\tfrac{1}{2}\theta^I(t_{k++}), 1]^T)\theta^I(t_{k++}) \\
&\approx T([-\tfrac{1}{2}\hat{\theta}^I(t_{k+}), 1]^T \otimes [\tfrac{1}{2}\theta^I(t_{k+}), 1]^T)(\theta^I(t_{k+}) - \hat{\theta}^I(t_{k+})) \\
&\approx T([-\tfrac{1}{2}\hat{\theta}^I(t_{k+}), 1]^T)(T([\tfrac{1}{2}\theta^I(t_{k+}), 1]^T)\theta^I(t_{k+}) - T([\tfrac{1}{2}\theta^I(t_{k+}), 1]^T)\hat{\theta}^I(t_{k+})) \\
&\approx T([-\tfrac{1}{2}T(\mathbf{p}^I)\hat{\theta}^{\hat{I}}(t_{k+}), 1]^T)(\theta^{\hat{I}}(t_{k+}) - \hat{\theta}^{\hat{I}}(t_{k+})) \\
&\approx T([-\tfrac{1}{2}\hat{\theta}^{\hat{I}}(t_{k+}), 1]^T)(\theta^{\hat{I}}(t_{k+}) - \hat{\theta}^{\hat{I}}(t_{k+}))
\end{aligned} \tag{43}$$

The last equation shows that $\theta^{\hat{I}}$ is not additive because the $\hat{I}$ frame changes by the rotation $T([-\frac{1}{2}\hat{\theta}^{\hat{I}}(t_{k+}), 1]^T)$ due to the feed back operations. In this case the covariance matrix must also be updated to reflect the frame change. Therefore, as a final step in the update routine, a discrete propagation of the covariance matrix must be done from t_{k+} to t_{k++} to change the frame of the attitude covariances. The discrete form of the EKF propagation equations in Table 1 are[9, 12]:

$$\hat{\mathbf{x}}(t_{k++}) = f(\hat{\mathbf{x}}(t_{k+})) \tag{44}$$

$$P(t_{k++}) = F_{k+}P(t_{k+})F_{k+}^T + Q_{k+} \tag{45}$$

$$F_{k+} \equiv F(\hat{\mathbf{x}}(t_{k+})) = [\partial f(\mathbf{x}(t))/\partial \mathbf{x}(t)]_{\mathbf{x}(t)=\hat{\mathbf{x}}(t_{k+})} \tag{46}$$

The f function does not change $\mathbf{r}^I$ or $\mathbf{v}^I$, but uses eq. (43) for the attitude states. From the Jacobian of eq. (43), the system matrix is:

$$F_{k+} = \begin{bmatrix} I_{3\times 3} & 0_{3\times 3} & 0_{3\times 3} \\ 0_{3\times 3} & I_{3\times 3} & 0_{3\times 3} \\ 0_{3\times 3} & 0_{3\times 3} & T([-\frac{1}{2}\hat{\theta}^{\hat{I}}(t_{k+}), 1]^T) \end{bmatrix} \tag{47}$$

and the process noise is $Q_{k+} = \mathbf{0}$. The EKF update is finally complete with these last steps.

A Monte Carlo of 100 simulations, like those discussed in section 4, was done for the closed loop $A1 = \hat{I}$ case to give a quantitative measure its performance. 100 simulations were run and the same set of initial conditions from section 4 are used. Figure 2 show the results. As with the other simulation results, the root mean square (RMS) of the velocity errors, attitude errors, and their corresponding 1σ values for all 100 simulations are plotted. The position errors are not shown because they are approximately an integral of the corresponding velocity errors and are therefore not interesting.

The filter performance for this case is similar to the CLB, OLI and OLB cases. The RMS errors are all near or below the $1\sigma_{RMS}$ boundaries.

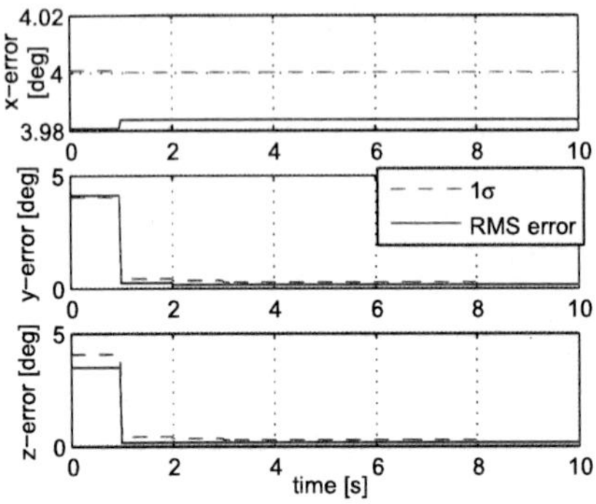

Fig. 2 RMS attitude error for closed loop EKF with $A1 = \hat{I}$ Monte Carlo.

7 Conclusion

The presented work has shown how the choice of the attitude error vector frame and filter type can effect filter performance. Four separate Monte Carlo simulations were done for a simplified navigation problem. The filter performance for the CLB, OLI and OLB cases was as expected, but the performance of the CLI case was relatively poor. In all cases, the attitude corrections were estimated indirectly via the velocity measurements with the $F_{v\theta}$ term in the EKF propagation equation. In the CLI case, the $\mathbf{a}^{\hat{I}}$ term in $F_{v\theta}$ changes after the EKF update but the attitude covariance states are unchanged. The poor performance of the CLI case can be attributed to this inconsistent modeling.

To avoid the problems with the CLI case, the $\hat{I}$ frame was used to represent the attitude states instead. This case has the same propagation equations as the CLI case, but uses an additional discrete filter propagation step immediately after the normal EKF update to rotate the attitude covariance states to the new $\hat{I}$ frame. In this case both the $\mathbf{a}^{\hat{I}}$ term in $F_{v\theta}$ and the attitude covariance states are always in the current $\hat{I}$ frame, which fixes the inconsistent modeling problem from the CLI case. Monte Carlo results for this case show similar performance to the CLB, OLI and OLB cases.

Most INS systems are much more complicated than the example used in this work. However, if a filter like the CLI case is used then the inconsistent modeling issues discussed in this work may degrade performance. For any system, all options should be considered to find the best choice for the application.

References

1. Crassidis, J.L., Markley, F.L., Cheng, Y.: Survey of Nonlinear Attitude Estimation Methods. Journal of Guidance, Control and Dynamics 30(1), 12–28 (2007)
2. Gray, C.W.: Star Tracker/IRU Attitude Determination Filters, vol. 107, pp. 459–476 (2001), AAS 01-039

3. Crassidis, J.L.: Sigma-Point Kalman Filtering for Integrated GPS and Inertial Navigation. In: AIAA Guidance, Navigation, and Control Conference, San Francisco, CA (August 2005), AIAA-2005-6052
4. Markley, F.L.: Attitude Error Representations for Kalman Filtering. Journal of Guidance, Control and Dynamics 26(2), 311–317 (2003)
5. Farrell, J.A.: Aided Navigation: GPS with High Rate Sensors. McGraw Hill, New York (2008)
6. Wendel, J.: Integrierte Navigationssysteme: Sensordatenfusion, GPS und Inertiale Navigation. Oldenbourg Wissenschaftsverlag, GmbH (2007)
7. Gai, E., Kevin Daly, J.H., Lemos, L.: Star-Sensor-Based Satellite Attitude/Attitude Rate Estimator. Journal of Guidance, Control and Dynamics 8(5), 560–565 (1985)
8. Thompson, I.C., Quasius, G.R.: Attitude Determination for the P80-1 Satellite. In: Proceedings of AAS Guidance and Control Conference (1980), AAS 80-001
9. Gelb, A. (ed.): Applied Optimal Estimation. The MIT Press, Cambridge (1974)
10. Brown, R., Hwang, P.: Introduction to Random Signals and Applied Kalman Filtering. John Wiley and Sons, Chichester (1997)
11. Grewal, M.S., Andrews, A.P.: Kalman Filtering - Theory and Practice using MATLAB, 2nd edn. John Wiley & Sons, Chichester (2001)
12. Kayton, M., Fried, W.R.: Avionics Navigation Systems, 2nd edn. John Wiley and Sons Inc., Chichester (1997)
13. Savage, P.G.: Strapdown Inertial Navigation Integration Algorithm Design Part 1: Attitude Algorithms. Journal of Guidance, Control, and Dynamics 21(1), 19–28 (1998)
14. Savage, P.G.: Strapdown Inertial Navigation Integration Algorithm Design Part 2: Velocity and Position Algorithms. Journal of Guidance, Control and Dynamics 21(2), 208–221 (1998)
15. Shuster, M.D.: A Survey of Attitude Representations. Journal of the Astronautical Sciences 41(4), 439–517 (1993)
16. McKern, R.A.: A Study of Transformation Algorithms for Use in a Digital Computer. Master's thesis, Massachusetts Institute of Technology (January 1968)

Nonlinear Filtering Using Sparse Grids

Carolyn Kalender and Alfred Schöttl

Abstract. This paper presents a new nonlinear filtering algorithm applicable in real-time. Nonlinear filtering problems are mostly solved with the Extended Kalman Filter which due to the nonlinearities is a suboptimal estimator. Optimal estimates are provided by Fokker-Planck-Equation in combination with Bayes rule. Conventional approaches for the numerical solution of this equation suffer from the "curse of dimension" and are therefore not applicable in higher dimensions. We use sparse grids for solving the Fokker-Planck-Equation and present a six dimensional nonlinear problem solved in real-time with this new approach.

1 Introduction

The estimation of the course (and further states) of a manoeuvring target based on measurements by one or more sensors such as radar stations are considered. This problem plays an important role in various applications such as in the guidance of an interceptor missile against an incoming threat. It is well-known that a precise estimation is crucial for the miss distance or, equivalently, the hit probability. Most problems of this kind are highly nonlinear.

Let $(\Omega, \mathscr{F}, P)$ be a probability space endowed with a right-continuous filtration $(\mathscr{F}_t)$ and let W and V be a d- and m-dimensional adapted Brownian motion.

The object's motion is modelled by an adapted stochastic process $X = (X_t)$, $X_t \in \mathbb{R}^d$, the dynamics of which are given as the strong solution of a nonlinear stochastic differential equation

$$dX_t = f_t(X_t)\,dt + \sigma_t(X_t)\,dW_t \quad . \tag{1}$$

Carolyn Kalender
LFK-Lenkflugkörpersysteme GmbH, Landshuter Straße 26,
85716 Unterschleißheim, Germany
e-mail: carolyn.kalender@mbda-systems.de

Alfred Schöttl
LFK-Lenkflugkörpersysteme GmbH, Landshuter Straße 26,
85716 Unterschleißheim, Germany
e-mail: alfred.schoettl@mbda-systems.de

In the continuous time setup, the measurement can be modelled by another stochastic process Y, again defined (up to versions) as a strong solution of the stochastic differential equation

$$dY_t = g_t(X_t)\,dt + \nu_t(X_t)\,dV_t \quad . \tag{2}$$

It is well-known (see [7], chapter 8.6) that under appropriate growth conditions such as (q denotes any of the measurable functions f, σ, g, ν)

$$\begin{aligned} \|q_t(x) - q_t(x')\|^2 &\leq K\|x - x'\|^2 \\ \|q_t(x)\|^2 &\leq K(1 + \|x\|^2) \end{aligned}$$

a strong solution exists.

As we will see, the filter may also be applied at the discrete update times

$$dY_{t_k} = h_t(X_{t_k}) + \nu_{t_k}(X_{t_k})V_{t_k} \quad . \tag{3}$$

By interpreting (1) as a system equation and (2) as a measurement equation, the problem of estimating the actual state X_t of the object only using measurements $Y_{s\leq t}$ can be seen as a filtering problem: Let $\mathscr{F}_t^Y$ be the filtration generated by Y. We are considering the problem of finding an optimal (in the $\mathscr{L}^2$ sense) $\mathscr{F}^Y$-adapted estimation of X. It can be easily seen that this problem is equivalent to finding the conditional expectation $E(X_t|\mathscr{F}_t)$.

There are various ways of approaching such a problem. The most usual method is to apply an extended Kalman filter (see e. g. [10]), a method suitable and highly efficient for systems with modest nonlinearities. Since the extended Kalman filter is based on linearisations of the system equation, divergence is possible. In addition, nonsymmetric or multimodal distributions cannot be treated since classical Gaussian theory is applied.

Another widely used method is the particle filter (see e. g. [10]) which uses a reasonable amount of state samples and propagates them through simulation of the system. The particles which may be viewed as a discrete distribution approximating the conditional probability distribution, are assigned normalized weights. An updated approximation is generated by changing the weights with respect to the measurements (e. g. using Bayes formula) . The quality of the approximation is heavily dependent on the number of particles. As a rule of thumb, the necessary number of particles grows exponentially with the number of dimensions of the system. This "curse of dimension" (see [4]) restricts the usage of particle filters to relatively small dimension numbers. In typical applications however, the dimension d of the system state is relatively high (5-10), while the dimension m of the measurement is moderate (3-6).

It is furthermore well known (e. g. [6]) that only in very special setups it is possible to calculate the estimate in a closed form with a finite-dimensional system of equations. Most importantly is the linear case with the Kalman filter which just needs to update d conditional expectations and $\frac{d(d-1)}{2}$ covariances. Since the linearity implies Gaussian conditional distribution, the complete distribution is specified by these parameters. Other so-called finite dimensional filters are the Beneš [2] and the Daum filters [4], which are of upmost theoretical interest but require artificial and restrictive conditions on f and g. In the vast majority of cases it is necessary to consider an infinite amount of numbers to specify the whole conditional distribution of $(X|\mathscr{F}^Y)$ or to be satisfied with only approximations.

We are assuming that the conditional probability density function (pdf) p_t

$$p_t(x) = \frac{\partial}{\partial x} P(X_t \leq x | \mathscr{F}_t^Y) \quad ,$$

which is a measurable function of (t,x), exists.

The analysis of the evolution of the conditional distribution is part of the general filter theory. Under very mild assumptions, filter formulas such as the Kushner–Stratonovich equation (see [1] Theorem 3.30) have been developed. The Kushner–Stratonovich equation is equivalent to a stochastic partial differential equation for the pdf if the solution of the differential equation exists (see [7], Theorem 8.6). For details about the existence of the conditional pdf see also Theorem 7.11 in [1]. A thorough analysis of the conditions and properties of the solution is contained in [8]. It holds

$$\begin{aligned} d_t p_t(x) = & \left(-\sum_k \frac{\partial}{\partial x_k} (f_{t,k}(x) p_t(x)) + \frac{1}{2} \sum_{jk} \frac{\partial^2}{\partial x_j \partial x_k} (b_{t,jk}(x) p_t(x)) \right) dt \\ & + p_t(x) \left(h_t(x) - E(h(X_t)|\mathscr{F}_t^Y) \right) \left(dY_{t_k} - E(h(X_t)|\mathscr{F}_t^Y)\, dt \right) \end{aligned}$$

with $b_t = \sigma_t \cdot \sigma_t^T$.

The right-hand side of this equation may be seen as the sum of a propagation part (the first line), containing a transport (or advection) term and a dissipation (or diffusion) term, and the innovation part (the second line) which handles measurements. While the transport term shifts the pdf according to the model, the diffusion term widens the pdf in time which inserts uncertainty into the estimation. The measurement term will in turn narrow the pdf due to the measurement update.

The discrete time case is similar. The innovation part (which can be seen as an abstract version of the Bayesian rule) is simply replaced by the classical Bayesian rule such that the propagation is done via the partial differential equation

$$d_t p_t(x) = \left(-\sum_k \frac{\partial}{\partial x_k} (f_{t,k}(x) p_t(x)) + \frac{1}{2} \sum_{jk} \frac{\partial^2}{\partial x_j \partial x_k} (b_{t,jk}(x) p_t(x)) \right) dt \tag{4}$$

while measurements y at time t are incorporated separately via

$$p_t^+(x) = \frac{p_t^-(x) p_{Y|X}(y|x)}{\int p_{Y|X}(y|z) p_t^-(z)\, dz} \quad . \tag{5}$$

Herein p_t^+ denotes the conditional pdf of X_t just after the measurement at time t has been considered, p_t^- denotes the pdf just before the measurement. For small measurement time intervals the discrete time and the continuous time formula become equivalent. In the sequel, we are considering the discrete time formulation.

2 Discretization

Conventional approaches for solving partial differential equations numerically suffer from the "curse of dimension". To achieve a given order of approximation the number of grid points grow exponentially with the dimension such that

$$||f - f_n|| = \mathcal{O}(n^{-r/d})$$

for a function f of smoothness r.

The dimension of the filtering problem ranges typically from 5 to 10 and is therefore out of reach for real-time applications with regular grid methods. Even the use of an adaptive grid as shown in [13] for only 4 dimensions could not be implemented in real time. However, the technique of sparse grids offers a possibility to drastically lower the number of necessary grid points from $\mathcal{O}(N^d)$ to $\mathcal{O}(N(\log N)^{d-1})$.

We suggest the use of sparse grids to make the full nonlinear problem in higher dimension treatable in real-time. The algorithm is split up into three parts. First the density function is propagated by solving equation (4) on sparse grids. Next, the density is updated with the actual measurements and then finally the expected values or other characteristic values (e. g. higher moments) of the conditional distribution are extracted from the sparse grids.

2.1 *Sparse Grids*

Sparse grids were first introduced by Zenger [12] and have in the meantime been widely used e. g. in the area of finance mathematics.

The basic idea is to decompose the space of piecewise multilinear functions in its hierarchical subspaces and then consider only those for which their contribution to the interpolation of smooth functions is significant.

For the multilinear basis functions

$$\phi_{l,i}(x) = \prod_{j=1}^{d} \phi_{l_j,i_j}(x_j) \text{ with } \phi_{l_j,i_j}(x_j) = \phi\left(\frac{x_j - i_j h_{l_j}}{h_{l_j}}\right),\ \phi(x) = \begin{cases} 1 - |x| & x \in [-1,1] \\ 0 & \text{otherwise} \end{cases},$$

where the gridwidth $h_{l_j} = 2^{-l_j}$ and the related grid points $x_{l,i}$ with $x_{l_j,i_j} = i_j \cdot h_{l_j}$ on the level l with the index i, the space of piecewise multilinear functions of level l in the interior of $[0,1]^d$ is given by

$$V_l = \text{span}\{\phi_{l,i} : 1 \leq i \leq 2^l - 1\}.$$

The hierarchical space

$$W_l = V_l \setminus \bigoplus_{j=1}^{d} V_{l-e_j} = \text{span}\{\phi_{l,i} : 1 \le i \le 2^l - 1, i_j \text{ odd } \forall j\} \quad (\Rightarrow V_l = \bigoplus_{k \le l} W_k)$$

is given as a set difference from a coarser to a finer grid. Hence for a functional decomposition

$$u(x) = \sum_{l,i} u_{l,i}\phi_{l,i}, \quad \phi_{l,i} \in W_l$$

the hierarchical surplus $u_{l,i}$ contains the difference in $x_{l,i}$ between the relative coarser and finer interpolation. While the number of grid points increases considerably with the level, it turns out (see Zenger [12]) that the gain in interpolation accuracy is getting comparatively small for smooth functions. The idea of sparse grids is instead of using a full grid with $\bigoplus_{|k|_\infty \le L} W_k$ only to use the lower level hierarchical subspaces and form $\bigoplus_{|k|_1 \le L+d-1} W_k$ (a tetraeder of subspaces is composed instead of a quader). L is called the level of the sparse grid. It should be noted that no grid points are located at the boundary of the domain.

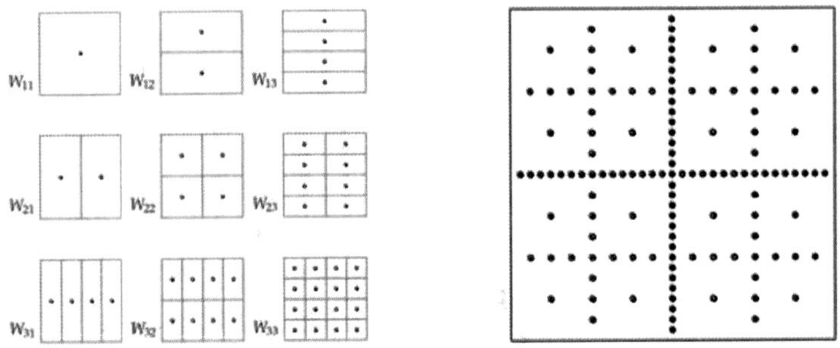

Fig. 1 Hierarchical subspaces $W_l, |l|_\infty \le 3$ and sparse grid of level $L = 5$

Instead of $\mathcal{O}(N^d)$ grid points in the regular grid, the sparse grids contain only $\mathcal{O}(N(\log N)^{d-1})$ grid points ($N = 1/h = 2^L$) by having only a slightly smaller order of interpolation accuracy for smooth functions in the maximum norm (respectively identical order of interpolation accuracy in the energy norm, for details see e. g. [3]).

Due to these promising properties of sparse grids we will use them in the sequel for solving the filtering problem numerically in real time for higher dimensions.

2.2 *Propagation*

Finite differences are used to discretize the propagation equation (4). The advection part, especially for highly agile targets, forces the localized density to move with time across a large region in the state domain. It would imply unnecessary high computational effort to discretize the whole region at every time step. As we will see, this can be avoided by introducing grid-tiling and grid-drifting.

2.2.1 Finite Differences

The following finite difference scheme is proposed

- A first order forward scheme for the time derivative

$$D_t^+ p_t = \frac{p_{t+\Delta t} - p_t}{\Delta t}, \quad \frac{\partial p_t}{\partial t} = D_t^+ p_t + \mathcal{O}(\Delta t)$$

- Upwind differences for the advection part

$$D_x^u p_t = \begin{cases} \frac{p_t(x+\Delta x) - p_t(x)}{\Delta x} & \text{for } f_t(x) < 0 \\ \frac{p_t(x) - p_t(x-\Delta x)}{\Delta x} & \text{for } f_t(x) > 0 \end{cases}, \quad \frac{\partial p_t}{\partial x} = D_x^u p_t + \mathcal{O}(\Delta x)$$

- Central differences for the diffusion part

$$D_x^2 p_t = \frac{p_t(x+\Delta x) - 2p_t(x) + p_t(x-\Delta x)}{\Delta x^2}, \quad \frac{\partial^2 p_t}{\partial x^2} = D_x^2 p_t + \mathcal{O}(\Delta x)$$

Mixed derivatives are not regarded here but could also be realized by finite differences. In contrast to regular grids however, there is not a natural grid neighbour in mixed directions but one can interpolate between different points. For usage peculiarities of finite differences on sparse grids see [9].

All together this gives an explicit first order scheme:

$$p_{t+\Delta t}(x) = p_t(x) - \Delta t \sum f_{t,k}(x) D_{x_k}^u p_t + \frac{1}{2} \Delta t \sum b_{kk} D_{x_k}^2 p_t \tag{6}$$

As an explicit scheme it has a timestep condition to be numerically stable. For

$$\Delta t \leq \frac{1}{\sum \frac{b_{kk} + |f_{t,k}(x)| \Delta x_k}{\Delta x_k^2}} \tag{7}$$

the following theorem holds

Theorem 1. *The discretization scheme (6) for which (7) holds is stable according to the von-Neumann stability analysis.*

Proof. Apply the argumentation of [11, p. 160] to the multidimensional case.

2.2.2 Tiling

Tiling helps in dealing with pdfs which move and widen due to drift and diffusion with time. Also Tiling restricts the computational effort to a possibly small subset of the state domain with a probability close to 1.

In contrast to regular grids it is not natural to expand sparse grids just by adding a few rows of grid points since this would contradict the hierarchical structure of those grids. If we want to add or delete grid points we do so by adding or removing an entire sparse grid.

To represent a more general area, we cover the relevant region of the domain with tiles each containing a sparse grid (which does not contain boundary points). Between the sparse grid tiles we have to add boundary layers as a connection. Those boundary layers are sparse grids themselves with a lower dimension.

A tile is removed if the integral of the pdf on this tile is below a given threshold. If on a boundary strip the integral part of the pdf is over a given threshold, a neighbour tile is then added.

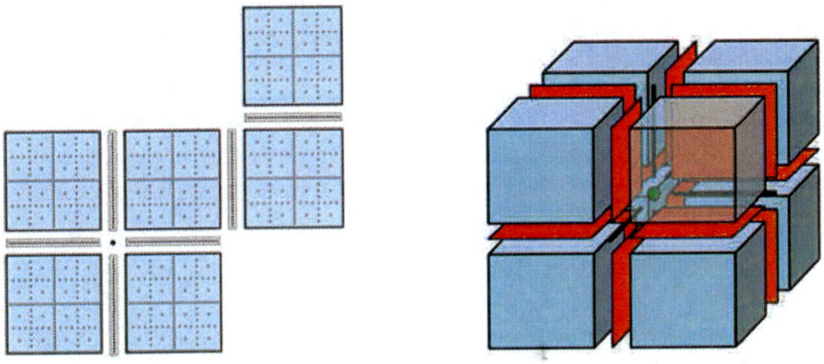

Fig. 2 Tiling examples in two and three dimensions

In regular intervals it is checked if still every tile is necessary and if new tiles are required.

2.2.3 Drift Compensation

As stated in section 2.2.1, the time stepsize must be limited in order to guarantee the stability of the scheme. Theorem 1 states that the condition (7) must be fulfilled for a stable scheme. Note that the magnitude of the drift $|f_{t,k}|$ determines the stability margin in (7).

In applications, a huge drift term in some dimensions (e. g. the position states in a tracking application) is likely. Fortunately, the pdf is often well-localized for these dimensions. A significant improvement can be achieved by modification of the drift

$$dX_t = (\bar{f}_t(X_t) + c)\,dt + \sigma_t(X_t)\,dW_t \tag{8}$$

with an appropriate choice of c. Obvious choices are $c = E(f_t(X_t)|\mathscr{F}_t^Y)$ or $c = f_t(E(X_t|\mathscr{F}_t^Y))$.

Rearranging the advection term part of equation (4) results in

$$d_t p_t(x) + \sum_k c_k \frac{\partial}{\partial x_k} p_t(x) = \left(-\sum_k \frac{\partial}{\partial x_k} (\bar{f}_{t,k}(x) p_t(x)) \right) dt$$

or, discretized by the upwind-scheme,

$$p_{t+\Delta t}(x + c\Delta t) = p_t(x,t) - \Delta t \sum_k \bar{f}_{t,k}(x) D_{x_k} p_t(x) \quad .$$

The spatial shift from x to $x + c\Delta t$ is realized by introducing a shift vector v which is attached to the grid. The numerically treated advection term is just the difference in drift $\bar{f}_t = f_t - c$ which should be absolutely smaller than the original drift. This has an influence on the stability behaviour: a significantly larger step size is then feasible.

In the implementation, the attached shift vector v is modified as soon as the calculated stability margin falls below the threshold.

2.3 Measurement Update

The pdf is updated using measurements according to Bayes' rule (5). The multiplication is simply done for every sparse grid point. The denominator is computed by a piecewise linear interpolation of the integrand which gives a coarse norm value of the updated pdf.

2.4 Expected Values

Usually, we are not interested in the whole pdf but only in its certain characteristic properties. For most cases it is reasonable to extract the expected values

$$Ex = \int x p_t(x) dx.$$

Instead of approximating the integrand with piecewise constant or linear functions as it is done in the normalization of Bayes' rule, the pdf itself is piecewise interpolated using Gaussian densities in every coordinate direction. With

$$y = a \exp\left(-\frac{1}{2} \left(\frac{x-\mu}{\sigma} \right)^2 \right),$$

the interpolation points $(x,y) = (x_i, p_t(x_i))$, $(x_j, p_t(x_j))$ and $(x_k, p_t(x_k))$, and building the quotients

$$h_{ij} = -2 \ln \frac{p_t(x_i)}{p_t(x_j)}, \quad r_{ijk} = \frac{h_{ij}}{h_{ik}}$$

we obtain for three neighbouring grid points $x_i = x_0$, $x_j = x_0 + \Delta x$ and $x_k = x_0 - \Delta x$, the parameters

$$\mu = x_0 + \frac{1}{2} \Delta x \frac{1 - r_{ijk}}{1 + r_{ijk}}, \quad \sigma^2 = -2 \frac{\Delta x^2}{h_{ij}} \frac{1}{1 + r_{ijk}}$$

and accept this interpolation for $\sigma^2 > 0$ and good conditioned values of μ. All together we find the expected values by averaging the piecewise values of μ.

3 Algorithm and Illustration

The algorithm consisting of the parts described above can be summarized as follows:

Algorithm 1. Filtering algorithm

```
Initialize sparse grid structure and pdf
for every time step do
    Compute drift compensation c and shift vector
    Solve (4) with difference drift f̄ using finite differences
    if measurement available then
        Measurement Update
    end
    Normalization
    Compute expected values
    if Tile Adapt Time then
        Adapt tiles
    end
end
```

For illustrating the algorithm, it is useful to apply it to a two-dimensional example. As a model problem we take the undamped pendulum with the following system equations

$$\dot{x}_1 = x_2$$
$$\dot{x}_2 = -c \cdot \sin(x_1)$$

and the measurement equation

$$y_1 = x_1$$

with the states angle and angular velocity: $x_1 = \varphi$, $x_2 = \dot{\varphi}$.

Figure 3 shows contour plots of the pdf at different time instances. The x-axes ranges from $\varphi = 2.73$ rad to 4.69 rad, the y-axes from $\dot{\varphi} = -0.38$ rad/s to 3.56 rad/s. The Gaussian distribution at $t = 0.0$ s is initialized with $\mu = (\pi, 0.45)^T$ and $\sigma = (0.1, 0.2)^T$. Afterwards, the pdf moves along its states with time according to (4). The position of the tiles is marked with squares, the interconnecting layers are depicted by lines. The grid drift, addition and removal of tiles can be observed from one time step to the next. Measurement updates were performed at $t = 0.3$ s and $t = 0.6$ s.

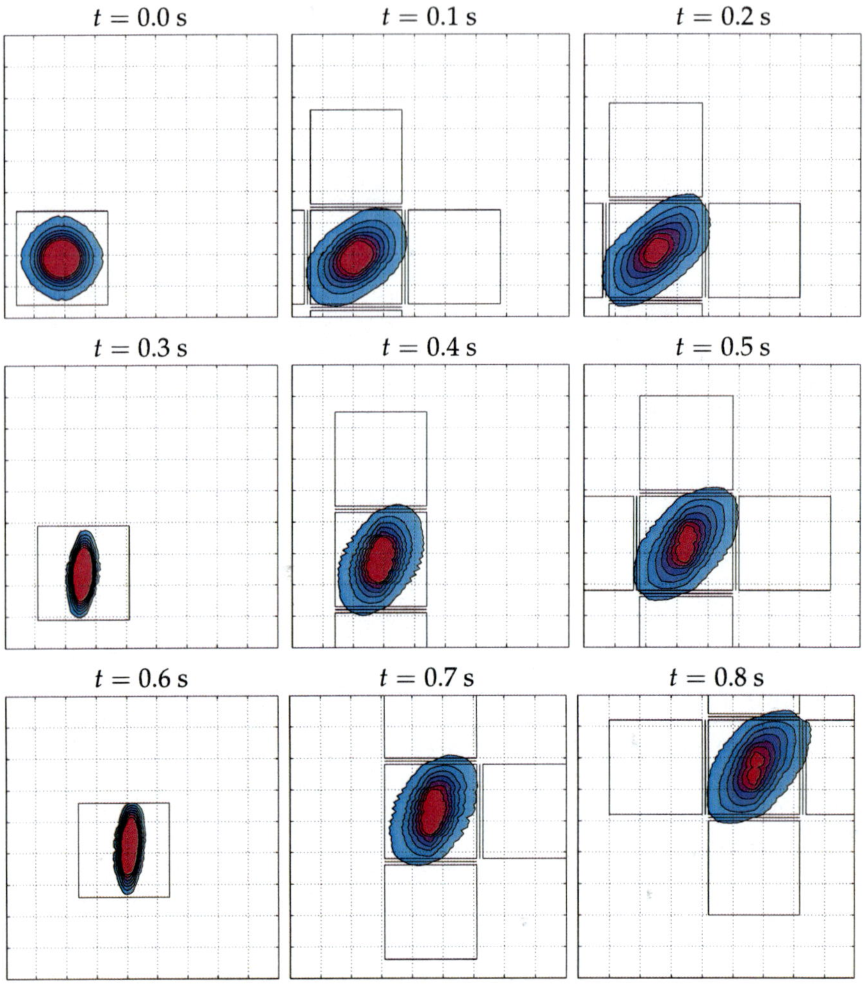

Fig. 3 Evolution of the pdf for the model problem

4 Example

To show the performance of the sparse grid tracking filter a target with high nonlinear dynamics is considered. In [5] a guidance for an anti-ship missile is proposed composed of a PN plus a barrel-roll distraction manoeuvre. The target dynamics can be described with the following nonlinear state equations:

$$\begin{aligned}
\dot{x}_1 &= -v_m \cos x_5 \cos x_4 \\
\dot{x}_2 &= -\frac{v_m}{x_1} \sin x_5 \\
\dot{x}_3 &= -\frac{v_m}{x_1 \cos x_2} \cos x_5 \sin x_4 \\
\dot{x}_4 &= \frac{v_m \sin x_4}{x_1 \cos x_5} \left(1 - N - \tan x_2 \sin x_5 \cos x_4 \cos x_5\right) - \omega \frac{\sin x_5 \cos x_4}{\cos x_5} \\
\dot{x}_5 &= \frac{v_m}{x_1} \left((1-N) \sin x_5 \cos x_4 + \tan x_2 \sin^2 x_4 \cos x_5\right) + \omega \sin x_4
\end{aligned}$$

with the states describing the distance $x_1 = r$, the line of sight angles $x_2 = \theta_L$, $x_3 = \psi_L$ and the angles of the missile velocity with respect to the line-of-sight $x_4 = \psi_M$, $x_5 = \theta_M$. There are three parameters in the model: the value of missile velocity v_m, the coefficient of the proportional navigation N and the angular velocity of the barrel roll ω. The dimension of the system ranges between five and eight depending on whether the parameters are known in advance or not. The results shown represent the case with an estimated ω with $\dot{\omega} = 0$ and we have therefore a six dimensional system.

A radar station shall measure the range and LOS angles

$$Y_{t_k} = \begin{pmatrix} 1 & & \\ & 1 & 0 \\ & & 1 \end{pmatrix} X_{t_k} + v_{t_k}(X_{t_k}) V_{t_k}$$

with

$$v_{t_k} = \begin{pmatrix} 100 & & \\ & 0.004 & \\ & & 0.004 \end{pmatrix}$$

at a frequency of 5 Hz. The system noise is given by

$$\sigma_{t_k} = \mathrm{diag}(10,\ 0.3,\ 0.3,\ 0.6,\ 0.6,\ 1).$$

The initial state values are assumed to be known except for the noise with a standard deviation of

$$\sigma = \mathrm{diag}(100,\ 0.004,\ 0.004,\ 0.1,\ 0.1,\ 0.25).$$

The missile system angles ψ_M and θ_M are only very poorly observable. These states diverge using an Extended Kalman Filter due to the strong nonlinearities. Simulation results computed in real time in comparison to an EKF are shown in figure 4 for ψ_L and ψ_M. The improved behaviour of the sparse grid filter is obvious and confirmed by 100 Monte-Carlo runs. Figure 5 shows the RMSE in ψ_L and ψ_M. The divergence of the EKF in ψ_M causes also bad estimates in the observed states as can be seen from the RMSE for ψ_L. The EKF tends to diverge and depends strongly on the measurement update while the sparse grid filter shows a good performance.

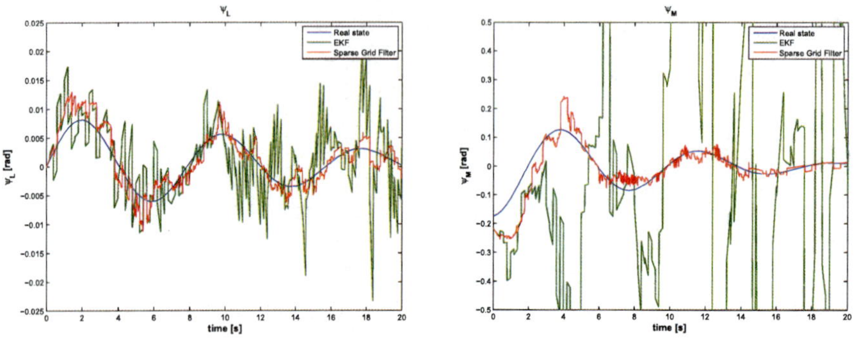

Fig. 4 Simulation results for ψ_L and ψ_M

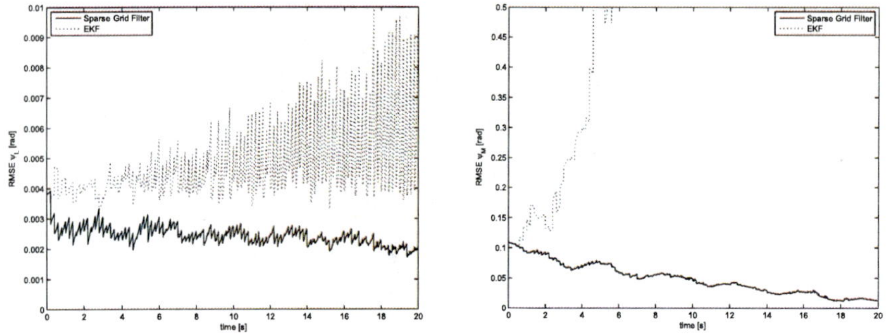

Fig. 5 RMSE for ψ_L and ψ_M

5 Conclusions

This paper presented a new approach for nonlinear filtering. The full nonlinear filtering problem is solved by a probability density function in which the time evolution is determined by a partial differential equation for prediction and Bayes' rule for measurements. The technique of sparse grids was used for the discretization of the continuous problem and provided an algorithm applicable in real-time also for higher dimensions. The discretization was done by finite differences on the sparse grids for which tiling and grid drifting to increase efficiency and accuracy was introduced. The estimated state values were extracted using piecewise interpolation with Gaussian densities. The performance of the new algorithm was shown on a six dimensional nonlinear problem in comparison to an EKF. The algorithm was applicable in real-time and showed a better performance than the EKF.

References

1. Bain, A., Crisan, D.: Fundamentals of Stochastic Filtering. Springer, New York (2009)
2. Beneš, V.E.: Exact finite-dimensional filters for certain diffusions with nonlinear drift. Stochastics 5(1-2), 65–92 (1981)

3. Bungartz, H.-J., Griebel, M.: Sparse Grids. Acta Numerica 13, 147–269 (2004)
4. Daum, F.E.: New exact nonlinear filters: Theory and applications. In: Proc. SPIE, vol. 2235, pp. 636–649 (1994)
5. Kim, Y.-H., Tahk, M.-J.: Guidance Synthesis for Evasive Maneuver of Anti-Ship Missiles. In: AIAA Guidance, Navigation and Control Conference and Exhibit, August 20-23 (2007)
6. Li, X.R., Jilkov, V.P.: A survey of maneuvering target tracking-Part VIa: density-based exact nonlinear filtering. In: Proc. 2010 SPIE Conf. Signal and Data Processing of Small Targets, Orlando (2010)
7. Lipster, R.S., Shiryaev, A.N.: Statistics of Random Processes: I General Theory. Springer, Berlin (2001)
8. Rozovskii, B.L.: Stochastic Evolution Systems. Kluwer, Dordrecht (1990)
9. Schiekofer, T.: Die Methode der finiten Differenzen auf dünnen Gittern zur Lösung elliptischer und parabolischer partieller Differentialgleichungen, Diss., Universität Bonn (1998)
10. Simon, D.: Optimal State Estimation. Wiley, Hoboken (2006)
11. Strikwerda, J.C.: Finite Difference Schemes and Partial Differential Equations. SIAM, Philadelphia (2004)
12. Zenger, C.: Sparse Grids. In: Hackbusch, W. (ed.) Parallel Algorithms for Partial Differential Equations. Notes on Numerical Fluid Mechanics, vol. 31, pp. 241–251. Vieweg, Braunschweig (1991)
13. Zhang, H., Laneuville, D.: Grid Based Solution of Zakai Equation with Adaptive Local Refinement for Bearings-only Tracking. In: IEEE Aerospace Conference, pp. 1–8 (2008)

Observability of Star Tracker / Gyro Based Attitude Estimation Considering Time-Variant Sensor Misalignment

Stefan Winkler

Abstract. The fusion of measurements from star trackers and gyroscopes within optimal estimators/filters is a common approach for spacecraft attitude determination. For applications where filter tuning is not sufficient to account for unmodelled deterministic errors, state augmentation is often the method of choice. So also here, where the focus is on deterministic time-variant misalignment between star tracker and gyroscope unit as this often occurs in missions with repetitive ecplipse and sun phases. Based on the derived filter dynamics and measurement equations, an observability analysis is performed. Different practical cases are distinguished to analyze: (1) which filter states are observable, (2) which only in linear combination and (3) which not at all.

1 Introduction

Information on the spacecraft attitude is essential for almost every space mission. A comprehensive survey of the most promising nonlinear methods that have been developed during the last 25 years is given in [2]. This paper here focuses on sensor data fusion of star tracker (STR) and gyros, also known as *gyro-stellar estimation.*

Star trackers and gyros have a number of complementary properties which makes them interesting for sensor data fusion. E.g. while a star tracker provides low-frequent long-term stable attitude measurements, the attitude computed from the high-frequent gyro measurements is only short-term stable. The motivation to fuse star tracker and gyro is to generate a long-term stable attitude estimate at gyro measurement frequency of attenuated noise compared to the star tracker measurement and, hence, to optimally bridge periods where no star tracker measurement is available. In addition to the improved attitude information, the estimator provides an

Dr. Stefan Winkler
AOCS/GNC and Flight Dynamics
Astrium GmbH - Satellites, 88039 Friedrichshafen, Germany
e-mail: `stefan.winkler.sat@astrium.eads.net`

estimate of the gyro angular rate bias. The typical extended Kalman filter has 6 states: 3 attitude errors, 3 gyro bias errors. For details see e.g. [1, 2, 5, 6].

Misalignment between star tracker and gyro unit and error in gyro scale factor always occur. If the spacecraft angular rate vector is constant[1] w.r.t. to the spacecraft, the filter will estimate constant misalignment and scale factor error as part of the gyro bias estimate. Hence, when the 6-state filter is in steady-state, both constant misalignment and scale factor error do not influence the attitude estimate anymore. Performing hereafter an attitude maneuver, where the angular rate vector is not constant anymore, will corrupt the attitude estimate. This fact is important for agile spacecraft that require high-precision attitude information.

Time-variant misalignment between star tracker and gyro unit is often caused by temperature changes between sun and eclipse phases during orbit. It can have a significant impact on the achievable attitude estimate performance. Hence, it is usually tried to limit it by specific constructional design, e.g. [3, 10, 11]. The remaining time-variant misalignment can be considered in the attitude estimation filter.

While time-correlated star tracker measurement noise can be well considered by appropriate filter tuning[2], the attitude estimate accuracy degradation due to (deterministic) time-variant star tracker / gyro unit misalignment can not sufficiently be compensated with this method for high-precision attitude determination. It is rather useful to augment the common 6-element filter state vector by the time-variant misalignment states [9, 10]. This paper focuses on the observability of such state augmented filters.

2 Attitude Difference Differential Equation

2.1 Attitude Representations

This section summarizes the attitude representations, their nomenclature and relations as used throughout this paper. For a thorough review see [8] and for one regarding attitude estimation with Kalman filtering [6].

2.1.1 Quaternion

An attitude or unit quaternion, $\mathbf{q}$, here shortly denoted as *quaternion*, describes a single-axis rotation about the Euler-axis (unit vector $\mathbf{e}$) about the rotation angle ψ

$$\mathbf{q} = \begin{bmatrix} \mathbf{e}\sin\frac{\psi}{2} \\ \cos\frac{\psi}{2} \end{bmatrix} = \begin{bmatrix} \boldsymbol{q} \\ q_4 \end{bmatrix} \tag{1}$$

where $\boldsymbol{q}$ is its vector element and q_4 its scalar element. A quaternion obey the unit length constraint $|\boldsymbol{q}|^2 + q_4^2 = 1$. The quaternion $\mathbf{q}_{21}$ describes the rotation from an 1- into a 2-frame and, hence, the attitude of a 2- w.r.t. an 1-frame.

[1] In amplitude and direction.

[2] Adaptation of the measurement error covariance matrix in the filter.

2.1.2 Gibbs Vector

The Gibbs vector[3], g, is defined using quaternion such that

$$\mathbf{g} = \boldsymbol{q}/q_4 = \mathbf{e}\tan(\psi/2) = \mathbf{a}/2 \tag{2}$$

where $\mathbf{a}$ is used throughout this paper as Gibbs vector *representation*. It clearly is equal to twice the Gibbs vector g. For small rotation angles ψ, one can write

$$\mathbf{a} \approx \mathbf{e}\psi = [\psi_x \; \psi_y \; \psi_z]^T \tag{3}$$

where ψ_x, ψ_y and ψ_z is the rotation angle about the x-, y- and z-axis, respectively. Hence, $\mathbf{a}$ becomes the rotation vector.

In this paper, the vector $\mathbf{a}_{21}$ describes the rotation from an 1- into a 2-frame and, hence, the attitude of a 2- w.r.t. an 1-frame.

2.1.3 Direction Cosine Matrix

Considering a sequential 3-2-1 rotation about the small angles $\delta\Psi$, $\delta\Theta$ and $\delta\Phi$, respectively, leads to the direction cosine matrix

$$\mathbf{T} = \mathbf{I} - \delta\mathbf{T} \tag{4}$$

where $\delta\mathbf{T}$ is a skew-symmetric matrix, namely

$$\delta\mathbf{T} = \begin{bmatrix} 0 & -\delta\Psi & \delta\Theta \\ \delta\Psi & 0 & -\delta\Phi \\ -\delta\Theta & \delta\Phi & 0 \end{bmatrix} = \begin{bmatrix} \delta\Phi \\ \delta\Theta \\ \delta\Psi \end{bmatrix}^{\times} = \delta\boldsymbol{\phi}^{\times}. \tag{5}$$

A direction cosine matrix $\mathbf{T}_{21}$ describes the rotation from an 1- into a 2-frame and, hence, the attitude of a 2- w.r.t. an 1-frame. A vector given in components of an 1-framce, $\mathbf{v}_1$, is transformed into a 2-frame, $\mathbf{v}_2$, by $\mathbf{v}_2 = \mathbf{T}_{21}\mathbf{v}_1$.

2.2 *Theoretical Derivation*

Considering the attitude between body- and inertial frame (subscript b and i, resp.), the attitude kinematics equation using quaternions is given by

$$\dot{\mathbf{q}}_{bi} = \frac{1}{2}\begin{bmatrix} \boldsymbol{\omega}_b^{ib} \\ 0 \end{bmatrix} \otimes \mathbf{q}_{bi} \tag{6}$$

$$= \frac{1}{2}\boldsymbol{\varpi}_b^{ib} \otimes \mathbf{q}_{bi} \tag{7}$$

where $\boldsymbol{\omega}_b^{ib}$ is the angular velocity of b- w.r.t. i- in the b-frame and $\otimes$ is the quaternion multiplication introduced in [5].

[3] Also known as Rodrigues parameters.

Analogous, the attitude kinematics between a second frame (subscript s) and the inertial frame is

$$\dot{\mathbf{q}}_{si} = \frac{1}{2}\,\boldsymbol{\varpi}_s^{is} \otimes \mathbf{q}_{si}. \tag{8}$$

For the attitude difference quaternion, $\mathbf{q}_{bs}$, which will reflect an attitude error later on in this paper, one can write

$$\mathbf{q}_{bi} = \mathbf{q}_{bs} \otimes \mathbf{q}_{si} \;\Rightarrow\; \mathbf{q}_{bs} = \mathbf{q}_{bi} \otimes \mathbf{q}_{si}^{-1}. \tag{9}$$

The time-derivative of this attitude difference is

$$\dot{\mathbf{q}}_{bs} = \dot{\mathbf{q}}_{bi} \otimes \mathbf{q}_{si}^{-1} + \mathbf{q}_{bi} \otimes \dot{\mathbf{q}}_{si}^{-1}. \tag{10}$$

Using the attitude kinematics equations from above leads to

$$\dot{\mathbf{q}}_{bs} = \frac{1}{2}\,\boldsymbol{\varpi}_b^{ib} \otimes \mathbf{q}_{bi} \otimes \mathbf{q}_{si}^{-1} + \mathbf{q}_{bi} \otimes \left(\frac{1}{2}\,\boldsymbol{\varpi}_s^{is} \otimes \mathbf{q}_{si}\right)^{-1} \tag{11}$$

$$= \frac{1}{2}\,\boldsymbol{\varpi}_b^{ib} \otimes \mathbf{q}_{bs} - \frac{1}{2}\mathbf{q}_{bs} \otimes \boldsymbol{\varpi}_s^{is}. \tag{12}$$

Introducing the Gibbs vector representation $\mathbf{a}_{bs}$, which corresponds to $\mathbf{q}_{bs}$, it follows

$$\dot{\mathbf{a}}_{bs} = \frac{1}{2}\left[\mathbf{I} - \frac{1}{2}\mathbf{a}_{bs}\right]\boldsymbol{\varpi}_b^{ib} \otimes \begin{bmatrix}\mathbf{a}_{bs}\\ 2\end{bmatrix} - \frac{1}{2}\left[\mathbf{I} - \frac{1}{2}\mathbf{a}_{bs}\right]\begin{bmatrix}\mathbf{a}_{bs}\\ 2\end{bmatrix} \otimes \boldsymbol{\varpi}_s^{is} \tag{13}$$

$$= \left(\mathbf{I} + \frac{1}{4}\mathbf{a}_{bs}\mathbf{a}_{bs}^T\right)\left(\boldsymbol{\omega}_b^{ib} - \boldsymbol{\omega}_s^{is}\right) - \frac{1}{2}\left(\boldsymbol{\omega}_b^{ib} + \boldsymbol{\omega}_s^{is}\right) \times \mathbf{a}_{bs}. \tag{14}$$

From now on, only small angular differences between s- and b-frame are assumed. Hence, $\mathbf{a}_{bs}$ is the small angular difference between both frames. Using the common small angle simplifications ($\sin\psi \approx \psi$, $\cos\psi \approx 1$, $\psi\psi \approx 0$) leads to $\mathbf{a}_{bs}\mathbf{a}_{bs}^T = \mathbf{0}$. Hence, Eq. (14) becomes

$$\dot{\mathbf{a}}_{bs} = \left(\boldsymbol{\omega}_b^{ib} - \boldsymbol{\omega}_s^{is}\right) - \frac{1}{2}\left(\boldsymbol{\omega}_b^{ib} + \boldsymbol{\omega}_s^{is}\right) \times \mathbf{a}_{bs}. \tag{15}$$

This is the attitude difference equation in Gibbs-vector notation for small attitude differences between b- and s-frame.

2.3 *Application to Star Tracker / Gyro Fusion*

A problem often considered in practice is the determination of the attitude of the body-frame w.r.t. inertial-frame. This attitude shall be expressed by the quaternion $\mathbf{q}_{bi}$. If $\boldsymbol{\omega}_b^{ib}$ is known, starting from an initial attitude, the attitude vs. time can be determined by integrating the kinematic equation (Eq. (6)). In reality $\boldsymbol{\omega}_b^{ib}$ is not exactly known but it can be measured with an inertial measurement unit (IMU)[4]. With the u-frame as the IMU-(fixed-)frame (subscript u), the measurement provided by the IMU can be modeled as

$$\tilde{\boldsymbol{\omega}}_u^{iu} = \boldsymbol{\omega}_u^{iu} + \mathbf{b}_u + \mathbf{w}_u \tag{16}$$

where $\mathbf{b}$ is the deterministic error and $\mathbf{w}$ the stochastic error on the IMU measurement.

If there is an angular velocity between IMU- and body-frame, $\boldsymbol{\omega}^{bu}$, which is the considered time-variant misalignment, one can write in more detail

$$\tilde{\boldsymbol{\omega}}_u^{iu} = \boldsymbol{\omega}_u^{ib} + \boldsymbol{\omega}_u^{bu} + \mathbf{b}_u + \mathbf{w}_u. \tag{17}$$

Obviously, there will be a difference, an error, between the attitude of body- w.r.t. inertial-frame and IMU- w.r.t. inertial-frame. The linear approximation of this attitude error in Gibbs-vector notation is given by Eq. (15). If this error, namely $\mathbf{a}_{bs}$, is known, the attitude obtained from integrating IMU measurements can be corrected. Hence, the goal must be to determine $\mathbf{a}_{bs}$. For this, Eq. (15) can be integrated. The expression for the ideal angular rate of the body- w.r.t. inertial frame, $\boldsymbol{\omega}_b^{ib}$, is given using Eq. (17) by

$$\boldsymbol{\omega}_b^{ib} = \mathbf{T}_{bu}\left(\tilde{\boldsymbol{\omega}}_u^{iu} - \boldsymbol{\omega}_u^{bu} - \mathbf{b}_u - \mathbf{w}_u\right). \tag{18}$$

In general, the term $\boldsymbol{\omega}_s^{is}$ in Eq. (15) could be the angular rate measured by the IMU. But to keep the attitude error $\mathbf{a}_{bs}$ as small as possible, all availabe information shall be used to correct the IMU measurement and to come with $\boldsymbol{\omega}_s^{is}$ as close as possible to $\boldsymbol{\omega}_b^{ib}$. If one would achieve $\boldsymbol{\omega}_s^{is} = \boldsymbol{\omega}_b^{ib}$ in practice, the attitude of the body-frame w.r.t. the inertial-frame could be reconstracted exactly using the IMU measurements (despite errors due to numerical integration etc.). To come with $\boldsymbol{\omega}_s^{is}$ as close as possible to $\boldsymbol{\omega}_b^{ib}$, it shall be

$$\boldsymbol{\omega}_s^{is} = \hat{\mathbf{T}}_{bu}\left(\tilde{\boldsymbol{\omega}}_u^{iu} - \hat{\boldsymbol{\omega}}_u^{bu} - \hat{\mathbf{b}}_u\right). \tag{19}$$

The term $\hat{\mathbf{T}}_{bu}\left(\tilde{\boldsymbol{\omega}}_u^{iu} - \hat{\boldsymbol{\omega}}_u^{bu} - \hat{\mathbf{b}}_u\right)$ is the best available estimate for the seeked angular rate $\boldsymbol{\omega}_b^{ib}$, namely $\hat{\boldsymbol{\omega}}_b^{ib}$. Hence, $\boldsymbol{\omega}_s^{is} = \hat{\boldsymbol{\omega}}_b^{ib}$. So one can write

$$\boldsymbol{\omega}_b^{ib} - \boldsymbol{\omega}_s^{is} = \mathbf{T}_{bu}\left(\tilde{\boldsymbol{\omega}}_u^{iu} - \boldsymbol{\omega}_u^{bu} - \mathbf{b}_u - \mathbf{w}_u\right) - \hat{\mathbf{T}}_{bu}\left(\tilde{\boldsymbol{\omega}}_u^{iu} - \hat{\boldsymbol{\omega}}_u^{bu} - \hat{\mathbf{b}}_u\right) \tag{20}$$

and

$$\boldsymbol{\omega}_b^{ib} + \boldsymbol{\omega}_s^{is} = \mathbf{T}_{bu}\left(\tilde{\boldsymbol{\omega}}_u^{iu} - \boldsymbol{\omega}_u^{bu} - \mathbf{b}_u - \mathbf{w}_u\right) + \hat{\mathbf{T}}_{bu}\left(\tilde{\boldsymbol{\omega}}_u^{iu} - \hat{\boldsymbol{\omega}}_u^{bu} - \hat{\mathbf{b}}_u\right). \tag{21}$$

With

$$\hat{\boldsymbol{\omega}}_u^{ib} = \tilde{\boldsymbol{\omega}}_u^{iu} - \hat{\boldsymbol{\omega}}_u^{bu} - \hat{\mathbf{b}}_u \tag{22}$$

[4] In this paper, IMU is equivalent to gyro unit.

and the general definition of the error δx of an arbitrary variable x, namely $\delta x = \hat{x} - x$, for Eq. (20) one can write

$$\boldsymbol{\omega}_b^{ib} - \boldsymbol{\omega}_s^{is} = (\mathbf{I} - \delta\mathbf{T})\,\hat{\mathbf{T}}_{bu}\left(\tilde{\boldsymbol{\omega}}_u^{iu} - \hat{\boldsymbol{\omega}}_u^{bu} + \delta\boldsymbol{\omega}_u^{bu} - \hat{\mathbf{b}}_u + \delta\mathbf{b}_u - \mathbf{w}_u\right) \\ -\hat{\mathbf{T}}_{bu}\hat{\boldsymbol{\omega}}_u^{ib}. \tag{23}$$

And considering small errors (linear approximation) it follows that

$$\boldsymbol{\omega}_b^{ib} - \boldsymbol{\omega}_s^{is} = \hat{\mathbf{T}}_{bu}\,\delta\boldsymbol{\omega}_u^{bu} + \hat{\mathbf{T}}_{bu}\,\delta\mathbf{b}_u - \delta\mathbf{T}\,\hat{\mathbf{T}}_{bu}\hat{\boldsymbol{\omega}}_u^{ib} - \hat{\mathbf{T}}_{bu}\mathbf{w}_u. \tag{24}$$

Furthermore, using Eq. (21), it can be written

$$\left(\boldsymbol{\omega}_b^{ib} + \boldsymbol{\omega}_s^{is}\right) \times \mathbf{a}_{bs} = \Big[(\mathbf{I} - \delta\mathbf{T})\,\hat{\mathbf{T}}_{bu}\left(\tilde{\boldsymbol{\omega}}_u^{iu} - \boldsymbol{\omega}_u^{bu} - \mathbf{b}_u - \mathbf{w}_u\right) \\ + \hat{\mathbf{T}}_{bu}\hat{\boldsymbol{\omega}}_u^{ib}\Big] \times \mathbf{a}_{bs}. \tag{25}$$

And considering small errors (linear approximation) it follows that

$$\left(\boldsymbol{\omega}_b^{ib} + \boldsymbol{\omega}_s^{is}\right) \times \mathbf{a}_{bs} = 2\hat{\mathbf{T}}_{bu}\hat{\boldsymbol{\omega}}_u^{ib} \times \mathbf{a}_{bs}. \tag{26}$$

So, finally Eq. (15) becomes

$$\dot{\mathbf{a}}_{bs} = -\left(\hat{\mathbf{T}}_{bu}\hat{\omega}_u^{ib}\right)^{\times}\mathbf{a}_{bs} + \hat{\mathbf{T}}_{bu}\,\delta\mathbf{b}_u + \hat{\mathbf{T}}_{bu}\,\delta\boldsymbol{\omega}_u^{bu} + \left(\hat{\mathbf{T}}_{bu}\hat{\omega}_u^{ib}\right)^{\times}\delta\boldsymbol{\phi} - \hat{\mathbf{T}}_{bu}\mathbf{w}_u \tag{27}$$

where $\delta\boldsymbol{\phi}$ is the misalignment error and $\delta\boldsymbol{\omega}_u^{bu}$ its rate of change. Clearly, the equal influence of $\delta\mathbf{b}_u$ and $\delta\boldsymbol{\omega}_u^{bu}$ on $\dot{\mathbf{a}}_{bs}$ can be recognized. Their separation must be realized with different dynamic models. If the underlying dynamic models are equal, a separation is not possible. However, there might be applications where the exact knowledge of $\delta\mathbf{b}_u$ and $\delta\boldsymbol{\omega}_u^{bu}$ is unimportant. There, the focus is on rejecting the sum of both from $\dot{\mathbf{a}}_{bs}$ no matter what the individual values are.

The state vector of the considered estimator shall be

$$\begin{bmatrix} \mathbf{a}_{bs} \\ \delta\mathbf{b}_u \\ \delta\boldsymbol{\phi} \\ \delta\dot{\boldsymbol{\phi}} \end{bmatrix} \tag{28}$$

which leads to the corresponding continuous-time system matrix

$$\mathbf{F} = \begin{bmatrix} -\left(\hat{\mathbf{T}}_{bu}\hat{\omega}_u^{ib}\right)^{\times} & \hat{\mathbf{T}}_{bu} & \left(\hat{\mathbf{T}}_{bu}\hat{\omega}_u^{ib}\right)^{\times} & \hat{\mathbf{T}}_{bu} \\ \mathbf{0} & \mathbf{0} & \mathbf{0} & \mathbf{0} \\ \mathbf{0} & \mathbf{0} & \mathbf{0} & \mathbf{I} \\ \mathbf{0} & \mathbf{0} & \mathbf{F}_{\ddot{\phi}\phi} & \mathbf{0} \end{bmatrix} \tag{29}$$

where $\mathbf{F}_{\ddot{\phi}\phi} = \partial(\delta\ddot{\boldsymbol{\phi}})/\partial(\delta\boldsymbol{\phi})$.

The star tracker measurement $\tilde{\mathbf{q}}_{bi}$ and the attitude estimate $\hat{\mathbf{q}}_{si}$ are used to formulate the filter innovation as quaternion

$$\delta\mathbf{q} = \tilde{\mathbf{q}}_{bi} \otimes \hat{\mathbf{q}}_{si}^{-1}. \tag{30}$$

It is converted into Gibbs vector formulation, $\mathbf{a}$, for usage in the estimator. Hence, the measurement matrix of the estimator is

$$\mathbf{H} = \begin{bmatrix} \mathbf{I}\ \mathbf{0}\ \mathbf{0}\ \mathbf{0} \end{bmatrix}. \tag{31}$$

3 Observability

3.1 Theoretical Background

This section goes beyond just presenting the well known equation for the observability matrix. It rather focuses on its interpretation to draw conclusions on the observability of each system state.

An observer or estimator (e.g. Kalman filter) is often used to determine system states that cannot or only very difficult be determined. Such states could be the time-variant misalignment between star tracker and gyro unit. The question rises if all states can be determined/distinguished using the available sensor and system information. This leads to the question of *observability* which was initially defined by Kalman [4].

Consider a system defined on some time interval T. An initial state $\mathbf{x}_0 = \mathbf{x}(t_0)$ with $t_0 \in T$ is said to be *observable* if it is possible to determine $\mathbf{x}_0$ using the measurements $\mathbf{z}$ over a finite time interval $t_0 \leq t \leq t_f$, $t \in T$ [7]. In other words, if the system is observable, the state $\mathbf{x}_0$ can be reconstructed using the measurements $\mathbf{z}$.

In the following, it shall be distinguished between *complete* and *non-complete* observability. In the former case, each single state of the state vector can be determined and, hence, distinguished from the other states. In the later, this is true only for some of the sates. The other states can only be determined in form of linear combinations of each other or not at all.

Consider the linear time in-variant system of n states with state vector $\mathbf{x}(t)$ and measurement $\mathbf{z}(t)$

$$\dot{\mathbf{x}}(t) = \mathbf{F}\,\mathbf{x}(t) \tag{32}$$

$$\mathbf{z}(t) = \mathbf{H}\,\mathbf{x}(t) \tag{33}$$

and state solution

$$\mathbf{x}(t) = e^{\mathbf{F}\,(t-t_0)}\,\mathbf{x}_0. \tag{34}$$

The available information to evaluate $\mathbf{x}_0$ can be summarized to

$$\begin{bmatrix} \mathbf{z}(t) \\ \dot{\mathbf{z}}(t) \\ \ddot{\mathbf{z}}(t) \\ \vdots \\ \mathbf{z}(t)^{(n-1)} \end{bmatrix} = \begin{bmatrix} \mathbf{H} \\ \mathbf{HF} \\ \mathbf{HF}^2 \\ \vdots \\ \mathbf{HF}^{n-1} \end{bmatrix} e^{\mathbf{F}\,(t-t_0)}\,\mathbf{x}_0 \tag{35}$$

$$= \mathcal{O} \quad e^{\mathbf{F}\,(t-t_0)}\,\mathbf{x}_0 \tag{36}$$

$$= \mathcal{O} \quad \mathbf{x}(t). \tag{37}$$

Essential for the determination/observation of $\mathbf{x}_0$ (and $\mathbf{x}(\mathbf{t})$) is $\mathcal{O}$, known as observability matrix. The system is completely observable if and only if $\mathrm{Rg}(\mathcal{O}) = n$. Then, Eq. (36) can be solved for $\mathbf{x}_0$ (and Eq. (37) for $\mathbf{x}(t)$).

3.2 *Application to Star Tracker / Gyro Fusion*

3.2.1 Observability Matrix

With the simplified expressions

$$\mathbf{T} = \hat{\mathbf{T}}_{bu} \tag{38}$$

$$\omega = \hat{\boldsymbol{\omega}}_u^{ib} \tag{39}$$

$$\mathbf{M} = \mathbf{F}_{\ddot{\phi}\phi} \tag{40}$$

the system matrix in Eq. (29) becomes

$$\mathbf{F} = \begin{bmatrix} -(\mathbf{T}\boldsymbol{\omega})^{\times}\,\mathbf{T} & (\mathbf{T}\boldsymbol{\omega})^{\times}\,\mathbf{T} \\ \mathbf{0} & \mathbf{0} & \mathbf{0} & \mathbf{0} \\ \mathbf{0} & \mathbf{0} & \mathbf{0} & \mathbf{I} \\ \mathbf{0} & \mathbf{0} & \mathbf{M} & \mathbf{0} \end{bmatrix}. \tag{41}$$

And assuming small differences between b- and u-frame, hence, $\mathbf{T} \approx \mathbf{I}$, it follows

$$\mathbf{F} = \begin{bmatrix} -\boldsymbol{\omega}^{\times} & \mathbf{I} & \boldsymbol{\omega}^{\times} & \mathbf{I} \\ \mathbf{0} & \mathbf{0} & \mathbf{0} & \mathbf{0} \\ \mathbf{0} & \mathbf{0} & \mathbf{0} & \mathbf{I} \\ \mathbf{0} & \mathbf{0} & \mathbf{M} & \mathbf{0} \end{bmatrix}. \tag{42}$$

With the measurement matrix, Eq. (31), one can write

$$
\begin{array}{lll}
\mathbf{HF} & = [& \begin{array}{c|c|c|c} -\boldsymbol{\omega}^{\times} & \mathbf{I} & \boldsymbol{\omega}^{\times} & \mathbf{I} \end{array}] \\
\mathbf{HF}^2 & = [& \begin{array}{c|c|c|c} (\boldsymbol{\omega}^{\times})^2 & -\boldsymbol{\omega}^{\times} & -(\boldsymbol{\omega}^{\times})^2+\mathbf{M} & \mathbf{0} \end{array}] \\
\mathbf{HF}^3 & = [& \begin{array}{c|c|c|c} -(\boldsymbol{\omega}^{\times})^3 & (\boldsymbol{\omega}^{\times})^2 & (\boldsymbol{\omega}^{\times})^3 & \mathbf{M} \end{array}] \\
\mathbf{HF}^4 & = [& \begin{array}{c|c|c|c} (\boldsymbol{\omega}^{\times})^4 & -(\boldsymbol{\omega}^{\times})^3 & -(\boldsymbol{\omega}^{\times})^4+\mathbf{M}^2 & \mathbf{0} \end{array}] \\
\mathbf{HF}^5 & = [& \begin{array}{c|c|c|c} -(\boldsymbol{\omega}^{\times})^5 & (\boldsymbol{\omega}^{\times})^4 & (\boldsymbol{\omega}^{\times})^5 & \mathbf{M}^2 \end{array}] \\
\vdots & \vdots [& \begin{array}{c|c|c|c} \vdots & \vdots & \vdots & \vdots \end{array}].
\end{array}
\tag{43}
$$

From now on, the observability analysis shall focus on spacecraft with angular rate

$$
\boldsymbol{\omega} = \begin{bmatrix} 0 \\ \omega \\ 0 \end{bmatrix}
\tag{44}
$$

which is a good approximation for e.g. Earth observation satellites despite mission/payload related steering. In this case, ω is equal to the orbital rate and constant. Hence, in addition to $\mathbf{H}$, also $\mathbf{F}$ and therefore the system is time in-variant.

So with

$$
\mathbf{M} = \begin{bmatrix} M_1 & 0 & 0 \\ 0 & M_2 & 0 \\ 0 & 0 & M_3 \end{bmatrix}
\tag{45}
$$

it can be written

$$
\begin{bmatrix} \mathbf{H} \\ \mathbf{HF} \\ \mathbf{HF}^2 \\ \mathbf{HF}^3 \\ \mathbf{HF}^4 \\ \mathbf{HF}^5 \end{bmatrix} =
\left[\begin{array}{ccc|ccc|ccc|ccc}
1 & 0 & 0 & 0 & 0 & 0 & 0 & 0 & 0 & 0 & 0 & 0 \\
0 & 1 & 0 & 0 & 0 & 0 & 0 & 0 & 0 & 0 & 0 & 0 \\
0 & 0 & 1 & 0 & 0 & 0 & 0 & 0 & 0 & 0 & 0 & 0 \\
\hline
0 & 0 & -\omega & 1 & 0 & 0 & 0 & 0 & \omega & 1 & 0 & 0 \\
0 & 0 & 0 & 0 & 1 & 0 & 0 & 0 & 0 & 0 & 1 & 0 \\
\omega & 0 & 0 & 0 & 0 & 1 & -\omega & 0 & 0 & 0 & 0 & 1 \\
\hline
-\omega^2 & 0 & 0 & 0 & 0 & -\omega & \omega^2+M_1 & 0 & 0 & 0 & 0 & 0 \\
0 & 0 & 0 & 0 & 0 & 0 & 0 & M_2 & 0 & 0 & 0 & 0 \\
0 & 0 & -\omega^2 & \omega & 0 & 0 & 0 & 0 & \omega^2+M_3 & 0 & 0 & 0 \\
\hline
0 & 0 & \omega^3 & -\omega^2 & 0 & 0 & 0 & 0 & -\omega^3 & M_1 & 0 & 0 \\
0 & 0 & 0 & 0 & 0 & 0 & 0 & 0 & 0 & 0 & M_2 & 0 \\
-\omega^3 & 0 & 0 & 0 & 0 & -\omega^2 & \omega^3 & 0 & 0 & 0 & 0 & M_3 \\
\hline
\omega^4 & 0 & 0 & 0 & 0 & \omega^3 & -\omega^4+M_1^2 & 0 & 0 & 0 & 0 & 0 \\
0 & 0 & 0 & 0 & 0 & 0 & 0 & M_2^2 & 0 & 0 & 0 & 0 \\
0 & 0 & \omega^4 & -\omega^3 & 0 & 0 & 0 & 0 & -\omega^4+M_3^2 & 0 & 0 & 0 \\
\hline
0 & 0 & -\omega^5 & \omega^4 & 0 & 0 & 0 & 0 & \omega^5 & M_1^2 & 0 & 0 \\
0 & 0 & 0 & 0 & 0 & 0 & 0 & 0 & 0 & 0 & M_2^2 & 0 \\
\omega^5 & 0 & 0 & 0 & 0 & \omega^4 & -\omega^5 & 0 & 0 & 0 & 0 & M_3^2
\end{array}\right].
\tag{46}
$$

This is the upper part of the observability matrix. The parts $\mathbf{HF}^k$ for $k > 5$ were left out. Because for the cases considered in sec. 3.2.2 they do not gain additional information.

3.2.2 Interpretation of Observability Matrix

The first 3 scalar lines of the observability matrix in Eq. (46) show that $\mathbf{a}_{bs}$ is always observable. This result is expected since $\mathbf{a}_{bs}$ is directly provided as measurement to the filter (see Eq. (30)). Writing Eq. (46) line by line gives

$$
\begin{array}{rcrrrr}
[\mathbf{HF}]_1 - c_{21} &=& \delta b_x \; + & & \omega\phi_z \; + & \dot{\phi}_x \\
[\mathbf{HF}]_2 - c_{22} &=& \delta b_y & & + & \dot{\phi}_y \\
[\mathbf{HF}]_3 - c_{23} &=& \delta b_z \; - & & \omega\phi_x \; + & \dot{\phi}_z \\
\left[\mathbf{HF}^2\right]_1 - c_{31} &=& -\omega\,\delta b_z \; + & (\omega^2 + M_1)\phi_x & & \\
\left[\mathbf{HF}^2\right]_2 - c_{32} &=& + & M_2\phi_y & & \\
\left[\mathbf{HF}^2\right]_3 - c_{33} &=& \omega\,\delta b_x \; + & (\omega^2 + M_3)\phi_z & & \\
\left[\mathbf{HF}^3\right]_1 - c_{41} &=& -\omega^2\,\delta b_x \; - & & \omega^3\phi_z \; + & M_1\dot{\phi}_x \\
\left[\mathbf{HF}^3\right]_2 - c_{42} &=& & & & M_2\dot{\phi}_y \\
\left[\mathbf{HF}^3\right]_3 - c_{43} &=& -\omega^2\,\delta b_z \; + & & \omega^3\phi_x \; + & M_3\dot{\phi}_z \\
\left[\mathbf{HF}^4\right]_1 - c_{51} &=& \omega^3\,\delta b_z \; + & (-\omega^4 + M_1^2)\phi_x & & \\
\left[\mathbf{HF}^4\right]_2 - c_{52} &=& & M_2^2\phi_y & & \\
\left[\mathbf{HF}^4\right]_3 - c_{53} &=& -\omega^3\,\delta b_x \; + & (-\omega^4 + M_3^2)\phi_z & & \\
\left[\mathbf{HF}^5\right]_1 - c_{61} &=& \omega^4\,\delta b_x \; + & & \omega^5\phi_z \; + & M_1^2\dot{\phi}_x \\
\left[\mathbf{HF}^5\right]_2 - c_{62} &=& & & & M_2^2\dot{\phi}_y \\
\left[\mathbf{HF}^5\right]_3 - c_{63} &=& \omega^4\,\delta b_z \; - & & \omega^5\phi_x \; + & M_3^2\dot{\phi}_z
\end{array}
$$

where c_{ij} is a known value. It is known since $\mathbf{a}_{bs}$ is always observable and, hence, computed from the first matrix line ($\mathbf{H}$-line) in Eq. (46). The c_{ij} correspond to the first matrix column in Eq. (46). Using these equations, now different cases shall be distinguish to analyze the observability of the estimator states.

Case 1: Time-constant STR/IMU misalignment

The time-*constant* misalignment leads to $\mathbf{M} = \mathbf{0}$. Cancelling the last vector column in Eq. (46) results in

- δb_y is fully observable
- δb_x and ϕ_z are only in combination observable
- δb_z and ϕ_x are only in combination observable
- ϕ_y is not observable

To determine the time-constant misalignment to extract its influence on the attitude estimate, calibration maneuvers are necessary. Such maneuvers must ensure the observability of the misalignment states.

Case 2: Harmonically oscillating STR/IMU misalignment with orbit frequency

This case is close to Earth observation spacecraft with approximately equal sun and eclipse duration. One can write $\mathbf{M} = -\omega^2\,\mathbf{I}$ where ω is the orbital rate. Hence,

- δb_x, δb_y and δb_z are fully observable
- ϕ_y is fully observable
- $\dot{\phi}_y$ is fully observable
- ϕ_z and $\dot{\phi}_x$ are only in combination observable
- ϕ_x and $\dot{\phi}_z$ are only in combination observable

The filter can not estimate each individual time-variant misalignment state. However, in [9] it was shown that the influence of the time-variant misalignment could almost completely extracted from the attitude estimates. So, if the focus is on precise attitude estimates and not on precise determination of the time-variant misalignment states, this seams to be a valid approach.

Case 3: Harmonically oscillating STR/IMU misalignment with doubled orbit frequency

This case is more of theoretical nature to study the observability of the system states. One can write $\mathbf{M} = -4\omega^2\,\mathrm{I}$. Hence,

- δb_x, δb_y and δb_z are fully observable
- ϕ_x, ϕ_y and ϕ_z are fully observable
- $\dot{\phi}_x$, $\dot{\phi}_y$ and $\dot{\phi}_z$ are fully observable

4 Conclusion

Deterministic time-variant misalignment between star tracker and gyro unit is mainly caused by temperature variation during orbit (e.g. sun/eclipse phases). It can significantly degrade the achievable attitude estimation performance when not considered by state augmentation within the sensor data fusion filter. The system dynamics equations required for state augmentation were derived. Considering typical nadir-oriented Earth observation missions, the observability analyses showed that often the misalignment states can only be observed as linear combination with other filter states. Nevertheless, the proposed state augmentation can significantly decrease the influence of misalignment on the attitude estimation performance as shown in [9, 10].

References

1. Crassidis, J.L., Junkins, J.L.: Optimal Estimation of Dynamic Systems. Chapman & Hall/CRC, Boca Raton (2004)
2. Crassidis, J.L., Markley, F.L., Cheng, Y.: Survey of Nonlinear Attitude Estimation Methods. Journal of Guidance, Control and Dynamics 30(1), 12–28 (2007)
3. Iwata, T., Hoshino, H., Yoshizawa, T., Kawahara, T.: Precision Attitude Determination for the Advanced Land Observing Satellite (ALOS): Design, Verification, and On-Orbit Calibration. In: AIAA Guidance, Navigation and Control Conference, Hilton Head, SC, August 20-23 (2007), AIAA-2007-6817
4. Kalman, R.E.: On the General Theory of Control Systems. In: Proceedings IFAC Moscow Congress, vol. 1, pp. 481–492. Butterworth Inc., Washington DC (1960)

5. Lefferts, E., Markley, F., Shuster, M.: Kalman Filtering for Spacecraft Attitude Estimation. Journal of Guidance, Control and Dynamics 5(5), 417–429 (1982)
6. Markley, F.L.: Attitude Error Representations for Kalman Filtering. Journal of Guidance, Control and Dynamics 26(2), 311–317 (2003)
7. Minkler, G., Minkler, J.: Theory and Application of Kalman Filtering. Magellan Book Company (1993)
8. Shuster, M.D.: A Survey of Attitude Representations. Journal of the Astronautical Sciences 41(4), 439–517 (1993)
9. Winkler, S., Wiedermann, G., Gockel, W.: Gyro-Stellar Attitude Estimation Considering Measurement Noise Correlation and Time-Variant Relative Sensor Misalignment. In: International Astronautical Congress, Glasgow, Scotland, September 29 - October 3 (2008), IAC-08-C1.7.4
10. Winkler, S., Wiedermann, G., Gockel, W.: High-Accuracy On-Board Attitude Estimation for the GMES Sentinel-2 Satellite: Concept, Design, and First Results. In: AIAA Guidance, Navigation and Control Conference, Honolulu, HI, August 18-21 (2008), AIAA-2008-7482
11. Wu, Y.-w.A., Li, R.K., Robertson, A.D.: Precision Attitude Determination for GOES N Satellite. In: 26th Annual AAS Guidance and Control Conference, Breckenridge, CO, February 5-9 (2003), AAS 03-002

Performance Comparison of Maneuver Detection Algorithms

Sebastian Bayerl, Georg Herbold, and Lorenzo Pettazzi

Abstract. This paper compares several algorithms for target maneuver detection. The algorithms are based on input estimation coupled with Kalman filter or IMM filtering. The performance of each method is analyzed computing a Pareto frontier, which is based on different quality criteria such as root mean square (rms) of state estimation error and time of maneuver detection. It is shown that computation of such Pareto frontiers allows to compare the performance of the different detectors in a systematic way.

1 Introduction

In target tracking applications a maneuver detector can be used to upgrade an existing model of a tracking filter. For example a filter modeling a constant velocity is working in a reliable manner, if a target is moving without any maneuver according to the filter assumptions. But, if the target motion differs to a constant velocity, the filter is estimating a false target state with a biased estimation error. The failure in modeling can be caused by any kind of maneuver. Thus a maneuver detection is used to give a corrected estimation of the target state, if the maneuver is according

Sebastian Bayerl
Student at Deggendorf University of Applied Sciences, Edlmairstraße 6+8, 94469 Deggendorf, Germany
e-mail: sebastian.bayerl@stud.fh-deggendorf.de

Georg Herbold
MBDA / LFK-Lenkflugkörpersysteme GmbH, Landshuter Straße 26, 85716 Unterschleißheim, Germany
e-mail: georg.herbold@mbda-systems.de

Lorenzo Pettazzi
European Organisation for Astronomical Research in the Southern Hemisphere, Karl-Schwarzschild-Straße 2, 85748 Garching, Germany
e-mail: lpettazz@eso.org

to the maneuver detector model. Additionally a statement can be made whether the target is moving with a constant motion or performing a maneuver. This paper will focus on the maneuver detection and state estimation performance comparison for five maneuvering detection algorithms:

- Open-loop input estimation with χ^2-based detector
- Open-loop input estimation with Gaussian significance based detector
- Closed-loop input estimation detector
- IMM detector without interaction
- IMM detector with interaction

The remainder of this paper is organized as follows. In section 2 the problem of Maneuvering Target Tracking is defined. Section 3 presents several algorithms to solve this problem. In section 4 these algorithms are compared. Conclusions are presented in section 5.

2 Problem Formulation

A linear system is used to describe the motion of a target:

$$\mathbf{x}(k+1) = \Phi(k) \cdot \mathbf{x}(k) + G(k) \cdot \mathbf{u}(k) + \mathbf{q}(k) \tag{1}$$

$$\mathbf{y}(k) = H(k) \cdot \mathbf{x}(k) + \mathbf{n}(k) \tag{2}$$

$\mathbf{x}(k)$ is the state vector of the target, $\mathbf{y}(k)$ is the measurement and $\mathbf{u}(k)$ is an unknown input, which is modeling a maneuver. The process noise $\mathbf{q}(k)$ and the measurement noise $\mathbf{n}(k)$ are assumed to be uncorrelated zero mean, white noise with known covariance matrices $Q(k)$ and $R(k)$. $H(k)$ is the measurement matrix.

Maneuver detection consists of two different tasks:

- On the one hand the existence of a maneuver has to be detected. Maneuver detection can be formulated as testing between two hypotheses:
  - Hypothesis H_1: The target is in non-maneuverig mode 1 ($\mathbf{u}(k) = 0$).
  - Hypothesis H_2: The target is in maneuverig mode 2 ($\mathbf{u}(k) \neq 0$).
- On the other hand the estimated state vector $\hat{\mathbf{x}}$ has to be corrected through the measurement vector $\mathbf{y}(k)$. In order to obtain an accurate estimation, the maneuver $\mathbf{u}(k)$ has to be estimated and used to correct the state estimation.

3 Detection Algorithms

The basic building block of the maneuver detection algorithms presented in this paper is a Kalman filter with a constant velocity model (CV filter). This Kalman filter is able to compute an estimation $\hat{\mathbf{x}}$ of the true target state $\mathbf{x}$ with associated covariance $\hat{P}$. The transition matrix Φ in (1) models a constant velocity without any maneuver.

$$\hat{\mathbf{x}}(k+1|k) = \Phi(k) \cdot \hat{\mathbf{x}}(k|k) \tag{3}$$
$$\hat{P}(k+1|k) = \Phi(k) \cdot \hat{P}(k|k) \cdot \Phi(k)^T + Q(k) \tag{4}$$
$$\mathbf{v}(k) = \mathbf{y}(k) - H(k) \cdot \hat{\mathbf{x}}(k+1|k) \tag{5}$$
$$S(k) = H(k) \cdot \hat{P}(k+1|k) \cdot H(k)^T + R(k) \tag{6}$$
$$K(k) = \hat{P}(k+1|k) \cdot H(k)^T \cdot S(k)^{-1} \tag{7}$$
$$\hat{\mathbf{x}}(k+1|k+1) = \hat{\mathbf{x}}(k+1|k) + K(k) \cdot \mathbf{v}(k) \tag{8}$$
$$\hat{P}(k+1|k+1) = [I - K(k) \cdot H(k)] \cdot \hat{P}(k+1|k) \tag{9}$$

The maneuver $\mathbf{u}$ can be detected and estimated by using the innovation $\mathbf{v}$ and its covariance matrix S.

A maneuver detection algorithm can be build with two different approaches. The first way is to handle a maneuver as an unknown input into a system with constant velocity. Therefore the CV filter is extended with an input estimation algorithm. The other way to build a maneuver detector is to use an IMM filter system including two filters with different motion models. One filter is modeling a non-maneuvering motion, the other one is modeling a maneuvering motion. The structure of the filter system allows to get information about the maneuvering mode of the target.

3.1 Input Estimation

3.1.1 Preliminaries

In [2] it is shown that the innovation $\mathbf{v}$ of a Kalman filter is a linear function of the unknown input $\mathbf{u}(k)$. This can be easily shown by comparing the residuals of two different filters:

- A false filter with non-maneuvering model:

$$\mathbf{x}(k+1) = \Phi(k) \cdot \mathbf{x}(k) + \mathbf{q}(k) \tag{10}$$

- A correct, but hypothetic filter with the maneuver as input:

$$\mathbf{x}^*(k+1) = \Phi(k) \cdot \mathbf{x}^*(k) + G(k) \cdot \mathbf{u}(k) + \mathbf{q}(k) \tag{11}$$

With use of the residuals of the false filter, the unknown maneuver of the hypothetic filter can be estimated. The innovations $\mathbf{v}^*$ of the correct, hypothetical filter are zero mean, white noise with covariance S^*. In comparison with the hypothetic innovations the real innovations $\mathbf{v}$ are biased by the maneuver. Hence the real innovations can be written as a function of $\mathbf{v}^*$, i.e.

$$\mathbf{v}(k) = \mathbf{v}^*(k) + \Lambda(k) \cdot \mathbf{u}(k) \tag{12}$$

It is then possible to conclude that the real innovations are a linear measurement of the maneuver. $\Lambda(k)$ is a measurement matrix defined with $\Lambda(k) = H(k) \cdot G(k)$.

3.1.2 Open-Loop Structure

Maneuver estimation

To estimate the unknown maneuver **u** and its covariance matrix $\hat{P}_u$ a derivation of the recursive least square method according to [2] is used.

$$\hat{P}_u(k) = \left[\hat{P}_u^{*-1}(k-1) + \Lambda^T(k) \cdot S^{-1}(k) \cdot \Lambda^T(k)\right]^{-1} \tag{13}$$

$$\hat{\mathbf{u}}(k) = \hat{\mathbf{u}}(k-1) + \Delta\hat{\mathbf{u}}(k) \tag{14}$$

$$\text{with} \quad \Delta\hat{\mathbf{u}}(k) = \hat{P}_u(k) \cdot \Lambda^T(k) \cdot S^{-1}(k) \cdot [\mathbf{v}(k) - \Lambda(k) \cdot \hat{\mathbf{u}}(k-1)] \tag{15}$$

This estimation differs to the standard recursive least square method. The estimation of the unknown input **u** with the standard least square method would lead to a smaller covariance matrix with each new time step until new measurements have almost no influence to the estimation. In order to mitigate such problem the old covariance matrix $\hat{P}_u(k-1)$ will be rescaled with a factor $\gamma \in [0,1]$.

$$\hat{P}_u^*(k-1) = \frac{1}{\gamma} \cdot \hat{P}_u(k-1) \tag{16}$$

Maneuver detection

In order to detect if the target is in maneuvering mode, tests for the statistical significance of the maneuver can be introduced, such as the χ^2 and the Gauss test. ([6]).

For the χ^2 test a variable $\varepsilon(k) = \hat{\mathbf{u}}(k)^T \cdot \hat{P}_u(k)^{-1} \cdot \hat{\mathbf{u}}(k)$ is calculated. Under assumption of H_1, $\varepsilon(k)$ is χ^2 distributed. This leads to a simple test.

$$\varepsilon(k) > \chi^2(n,\alpha) \ \Rightarrow H_2, \text{ target is in maneuvering mode} \tag{17}$$

where $(1-\alpha)$ is the confidence level of the test. This means H_1 will be rejected with confidence $(1-\alpha)$, if $\varepsilon(k)$ exceeds the corresponding threshold. n is the degree of freedom of the estimated maneuver.

Similar to the χ^2 test is the Gauss test. According to the Gauss test a maneuver will be declared, if one single component $\hat{\mathbf{u}}_i$ of the estimated maneuver is statistical significant. λ is determined from the standard Gaussian distribution.

$$max\left(\frac{|\hat{\mathbf{u}}_i(k)|}{\sqrt{\hat{P}_{u\,i,i}(k)}}\right) > \lambda \ \Rightarrow H_2, \text{ target is in maneuvering mode} \tag{18}$$

State correction

If the estimated maneuver has been estimated and it is statistical significant, the state and covariance must be corrected. In the open-loop structure only the output of the system is being corrected so that the corrected state estimation is not fed back to the filter. In order to characterize the effect of the maneuver onto the state, a time window is being built continuously. The matrix Ψ describes the windowed relation between the target state and the maneuver. It is calculated as follows:

$$\Psi(k) = [I - K(k-1) \cdot H(k-1)] \cdot \Phi(k-1) \cdot \Psi(k) + G(k) \tag{19}$$

According to [2] the estimated state and its covariance can be corrected in the following way.

$$\begin{aligned} \hat{\mathbf{x}}_{corr}(k|k) &= \hat{\mathbf{x}}(k|k) + \Delta\hat{\mathbf{x}}(k) \\ &= \hat{\mathbf{x}}(k|k) + [I - K(k) \cdot H(k)] \cdot \Psi(k) \cdot \hat{\mathbf{u}}(k) \end{aligned} \tag{20}$$

$$\begin{aligned} \hat{P}_{corr}(k|k) &= \hat{P}(k|k) + \Delta\hat{P}(k) \\ &= \hat{P}(k|k) + [I - K(k) \cdot H(k)] \cdot \Psi(k) \cdot \hat{P}_u(k) \cdot \Psi^T(k) \cdot [I - K(k) \cdot H(k)]^T \end{aligned} \tag{21}$$

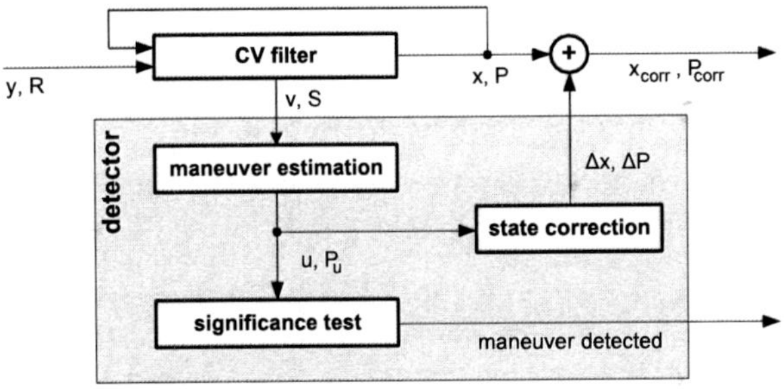

Fig. 1 Maneuver detection and state correction with open-loop structure

3.1.3 Closed-Loop Structure

In the open-loop structure the detector does not share the information about the estimated maneuver with the filter. The corrected state and covariance is only available at the output. According to [4] the optimal estimation is possible, if a closed-loop structure is used, where the feedback of the filter is corrected.

Through this closed-loop structure it is not possible to make a windowed estimate of the maneuver, because the filter state is being corrected each time step. Instead

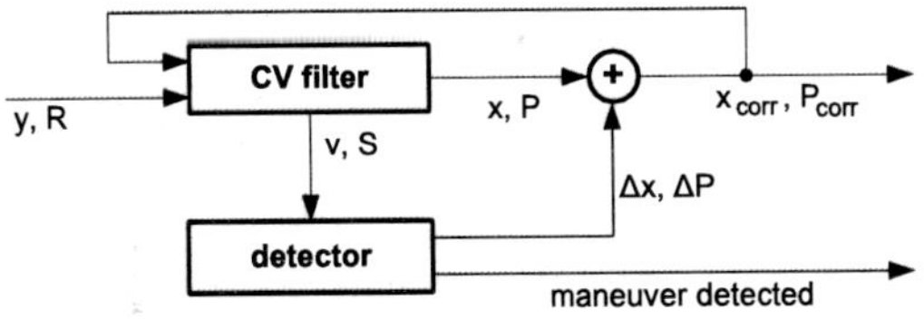

Fig. 2 Maneuver detection and state correction with closed-loop structure

of a time window each single measurement has to be used separately to estimate the maneuver.

Maneuver estimation

At the beginning of this section it is shown, that the residuals are a linear measurement of the maneuver. Thus the maneuver and its covariance can be estimated via least square methods according to [3].

$$\hat{\mathbf{u}}(k) = \left[\Lambda^T(k)\cdot S^{-1}(k)\cdot \Lambda(k)\right]^{-1}\cdot \Lambda(k)^T\cdot S(k)^{-1}\cdot \mathbf{v}(k) \quad (22)$$

$$\hat{P}_u(k) = \left[\Lambda^T(k)\cdot S^{-1}(k)\cdot \Lambda(k)\right]^{-1} \quad (23)$$

The estimation of the maneuver, out of the innovation without windowing, means to loose information about the composition of the innovations e.g. it is not possible to extract the acceleration from the innovation noise. This results in an estimation of a maneuver caused by noise. In order to reduce this effect several algorithms were tested. The best results were reached by filtering the estimations with an filtering factor $\alpha \in [0,1]$.

$$\hat{\mathbf{u}}(k) = (1-\alpha)\cdot \hat{\mathbf{u}}(k-1) + \alpha\cdot \hat{\mathbf{u}}(k) \quad (24)$$

$$\hat{P}_u(k) = (1-\alpha)^2\cdot \hat{P}_u(k-1) + \alpha^2\cdot \hat{P}_u(k) \quad (25)$$

Maneuver detection

The maneuver is detected by an absolute test. This means one single component $\hat{\mathbf{u}}_i$ of the estimated acceleration has to exceed a threshold. This threshold is defined in adaptive fashion through the process noise of the Kalman filter. Δv_{filter} is the velocity noise of the process noise Q and dt is the elapsed time since the last detection.

$$\hat{\mathbf{u}}_i > u_{threshold} \quad \rightarrow H_2,\ \text{target is in maneuvering mode} \quad (26)$$

with $u_{threshold} = \frac{\Delta v_{filter}}{dt}$

State correction

If the maneuver is detected, the filtered state and state covariance can be corrected. The correcting term is calculated similar to the open-loop structure.

$$\hat{\mathbf{x}}_{corr}(k|k) = \hat{\mathbf{x}}(k|k) + \Delta\hat{\mathbf{x}}(k) \quad (27)$$

$$\hat{P}_{corr}(k|k) = \hat{P}(k|k) + \Delta\hat{P}(k) \quad (28)$$

with $\Delta\hat{x}(k) = [I - K(k)\cdot H(k)]\cdot G(k)\cdot \hat{u}(k) \quad (29)$

$$\Delta\hat{P}(k) = [I - K(k)\cdot H(k)]\cdot G(k)\cdot \hat{P}_u(k)\cdot G^T(k)\cdot [I - K(k)\cdot H(k)]^T \quad (30)$$

3.2 IMM Filtering

The algorithms presented in section 3.1.2 and 3.1.3 are both based on a combination of a filter which is modeling a constant velocity, and a detector. In [1] an algorithm is described which is using several tracking filters with different motion models. Hence the name interactive multiple model filtering (IMM filtering). The IMM algorithms used in this paper is using two different filter:

- One filter with a model describing a non-maneuvering motion
- One filter with a model describing a maneuvering motion

The maneuvering filter model can describe several maneuvers. For the comparison in this paper, a maneuver with constant acceleration is modeled. There is a fixed initial switching probability $P_{T_{i,j}}$, which describes the probability that the target state will switch from mode i to mode j in the next time step. Furthermore there are other conditional probabilities which are being updated by the innovations of the filters:

- The switching probability $\mu_{i,j}(k)$ describes the probability that the target has changed from mode i at time step (k-1) into mode j at the actual time step (k).
- The fixed switching probability $P_{T_{i,j}}$ describes the probability that the target will change from mode i at the actual time step (k) and to the mode j in the next time step (k+1).
- The state probability $\mu_i(k)$ describes the probability that the target is in mode i at time (k).
- The predicted state probability $C_i(k)$ describes the probability that the target will be in mode i at time (k+1).

3.2.1 Probability Update

As in the χ^2 test, statistical deviations χ_i^2 of each filter are calculated to update the probabilities of each filter i:

$$\chi_i^2(k) = \mathbf{v_i}^T(k) \cdot S_i^{-1}(k) \cdot \mathbf{v_i}(k) \tag{31}$$

The probability density can be then obtained with $\Lambda_i(k) = \frac{\exp(-0.5 \cdot \chi_i^2(k))}{\sqrt{(2\pi)^N \cdot |S_i(k)|}}$.

Where N is the degree of freedom of the innovations. Finally the probabilities can be updated as follows:

$$\mu_i(k) = \frac{\Lambda_i(k) \cdot C_i(k)}{C(k)} \quad \text{and} \quad \mu_{i,j}(k) = \frac{P_{T_{i,j}} \cdot \mu_i(k)}{C_j(k)} \tag{32}$$

$$\text{with} \quad C(k) = \sum_{j=1}^{2} \Lambda_j(k) \cdot C_j(k) \quad \text{and} \quad C_j(k) = \sum_{i=1}^{2} P_{T_{i,j}} \cdot \mu_i(k-1) \tag{33}$$

3.2.2 IMM Interaction

It is possible to mix the states and covariances in the feedback of all filters. This interaction leads to an exchange of information between the filters.

$$\hat{\mathbf{x}}_{j-mix}(k-1|k-1) = \sum_{i=1}^{2} \mu_{ij}(k-1) \cdot \hat{\mathbf{x}}_i(k-1|k-1) \tag{34}$$

$$\hat{P}_{j-mix}(k-1|k-1) = \sum_{i=1}^{2} \mu_{ij}(k-1) \cdot [\hat{P}_i(k-1|k-1) + \hat{DP}_{ij}(k-1)] \tag{35}$$

$$\text{with} \quad \hat{DP}_{ij}(k-1) = \mathbf{Dx}_{ij}(k-1) \cdot \mathbf{Dx}_{ij}^T(k-1) \tag{36}$$

$$\mathbf{Dx}_{ij}(k-1) = \hat{\mathbf{x}}_i(k-1|k-1) - \hat{\mathbf{x}}_{j-mix}(k-1|k-1) \tag{37}$$

3.2.3 Maneuver Detection

The state probabilities μ_i of each filter i contain information about the target maneuvering mode. If the state probability of the maneuvering filter exceeds the probability of the non-maneuvering filter, then the hypothesis H_2 is correct. This leads to a simple test:

$$\mu_{maneuveringfilter} > 50\% \Rightarrow H_2, \text{ target is in maneuvering mode} \tag{38}$$

3.2.4 Combination of the System Output States and Covariances

Instead of correcting the state and covariance of the non-maneuvering filter, the filtered data will be combined. Thus, a weighted sum of the states and covariances is being built. Furthermore, it is not necessary to estimate the maneuver because it is already estimated in the maneuvering filter. These combined values are not significant for the IMM algorithm itself, but they are used for a possible user at the system output.

$$\hat{\mathbf{x}}(k|k) = \sum_{i=1}^{2} \mu_i(k) \cdot \hat{\mathbf{x}}_i(k|k) \tag{39}$$

$$\hat{P}(k|k) = \sum_{i=1}^{2} \mu_i(k) \cdot \left\{ \hat{P}_i(k|k) + [\hat{\mathbf{x}}_i(k|k) - \hat{\mathbf{x}}(k|k)] \cdot [\hat{\mathbf{x}}_i(k|k) - \hat{\mathbf{x}}(k|k)]^T \right\} \tag{40}$$

4 Comparison of the Algorithms

4.1 Criteria of Quality

Three criteria of quality are defined to compare the different algorithms.

- The first criterion is the delay time. It is defined as the arithmetic mean of the time the detector needs to detect the maneuver onset and the time the detector needs to detect the maneuver termination.
- The second criterion is the number of false detections. A false detection is a false statement of the detector about the maneuver. For example, a detector detects no maneuver during the period the target is maneuvering.
- The last criterion is the estimation error which is a measurement of the noise affecting the maneuver estimation and state correction. In order to obtain the estimation error a time difference between the true maneuver and the mean estimation of several Monte Carlo simulations is calculated via cross-correlation. This time difference is used to shift the maneuver estimation of one single Monte Carlo simulation towards the true maneuver. The estimation error is the difference between the shifted estimation and the truth. The root mean square of this difference over time is built to get the estimation error. Important is that areas of Δt around the maneuver start and end will not be considered. These manipulations are needed in order to avoid the influence of the estimation delay into the calculation of the estimation error. Finally, this time averaged error is additionally averaged over all Monte Carlo simulations.

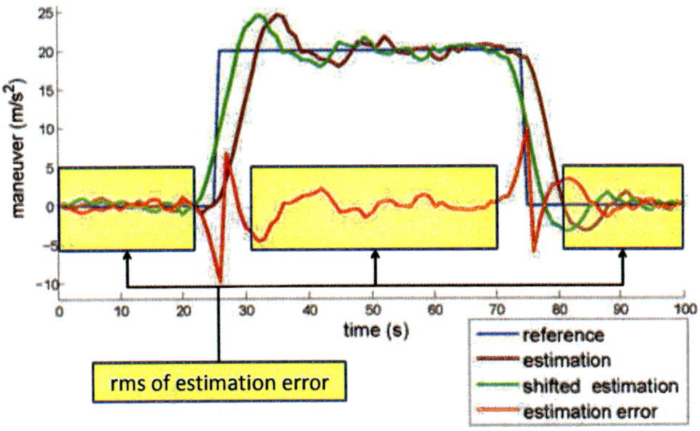

Fig. 3 Calcluation of the error in maneuver estimation

4.2 *Pareto Frontier*

Comparing two or more algorithms is difficult because the algorithms differ in more than one criterion of quality. Thus, a Pareto frontier according to [5] is used to compare the detection algorithms in a plane with estimation error and the detection delay as axis. The Pareto frontier is the connection line between all dominant points of one algorithm. The attribute dominant is declared, if there is no other point which has a lower detection delay and a smaller estimation error. This means that one point on the Pareto frontier describes the best estimation error at a fixed delay time and vice versa for a special scenario. To get one point in the plane a simulation with fixed input parameters $[p_1, \ldots, p_n]$ will be calculated. These input parameters have influence on the behavior of the detection algorithms. After the simulation is done, all criteria of quality are calculated and used to localize a point in the Pareto plane. This point is representing the behavior of the algorithm with the input parameters $[p_1,\dots,p_n]$. To approximate the Pareto frontier each input parameter p_i will be sampled with $[p_i^{min},\dots,p_i^{max}]$. In this paper the input parameters for the input estimation based detectors are the process noise Q_{CV} and the filtering factors α / γ. For approximating the Pareto frontier of the IMM algorithm the two process noise matrices Q_i of each filter i and the initial switching probability P_T are varied.

4.3 *Comparison of the Pareto Frontiers*

A scenario has to be defined to compare the algorithms. In the simulated scenario a target is moving with constant velocity $\mathbf{v}$ until a maneuver with constant turn begins. The acceleration of the target during the maneuver is defined with $\mathbf{a}(k) = \omega \times \mathbf{v}(k)$. The simulated trajectory has duration of 100 seconds. The maneuver starts at 25 seconds and ends at 75 seconds. The radar measures in radius r, azimuth ϕ and elevation θ. Radar noise is added correspondingly. The following parameters are used to simulate the trajectory:

- radar position $\mathbf{p}_0 = \begin{bmatrix} 0\,\text{m} \; 0\,\text{m} \; 0\,\text{m} \end{bmatrix}^T$
- error of measurement $\sigma_r = 20\,\text{m}, \quad \sigma_\phi = 3.4\,\text{mrad}, \quad \sigma_\theta = 1.7\,\text{mrad}$
- initial target position $\mathbf{p}_0 = \begin{bmatrix} -10000\,\text{m} \; 0\,\text{m} \; 1000\,\text{m} \end{bmatrix}^T$
- initial target velocity $\mathbf{v}_0 = \begin{bmatrix} 250\,\frac{\text{m}}{\text{s}} \; 0\,\frac{\text{m}}{\text{s}} \; 0\,\frac{\text{m}}{\text{s}} \end{bmatrix}^T$
- turn rate $\omega = \begin{bmatrix} 0\,\frac{1}{\text{s}} \; 0\,\frac{1}{\text{s}} \; 0.1\,\frac{1}{\text{s}} \end{bmatrix}^T$

Figure 4 shows the Pareto frontier of this constant turn motion with a maximum part of 25 % false detections.

- Maneuver detection by IMM-filtering has best performance. It has the best detection delay at a fixed maneuver estimation error. The interaction of the filters does not significantly improve the performance of the detector.
- Maneuver detection with input estimation has a inferior quality to the IMM detection. With open-loop structure the χ^2 and Gauss test have almost the same results.
- The maneuver detection with closed-loop structure shows nearly the same performance as the open-loop structure. The advantage of the information interchange between filter and detector is nullified by the fact that the estimation has to be filtered strongly in order to avoid a feed through of the measurement noise to the maneuver estimation. The closed-loop estimation does not achieve a low noise estimation, because the strong filtering, which is needed, would cause an enormous detection delay.

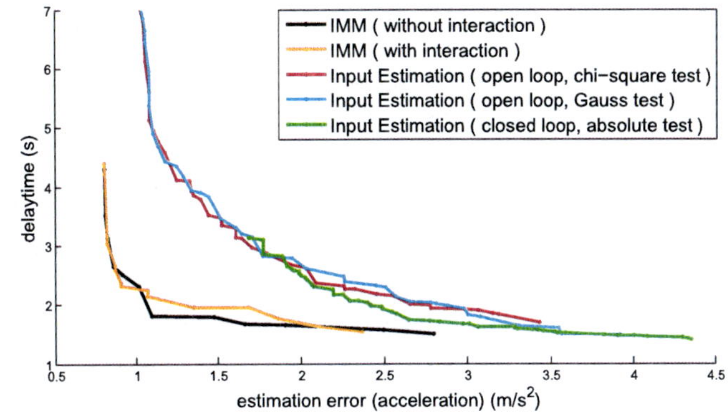

Fig. 4 Pareto frontier for a constant turn motion

4.4 *Comparison of the Computing Time*

In this section the maneuver detection algorithms are analyzed with respect to the computing time. This is also a critical quality criterion for real applications. The computing time is calculated as follows. With each maneuver detection step the calculation time is measured and saved. After the simulation all these values are taken and averaged in time. This average is additionally averaged above several Monte Carlo runs. The given data in figure 5 are scaled to the longest computing time in order to avoid absolute values.

- The IMM algorithms need the most calculating time. The reason lies in the two parallel running Kalman filters and the complexity of the algorithm structure. The IMM algorithm without interaction can be processed faster than the IMM algorithm with interaction.
- The calculation time of open-loop input estimation with χ^2 test is a bit lower than those of the version with Gauss test.
- The closed-loop input estimation has the lowest computing time.

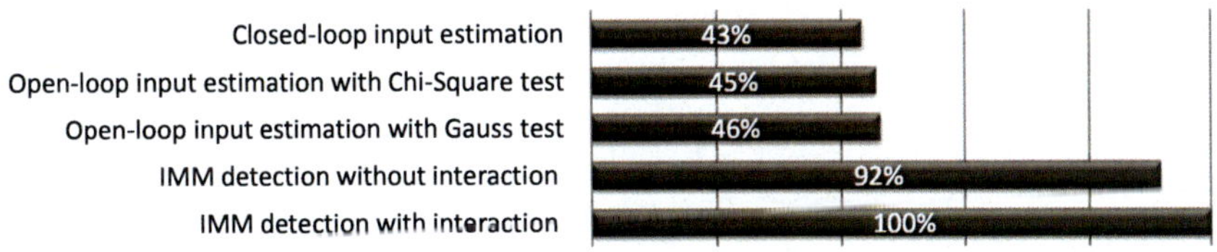

Fig. 5 Relative computing time of the maneuver detection algorithms

5 Conclusion

Several maneuver detectors are developed to detect a maneuver and to estimate the target state. With the use of one Kalman filter with non-maneuvering motion model the state can be estimated and corrected via least square methods. A state estimation and maneuver detection is also possible by using two Kalman filters with different motion models in an IMM algorithm. Also a special scenario was defined, in which the maneuver detectors are compared. Note that the performance of the maneuver detectors has been assessed against a maneuvering scenario that does not match with the one assumed in the input estimation and the IMM algorithm. It is shown, that different maneuver detection algorithms with several criteria of quality can be compared with the use of a Pareto frontier in a systematic way. The results of this comparison show, that IMM filtering has on the one hand best maneuver detection and the best estimation behavior but on the other hand the highest computing time.

References

1. Blackman, S., Popoli, R.: Design and Analysis of Modern Tracking Systems. Artech House, Boston (1999)
2. Bar-Shalom, Y., Rong Li, X., Kirubarajan, T.: Estimation with Applications to Tracking and Navigation: Theory, Algorithms and Software. Wiley, Chichester (2001)
3. Gillijns, S.: Kalman Filtering Techniques for System Inversion and Data assimilation. PhD thesis, Katholieke Universiteit Leuven (2007)
4. Gillijns, S., De Moor, B.: Unbiased minimum-variance input and state estimation for linear discrete-time systems. Automatica 43, 111–116 (2007)
5. Godfrey, P., Shipley, R., Gryz1, J.: Algorithms and analyses for maximal vector computation. VLDB Journal (August 2006)
6. Ru, J., Bashi, A., Rong Li, X.: Performance comparison of target maneuver onset detection algorithms. In: SPIE Conf. on Signal and Data Processing of Small Targets (2004)

Spacecraft Attitude Estimation and Gyro Calibration via Stochastic H_∞ Filtering

Daniel Choukroun, Lotan Cooper, and Nadav Berman

Abstract. A filter for estimating spacecraft attitude quaternion and gyro drift from vector measurements in the presence of white noises in the gyro error, in the drift dynamics, and in the line-of-sight measurement error is developed. The variance parameters of the white noises are unknown, and are modeled as non-anticipative second order stochastic processes. The approach taken in this work consists in estimating the attitude quaternion and the gyro drift while attenuating the transmission from the unknown variances to the estimation error. The resulting H_∞ filter involves the solution of a set of (differential) linear matrix inequalities. In the case of gyro white noises, extensive Monte-Carlo simulations were run showing that the proposed filter performs well from the standpoint of attitude estimation per se, in a wide range of gyro noise and line-of-sight noise intensities. The guaranteed disturbance attenuation level seems to be slightly dependent on the noises intensities. The actual level of disturbance attenuation is improving when the noises levels increase and admits as worst scenario the case of(ideal) noise-free sensors, as expected from the analysis. When compared with a matched quaternion Kalman filter, the H_∞ filter produces higher Monte-Carlo standard deviations of the estimation error, but lower Monte-Carlo means. The higher the level of noises are, the less obvious the advantage of the Kalman filter is. When estimating the quaternion only, and as opposed to standard quaternion Kalman filters, the H_∞ filter's gains can be computed independently from the quaternion estimates, which makes it insensitive to estimation errors.

Daniel Choukroun
Chair of Space Systems Engineering
Delft University of Technology
2629, NS, Netherlands
e-mail: D.Choukroun@TUDelft.NL

Lotan Cooper · Nadav Berman
Mechanical Engineerting Department
Ben-Gurion University
84105, Beer-Sheva, Israel
e-mail: {lotanco,nadav}@bgu.ac.il

This favorable feature is further emphasized when comparing its performances with those of unmatched Kalman filters. When provided with too high or too low noise covariances, the Kalman filter is outperformed by the H_∞ filter, which delivers essentially identical levels of errors within a wide range of noise intensities.

1 Introduction

The attitude quaternion [1], $\mathbf{q}$, is a very popular spacecraft attitude parametrization and the mathematical modeling and filtering have been ongoing topics of research for more than four decades [2]. Classical techniques for quaternion stochastic estimation belong to the realm of Kalman filtering, i.e., optimal filtering, and numerous successful quaternion estimators have been developed along that approach (e.g. [3], [4]). An inherent drawback to the optimal filtering approach consists of being sensitive to the model parameters' inaccuracies, in particular, the estimator's performances rely on the adequate knowledge of the measurement and process noises variances. Although adaptive noise estimation might provide satisfactory performances for some cases [5], the designer may be willing to tackle the parameters uncertainty pitfall by a less sensitive approach: rather than trying to estimate the uncertainty, the filter will attenuate their effect on the estimation error down to an arbitrary transmission level. In that case, the system satisfies the L_2-gain property between the perturbation and the estimation error. That approach was applied in [6], where an H_∞ spacecraft quaternion and gyro bias estimator was developed. The point of view of that work was deterministic, in the sense that the measurement and process noises were modeled as deterministic functions.

In this work, building on a previous work [7], a stochastic H_∞ quaternion filtering problem is proposed. We consider the case where a single continuous-time noisy vector measurement is being acquired and a triad of body-mounted gyros are providing a measurement of the spacecraft angular velocity with drift errors and white noise. We assume that the drift behaves like a non-zero mean random walk and that the variances of the line-of-sight measurements, of the gyro errors, and of the drift dynamics are unknown. The proposed filter is designed such as to estimate the quaternion and the gyro drift while attenuating these unknown perturbations. Motivated by recent works on stochastic H_∞ filtering and control for nonlinear systems [9], [10], we follow an approach based on the dissipativity theory.

As a first step, an H_∞ filter is developed where the variance of the gyro noises only is modeled as a perturbation. The measurement noise level is known and the drifts are assumed to be zero. The filtering problem solution hinges on solving a (differential) linear matrix inequality (DLMI) in order to compute the filter's time-varying gain. The DLMI provides a sufficient condition for the sought L_2-gain property. The development avoids linearization, and the structure of the quaternion state-space equations is exploited such that the resulted DLMI does not depend on the estimated process. As a consequence, computations can be performed off-line for a priori known histories of the line-of-sight measurements and of the angular velocity. Further, another H_∞ filter is developed along the same approach, but assuming that both the gyro white noise variance and the attitude sensing noise variance are unknown and modeled as disturbances. The resulting estimator shows similar features to the first one. Extensive Monte-Crlo simulation were run in order to check the H_∞ filter's performances, *as a quaternioin estimator*, and to compare the performances of this type of estimator with another type of quaternion Kalman filter. The Kalman filter is a variant of the classical Multiplicative EKF for quaternion estimation [3]. The case of quaternion and gyro drift H_∞ estimation was also addressed analytically.

Section 2 presents the mathematical formulation of the problem. Section 3 includes the development of the quaternion H_∞ estimators. Section 4 provides the basis for the development of a quaternion-drift H_∞ estimator. Section 5 presents the results of the Monte-Carlo simulations. Conclusions are proposed in the last section.

2 Statement of the Problem

Consider the following stochastic dynamical system in Itô form:

$$\mathbf{dq}_t = \frac{1}{2}\Omega(\boldsymbol{\omega}_t - \mathbf{c}_t)\,\mathbf{q}_t\,dt - \frac{1}{2}\Xi(\mathbf{q}_t)\sigma_\epsilon(t)\mathbf{d}\boldsymbol{\beta}_t; \qquad \mathbf{q}(0) \overset{a.e.}{=} \mathbf{q}_0; \qquad t \in [0,T] \quad (1)$$

$$\mathbf{dc}_t = \sigma_c(t)\mathbf{d}\boldsymbol{\nu}_t; \qquad \mathbf{c}(0) \overset{a.e.}{=} \mathbf{c}_0 \quad (2)$$

$$\mathbf{dy}_t = H_t\,\mathbf{q}_t\,dt - \frac{1}{2}\Xi(\mathbf{q}_t)\,\sigma_b(t)\mathbf{d}\boldsymbol{\eta}_t \quad (3)$$

where $\mathbf{q}_t$ denotes the attitude quaternion, Ω_t is the following matrix function of the measured angular velocity $\boldsymbol{\omega}_t$,

$$\Omega_t = \begin{bmatrix} -[\boldsymbol{\omega}\times] & \boldsymbol{\omega} \\ -\boldsymbol{\omega}^T & 0 \end{bmatrix} \quad (4)$$

where $\boldsymbol{\omega}_t$, which is acquired by a triad of body-mounted gyroscopes, is corrupted by an additive drift, $\mathbf{c}_t$, and by an additive standard Brownian motion, $\boldsymbol{\beta}_t$, with infinitesimal independent increments $\mathbf{d}\boldsymbol{\beta}_t$ such that $E\{\mathbf{d}\boldsymbol{\beta}_t\mathbf{d}\boldsymbol{\beta}_t^T\} = I_3\,dt$. $\sigma_\epsilon(t)$ denotes the variance parameter of the gyro output noise $\boldsymbol{\beta}_t$. The gyro drift is modeled as a random walk process with mean $\mathbf{c}_0$ and variance parameter $\sigma_c(t)$. In Eq. (2), $\boldsymbol{\nu}_t$ denotes a standard Brownian motion that is independent from $\boldsymbol{\beta}_t$. The system of equations (1), (2) is a straightforward extension of the quaternion stochastic differential equation (SDE) developed in [8] to the case of drifts in the gyro output error. Equation (3) is the continuous-time equivalent of the quaternion measurement equation in [5], where the measurement value is identically zero. Hence,

$$\mathbf{dy}_t = \mathbf{0} \quad (5)$$

and the measurement matrix H_t is constructed from vector measurements. Let $\mathbf{b}_t$ and $\mathbf{r}_t$ denote the projections of a measured line-of-sight (LOS) in the spacecraft body frame axes and in a reference frame, respectively, then H_t is computed as follows

$$\mathbf{s} = \frac{1}{2}(\mathbf{b} + \mathbf{r}) \tag{6}$$

$$\mathbf{d} = \frac{1}{2}(\mathbf{b} - \mathbf{r}) \tag{7}$$

$$H = \begin{bmatrix} -[\mathbf{s}\times] & \mathbf{d} \\ -\mathbf{d}^T & 0 \end{bmatrix} \tag{8}$$

The matrix, Ξ, which appears both in the process and measurement multiplicative noises, is the following linear matrix function of the quaternion $\mathbf{q} = [\mathbf{e}^T\, q]^T$:

$$\Xi(\mathbf{q}) = \begin{bmatrix} q I_3 - [\mathbf{e}\times] \\ \mathbf{e}^T \end{bmatrix} \tag{9}$$

The measurement noise is modeled as a standard Brownian motion, η_t, with variance parameter $\sigma_b(t)$.

The filtering problem consists in estimating the quaternion, $\mathbf{q}_t$, and the gyro drift, $\mathbf{c}_t$, from the LOS measurements in the presence of unknown intensities in the system noises, $\sigma_\epsilon(t)$, $\sigma_b(t)$, and $\sigma_c(t)$. The filtering problem is formulated as a stochastic disturbance attenuation problem via the H_∞ approach in the following way. Assuming that $\sigma_\epsilon(t)$, $\sigma_b(t)$, and $\sigma_c(t)$ are stochastic non-anticipative processes with finite second-order moments, we consider the following estimator:

$$d\widehat{\mathbf{q}}_t = \frac{1}{2}\Omega(\omega_t - \widehat{\mathbf{c}}_t)\widehat{\mathbf{q}}_t\, dt + K_q(\widehat{\mathbf{q}}_t, \widehat{\mathbf{c}}_t)\,(d\mathbf{y}_t - \widehat{\mathbf{q}}_t\, dt) \tag{10}$$

$$d\widehat{\mathbf{c}}_t = K_c(\widehat{\mathbf{q}}_t, \widehat{\mathbf{c}}_t)\,(d\mathbf{y}_t - \widehat{\mathbf{q}}_t\, dt) \tag{11}$$

$$\widehat{\mathbf{q}}(0) = \widehat{\mathbf{q}}_0,\ \widehat{\mathbf{c}}(0) = \widehat{\mathbf{c}}_0 \tag{12}$$

Let $\tilde{\mathbf{q}}_t$ and $\tilde{\mathbf{c}}_t$ denote the additive quaternion and biases estimation error, i.e.,

$$\tilde{\mathbf{q}}_t = \mathbf{q}_t - \widehat{\mathbf{q}}_t \tag{13}$$

$$\tilde{\mathbf{c}}_t = \mathbf{c}_t - \widehat{\mathbf{c}}_t \tag{14}$$

Given a scalar $\gamma > 0$, we seek for a gain process $K(\widehat{\mathbf{q}}_t)$ such that the following H_∞ criterion is satisfied:

$$E\{\int_0^T (\|\tilde{\mathbf{q}}_t\|^2 + \|\tilde{\mathbf{c}}_t\|^2)\, dt\} \le \gamma^2 E\{\|\tilde{\mathbf{q}}_0\|^2 + \|\tilde{\mathbf{c}}_0\|^2 + \int_0^T \|\mathbf{v}_t\|^2\, dt\} \tag{15}$$

under the constraints (1)-(3), and where $\mathbf{v}_t$ denotes the augmented process of admissible disturbance functions, i.e., $\mathbf{v}_t = \{\sigma_\epsilon(t), \sigma_b(t), \sigma_c(t)\}$. Whenever Eq. (15) is true, it is said that the L_2-gain property is satisfied from $\{\tilde{\mathbf{q}}_0, \mathbf{v}_t\}$ to $\tilde{\mathbf{q}}_t$, for $0 \le t \le T$.

3 Quaternion H_∞ Estimation

3.1 Attenuation of Process Noise with Unknown Intensity σ_ϵ

As in [7], we will first consider the simpler case where there are no biases and the intensity parameter σ_b is known.

Augmented stochastic process

Following standard techniques [9] , we define an augmented process as follows:

$$\mathbf{q}^a_t = \begin{bmatrix} \widehat{\mathbf{q}}_t \\ \tilde{\mathbf{q}}_t \end{bmatrix} \tag{16}$$

It is straightforward to develop the following SDE for the augmented process $\mathbf{q}^a_t$:

$$\begin{bmatrix} \mathbf{d}\widehat{\mathbf{q}}_t \\ \mathbf{d}\tilde{\mathbf{q}}_t \end{bmatrix} = \begin{bmatrix} (\frac{1}{2}\Omega_t - \widehat{K}_t H_t)\widehat{\mathbf{q}}_t \\ (\frac{1}{2}\Omega_t - \widehat{K}_t H_t)\tilde{\mathbf{q}}_t \end{bmatrix} dt + \sum_{i=1}^{3} \begin{bmatrix} \mathbf{0}_{4\times 1} \\ -\frac{1}{2}\Xi_{Ci} \end{bmatrix} \sigma_\epsilon(t) d\beta_i + \begin{bmatrix} O_{4\times 3} \\ \frac{1}{2}\widehat{K}_t \Xi \end{bmatrix} \mathbf{d}\eta_t \tag{17}$$

where Ξ_{Ci}, $i = 1, 2, 3$, denote the columns of the matrix Ξ, and the scalar processes β_i, $i = 1, 2, 3$, are the components of the vector Brownian motion $\boldsymbol{\beta}_t$. Notice that Ξ is a function of the augmented process $\{\widehat{\mathbf{q}}_t, \tilde{\mathbf{q}}_t\}$ since it is a function of the quaternion $\mathbf{q}$. Equation (17) can be re-written in the following compact form:

$$\mathbf{d}\mathbf{q}^a_t = F^a \mathbf{q}^a_t\, dt + \sum_{i=1}^{3} \mathbf{g}^i_2(\mathbf{q}^a_t)\sigma_\epsilon(t) d\beta_i + G(\mathbf{q}^a_t)\mathbf{d}\eta_t \tag{18}$$

where F^a , $\mathbf{g}^i_2(\mathbf{q}^a_t)$ and $G(\mathbf{q}^a_t)$ are effectively defined from Eq. (17).

Hamilton-Jacobi-Bellman inequality

Following [9], the desired L_2-gain property will be satisfied if and only if the augmented system (18) is dissipative with respect to the supply rate $S(\sigma_\epsilon(t), \mathbf{q}^a_t) = \gamma^2[\sigma_\epsilon(t)]^2 - \|\tilde{\mathbf{q}}_t\|^2$, for a given positive scalar γ. We, thus, seek for a non-negative scalar-valued function, $V(\mathbf{q}^a, t)$, that satisfies the fundamental property [10]:

$$E\{V(\mathbf{q}^a_t, t)\} \leq E\{V(\mathbf{q}^a_s, s) + \int_s^t (\gamma^2\|\sigma_\epsilon(\tau)\|^2 - \|\tilde{\mathbf{q}}_\tau\|^2) d\tau\} \quad \forall\, 0 \leq s \leq t \leq T \tag{19}$$

$$V(\mathbf{q}^a_0, 0) \leq \gamma^2 \|\tilde{\mathbf{q}}_0\|^2 \tag{20}$$

for all $\mathbf{q}^a$ and for all admissible $\sigma_\epsilon(t)$. When Eq. (19) is satisfied, the function V is called a storage function with respect to the supply rate S. A sufficient condition for Eq. (19) is:

$$E\{dV(\mathbf{q}^a,t)\} \leq E\{\gamma^2\|\sigma_\epsilon(t)\|^2 - \|\tilde{\mathbf{q}}_t\|^2\} \quad \forall\, 0 \leq t \leq T \tag{21}$$

for all $\mathbf{q}^a$ and for all admissible $\sigma_\epsilon(t)$, where dV is the Itô differential of the function V. Using Itô differentiation rule [11] and dropping the expectation operator on both sides of Eq. (21) yields the following sufficient condition [Hamilton-Jacobi-Bellman (HJB) equation] for $V(\mathbf{q}^a,t)$:

$$\begin{aligned}\frac{\partial V}{\partial t} + \frac{\partial V}{\partial \mathbf{q}^{aT}} F^a \mathbf{q}^a + \frac{1}{2}\,\sigma_b^2 \operatorname{tr}\left\{GG^T \frac{\partial^2 V}{\partial \mathbf{q}^a \partial \mathbf{q}^{aT}}\right\} \\ + \frac{1}{2}\sigma_\epsilon^2(t)\sum_{i=1}^{3} \mathbf{g}_2^{i\,T} \frac{\partial^2 V}{\partial \mathbf{q}^a \partial \mathbf{q}^{aT}} \mathbf{g}_2^i \leq \gamma^2[\sigma_\epsilon(t)]^2 - \|\tilde{\mathbf{q}}_t\|^2\end{aligned} \tag{22}$$

for all $0 \leq t \leq T$, $\mathbf{q}^a$, and for all admissible $\sigma_\epsilon(t)$, where $G = G(\mathbf{q}^a)$. Bringing all terms to the left-hand-side (LHS) of Eq. (22) yields

$$\begin{aligned}\frac{\partial V}{\partial t} + \frac{\partial V}{\partial \mathbf{q}^{aT}} F^a \mathbf{q}^a + \frac{1}{2}\,\sigma_b^2 \operatorname{tr}\left\{GG^T \frac{\partial^2 V}{\partial \mathbf{q}^a \partial \mathbf{q}^{aT}}\right\} \\ + \mathbf{q}^{aT} L \mathbf{q}^a + \left[\frac{1}{2}\sum_{i=1}^{3} \mathbf{g}_2^{i\,T} \frac{\partial^2 V}{\partial \mathbf{q}^a \partial \mathbf{q}^{aT}} \mathbf{g}_2^i - \gamma^2\right]\sigma_\epsilon^2(t) \leq 0\end{aligned} \tag{23}$$

and where

$$L = \begin{bmatrix} O_4 & O_4 \\ O_4 & I_4 \end{bmatrix} \tag{24}$$

For a solution to exist for all $\mathbf{q}^a$ and for all admissible $\sigma_\epsilon(t)$, the coefficient multiplying the arbitrary disturbance function, $\sigma_\epsilon(t)$, in Eq. (23) must be negative, which yields the following condition:

$$\frac{1}{2}\sum_{i=1}^{3} \mathbf{g}_2^{i\,T} \frac{\partial^2 V}{\partial \mathbf{q}^a \partial \mathbf{q}^{aT}} \mathbf{g}_2^i - \gamma^2 \leq 0 \tag{25}$$

for all $(\mathbf{q}^a, t)$. Given Eq. (25), any non-zero disturbance will only add a negative term to the LHS, increasing the system's dissipativity with respect to the chosen supply rate. The *worst-case* scenario will, thus, consists in a vanishing disturbance, $\sigma_\epsilon^*(t)$, e.g.,

$$\sigma_\epsilon^*(t) = 0 \tag{26}$$

Inserting Eq. (26) into Eq. (23) yields the following sufficient condition:

$$\frac{\partial V}{\partial t} + \frac{\partial V}{\partial \mathbf{q}^{a\,T}} F \mathbf{q}^{a} + \frac{1}{2}\,\sigma_{b}^{2}\,\mathrm{tr}\left\{ G G^{T} \frac{\partial^2 V}{\partial \mathbf{q}^{a} \partial \mathbf{q}^{a\,T}} \right\} + \mathbf{q}^{a\,T} L \mathbf{q}^{a} \leq 0 \tag{27}$$

for all $0 \leq t \leq T$ and $\mathbf{q}^{a}$.

Candidate storage function V

A standard approach in order to circumvent the formidable task of solving the partial differential inequality for V [Eq. (27)] consists in guessing the solution in a parameterized form and developing sufficient conditions for its parameters. We use here the classical quadratic form for V, i.e., we assume

$$V(\mathbf{q}^{a}, t) = \mathbf{q}^{a\,T} P_{t} \mathbf{q}^{a} \tag{28}$$

with the further assumption that P_t is symmetric, positive, and block diagonal, i.e.,

$$P_{t} = \begin{bmatrix} \widehat{P}_{t} & O_4 \\ O_4 & \widetilde{P}_{t} \end{bmatrix} \tag{29}$$

Convexity condition with respect to $\mathbf{q}_{\epsilon}(t)$

The existence of a solution for the filtering problem is conditioned upon Eq. (25), which is re-written here:

$$\frac{1}{2} \sum_{i=1}^{3} \mathbf{g}_2^{i\,T} \frac{\partial^2 V}{\partial \mathbf{q}^{a} \partial \mathbf{q}^{a\,T}} \mathbf{g}_2^{i} - \gamma^2 \leq 0 \tag{30}$$

Using Eqs. (28),(29), and the following known property of the matrix Ξ

$$\Xi \Xi^{T} = \mathbf{q}_{t}^{T} \mathbf{q}_{t}\, I_4 - \mathbf{q}_{t} \mathbf{q}_{t}^{T} \tag{31}$$

without simplification via the quaternion unit-norm constraint, yields

$$\mathbf{q}_{t}^{T} \left[\frac{1}{4} \left(\mathrm{tr} \widetilde{P}_{t}\, I_4 - \widetilde{P}_{t} \right) - \gamma^2\, I_4 \right] \mathbf{q}_{t} \leq 0 \tag{32}$$

Since Eq. (32) must be satisfied for all $\mathbf{q}_t$, we deduce the following condition on γ:

$$\frac{1}{4} \left(\mathrm{tr} \widetilde{P}_{t}\, I_4 - \widetilde{P}_{t} \right) - \gamma^2 I_4 \leq 0 \tag{33}$$

Sufficient condition on the matrices K and P

Using Eqs. (28),(29) in Eq. (27), straightforward computations yield the following identity:

$$\frac{\partial V}{\partial t} + \frac{\partial V}{\partial \mathbf{q}^{aT}} F^a \mathbf{q}^a = [\widehat{\mathbf{q}}_t^T \; \tilde{\mathbf{q}}_t^T] \begin{bmatrix} \frac{d\widehat{P}_t}{dt} + \widehat{F}_t^T \widehat{P}_t + \widehat{P}_t \widehat{F}_t & O_4 \\ O_4 & \frac{d\widetilde{P}_t}{dt} + \widehat{F}_t^T \widetilde{P}_t + \widetilde{P}_t \widehat{F}_t \end{bmatrix} \begin{bmatrix} \widehat{\mathbf{q}}_t \\ \tilde{\mathbf{q}}_t \end{bmatrix} \tag{34}$$

where $\widehat{F}_t$ is defined as follows:

$$\widehat{F}_t = \frac{1}{2} \Omega_t - \widehat{K}_t H_t \tag{35}$$

Relying on Eq. (31), and using the expression for G, as given in Eq. (17), further algebraic manipulations yield the following identity:

$$\begin{aligned} \frac{1}{2} \sigma_b^2 \operatorname{tr} \left\{ G G^T \frac{\partial^2 V}{\partial \mathbf{q}^a \partial \mathbf{q}^{aT}} \right\} &= \mathbf{q}_t^T \left\{ \frac{\sigma_b^2}{4} \left[\operatorname{tr}(\widehat{K}_t^T \widetilde{P}_t \widehat{K}_t) I_4 - \widehat{K}_t^T \widetilde{P}_t \widehat{K}_t \right] \right\} \mathbf{q}_t \\ &= [\widehat{\mathbf{q}}_t^T \; \tilde{\mathbf{q}}_t^T] \begin{bmatrix} I_4 \\ I_4 \end{bmatrix} \left\{ \frac{\sigma_b^2}{4} \left[\operatorname{tr}(\widehat{K}_t^T \widetilde{P}_t \widehat{K}_t) I_4 - \widehat{K}_t^T \widetilde{P}_t \widehat{K}_t \right] \right\} \\ & [I_4 \; I_4] \begin{bmatrix} \widehat{\mathbf{q}}_t \\ \tilde{\mathbf{q}}_t \end{bmatrix} \end{aligned} \tag{36}$$

Using Eqs. (34) and (36), we are able to express the LHS of Eq. (27) as a quadratic form in $\mathbf{q}_t^a$, and thus to re-write Eq. (27) as follows:

$$[\widehat{\mathbf{q}}_t^T \;\; \tilde{\mathbf{q}}_t^T] \begin{bmatrix} \frac{d\widehat{P}_t}{dt} + \widehat{F}_t^T \widehat{P}_t + \widehat{P}_t \widehat{F}_t + \frac{\sigma_b^2}{4} \left[\operatorname{tr}\widehat{M}_t I_4 - \widehat{M}_t \right] & \frac{\sigma_b^2}{4} \left[\operatorname{tr}\widehat{M}_t I_4 - \widehat{M}_t \right] \\ \frac{\sigma_b^2}{4} \left[\operatorname{tr}\widehat{M}_t I_4 - \widehat{M}_t \right] & \frac{d\widetilde{P}_t}{dt} + \widehat{F}_t^T \widetilde{P}_t + \widetilde{P}_t \widehat{F}_t + \frac{\sigma_b^2}{4} \left[\operatorname{tr}\widehat{M}_t I_4 - \widehat{M}_t \right] + I_4 \end{bmatrix} \begin{bmatrix} \widehat{\mathbf{q}}_t \\ \tilde{\mathbf{q}}_t \end{bmatrix} \leq 0 \tag{37}$$

for all $(\widehat{\mathbf{q}}_t, \tilde{\mathbf{q}}_t, t)$, where $\widehat{M}_t = \widehat{K}_t^T \widetilde{P}_t \widehat{K}_t$. Inequality (37) yields the following differential matrix inequality

$$\begin{bmatrix} \frac{d\widehat{P}_t}{dt} + F^T \widehat{P}_t + \widehat{P}_t F + \frac{\sigma_b^2}{4} \left[(\operatorname{tr} M) I_4 - M \right] & \frac{\sigma_b^2}{4} \left[(\operatorname{tr} M) I_4 - M \right] \\ \frac{\sigma_b^2}{4} \left[(\operatorname{tr} M) I_4 - M \right] & \frac{d\widetilde{P}_t}{dt} + F^T \widetilde{P}_t + \widetilde{P}_t F + \frac{\sigma_b^2}{4} \left[(\operatorname{tr} M) I_4 - M \right] + I_4 \end{bmatrix} \leq 0 \tag{38}$$

for all $0 \leq t \leq T$, where

$$F = \frac{1}{2} \Omega_t - K_t H_t \tag{39}$$

$$M = K_t^T \widetilde{P}_t K_t \tag{40}$$

and we dropped the symbols $\widehat{\cdot}$ in order to emphasize that the gain matrix, K_t, will not be a function of the estimate.

3.2 Attenuation of Process Noises with Unknown Intensity σ_ϵ and of Measurement Noises with Unknown Intensity σ_b

In the present case, the intensities of both the gyro white noise, σ_ϵ, and of the attitude sensor's noise, σ_b, are assumed unknown and considered perturbations in the H_∞ estimation problem. The filtering problem is formulated, as previously, as a stochastic disturbance attenuation problem via the H_∞ approach, where the intensities are modeled as random non-anticipative processes with finite second-order moments on $[0, T]$. For the sake of brevity, and thanks to the similarity between the current case and the previous one, the filter development details will be omitted. We consider the following estimator:

$$d\widehat{\mathbf{q}}_t = \left[\frac{1}{2}\Omega_t - \widehat{K}_t H_t\right]\widehat{\mathbf{q}}_t\,dt \tag{41}$$
$$\widehat{\mathbf{q}}_t(0) = \widehat{\mathbf{q}}_0$$

where $\widehat{K}_t$ denotes $K(\widehat{\mathbf{q}}_t)$. Let $\tilde{\mathbf{q}}_t$ denotes the additive estimation error, i.e., $\tilde{\mathbf{q}}_t = \mathbf{q}_t - \widehat{\mathbf{q}}_t$. Given a positive scalar, $\gamma > 0$, we seek for a gain process, $\{\widehat{K}_t, 0 \le t \le T\}$, such that the following H_∞ criterion is satisfied:

$$E\{\int_0^T \|\tilde{\mathbf{q}}_t\|^2\,dt\} \le \gamma^2 E\{\|\tilde{\mathbf{q}}_0\|^2 + \int_0^T \|\mathbf{v}_t\|^2\,dt\} \tag{42}$$

where $\mathbf{v}_t$ denotes the augmented process of admissible disturbance functions, i.e., $\mathbf{v}_t = \{\sigma_\epsilon(t),\ \sigma_b(t)\}$. Whenever Eq. (42) is true, it is said that the L_2-gain property is satisfied from $\{\tilde{\mathbf{q}}_0, \sigma_\epsilon(t),\ \sigma_b(t)\}$ to $\tilde{\mathbf{q}}_t$, $0 \le t \le T$. Simply stated, the proposed estimator is designed such as to provide an attitude estimate, with a given level of attenuation from the initial error and the process and measurement noise levels to the estimation error. An attractive feature is that no a priori knowledge of the noise intensities is necessary and the attenuation performance is, thus, guaranteed for a very wide class of noise levels, which may be random and time-varying, provided that they have a second-order moment.

Sufficient conditions on the matrices K, $\widetilde{P}$, $\widehat{P}$

The sought sufficient conditions on K, $\widehat{P}_t$, and $\widetilde{P}_t$ are as follows:

$$\frac{d\widehat{P}_t}{dt} + F_t^T\widehat{P}_t + \widehat{P}_t F_t \le 0 \tag{43}$$
$$\frac{d\widetilde{P}_t}{dt} + F_t^T\widetilde{P}_t + \widetilde{P}_t F_t + I_4 \le 0 \tag{44}$$
$$\frac{1}{4}(\mathrm{tr}\widetilde{P}_t\, I_4 - \widetilde{P}_t) - \gamma^2 I_4 \le 0 \tag{45}$$
$$\frac{1}{4}(\mathrm{tr}M_t\, I_4 - M_t) - \gamma^2 I_4 \le 0 \tag{46}$$

where

$$F_t = \frac{1}{2}\,\Omega_t - K_t H_t \tag{47}$$

$$M_t = K_t^T \widetilde{P}_t K_t \tag{48}$$

For simplification purposes, it is assumed that the matrices $\widehat{P}_t$ and $\widetilde{P}_t$ are identical, i.e., we seek for a single positive definite matrix, $\widetilde{P}_t$, that satisfies Eqs. (43) to (46). Extensive simulations showed that this assumption did not significantly impair the estimator's performances. Moreover, this allows dropping Eq. (43), which is automatically satisfied if Eq. (44) is satisfied. The simplified sufficient conditions under the assumption of a single decision matrix variable $\widetilde{P}_t$ are

$$\frac{d\widetilde{P}_t}{dt} + F_t^T \widetilde{P}_t + \widetilde{P}_t F_t + I_4 \le 0 \tag{49}$$

$$\frac{1}{4}\,(\,\mathrm{tr}\widetilde{P}_t\, I_4 - \widetilde{P}_t) - \gamma^2\, I_4 \le 0 \tag{50}$$

$$\frac{1}{4}\,(\,\mathrm{tr}M_t\, I_4 - M_t) - \gamma^2\, I_4 \le 0 \tag{51}$$

where

$$F_t = \frac{1}{2}\,\Omega_t - K_t H_t \tag{52}$$

$$M_t = K_t^T \widetilde{P}_t K_t \tag{53}$$

Since the LHS of the above inequalities is independent from the quaternion estimate, the sought filter gain, K_t, will be independent from $\widehat{\mathbf{q}}_t$.

Sufficient condition in form of Linear Matrix Inequalities (LMI)

Since the above inequalities are not linear with respect to $\widetilde{P}$ and K, some manipulations are required in order to bring them to an LMI structure. The bilinear dependence with respect to $\widetilde{P}$ and K is readily coped with via a standard parametrization approach. Let $\widetilde{Y}_t$ denote the following four-dimensional matrix:

$$\widetilde{Y}_t = \widetilde{P}_t K_t \tag{54}$$

then, using Eq. (54) in Eq. (49) yields

$$\frac{d\widetilde{P}_t}{dt} + \frac{1}{2}\,(\Omega_t^T \widetilde{P}_t + \widetilde{P}_t \Omega_t) - (H_t^T \widetilde{Y}_t^T + \widetilde{Y}_t H_t) + I_4 \le 0 \tag{55}$$

In order to circumvent the difficulty arising from the quadratic structure of M_t with respect to $\widetilde{P}_t$ and K, we seek for a symmetric positive definite matrix W_t, such that

$$M_t - W_t = \widetilde{Y}_t^T \widetilde{P}_t^{-1} \widetilde{Y}_t - W_t \le 0 \tag{56}$$

Notice that $\widetilde{P}_t^{-1}$ exists since $\widetilde{P}_t$ is assumed to be positive definite. Then, we use the following bounds on the LHS of Eq. (51):

$$\frac{1}{4}\,(\,\mathrm{tr}M_t\,I_4 - M_t) - \gamma^2\,I_4 \leq (\frac{1}{4}\,\mathrm{tr}M_t - \gamma^2)\,I_4 \leq (\frac{1}{4}\,\mathrm{tr}W_t - \gamma^2)\,I_4 \leq 0 \quad (57)$$

and replace Eq. (49) by the following sufficient condition on W:

$$\frac{1}{4}\,\mathrm{tr}W_t - \gamma^2 \leq 0 \quad (58)$$

where W, $\widetilde{Y}$, and $\widetilde{P}$ satisfies Eq. (56), which by the Schur complement can be written as the following LMI :

$$\begin{bmatrix} -W_t & -\widetilde{Y}_t \\ -\widetilde{Y}_t^T & -\widetilde{P}_t \end{bmatrix} \leq 0 \quad (59)$$

Notice that the successive bounds in Eq. (57) may yield higher feasible values for γ, i.e., a worse level of guaranteed disturbance attenuation.

Summary of the continuous-time H_∞ filter

Given $\widehat{\mathbf{q}}_0$, choose $\widetilde{P}(0)$ such that Eq. (20) is satisfied. Solve for $\widetilde{P}_t = \widetilde{P}_t^T > 0 \in \mathbb{R}^4$, $\widetilde{Y}_t \in \mathbb{R}^4$, and $W_t = W_t^T > 0 \in \mathbb{R}^4$, the following set of (differential) LMIs:

$$\frac{d\widetilde{P}_t}{dt} + \frac{1}{2}\,(\Omega_t^T\widetilde{P}_t + \widetilde{P}_t\Omega_t) - (H_t^T\widetilde{Y}_t^T + \widetilde{Y}_t H_t) + I_4 \leq 0 \quad (60)$$

$$\begin{bmatrix} -W_t & -\widetilde{Y}_t \\ -\widetilde{Y}_t^T & -\widetilde{P}_t \end{bmatrix} \leq 0 \quad (61)$$

$$\frac{1}{4}\,(\,\mathrm{tr}\widetilde{P}_t\,I_4 - \widetilde{P}_t) - \gamma^2\,I_4 \leq 0 \quad (62)$$

$$\frac{1}{4}\,\mathrm{tr}W_t - \gamma^2 \leq 0 \quad (63)$$

For any pair of matrices $(\widetilde{Y}_t, \widetilde{P}_t)$, compute the gain K_t using

$$K_t = \widetilde{P}_t^{-1}\widetilde{Y}_t \quad (64)$$

and compute the estimated quaternion via the estimator differential equation

$$\dot{\widehat{\mathbf{q}}}_t = \left[\frac{1}{2}\,\Omega_t - K_t H_t\right]\widehat{\mathbf{q}}_t \quad (65)$$

Remark 1: The estimator equation, (65), is not designed to preserve the quaternion unit-norm property. For that purpose, a normalization stage of the estimate is performed along the estimation process, i.e., $\widehat{\mathbf{q}}_t(0)$ is continuously divided by its norm.

Remark 2: Inspired by previous works [12], the discrete approximation of the differential LMI (60), is developed via a finite-difference formula, which is here:

$$\frac{\widetilde{P}_{k+1}-\widetilde{P}_k}{\Delta t}+\frac{1}{2}(\Omega_{k+1}^T\widetilde{P}_{k+1}+\widetilde{P}_{k+1}\Omega_{k+1})-(H_{k+1}^T\widetilde{Y}_{k+1}^T+\widetilde{Y}_{k+1}H_{k+1})+I_4\leq 0 \tag{66}$$

where Δt denotes the time increment, and $k=0,1,...,N=T/\Delta t$.

4 Quaternion and Gyro Drift H_∞ Estimation

In this section, the problem of estimating the attitude quaternion and the gyro drift under the assumption of unknown intensities in the system's noises is addressed. The quaternion-drift system is governed by the following SDE:

$$\begin{bmatrix}d\mathbf{q}_t\\ d\mathbf{c}_t\end{bmatrix}=\begin{bmatrix}\frac{1}{2}\Omega(\boldsymbol{\omega}_t-\mathbf{c}_t)\mathbf{q}_t\\ O_{3\times 1}\end{bmatrix}dt+\begin{bmatrix}-\frac{1}{2}\Xi(\mathbf{q}_t) & O_{4\times 3}\\ O_3 & I_3\end{bmatrix}\begin{bmatrix}\sigma_\epsilon(t)I_3 & O_3\\ O_3 & \sigma_c(t)I_3\end{bmatrix}\begin{bmatrix}d\boldsymbol{\beta}_t\\ d\boldsymbol{\nu}_t\end{bmatrix} \tag{67}$$

where $\boldsymbol{\beta}_t,\boldsymbol{\nu}_t$ are vector standard Brownian motion with unknown variance parameters σ_ϵ and σ_c respectively. The filtering problem is formulated as a stochastic disturbance attenuation problem via the H_∞ approach where all the intensities σ_ϵ, σ_b and σ_c are modeled as random non-anticipative processes with finite second-order moments on $[0,T]$. Consider the following estimator:

$$\begin{bmatrix}d\widehat{\mathbf{q}}_t\\ d\widehat{\mathbf{c}}_t\end{bmatrix}=\begin{bmatrix}[\frac{1}{2}\Omega(\boldsymbol{\omega}_t-\widehat{\mathbf{c}}_t)-\widehat{K}_qH_t]\\ -\widehat{K}_cH_t\end{bmatrix}\widehat{\mathbf{q}}_t\,dt \tag{68}$$
$$\widehat{\mathbf{q}}_t(0)=\widehat{\mathbf{q}}_0;\quad \widehat{\mathbf{c}}_t(0)=\widehat{\mathbf{c}}_0$$

where $\widehat{K}_q$ and $\widehat{K}_c$ are the matrix blocks of $\widehat{K}_t$, with the apropriate dimensions. Let $\tilde{\mathbf{q}}_t,\tilde{\mathbf{c}}_t$ denotes the additive estimation error, i.e., $\tilde{\mathbf{q}}_t=\mathbf{q}_t-\widehat{\mathbf{q}}_t$ and $\tilde{\mathbf{c}}_t=\mathbf{c}_t-\widehat{\mathbf{c}}_t$. Given a positive scalar, $\gamma>0$, we seek for a gain process, $\{\widehat{K}_t, 0\leq t\leq T\}$, such that the following H_∞ criterion is satisfied

$$E\{\int_0^T\|\tilde{\mathbf{q}}_t\|^2\,dt\}\leq\gamma^2E\{\|\tilde{\mathbf{q}}_0\|^2+\int_0^T[\sigma_\epsilon^2(t)+\sigma_b^2(t)+\sigma_c^2(t)]\,dt\} \tag{69}$$

Whenever Eq. (69) is true, it is said that the L_2-gain property is satisfied from $\{\tilde{\mathbf{q}}_0, \sigma_\epsilon(t), \sigma_b(t), \sigma_c(t)\}$ to $\{\tilde{\mathbf{q}}_t, \tilde{\mathbf{c}}\}$, $0 \leq t \leq T$. The augmented process $\{\widehat{\mathbf{q}}_t, \widehat{\mathbf{c}}_t, \tilde{\mathbf{q}}, \tilde{\mathbf{c}}\}$ is governed by the following stochastic Itô differential equation

$$\begin{bmatrix} \mathbf{d}\widehat{\mathbf{q}}_t \\ \mathbf{d}\widehat{\mathbf{c}}_t \\ \mathbf{d}\tilde{\mathbf{q}}_t \\ \mathbf{d}\tilde{\mathbf{c}}_t \end{bmatrix} = \begin{bmatrix} \frac{1}{2}\,\Omega_t - \widehat{K}_q H_t - \frac{1}{2}\,\Xi(\widehat{\mathbf{q}}_t) & O & O \\ -\widehat{K}_c H_t & O & O & O \\ O & O & \frac{1}{2}\,\Omega_t - \widehat{K}_q H_t - \frac{1}{2}\,\Xi(\widehat{\mathbf{q}}_t) \\ O & O & -\widehat{K}_c H_t & O \end{bmatrix} \begin{bmatrix} \widehat{\mathbf{q}}_t \\ \widehat{\mathbf{c}}_t \\ \tilde{\mathbf{q}}_t \\ \tilde{\mathbf{c}}_t \end{bmatrix} dt \quad (70)$$

$$+ \begin{bmatrix} O_{4\times 3} \\ O_3 \\ -\frac{1}{2}\,\Xi(\widehat{\mathbf{q}}_t) \\ O_3 \end{bmatrix} \sigma_\epsilon(t)\mathbf{d}\boldsymbol{\beta}_t + \begin{bmatrix} O_{4\times 3} \\ O_3 \\ O_{4\times 3} \\ I_3 \end{bmatrix} \sigma_c(t)\mathbf{d}\boldsymbol{\nu}_t + \begin{bmatrix} O_{4\times 3} \\ O_{4\times 3} \\ \widehat{K}_q\,\frac{1}{2}\,\Xi(\widehat{\mathbf{q}}_t) \\ \widehat{K}_c\,\frac{1}{2}\,\Xi(\widehat{\mathbf{q}}_t) \end{bmatrix} \sigma_b(t)\mathbf{d}\boldsymbol{\eta}_t$$

where second-order terms with respect to the noises $\boldsymbol{\beta}_t, \boldsymbol{\nu}_t, \boldsymbol{\eta}_t$ and to the estimation errors $\tilde{\mathbf{q}}$, $\tilde{\mathbf{c}}$ have been neglected. Equation (4) may be re-written in the following compact form:

$$\mathbf{d}\mathbf{q}^a_t = F^a \mathbf{q}^a_t\, dt + G_1(\mathbf{q}^a_t)\sigma_\epsilon(t)\mathbf{d}\boldsymbol{\beta}_t + G_2(\mathbf{q}^a_t)\,\sigma_c(t)\mathbf{d}\boldsymbol{\nu}_t + G(\mathbf{q}^a_t)\,\sigma_b(t)\mathbf{d}\boldsymbol{\eta}_t \quad (71)$$

The remainder of the filter development is straightforward and is omitted for the sake of brevity.

5 Numerical Simulation

Consider a spacecraft rotating around its center of mass with the following time-varying inertial angular velocity vector, $\boldsymbol{\omega}^o(t)$:

$$\boldsymbol{\omega}^o(t) = [1\ \ -1\ 1]^T\ \sin(2\pi t/150)\quad [deg/sec] \quad (72)$$

The measured angular velocity is computed according to

$$\boldsymbol{\omega}(t) = \boldsymbol{\omega}^o(t) + \sigma_\epsilon\,\boldsymbol{\epsilon}(t) \quad (73)$$

where $\boldsymbol{\epsilon}(t)$ is a standard zero-mean white Gaussian noise, e.g., $E\{\,\boldsymbol{\epsilon}(t)\,\boldsymbol{\epsilon}(\tau)^T\} = I_3\delta(t-\tau)$. Typical values of low-grade gyros will be used in the ensuing, i.e., $\sigma_\epsilon \in [10^{-3}, 10^{-1}]\,[rad/\sqrt{sec}]$. A continuous-time single line-of-sight measurement, $\mathbf{b}(t)$, is assumed to be acquired. It is computed via the classical vector measurement model:

$$\mathbf{b}(t) = A[\mathbf{q}(t)]\,\mathbf{r}(t) + \sigma_b\,\boldsymbol{\delta b}(t) \quad (74)$$

where $\boldsymbol{\delta b}(t)$ is a zero-mean white Gaussian noise and $E\{\,\boldsymbol{\delta b}(t)\,\boldsymbol{\delta b}(\tau)^T\} \simeq (I_3 - \mathbf{b}\,\mathbf{b}^T)\delta(t-\tau)$, which stems from the unit norm property of $\mathbf{b}$. The time history of the reference line-of-sights, $\mathbf{r}(t)$, is chosen arbitrarily at this stage of the simulation study. Typical values of coarse and fine attitude sensors will be used in the ensuing, i.e., $\sigma_b \in [10^{-5}, 10^{-1}]\,[rad]$. The simulated sampling time for the gyro and attitude sensor (continuous) processes is $\Delta t = 0.1sec$. The true initial quaternion is $\mathbf{q}(0) = [1,1,1,1]^T/\sqrt{4}$. The filter is usually initialized with $\widehat{\mathbf{q}}(0) = [0,0,0,1]$ and $\widetilde{P}(0) = \widehat{P}(0) = 10\,I_4$ unless stated otherwise. Monte-Carlo simulations (50 runs) were run over time spans varying from 500 to 6000 seconds. This shows the estimation performances over several periods of the angular velocity dynamics, up to a typical revolution time of a Low-Earth-Orbit satellite around the Earth. For the purpose of comparison, the quaternion Kalman filter of [5], which will be referred to as QKF (where the linear measurement model (73) is used), and a typical Multiplicative EKF [3], referred to as MEKF (where the nonlinear measurement model (74) is linearized) were also implemented. The novel stochastic H_∞ quaternion estimator will be denoted as QHF.

The performances of the quaternion $H\infty$ filter that achieves a γ-level attenuation from the intensity σ_ϵ (and the initial estimation error) to the estimation error were presented in [7].

Unknown gyro noise and attitude sensor intensities, σ_ϵ and σ_b

Next, we summarize the performances of the quaternion $H\infty$ filter that achieves disturbance attenuation from the unknown σ_ϵ and σ_b (and the initial estimation error) to the estimation error. Extensive Monte-Carlo simulations were run in order to compute the actual attenuation ratio $AR(T)$ at the final time T, with $T = 500$ sec. The ratio is defined as follows:

$$\frac{E\{\int_0^T \|\tilde{\mathbf{q}}(t)\|^2\,dt\}}{E\{\|\tilde{\mathbf{q}}(0)\|^2 + \int_0^T (\sigma_\epsilon^2 + \sigma_b^2)\,dt\}} \tag{75}$$

where the integral is numerically computed using a time step $\Delta t = 0.1$ sec, and the expectations are computed as MC averages. Table 1 shows values of the $AR(500sec)$ for various values of σ_ϵ and σ_b. Also, the MC means of the best guaranteed level of attenuation, γ^2_{QHF}, were computed along the filtering process. Table 1 shows their values in steady-state, which is in general reached after 100 sec. Disturbance attenuation ratios from 0.45 down to 0.06 are achieved by the H_∞ filter. The filter has a better (attenuation) efficiency for higher levels of disturbances. For small noise intensities, the $AR(500)$ are similar and show a strong variation in the range of $(10^{-2}, 10^{-1})$. This hints at the fact that the H_∞ filtering approach can be exploited when using low-grade sensors, which are characterized by high levels of noises. As expected from the theory, the $AR(500)$ are all smaller than the corresponding guaranteed levels of attenuation, i.e., $AR(500) \leq \gamma^2_{QHF}$. The discrepancy between both quantities, however, depends on the noise intensities. For instance, when $(\sigma_\epsilon,\ \sigma_b\})$ are equal to $(0.1, 0.1)$, the actual attenuation ratio (0.08)is 30 times lower than the guaranteed level. On the other end of the table, in the vanishing noises case, the AR(500) is only six times lower. Therefore, it seems that the performance bound is more efficient, e.g., closer to the actual performance, for low noise intensities. Moreover, one can observe that the maximal values for γ^2_{QHF} as well as for the attenuation ratio $AR(500)$ occurs when σ_ϵ and σ_b are both zero. This, indeed, stands in good agreement with the analysis previously presented, where the case of vanishing disturbances is the *worst case* scenario for the H_∞ filter.

Figure 1-a depicts the time histories of the MC-means (50 runs)for $AR(t)$ and for γ^2_{QHF}, $0 \leq t \leq 2000$ sec. The best guaranteed bound, $\gamma^2_Q HF$ is quickly reaching a steady-state around 2.25, while the attenuation ratio is more slow to settle on a steady-state at 0.06.

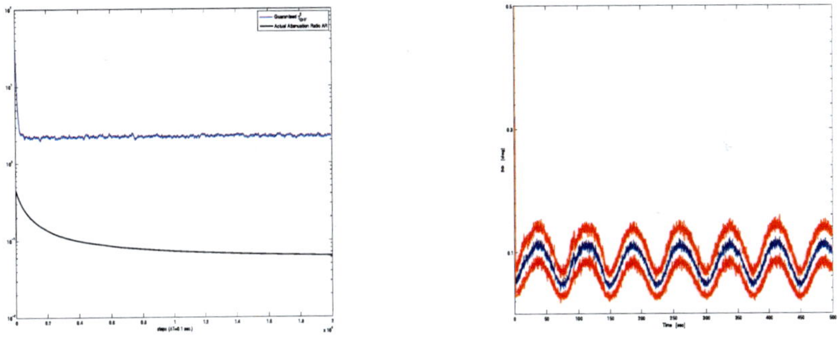

(a) AR(t) and γ^2_{QHF} (b) $\delta\phi$ MC-mean and MC-σ

Fig. 1 Performances of the QHF filter. 50 MC runs. $(\sigma_\epsilon, \sigma_b) = (0.001, 0.1)$.

Figure 1-b presents the time histories of the MC-means and MC-standard deviation envelop of the angular estimation error, $\delta\phi$. Albeit oscillating with an amplitude of 0.06 deg around 0.08 deg, $\delta\phi$ shows good performances: recalling the

Table 1 Attenuation Ratios $AR(500sec)$ (50 MC runs) between the quaternion estimation error, $\tilde{\mathrm{q}}$, and the disturbances $\{\tilde{\mathrm{q}}(0), \sigma_\epsilon, \sigma_b\}$ for various values of the intensities $\{\sigma_\epsilon, \sigma_b\}$. (Value of the steady-state MC-mean for the smallest reachable γ^2_{QHF}, as computed in the filter.)

$\sigma_\epsilon[\frac{rad}{\sqrt{sec}}]$ / $\sigma_b\,[rad]$	0	10^{-3}	10^{-2}	10^{-1}
0	(2.89) 0.45	(2.89) 0.45	(2.79) 0.44	(2.32) 0.06
10^{-3}	(2.89) 0.45	(2.65) 0.45	(2.60) 0.44	(2.32) 0.07
10^{-2}	(2.78) 0.44	(2.53) 0.43	(2.46) 0.41	(2.31) 0.07
10^{-1}	(2.41) 0.16	(2.40) 0.15	(2.39) 0.15	(2.25) 0.08

Table 2 Time-averages (and time standard deviations) of the quaternion estimation errors in QHF (right) and in MEKF (left) for various values of the intensities $\{\sigma_\epsilon,\ \sigma_b\}$. Single run. The final time is 2000 sec.

$\sigma_b[rad]$ / $\sigma_\epsilon[\frac{rad}{\sqrt{sec}}]$	10^{-3}	10^{-2}	10^{-1}
	MEKF | QHF	MEKF | QHF	MEKF | QHF
10^{-7}	$(1.2\,10^{-5} \mid 1.4\,10^{-3})$ $3\,10^{-5} \mid 2\,10^{-5}$	$(1.0\,10^{-4} \mid 1.4\,10^{-3})$ $0.003 \mid 0.003$	$(0.015 \mid 0.080)$ $0.020 \mid 0.017$
10^{-5}	$(2.1\,10^{-4} \mid 1.4\,10^{-3})$ $5.1\,10^{-5} \mid 1.3\,10^{-5}$	$(9.0\,10^{-4} \mid 1.4\,10^{-3})$ $0.008 \mid 0.007$	$(0.015 \mid 0.080)$ $0.020 \mid 0.016$
10^{-4}	$(1.2\,10^{-4} \mid 1.4\,10^{-3})$ $1.5\,10^{-5} \mid 1.3\,10^{-5}$	$(2.0\,10^{-3} \mid 2.1\,10^{-3})$ $0.0001 \mid 0.0003$	$(0.015 \mid 0.082)$ $0.020 \mid 0.015$
10^{-3}	$(1.2\,10^{-3} \mid 1.4\,10^{-3})$ $7.9\,10^{-5} \mid 2.0\,10^{-5}$	$(3.7\,10^{-3} \mid 1.2\,10^{-2})$ $0.0196 \mid 0.0008$	$(0.015 \mid 0.081)$ $0.020 \mid 0.002$
10^{-2}	$(1.6\,10^{-3} \mid 2.0\,10^{-3})$ $1.3\,10^{-5} \mid 1.2\,10^{-5}$	$(0.0120 \mid 0.0124)$ $2.8\,10^{-5} \mid 1.4\,10^{-4}$	$(0.019 \mid 0.080)$ $0.020 \mid 8.4\,10^{-4}$
10^{-1}	$(1.6\,10^{-3} \mid 1.6\,10^{-2})$ $4.80\,10^{-5} \mid 3.30\,10^{-6}$	$(0.0172 \mid 0.0192)$ $0.0001 \mid 0.0001$	$(0.116 \mid 0.084)$ $2.60\,10^{-3} \mid 6.80\,10^{-4}$

measurement noise level σ_b, equal to 5 deg, the QHF provides an attenuation ratio of order $0.02 \simeq 1/50$.

Extensive simulations were performed in order to compare the performances of the QHF and of a standard quaternion Multiplicative EKF (MEKF). In the MEKF, the measurement equation model, which is quadratic in $\mathbf{q}_t$ is linearized around the estimated trajectory. The MEKF is matched to the true levels of noises. For comparative purposes, we consider the additive estimation errors in both filters. Table 2 shows the time averages, computed on single runs of 2000 sec, of the quaternion estimation error norm in QHF (right) and in MEKF (left). In addition, the values of the time standard deviations are provided for both filters (in parenthesis). In general, both filters provide similar time averages of the estimation error. If σ_b remains constant and σ_ϵ is increased, then no significant changes appear. However, when increasing σ_b, while keeping σ_ϵ constant, there is a proportional increase in the time-average of the error. As expected, the MEKF provides smaller standard deviations of the quaternion error than QHF, since MEKF approximates the minimum variance estimator. Increasing σ_ϵ, while keeping σ_b constant, does not affect the performances (standard deviation's level) of QHF. On the other hand, the dispersion of the estimation error in MEKF grows proportionally with σ_ϵ. The same fact can be observed when increasing σ_b while keeping σ_ϵ constant. This property goes is readily explained via the approach used for the filter's development.

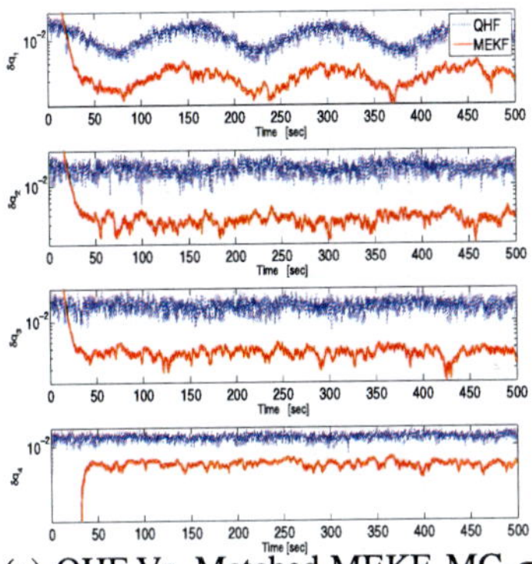

(a) QHF Vs. Matched MEKF. MC-σ.

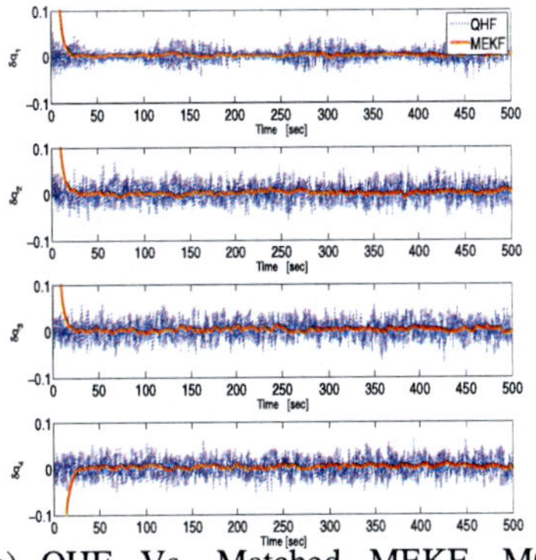

(b) QHF Vs. Matched MEKF. MC-means.

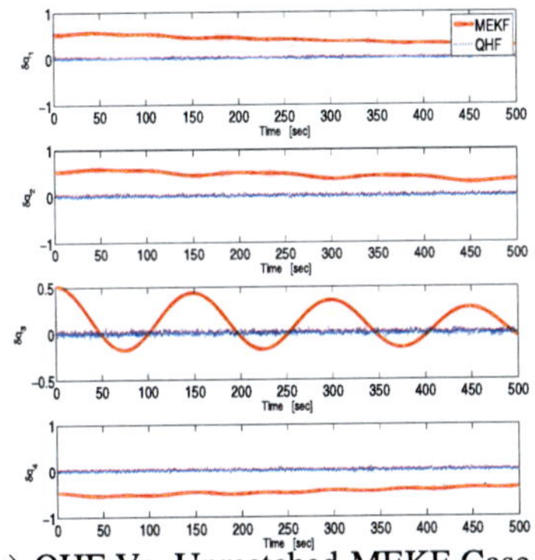

(c) QHF Vs. Unmatched MEKF Case A. MC-means.

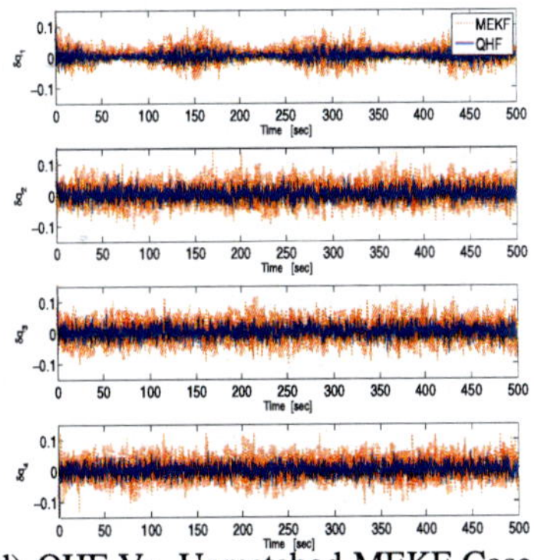

(d) QHF Vs. Unmatched MEKF Case B. MC-means.

Fig. 2 Quaternion estimation errors in QHF (dashed blue) and in MEKF (full red). 50 MC runs. $(\sigma_\epsilon, \sigma_b) = (0.01, 0.1)$.

Additional Monte-Carlo simulations were performed for specific levels of the noises. Figure 2-a depicts the MC-standard deviations of the additive quaternion estimation errors in the QHF and in the MEKF, for the case $(\sigma_\epsilon, \sigma_b) = (0.01, 0.1)$. As expected, the values of σ in the MEKF (around $2\,10^{-3}$ are significantly lower than in the QHF (around $2\,10^{-2}$. This is consistent with the property that the MEKF is approximating the minimum-variance estimator. Figure 2-b shows the time histories of the MC-means of the quaternion estimation errors, for $(\sigma_\epsilon,) = (0.001, 0.1)$. The errors appear to be unbiased in both filters. The MEKF plots, however, are smoother than those of the QHF. Further, we tested both filters in cases where the true levels of noises where not accurately known. This may be the result of on-board failures of gyros or attitude sensing devices, which, while undetected, render the MEKF unmatched and probe to divergence. In Case A, the filter level of σ_b was fixed to ten times its true value. As a consequence, as can be seen from Fig.2-c, the MEKF is very slow to converge. On the other hand, the QHF, which design is essentially independent from the actual values of the noise intensities, produces means that are close to the levels obtained in the previous simulation (Fig. 2-a). In addition, we tested in Case B an unmatched MEKF where the filter σ_b is ten times lower than the true one. As a result, the estimation errors in the MEKF converge quicker but to a noisier steady-state, which is expected from the Kalman filtering theory. The QHF, on the other hand, provides essentially the same performances as in Case A. The slight differences arise from the fact that the QHF implement data that are noisy, and, thus, the QHF performances are indirectly affected by the level of noises.

6 Conclusion

In this work, a novel quaternion attitude estimator was developed in the framework of stochastic H_∞ filtering using gyro measurements and line-of-sights (LOS) measurements. The proposed design hinges on a model where the white noises in gyro and the LOS measurements have unknown possibly random intensities. The case of quaternion and gyro drift H_∞ estimation was also addressed analytically. The estimator computes a quaternion while attenuating the effect of the noise intensities on the estimation error. The H_∞ filter involves the solution of a set of (differential) linear matrix inequalities, which do not depend on the estimated quaternion. Extensive Monte-Carlo simulations were run, for the case of guro white noise, showing that the proposed filter performs well from the standpoint of attitude estimation, *per se*, in a wide range of gyro and LOS intensities. The guaranteed disturbance attenuation level seems to be slightly dependent on the noises intensities, which may be due to the fact that the filter's parameters are noisy. The actual level of disturbance attenuation between the noises levels and the estimation error is improving when the noises levels increases, and is minimal for (ideal) noise-free sensors. This was expected by the analysis and illustrates the conservative nature of the H_∞ filter. When compared with two different matched quaternion Kalman filters (KF), the $H\infty$ filter produces higher Monte-Carlo standard deviations of the estimation error, but lower Monte-Carlo means. The higher the level of noises are, the less obvious the advantage to the Kalman filter is. Besides, as opposed to standard quaternion Kalman filters, this filter's gain process can be computed independently from the quaternion estimate process, which makes it insensitive to estimation errors. This nice feature is further emphasized when comparing the H_∞ performances with those of unmatched Kalman filters. When provided with too high or too low noise covariances, the KF is outperformed by the H_∞ filter, which delivers essentially identical levels of errors within a wide range of noise intensities.

Acknowledgements

This research was supported by THE ISRAEL SCIENCE FOUNDATION (grant No. 1546/08).

References

1. Wertz, J.R. (ed.): Spacecraft Attitude Determination and Control. D. Reidel, Dordrecht (1984)
2. Crassidis, J., Markley, F.L., Cheng, Y.: Nonlinear Attitude Filtering Methods. Journal of Guidance, Control and Dynamics 30(1), 12–28 (2007)
3. Lefferts, E.J., Markley, F.L., Shuster, M.D.: Kalman Filtering for Spacecraft Attitude Estimation. Journal of Guidance, Control and Dynamics 5, 417–429 (1982)
4. Bar-Itzhack, I.Y., Oshman, Y.: Attitude Determination from Vector Observations: Quaternion Estimation. IEEE Transactions on Aerospace and Electronic Systems AES-21, 128–136 (1985)
5. Choukroun, D., Oshman, Y., Bar-Itzhack, I.Y.: Novel Quaternion Kalman Filter. IEEE Transactions on Aerospace and Electronic Systems AC-42(1), 174–190 (2006)
6. Markley, F.L., Berman, N., Shaked, U.: Deterministic EKF-like Estimator for Spacecraft Attitude Estimation. In: Proceedings of the American Control, WA10, Baltimroe, MD (June 1994)
7. Cooper, L., Choukroun, D., Berman, N.: Spacecraft Attitude Estimation via Stochastic Filtering. In: Proceedings of the AIAA Guidance Navigation and Control Conference, Toronto, CA (August 2010)
8. Choukroun, D.: On Continuous-Time Ito Stochastic Modeling of the Attitude Quaternion. In: Proceedings of the AAS/F. Landis Markley's Astronautics Symposium, AAS-2008-288, Cambridge, MD (June 2008); Journal of Astronautical Sciences (in Press)
9. Berman, N., Shaked, U.: H_∞ Filtering for Nonlinear Stochastic Systems. In: Proceedings of the 13th Mediteranean Conference on Control and Automation, TuA05-3, Limassol, Cyprus (June 2005)
10. Berman, N., Shaked, U.: H_∞-like Control for Nonlinear Stochastic Systems. Systems and Control Letters 55, 247–257 (2006)
11. Jazwinski, A.H.: Stochastic Processes and Filtering Theory. Academic, New York (1970)
12. Gershon, E., Shaked, U., Yaesh, I.: H_∞ Control and Estimation of State-multiplicative Linear Systems. LNCIS, vol. 318. Springer, Heidelberg (2005)

Part IV
Space Applications

Advanced Optical Terrain Absolute Navigation for Pinpoint Lunar Landing

Marco Mammarella, Marcos Avilés Rodrigálvarez, Andrea Pizzichini, and Ana María Sánchez Montero

Abstract. Pin-point landing can only be achieved developing precise Absolute Navigation systems. Craters, for their intrinsic properties, are one of the most suitable and robust features identifiable in lunar landscape. The Optical Terrain Absolute Navigation (OTAN) system provides absolute navigation features and is composed by two main parts: the off-line part, focused on the extraction of the Landmark Database; the on-line part instead is focused on the Real Time identification of the craters and the Orbit Determination process. The presented vision-based approach uses Real-Time crater identification in order to extract relevant features from on-board captured images of the lunar surface. The detected craters are fitted with ellipses and matched to a Lunar Crater Database previously created. The matching of the two sets permits the computation of the absolute position of the camera. A Kalman Filter uses this information and the IMU measurements in order to provide precise complete state space information of the vehicle. In this paper, a detailed description of the complete structure of the Optical Terrain Absolute Navigation system based on craters detection and recognition is provided.

1 Introduction

One of the most interesting challenges in descent and landing missions is PinPoint Landing (PPL). Considering a probe landing on a planet, the accuracy level that is nowadays guaranteed, in terms of distance of the achieved landing site with respect to the target site, is not smaller than the order of magnitude of tens (or even hundreds) of kilometers [1]. Instead, PPL requires a few hundreds of meters wide error ellipse, that is to say about two-three orders of magnitude smaller than the one achieved.

Marco Mammarella · Marcos Avilés Rodrigálvarez · Andrea Pizzichini · Ana María Sánchez Montero
GMV, Tres Cantos, Spain
e-mail: mmammarella@gmv.com, maaviles@gmv.com, andrea.pizzichini1@gmail.com, amsmontero@gmv.com

Various PPL solutions have been proposed in recent years for example: VISINAV (vision aided inertial navigation system), which its main drawback is its high memory requirement due to the usage of image correlation in finding mapped landmarks [2]; in [3], authors use SIFT features (Scale Invariant Feature Transform) as mapped landmark to estimate the spacecraft's global position. SIFT features only prove to be invariant to an affine transformation in illumination change. Thus, they are quite vulnerable in case of important illumination changes between the orbiter and descent images, such as caused by the difference of the sun's direction. [4] relies on a Lidar sensor and matching of landmark constellations. Where either maximum or minimum of the surface's height and slope are extracted as landmarks from the surface elevation image provided by the Lidar sensor. The metrics distances of a landmark with its neighbors are used as it signature. Then a matching is performed between the surface and the DEMs landmarks. The main problem to this solution is that the system requires the resolution of the surface image obtained with the Lidar sensor to be equivalent to that of the map image.

In [5], modification Harris corner detection is used to extract landmarks from orbital and camera images. Landmarks from orbital images are stored onboard in a database. The two sets of landmarks are matched through constellation method and the absolute position is extracted. The main drawback of this method is the requirement of prior orbital images with similar environmental conditions (and not yet clearly specified). Furthermore, the number of stored landmarks and associate information can be a serious issue to handle since in each image a relatively big number of landmarks need to be stored. In order to rapidly access the database in real time, the database must be handled carefully.

All of the PPL solutions have at least the following functionalities: landmarks extraction, landmarks matching and position estimation. The present paper presents a complete description of an Automated Lunar Lander system, which has been designed to be compliable with different solutions but has been implemented using craters as landmarks.

Craters have been claimed by several authors [5] to be excellent landmarks due to its scale, rotation and illumination invariance properties. Crater shapes commonly correspond to a known geometric model (ellipse), invariant to scale changes and orientation between the spacecraft camera and the surface. The main drawback of craters, instead, is the not easily method of identification especially when the Image Processing algorithm has to deal with irregular and overlapped craters. In the developed method these issues are limited through appropriate pre-filtering function and discarding ambiguous craters directly in the database creation.

Optical Terrain Landmark Navigation (OTAN) is an absolute navigation system conceived to perform PPL in a fully autonomous way. It exploits optical navigation techniques in order to support and, if necessary, to correct, the propagation data of an Inertial Measurement Unit (IMU) in conjunction with Ground Tracking (GT) measurements.

The present paper describes the complete structure of the OTAN applied to the ESA NEXT-MOON [6] mission. In particular, after giving an overview of the mission scenario and explain the necessity of optical navigation, a complete description of the OTAN system is provided.

2 Mission Overview

The ESA NEXT-MOON mission [6] is the mission scenario in which OTAN technology is applied and tested. A typical GNC design for the NEXT-MOON Lunar Lander mission may consider the Descent and Landing trajectory as divided in four principal phases: Orbital, Main Braking, Visual and Terminal Descent Phases. A synthetic overview of the four phases is given in Table 1.

Table 1 Mission Overview

Phase name	**ORBITAL**		**MAIN BRAKING**	**VISUAL**	**TERMINAL DESCENT**
	LLO	**DESCENT ORBIT**			
Characteristics	Free-flight orbit.	Free-flight orbit.	Maximum thrust.	Landing site inside FOV, reduction of thrust level.	Pure vertical descent
Altitude range (approximate)	100 km	100 to 15 km	15 to 4-3 km	4-3 km to 15-5 m	15-5 m– until Touchdown

It can seem obvious that the most critical phases in a PPL mission are relative to visual and terminal phases. This is actually only partially true because if excessive dispersion in the navigation solution is accumulated in the first phases, these errors cannot be recovered during the last phases because of mass-budget and maneuverability limits and, also, due to the lack of a system able to recognize the landing site [6]. These limits are such that the only way to achieve PPL is to limit in all phases the dispersion as much as possible. The here described OTAN system has been conceived for providing a precise navigation solution during the orbital and main braking phases. It has to permit to reach the beginning of the visual phase well below the typical dispersion due to inertial navigation.

3 Optical Terrain Absolute Navigation System

Figure 1 shows the proposed Optical Terrain Absolute Navigation system able to recognize craters. The sources can be either geo-referenced images or DEMs. From these sources, the landmark extraction shall extract the coordinate and some geometrical properties of the craters off-line. Consequently, the resulting Landmark Database will be extracted off-line and stored on-board in order to be used. It is required that the spacecraft mounts a Camera providing image data, a Star Tracker providing the attitude data and IMU providing acceleration data. The image will be processed by the Lander Acquired Image Landmarks Extraction function, which is an IP function able to extract the craters on-line. This function can use data from the navigation filter in order to reduce computational time. The

Database projection function uses attitude data and position estimation data provided by the filter in order to predict which landmarks are expected to be seen in the current image. The landmark matching function finds the correspondence between the Lander acquired image landmarks and the expected Landmarks. The Absolute Position Estimation uses the knowledge of the attitude data, Landmarks Database and matched landmarks for computing Absolute Estimated Position. The matched landmarks are then checked by the Landmark Integrity function in order to select the most appropriate landmarks that have to be used for the position computation. The position is next provided to the Navigation filter with the landmark integrity in order to estimate the complete state vector. In next sections, each block of the OTAN system is described in more detail.

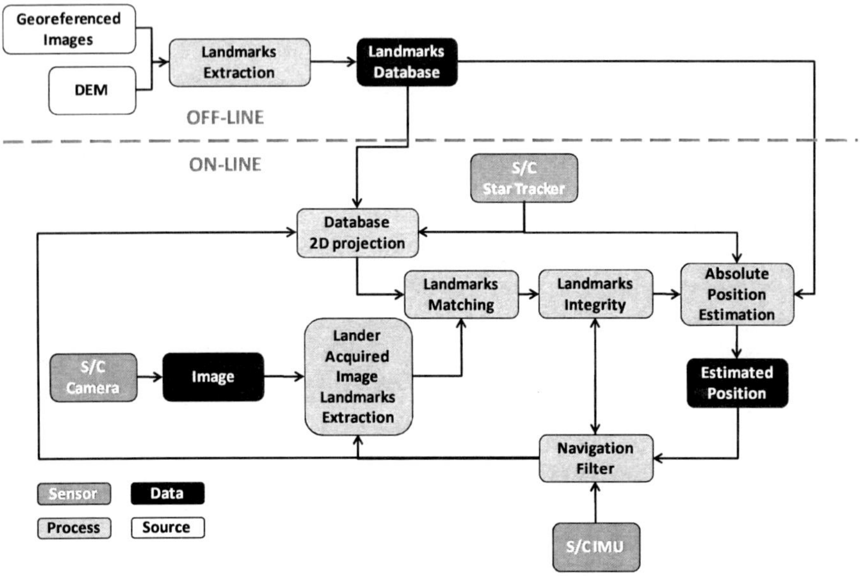

Fig. 1 Optical Terrain Absolute Navigation System

4 Landmark Database Creation

Creating a reliable landmarks database is a fundamental step in the system because the spacecraft positioning will be relative to this database. However, after decades of research and manual efforts, only tens of thousands of the millions of craters on the Moon have been catalogued, mainly those with diameters ≥ 2 km. Automated techniques for crater detection and cataloguing are therefore necessary. They take advantage of the vast quantities of remotely sensed data now available, especially now that precise 3D information is becoming available [7].

Crater detection techniques in DEMs are not very different from those employed in optical images (described in the following section). The edge detector is replaced with an extractor of high slope areas, whose binary output feeds the feature extraction stage. Template matching or voting schemes are the typical options for this second step as other alternatives that take into account the lighting effects cannot be considered for elevation maps.

Crater rims in a DEM can be effectively marked by searching for the zero crossings of the second-order derivative. After applying this process, the zero-crossings are identified and those above a certain threshold are marked.

The feature extraction stage mainly consists in performing the circle detection of the image obtained in the previous stage. The circle finding has been broken in two parts: center detection and radius estimation. A final step of ellipse fitting and refinement in the DEM image domain is recommended to fine tune the estimated parameters, deal with elliptical craters and remove errors introduced in the previous steps. When optical images corresponding to the DEM are also available, a parallel branch can be added to the system to extract a new set of craters and match its output with the one coming from the DEM. Figure 2 shows the results of this process in a section of the DEM obtained by LRO [7] and the visualization of a portion of the DEM using PANGU [8] with the obtained craters.

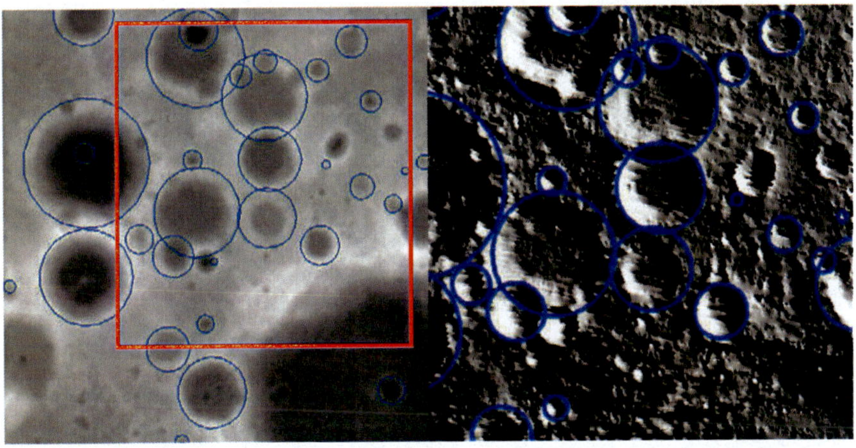

Fig. 2 LRO DEM [7] and identified craters (left), projection of the craters of DEM section using PANGU (right)

5 Lander Acquired Image Landmarks Extraction

The Lander Acquired Image Landmarks Extraction [9][10] has been designed following the approach presented in [1][11] and summarized in Figure 3. Namely, three stages can be identified in the algorithm: Border Detection (divided in Edge Detection and preliminary crater rim selection), Rim Grouping (composed of curvature check and crater border coupling) and Ellipse Fitting.

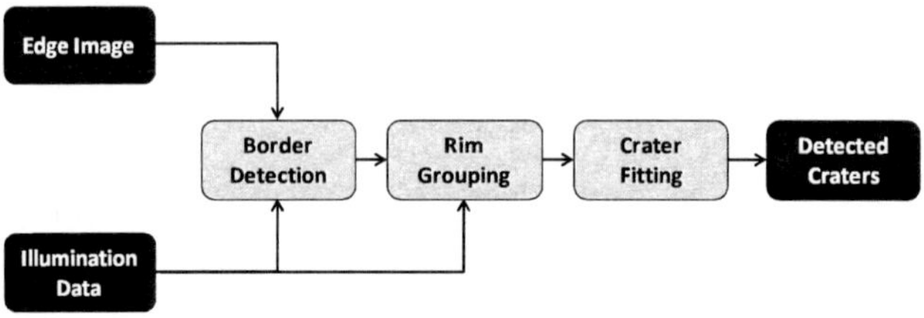

Fig. 3 Lander Acquired Image Landmarks Extraction scheme

5.1 Border Detection

The first step of the Border Detection process is the Edge Detection. The Edge Detection is a standard image processing technique [12] by means of which object outlines are identified in an image. Considering a picture as a brightness function depending on the two spatial coordinates (vertical and horizontal), edges can be defined as local maxima of the just mentioned function. The concept can be quantified recurring to the gradient of the brightness function, giving two kinds of information:

1) Gradient magnitude. Given a generic picture-element (pixel), higher is the gradient magnitude higher is the probability to encounter an object border.
2) Gradient direction. It gives information about the border direction, which is always perpendicular to the gradient vector.

Edge detection is the basis of the following steps. In fact, the aim of the whole algorithm is to identify the crater shapes among the other terrain features. Recognizing an object means, in this case, to associate its outline to a mathematical model, like an ellipse. Therefore, Edge Detection gives a first set of borders, in which the crater rims have to be selected among the other surface characteristics. Several different techniques are available in literature. The Border Detection process uses Sobel edge detector, which allows a sufficient level of accuracy with a non-excessive computational load.

The Edge Detection output image is a binary image in which each identified edge pixel is set to the '1' value. Non-edge pixels are, instead, set to '0'. In general, apart from edges actually belonging to craters, many other object outlines (e.g. mountains) are detected by the Edge Detection algorithm. The next step of the Lander Acquired Image Landmarks Extraction process operates a selection in order to keep as many crater pixels as possible, discarding the ones belonging to other objects. Craters, due to their particular geometric shape, have also a well-defined illumination profile, which can be exploited in order to operate the abovementioned pixel selection. As can be seen in the left part of Figure 4, the regions with the highest contrast are the ones perpendicular to the light direction (directed horizontally from left to right in the Figure). Moreover, considering the crater edges, such regions present themselves as curved segments, easily recognizable by the algorithm. The preliminary crater rim selection consists in the

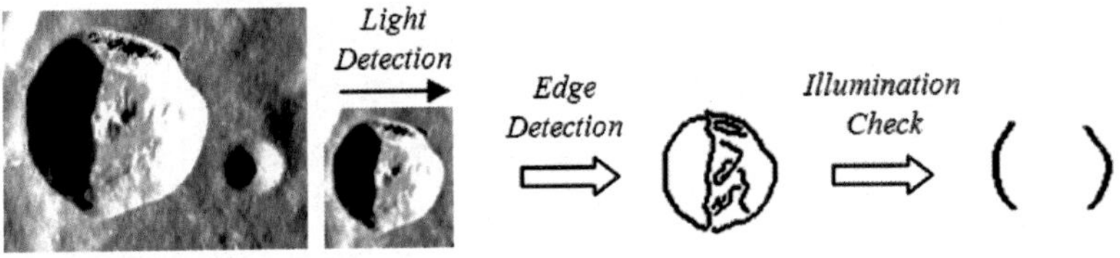

Fig. 4 Typical crater illumination profile (left), overview of the border detection process (right)

evaluation, at each edge point identified by the Edge Detector, of the gradient direction (perpendicular to the edge direction) with respect to the illumination unit-vector.

5.2 Rim Grouping

After performing the Border Detection stage (Right part of Figure 4 gives an overview of the whole process), crater edges present themselves as typical curve segments that are called crater signatures. Notice that each crater is characterized by a couple of such segments, one related to the lit side and the other one to the shaded side (see Figure 4). Both sides of the crater signature are to be taken together in order to be successfully fitted by the ellipse-specific fitting algorithm. This can be quantitatively translated in assigning a single label value to the pixels belonging to the whole crater signature. It is what is performed by the Rim Grouping process, which is divided in the Curvature Check and Crater Border Coupling steps [9][10]. The output sets of rim grouped are provided as input to the Ellipse Fitting.

5.3 Ellipse Fitting

Ellipse Fitting is the last stage of the Lander Acquired Image Landmarks Extraction. It applies an ellipse-specific fitting algorithm [13] to the border couples provided by the Rim Grouping block. Each crater candidate is fitted with an ellipse in order to take into account off-nadir Camera attitudes, which cause craters to be represented in the image as ellipses instead of circles. Although two selection checks have been performed in the previous stages, some non-crater borders can be mistakenly considered as crater signatures. In particular, false detections may present very high eccentricities. Therefore, a first check is done discarding too eccentric ellipses. Moreover, a further and more accurate check is performed using the fitting error and discarding the ellipses, which are characterized by the worse fitting. At the end of the process, each crater is associated with five parameters (i.e. the ellipse parameters): 2-D center coordinates, semi-axes lengths (both major and minor) and orientation. The list of crater ellipse parameters is the final output of the Lander Acquired Image Landmarks Extraction process. Figure 5 shows craters detected in images captured from LRO and Kaguya.

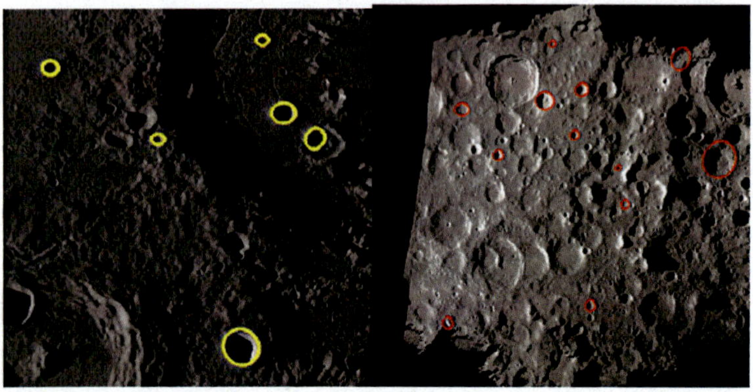

Fig. 5 Craters Detected in images captured from LRO (left) and Kaguya (right)

6 Database 2D Projection

The Database 2D projection algorithm projects the Landmark Database craters generating a crater distribution as would be seen by a hypothetical camera sharing the Lander attitude and estimated position. The output of the algorithm is a list of parameters, similar to the one generated by Lander Acquired Image Landmarks Extraction process. In fact, the list created by the Database 2D projection can be considered as the output of a "perfect" Lander Acquired Image Landmarks Extraction process if the estimated position and attitude were correct.

7 Landmarks Matching

Once the output of Lander Acquired Image Landmarks Extraction and the output of the Database 2D Projection are available, the problem becomes a typical Point Matching problem [14] where two sets of points with different dimension have to be matched together. In this case, four-dimensional Point Matching (PM) can be applied in order to give more robustness to the solution. The four dimensions available to the process are the two coordinate in the camera plane and the length of the both semi-axes of the ellipse for each crater, in substitution the orientation of the crater can be used instead the length of one semi-axis. The process of Point Matching has been extensively described in [14].

Figure 6 aims to illustrate the complexity faced during the Landmarks Matching process. The figure shows the analysis performed on a captured 512 x 512 frame. The black ellipses represent the projection of the database craters, while the red ellipses are the craters detected on the captured frame. The landmarks matching phase will associate the detected craters (in red) to the database craters (in black), using a four-dimensional Point Matching. As a result, the Landmarks Matching will return pairs of craters (one from the database and one from the captured frame). Matches are marked in the figure as a blue crosses joined by a blue line, where the crosses are located in the center of the database craters matched and in the center of the detected craters.

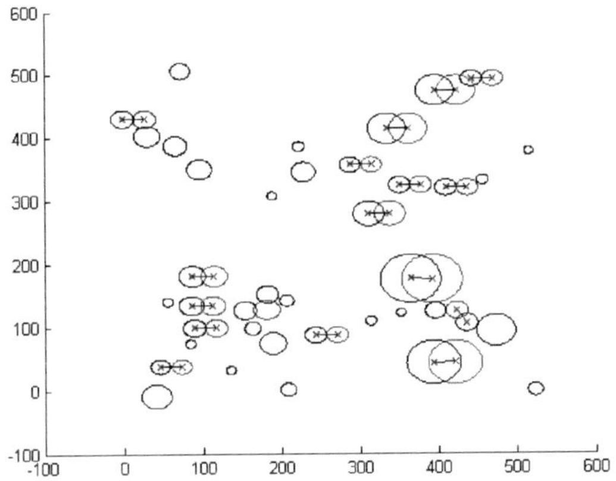

Fig. 6 Complexity of the Landmarks matching

8 Landmark Integrity

The Landmark Integrity function is an ulterior check performed before the Absolute Position Estimation. Since the Landmark Matching can be wrong, this function has the purpose of eliminating false matching or at least informing the Navigation filter that the new measurement is corrupted. The landmark integrity is performed using the Database 2D Projection data and the matched craters, the two sets of points are compared using proximity and rigidity criteria. If one or more craters are considered as an invalid matching the craters are removed from the subset whenever the number of craters is still larger than four. If the number is less than four the covariance matrix of the filter will be modified in order to provide the filter the information that the actual measurement is corrupted.

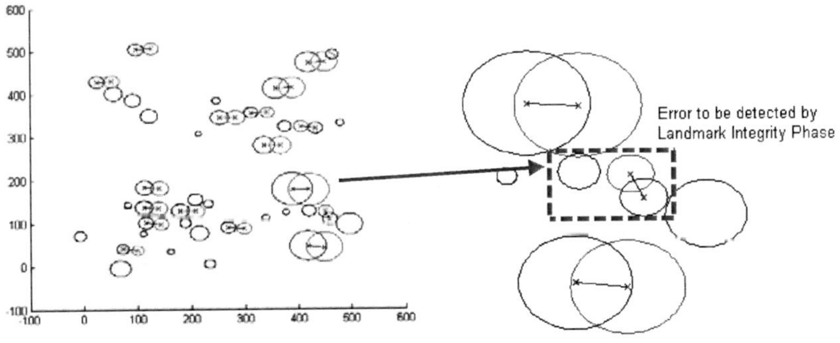

Fig. 7 Landmark Integrity process

Figure 7 shows the operation performed by the Landmark Integrity block. A false match can be detected in the figure on coordinates (450,105), this wrong matching is detected by the Landmark Integrity algorithm and the matching is deleted in the Absolute Position Estimation process.

9 Absolute Position Estimation

The Absolute Position Estimation algorithm uses the information of the Crater Database, the attitude information provided by the Star Tracker and the Landmark Integrity output for computing the Absolute Position Estimation [15]. A minimum of four craters are requested to be provided by the Landmark Integrity process in order to reach the requested level of accuracy. The output of this block is usually the distance of the camera to the origin of the reference frame in which the Crater Database is expressed. Since the position and orientation of the camera in the vehicle is known, the vehicle position can be computed.

10 Navigation Filter

The navigation filter includes a complete model of the Moon gravity containing the central term and the LP165 error gravity model. The filter is a Linear Kalman Filter (KF) able to integrate different sensors working with different frequency. This is particularly important when the measurement of different sensors have to be integrated / fused in order to obtain a more precise quantity. In the following figure is represented a schematic version of the Navigation filter. It requires the gravity components of the acceleration and the measurements provided by the OTAN system and the IMU. The sensor weight logic decides how much and when a sensor has to be trusted. Namely, it receives information of the Landmark integrity in order to transmit the information that the measurement can be corrupted or if more than four craters are recognized the covariance matrix is properly adjusted. In addition, when the vehicle is in the orbiting phase the accelerometer is discarded from the measurement whenever the acceleration is inside a certain threshold derived from the IMU noise parameters.

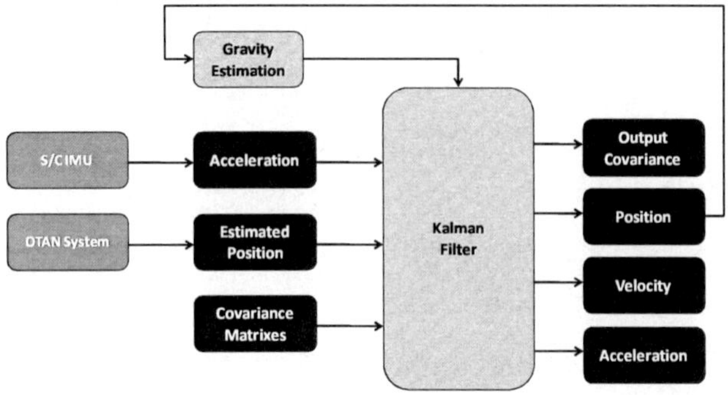

Fig. 8 Navigation Filter structure

The Kalman gain K_F is computed applying the formulas described in [16]. The matrixes Q and R used in the K_F computation contain the covariance of the state and measurement noises. Q and R are time variable and they determine when each sensor has to be considered and how much it has to be trusted. The initialization of the filter is performed using Ground Tracking measurements and the filter has to be able to start from almost lost in space condition.

11 Conclusion and Future Works

This paper has presented the Optical Terrain Absolute Navigation (OTAN) system being developed at GMV. Different parts of the system have been tested in simplified simulation environment [9][10] providing encouraging results in line with the expected accuracy. In particular, the on-line part of the system has shown robustness to illumination condition and to error in attitude less than 0.1 degrees. Information derived by the mentioned analysis has been used in an end-to-end GNC simulator showing that pinpoint landing is only achievable when precise absolute navigation is present in the orbital and main braking phases of the mission [6].

Extensive test campaigns are planned to be performed in the close future in the frame of ANTARES project, where the presented system will be extensively compared with [5]. In the project both systems will be adapted to a common architecture and the same test will be performed in both systems making the comparison straightforward. The objective is to bring the technology to TRL 4.

References

1. Cheng, Y., Ansar, A.: A Landmark Based Position Estimation for Pinpoint Landing on Mars. In: Proceedings of the 2005 IEEE International Conference on Robotics and Automation (ICRA), Barcelona, Spain, pp. 4470–4475 (2005)
2. Trawny, N., Mourikis, A.I., Roumeliotis, S.I.: Coupled Vision and Inertial Navigation for Pin-point Landing. In: NASA Science Technology Conference (2007)
3. Trawny, N., Mourikis, A.I., Roumeliotis, S.I., Johnson, A.E., Montgomery, J.: Vision-Aided Inertial Navigation for Pin-Point Landing using Observations for Mapped Landmarks. Journal of Fields Robotics (2006)
4. Hamel, J.F., Neveu, D., Jean, L.: Feature Matching Navigation Techniques for Lidar-Based Planetary Exploration. In: AIAA Guidance, Navigation and Control Conference and Exhibit (2006)
5. Pham, B.V., Lacroix, S., Devy, M., Drieux, M., Voirin, T.: Landmarks Constellation Matching for Planetary Lander Absolute Localization. In: International Conference on Computer Vision Theory and Applications (VISAPP 2010), Anger, France, May 17-21 (2010)
6. Melloni, S., Mammarella, M., Gil-Fernández, J., Colmenarejo, P.: GNC solution for Lunar Pinpoint and Soft Landing. In: Global Lunar Conference, 11th ILEWG Conference on Exploration and Utilisation of the Moon, Beijing, China, May 31 - June 3 (2010)
7. Lunar Orbital Deta Explorer, `http://ode.rsl.wustl.edu/moon/index.aspx`

8. PANGU. Planet and Asteroid Natural scene Generation Utility, University of Dundee, UK, http://spacetech.dundee.ac.uk/research/planetary-lander-technology/pangu/projects/pangu-enhancement
9. Pizzichini, A., Mammarella, M., Colmenarejo-Matellano, P., Melloni, S., Graziano, M., Curti, F.: Inertial Vision Based Navigation Technique For Low Lunar Orbit Position Determination. In: 4th International Conference on Astrodynamics Tools and Techniques (ICATT), Madrid, Spain, May 3-6 (2010)
10. Pizzichini, A., Mammarella, M., Colmenarejo-Matellano, P., Graziano, M., Curti, F.: Known Landmark Navigation for Precise Position Estimation in Lunar Landing Mission. In: Global Lunar Conference, 11th ILEWG Conference on Exploration and Utilisation of the Moon, Beijing, China, May 31 - June 3 (2010)
11. Cheng, Y., Johnson, A.E., Matthies, L.H., Olson, C.F.: Optical Landmark Detection for Spacecraft Navigation. In: 13th Annual AAS/AIAA Space Flight Mechanics Meeting (2003)
12. Parker, J.R.: Algorithms for Image Processing and Computer Vision, pp. 23–29. John Wiley & Sons, Inc., New York (1997)
13. Fitzgibbon, A.W., Pilu, M., Fisher, R.B.: Direct Least Squares Fitting of Ellipses. In: International Conference on Pattern Recognition, Vienna (August 1996)
14. Mammarella, M., Campa, G., Napolitano, M.R., Fravolini, M.L.: Comparison of Point Matching Algorithms for the UAV Aerial Refueling Problem. Journal of Machine Vision and Applications (in press)
15. Lu, C.P., Hager, G.D., Mjolsness, E.: Fast and Globally Con-vergent Pose Estimation from Video Images. IEEE Transactions on Pattern Analysis and Machine Intelligence 22(6), 610–622 (2000)
16. Brison, A.E., Ho, Y.C.: Applied Optimal Control, ch. 2. Hemisphere Publishing Corp., Washington DC (1975)

Methodologies to Perform GNC Design and Analyses for Complex Dynamic Systems Using Multibody Software

G. Baldesi, T. Voirin, A. Martinez Barrio, and M. Claeys

Abstract. Virtual simulation is currently a key activity that supports the specification, design, verification and operations of space systems. Taking advantage of dynamics research activities, ESA has currently been developing a multibody software (DCAP) together with industry for more than 30 years heritage in space applications. This software is a suite of fast, effective computer programs that provides the user with capability to model, simulate and analyze the dynamics and control performances of coupled rigid and flexible structural systems subjected to possibly time varying structural characteristics and space environmental loads. With the implementation of dedicated interfaces to other specialised software (such as NASTRAN, CATIA or Matlab/Simulink), it is possible to reproduce, with a quite good level of details, most of the key subsystems (such as trajectory, structures, configuration, mechanisms, aerodynamics, GNC and propulsion) of the launch vehicle and/or spacecraft in a single simulation.

In this paper, an overview of two GNC studies, a generic launch vehicle feasibility study and a spacecraft with no-conventional configuration, is presented.

1 Introduction

Rigid modelling of systems has proved to be efficient in many applications for AOCS/GNC designs. Nevertheless, in some specific cases, such as high resolution imaging satellites, high accuracy pointing space telescopes, or launchers, the flexible dynamics content of the system become a driver for the AOCS/GNC design, requiring ad-hoc dynamics model at the very first start of the project (feasibility phase, or phase 0).

To cope with this problematic, ESA/ESTEC has developed an in-house capability and methodology aiming at improving the flexible dynamics modelling and the subsequent GNC design. The purpose of the paper is to explain this strategy and

G. Baldesi · T. Voirin · A. Martinez Barrio · M. Claeys
ESA/ESTEC, Keplerlaan 1, 2201AZ Noordwijk ZH, The Netherlands
e-mail: Gianluigi.Baldesi@esa.int, Thomas.Voirin@esa.int, Alvaro.Martinez.Barrio@esa.int, Mathieu.Claeys@esa.int

its adaptation to two real applications, namely the generic launcher, and the IXO X-ray space telescope, focusing more on the tools and methodology adopted in that frame than in the actual discussion of the results.

2 Multibody Dynamics in Aerospace Applications

In recent years, space-system design has shown a clear trend towards increasing complex configurations. Typical examples are the use of several flexible components (antennas and solar arrays), the need for deployment and retrieval mechanisms, the demand for high precision pointing systems, and the increase in mission scenarios implying the assembly of large structures in space. This trend has also caused an evolution towards a multi-disciplinary design approach, particularly in the area of dynamics and control [1].

In order to study the performance of generic controlled dynamic systems, it is essential to have a dedicated tool, which allows the user to model, in a short time, the complex behaviour of the dynamic systems and their interactions with the control. In fact, some systems require a model with more than one body in order to take into account their different characteristics and their mutual dynamic interactions. This task is complex and requires dedicating some time to understand, code and validate the dynamic behaviour of the system.

A lot of research has gone into the development and improvement of multibody software, with the aim of reducing the time of modelling a system and the computation time required to run an analysis. Multibody software involve the derivation of the equations of motion for multibody systems, which are systems characterized by several bodies connected by hinges that permit relative motion across them. Robots, launchers and spacecraft including articulated appendages like solar arrays are typical examples of this kind of complex dynamic systems.

ESA has a long experience in using multibody software in order to analyse complex dynamic systems [2]. For this purpose, several software (such as DCAP, Adams and Samcef/MECANO) are currently used in different studies [3].

2.1 DCAP Software

Early approaches to the dynamics formulation for multibody systems lead to the equations of motion, for open-loop tree topologies, of the form:

$$F = M\ \ddot{q}$$

where, M is an $(n\ x\ n)$ mass matrix, $q = [q^1\ q^2\ \ldots\ q^n]^T$ is an (n x 1) column matrix representing the generalised coordinates and F is the column matrix containing the contributions from centrifugal, Coriolis and external forces.

For a numerical simulation of such a system, the mass matrix must be inverted. Since the inversion of an (n x n) matrix involves operations proportional to the cubic power of n, this is called an Order(n^3) approach. As the number of degrees of freedom increases, this matrix inversion for every integration step, becomes computationally expensive. Thus, researchers have sought methodologies to circumvent the mass matrix inversion and to improve computational efficiency. The research into improvements in formulations that increase computational speed resulted in - what are today called - Order(n) algorithms. The reason for this nomenclature is that the computational burden in these schemes increases only linearly with n. More details have already been presented in [4].

DCAP, which has currently been used in the Structures & Mechanisms Division at ESTEC (ESA), has the possibility to simulate the same problem performing the analysis based on *Order(n_3)* approach, derived by Lagrange method, and on *Order(n)* one using Kane's method of generalized speeds.

The development work that was undertaken during years by Alenia Spazio under ESA contract, culminated in the production of DCAP the Dynamics and Control Analysis Package [5]. DCAP is a suite of fast, effective computer programs that provides the spacecraft analyst with a powerful tool for designing and verifying the dynamics and control performance of coupled rigid and flexible structural systems. For the latter a dedicated symbolic manipulation pre-processor, used in the coding optimisation, has been coded in order to compute the minimum set of equations of motion for each particular problem.

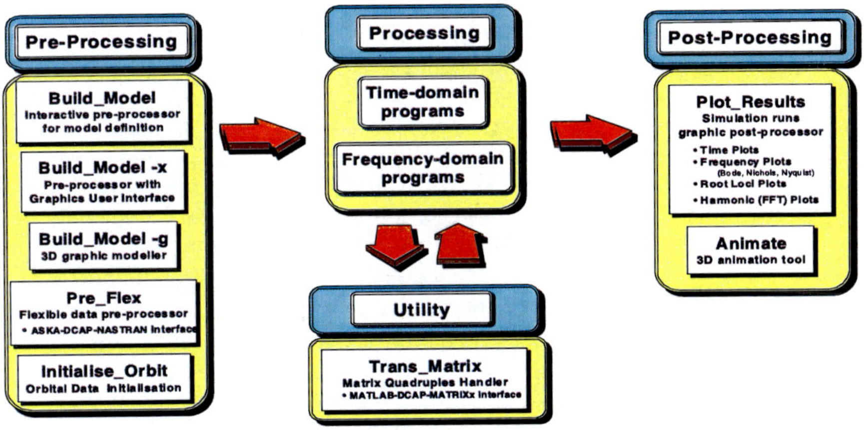

Fig. 1 DCAP features overview

The modelling capability is completed with the possibility of user-defined environment, allowing for modelling of specific control feature not directly included in the dynamic package's library. For the latter control modelling, straightforward interface with Matlab/Simulink exists.

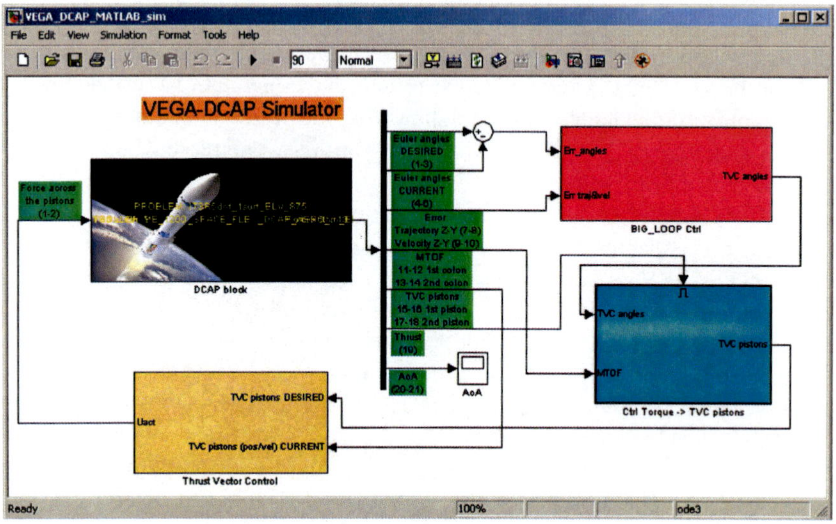

Fig. 2 DCAP IF to MATLAB-Simulink (co-simulation)

3 ESA Launch Flight Dynamics Simulator

Since 2001, ESA Structures & Mechanisms division (TEC-MS) has dedicated an important amount of energy in developing a solid knowledge in launcher flight dynamics [6]. Using multibody software, which can model the dynamic behaviour of interconnected rigid and/or flexible bodies, each of which may undergo large translational and rotational displacements, it is possible to assess and verify performances at system level. More in detail, launcher flight dynamics analysis allows to verify whether the selected launch vehicle design is able to accomplish the mission objectives taking into account the output information provided by all other subsystems such as Trajectory, Structures, Mechanisms, Aerodynamics, Propulsion, GNC etc. In particularly, non-linear dynamic simulations have been run in order to assess stability in nominal and off-nominal condition, TVC angular deflection and general loads.

Taking advantage of the strong know-how in dynamics, coming from long R&D activities in Multibody Dynamics, specific launcher flight dynamics tools (VEGA-DCAP-sim in Fig.1) was developed for different ESA projects in order to investigate not only flight dynamic effects for VEGA Support and CDF LV studies, but also local analyses (such as gust response, lift off, multi-payload separations) for GSTP3/GSTP4 Ariane 5 TVC, Swarm Support, Galileo IOV & IOF Support, IXV support. Furthermore, multibody software is gaining importance also in the difficult area of coupled load analysis [7], because it allows a much faster and malleable approach when compared to the classic FEM based procedures.

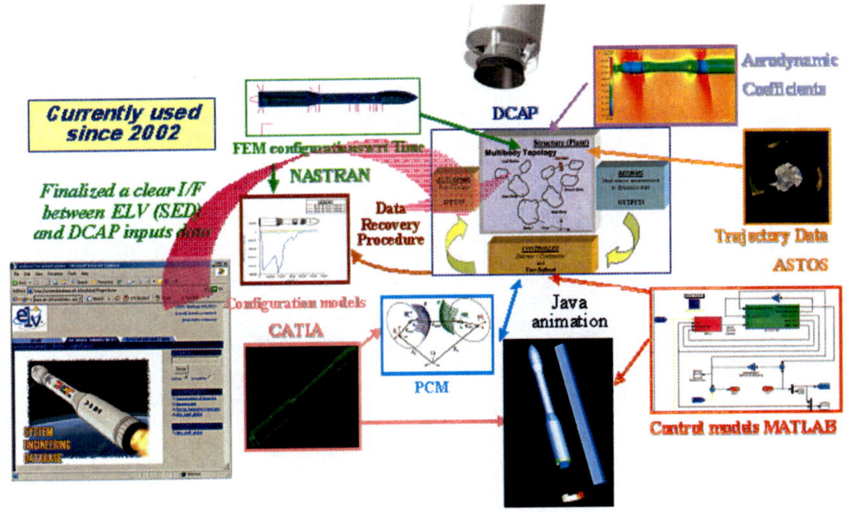

Fig. 3 VEGA-DCAP-sim overview

3.1 What Makes It Unique

In a nutshell, the main characteristics of ESA launcher flight dynamic simulator (VEGA-DCAP-sim) are:

- Multibody-based (easy to model the dynamics of "complex" launchers) [6]
- Flexible or rigid bodies both with time varying characteristics [1]
- LV flight Dynamics-Control interaction (directly imported in Matlab) [1]
- Environment modelling for Flexible launcher including external disturbances [8]
- Easy modelling non-linear effects (as misalignment, inertia influences, crosstalk effects, sloshing.)
- Easy modelling of dynamic transitions (lift-off analysis, stage and payload separations, …) [8][9]
- High degree of compatibilities with different software packages.

3.2 GNC Study Applications

3.3 Generic Launch Vehicle Feasibility Study

3.3.1 Launcher Design and Trajectory Optimization

The conceptual design of an expendable launch vehicle is dominated by interactions between many engineering disciplines: optimal trajectories, aerodynamics, propulsion, structures and mass estimations. All of these disciplines are usually highly coupled and several iterations between them are needed to produce a complete vehicle design.

In the field of trajectory optimization, ESA has developed, improved, and commercialized a trajectory optimization tool called ASTOS. The entire optimization scheme used in ASTOS has its foundations in the method of direct optimization which requires a non-linear programming solver (NLP).

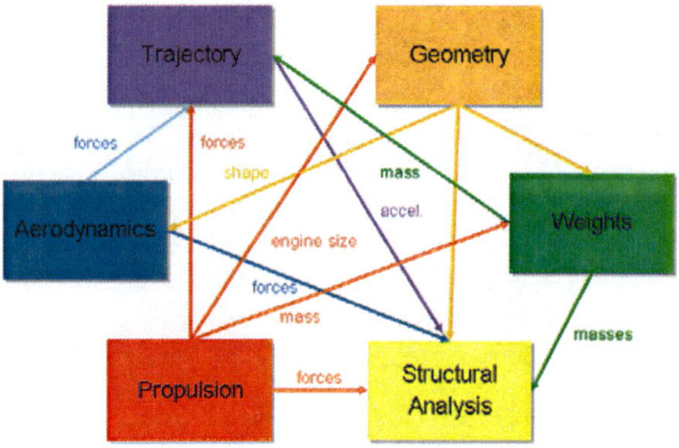

Fig. 4 Data exchange in ASTOS multidisciplinary design optimization environment

Recently, the trajectory optimization capability of ASTOS has been enhanced to include launch vehicle design optimization. In this way the tool is now capable of performing multidisciplinary design optimization (MDO) linking several discipline models. The data exchange between these disciplines is considerable as many disciplines depend on input from many others (see **Error! Reference source not found.**).

Nevertheless by means of adequate formulation of the problem and transcription methods, the optimal design of the vehicle together with the trajectory design can be obtained by means of NLP solvers using the MDO all-at-once approach. [10,11].

As a result the launcher scenario has been extended by a geometry model for all stages and the fairing, where lengths and diameters are optimisable. Interstages manage different stage diameters, which allow the modeling of hammerhead configurations. The aerodynamics are computed by Missile Datcom, where the shape is used from the geometry model. Tank models for separate, common bulkhead and enclosed tanks are provided to support the geometry model and the mass estimation.

The mass estimation is based on regression formulas for most important components. Alternatively, the mass estimation can be refined with a One-Beam Approximation (OBAX), which performs a structural analysis based on external and internal forces and weights and which results in a minimum wall thickness and mass estimation. Finally, the liquid engine model computes the combustion at equilibrium conditions with the NASA tool CEA and adds efficiency factors based on regression.

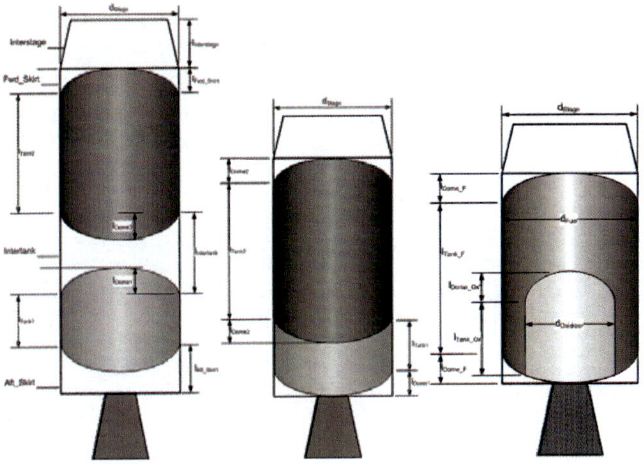

Fig. 5 Different tank configurations available

The ASTOS tool has been used in several ESA Concurrent Design Facility (CDF) studies improving the design process by obtaining a more realistic initial guess for a launcher system design. This initial guess is then used as the starting point for more detail subsystem design iterations. Once more accurate data is available from all the other subsystem, ASTOS can also be used as a simple trajectory optimization tool.

3.3.2 Dynamic Part

Lately, based on past successful experiments [12], the advantage of multibody software has also been made available to ESA Concurrent Design Facility (CDF)

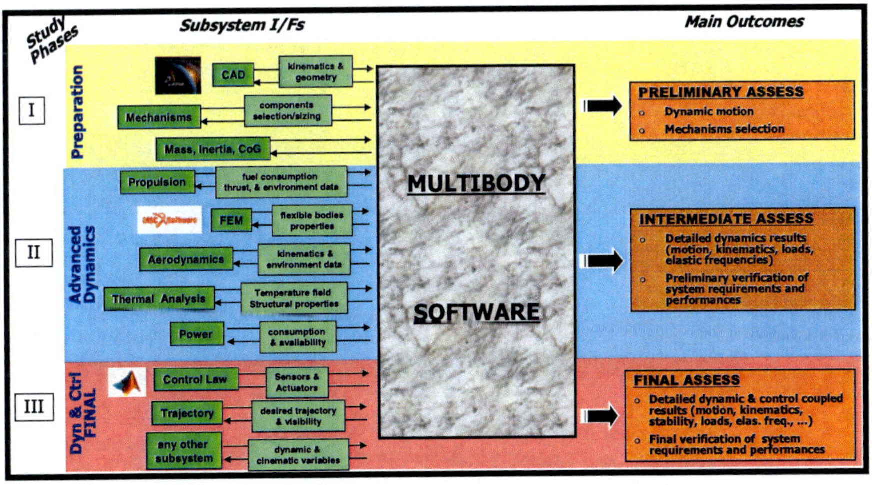

Fig. 6 Multibody expertises in CDF environment

studies (Figure 6). Indeed, since CDF applies the concurrent engineering method to the design of future space missions [13], the use multibody software is a key tool for analysing performances at system level, while taking into account simultaneously several subsystems. In the next figures launch flight dynamics simulation applications are briefly presented, which highlight the interfaces with other subsystems and some typical results.

Launch vehicle dynamics having multiple nozzles can be easily simulated using multibody capabilities with minimal workload. An example is reported in Fig.4, which represents an example configuration of future launcher study. The first stage consists of 4 boosters and a main motor controls 5 nozzles, which can be tilted independently. Despite being a rather complex to represent, the system proved relatively straightforward to model using the multibody tool, Therefore, more time was invested to investigate the best philosophy to command the individual nozzles, which could meet the strict requirements placed on the control part.

Fig. 7 Launcher with multiple nozzles

3.3.3 GNC Part

The first step has been the elaboration of a linear model for a rigid launch vehicle. The resulting model is presented hereafter:

$$\begin{bmatrix} \Delta\dot{\theta} \\ \Delta\ddot{\theta} \\ \Delta\ddot{z} \end{bmatrix} = \begin{bmatrix} 0 & 1 & 0 \\ a_6 & 0 & \frac{a_6}{U} \\ -a_1 & U & -a_2 \end{bmatrix} \begin{bmatrix} \Delta\theta \\ \Delta\dot{\theta} \\ \Delta\dot{z} \end{bmatrix} + \begin{bmatrix} 0 & 0 \\ -\frac{a_6}{U} & -k_1 \\ a_2 & -a_3 \end{bmatrix} \begin{bmatrix} W \\ \Delta\delta \end{bmatrix}$$

where U is the launch vehicle flight path speed and $\Delta\theta$ represents the perturbed pitch angle, i.e. the difference between the reference trajectory pitch angle and the actual pitch angle. Similarly, $\Delta\dot{z}$ represents the perturbed vertical component of the LV velocity. W is the wind velocity acting perpendicular to the launch vehicle and $\Delta\delta$ is the perturbed TVC deflection angle, measured with respect to the longitudinal axis of the LV.

Additionally, the different coefficients present in the matrices are detailed hereafter:

$$a_1 = \frac{L_\alpha + T_S + T_C - D}{m}$$

$$a_2 = \frac{L_\alpha}{m \cdot U}$$

$$a_3 = \frac{T_C}{m \cdot U}$$

$$a_6 = \frac{L_\alpha \cdot L_{CP2CoG}}{J_{yy}}$$

$$k_1 = \frac{T_C \cdot L_{Pivot2CoG}}{J_{yy}}$$

$$L_\alpha = Q \cdot S \cdot CN_\alpha$$

Where:

L_α is the derivative of the lift force with respect to the angle of attack, Q the dynamic pressure, S the aerodynamic reference area and CN_α the derivative of the aerodynamic normal coefficient (body axes) with respect to the angle of attack.

T_S is the static thrust, which can not be deflected by the nozzle (applicable for boosters with no TVC)

T_S is the "controllable" part of the thrust, deflectable by the TVC.

D is the drag force and U is the velocity of the LV

m and J_{yy} are the current mass and moment of inertia of the launcher at the given time.

$L_{Pivot2CoG}$ is the distance from the nozzle pivot point to the current LV center of mass.

L_{CP2CoG} is the distance from the current aerodynamic center of pressure to the current LV center of mass.

The second step has been to design a linear quadratic regulator (LQR) at different times of the flight, taking into account the time and frequency domain specifications. The different linearized models have been computed every 10 seconds of the flight.

Once the different control laws are computed for each time slice, the gains are linearly interpolated resulting in the final values shown in Figure 8:

Finally, the complete GNC system performance has been assessed using the DCAP simulator, which takes into account the elastic modes of the launcher. In order to reduce the effect of this vehicle elastic dynamics, a low-pass filter has been added.

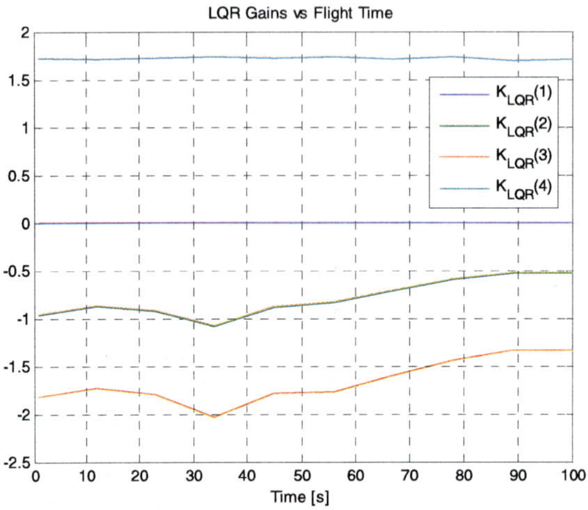

Fig. 8 LQR gains versus time

As mentioned before, the desired trajectory has been provided by the tool ASTOS. An overall overview of the DCAP simulator is shown in the following picture.

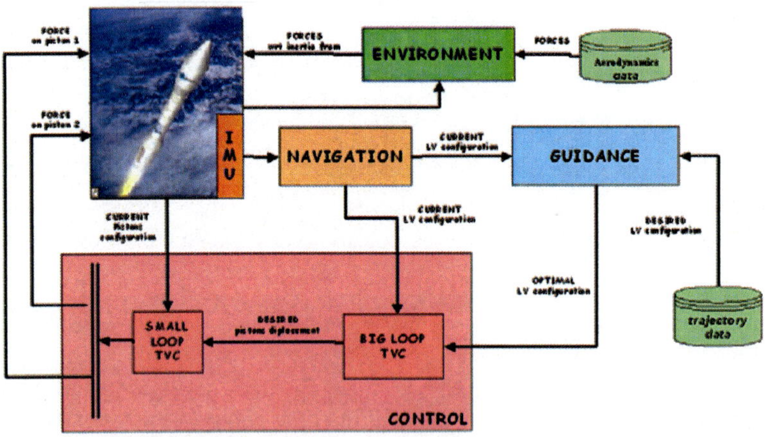

Fig. 9 Launch DCAP simulator flowchart

The results, provided in the following Figure 10, shows that the sample launcher analyzed is stable, the desired trajectory has been followed within 3.5 % of trajectory error and TVC does not exceed the design limit of 6 degrees.

It should be noted that no active roll control was implemented in the simulation presented hereafter. Nevertheless the roll increase occurred during the first stage flight of this particular launcher is relatively small and it is not expected to cause major concerns in future steps of the control design.

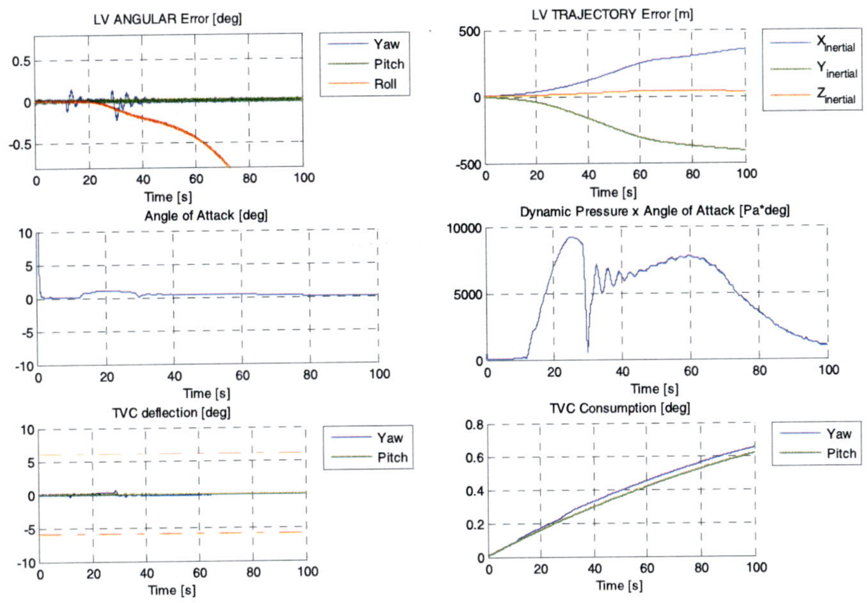

Fig. 10 Sample results of the launch vehicle simulator

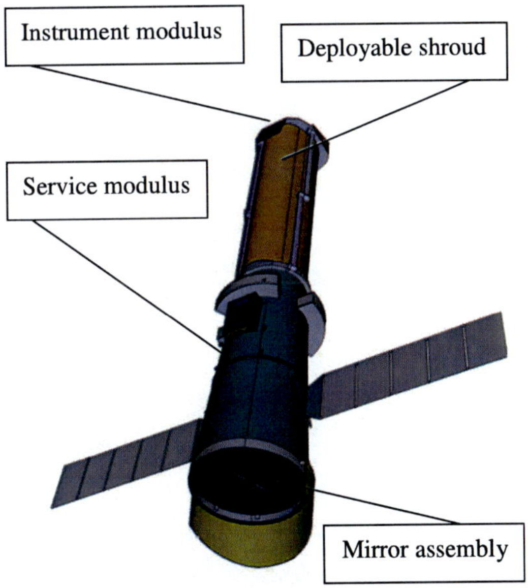

Fig. 11 IXO General Configuration (deployed). ESA CDF design

3.4 Spacecraft with No-Conventional Configuration

The International X-ray Observatory (IXO) is a Large-class mission candidate within ESA Cosmic Vision 2015-2025 program, being studied in a cooperative way by ESA, NASA and JAXA. Its main feature, which is of interest for this paper, is the need to extend the telescope structure from about 8 meters, as configured inside the launcher fairing, up to (at least) 20 meters (to achieve required operational focal length). Peculiar to the IXO telescope is the need to extend its focal length by a large amount (about 10 m) while ensuring extreme pointing accuracy and stability of the different telescope modules. The deployment principle is based on an innovative "articulated booms" concept, which is currently being investigated within (and outside) ESA. The elastic behaviour of the booms and the non-linear hinge characteristics (friction, backlash, hysteresis ...) are modelled to simulate the system deployment dynamics. Application of advanced multibody software techniques permit the investigation of coupled structural dynamics / control system, aiming at pointing accuracy / stability performance verification. [14]

3.5 Modelling

3.5.1 Dynamic Part

The dynamic behaviour of IXO spacecraft was modelled as a set of interconnected rigid or flexible bodies, each of which may undergo large translational and rotational displacements. The service and the instrument modules are considered as

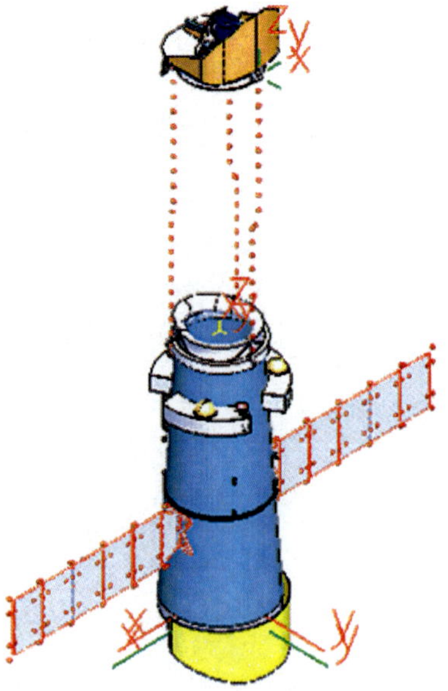

Fig. 12 IXO multibody model

rigid bodies as well as the motorised and passive hinges. On the other side, the solar arrays and the articulated booms are modelled as flexible taking advantage of the direct interface with MSC.NASTRAN. In addition of the main spacecraft free motion in space, "parallel kinematics" approach, uses nine revolute type of joints, which can benefit from the utilisation of very well known (to the space community) building blocks.

In order to provide mathematical models to Attitude and Orbit Control Systems (AOCS) and Guidance Navigation and Control (GNC) specialists (TEC-ECN) in ESA, IXO multibody model has been slightly modified by including system sensors (accelerometers, gyros,...), actuators (reaction wheels, thrusters,...) and additional output (lateral displacement of the instrument modules with respect to the service one, focal length,...).

For this purpose two models have been generated in "fully deployed" condition: a linear and a non linear model. Taking advantage of numerical linearization capability of most multibody software, it is possible to obtain linear model, described by means of the quadruplet (A, B, C, D). This mathematical model can be easily imported in Matlab/Simulink environment having as inputs the actuators/external force & torque disturbances and as outputs the sensor/additional dynamics outputs. On the other side, once the control law has been setup, it is compulsory to verify whether the assumptions are still validated. Furthermore, a "non linear" mathematical model is also generated and exported in Matlab/Simulink environment. The nonlinearities might be due not only to stiffness characteristics, transition events,... but also to the rigid motion of the whole system.

3.5.2 GNC Part

The size of the IXO telescope (20m) coupled with the need for relatively high agility (85% observation efficiency), and high accuracy pointing (10 arcs APE, 1 arcs AME around lateral axes), lead to unusual AOCS design, relying on a swarm of 5 "big" reaction wheels (150 Nms – 0.4 Nm class) to control the attitude of the spacecraft. The coupling of AOCS with flexible structure has to be analyzed carefully considering the very demanding lateral stability required by the payload (0.1 mm lateral displacement requirement).

To perform this assessment in the frame of IXO, a high-fidelity dynamics model of the Spacecraft is therefore required to model the flexibility inherent to this type of design, and in particular the induced motion of the detector with respect to the mirror. The DCAP-based linear flexi-body model detailed above has been used in that purpose, with as main objective to address the feasibility of and to characterize the AOCS performance for the nominal AOCS mode of IXO, that is to say the "Fine Pointing and Slew Mode" (FPS).

Design Approach

The following step-by-step approach for the control design and validation has been adopted:

1. First, a simplified dynamics model has been used, namely a cantilever beam attached to a rigid body, roughly representative of the expected MCI properties of the spacecraft ; a simplified controller has been then synthesized to characterize the achievable pointing performance of the spacecraft (Proportional Integrator & Derivative PID controller). This step is mandatory to initiate the preliminary AOCS design when no dynamics model is yet available. This step enables also to provide to the dynamics modelling expert the AOCS actuators characteristics (Reaction Wheels and Thrusters) and to specify the outputs of the dynamics model that are required to characterize the AOCS/GNC performance. For IXO these were mainly:
    a. the attitude and attitude derivative of the S/C at mirror node. This is obviously required to assess the pointing and pointing stability performance.
    b. the 3-axis translation of the detector with respect to the mirror at focus point of the telescope. This is required to characterize the blurring of the X-ray image induced by the detector relative motion wrt mirror.
2. Once the first high-fidelity Linear Dynamics Model has been made available by the dynamics modelling expert, a robust controller has been designed, based on the $H_{infinity}$ formalism. The use of the linear dynamics model at this stage enables to validate extensively the behaviour of the control loop on the various IXO operation scenarios.
3. Finally, the AOCS performance is cross-checked on reference cases using the final dynamics model, namely the Non-linear Dynamics Model with the robust controller previously designed.

This approach has many advantages:

1. The AOCS expert can start working on the design without having the fully representative dynamics model, which is by definition often the case at the start of a feasibility study.
2. A first iteration on the AOCS design can be performed ahead, and specifications for a representative dynamics model can be derived.
3. When the linear dynamics model is delivered, it is easy to perform a first validation of this dynamics model with the simplified model.
4. The non-linear model enables to validate the performance and robustness of the proposed controller.

This process is summarized in Figure 13

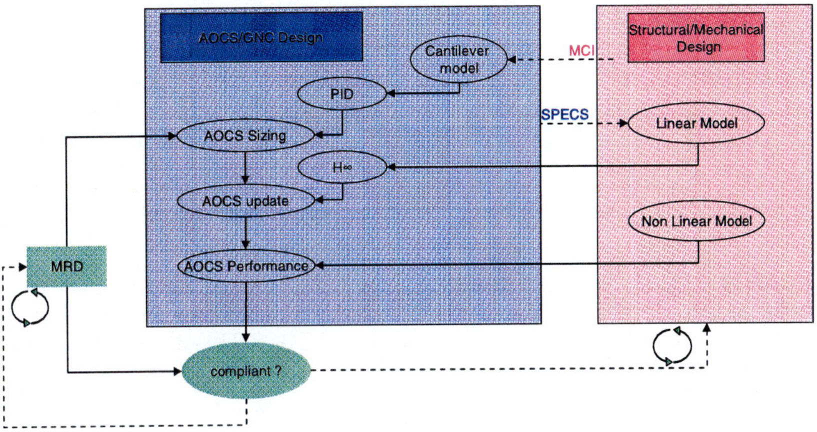

Fig. 13 IXO AOCS/GNC co-design process

AOCS Modes

The Fine Pointing and Slew Mode (FPS) of IXO includes four different operation modes, as identified in **Figure 14**:

- the Inertial Pointing Mode
- the dithering mode, where the boresight follows a Lissajous pattern
- the Raster mode, where the telescope follows a mosaic pattern
- the Slew mode, corresponding to a retargeting manoeuvre of the telescope from target n to target n+1.

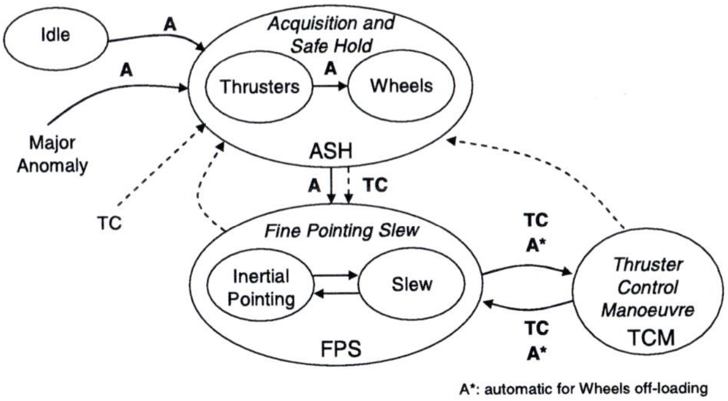

Fig. 14 IXO AOCS modes

Robust controller design

A single robust controller has been synthesized for the FPS, based on the $H_{infinity}$ framework. The IXO science requirements have been translated in the standard problem description for robust control as described in [15] [16]. The generated controller covers the whole operation modes of the FPS, with, for the slew operation mode, the use of a feed-forward scheme for the Reaction Wheels command.

The synthesized controller has a low bandwidth of 1 mHz to avoid the excitation of the S/C flexible modes. Standard robustness criteria of 6 dB margin on the gain and 30° margin on the phase have been applied.

The controller performance has then been assessed with a 3-dof high-fidelity relying on the linear dynamics model as described above, with the following scenarios:

- **Raster pointing scenario:** two consecutive steps of 10 arcs around the Y and X axis respectively, with 3000s between the two steps.
- **Dithering pattern scenario :** the Lissajous is 20 arcsec wide, with an X-axis frequency of 1/1200 Hz and a Y-axis frequency of 1/1800 Hz
- **Slews around X, Y and Z :** Amplitudes up to 180 ° have been considered to cover the whole spectrum of possible slews.

Results

The results obtained have confirmed the feasibility of the baseline IXO AOCS design, and has even shown that margins with respect to some requirements were important. In particular, the lateral stability of the detector platform with respect to mirror is always much below the requirement of 0.1mm, with a worst case excitation by AOCS smaller than 1 μm as shown in Figure 15. This proves that the AOCS might have coped with much more flexible (and therefore lighter) designs. It shall be noted that higher deviations are expected from thrusters impulses, used in TCM (required for example for wheels offloading or orbit control manoeuvres) or ASH modes. Future work dealing with the design of the other IXO AOCS modes will address this point, which is not expected to be critical, the lateral deviation requirement being not as critical in such phases.

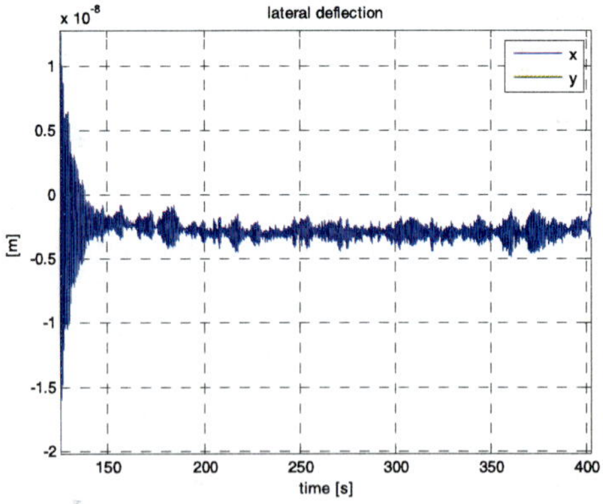

Fig. 15 IXO worst case lateral deviation (detector vs Mirror) 180° slew around sun axis

The overall pointing performance is in line with the 10 arcs APE requirement for all the observation modes. The dithering mode required by the Mission Requirements revealed to be the most challenging one for the AOCS, the frequency of the lissajous pattern being of the same order as the controller bandwidth. As a consequence, the robust controller has difficulties in tracking the Lissajous pattern as presented Figure 16, with on oscillatory tracking error of ~10 arcs amplitude If acceptable by the mission, a slight relaxation of the Lissajous frequency would solve this issue.

The slew mode is performed without saturating the wheels, and with a mean tranquillization time of ~1000s (worst case presented Figure 17 is for a 180° slew). The tranquillization time has proven to be for IXO a significant contributor to the overall observation efficiency, with an impact of ~2% on the overall telescope availability, making at the end the observation efficiency requirement only marginally met (86.6% obtained for 85% specified).

Finally, the maximal agility reached is close to 75°/hr, including tranquillization time. Realistic agility figures are an important input to scientists when establishing the telescope observation plan.

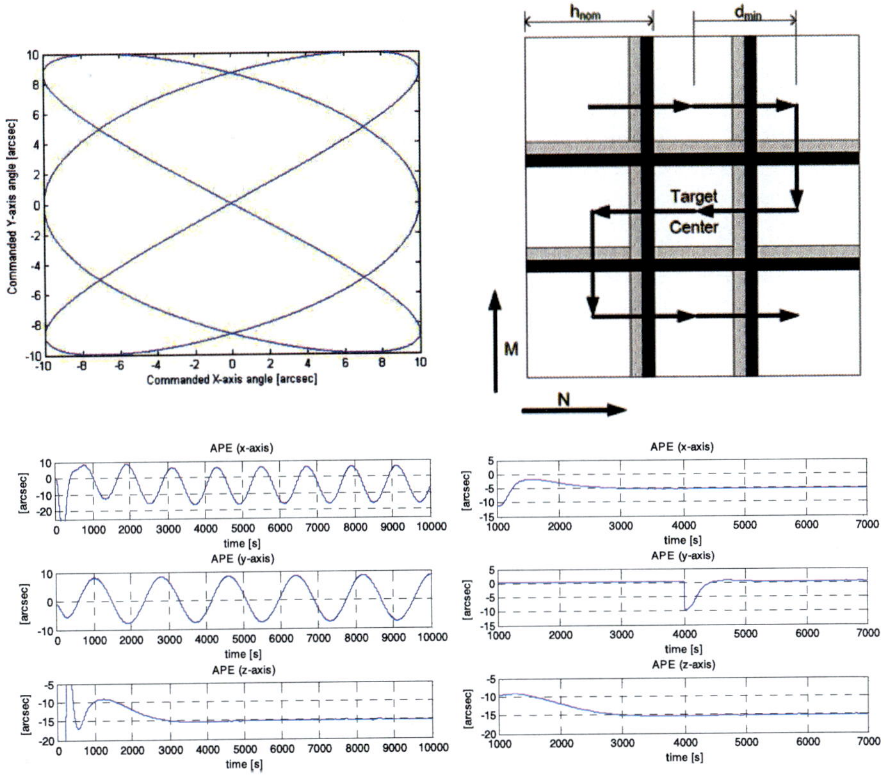

Fig. 16 IXO Lissajous and Raster guidance patterns (top), and simulated pointing performance (bottom)

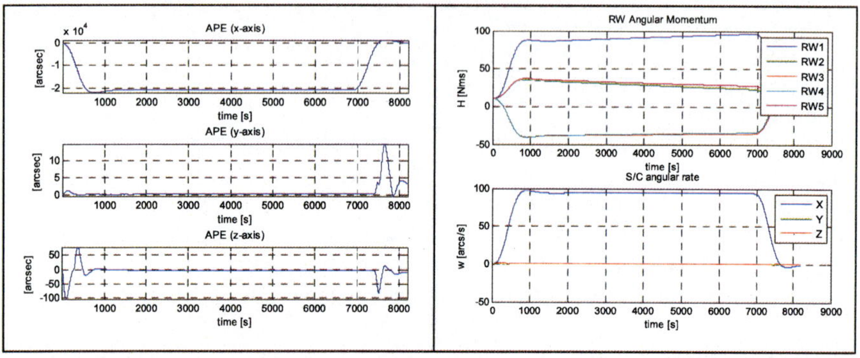

Fig. 17 IXO Slew of 180° amplitude wrt X-axis. Pointing error (left) and Reaction Wheels angular momentum load (right)

Conclusion

Overall, the AOCS analysis performed in the frame of IXO based on the DCAP dynamics model, pave the way for an even more ambitious approach for the assessment and pre-design of flexible spacecrafts: an integrated approach performed at phase 0 (CDF) in co-design between structure-mechanisms and AOCS/GNC subsystems would enable to improve the S/C design process by adjusting the structure flexibility to the AOCS capability, and vice & versa, for given science requirements.

4 Conclusions

In conclusion, the ESA Launcher Flight Dynamics Simulator represents an important successful story of a European software package, internally developed by the European Space Agency with the support of Thales Alenia Space Italy. Its peculiar generic modelling capabilities and computational speed are strong assets of the simulator, enabling real time comprehensive simulation of an entire multi-stage launcher along all the different phases of the launch. The high degree of interface with different specific tool packages allows for exchanging data and co-operating with other disciplines at the same time in a reliable and user-friendly environment.

The International X-ray Observatory spacecraft has been considered as a challenging test case to evaluate advanced multi-body simulation packages. The simulation strategies, tools and main results have been presented in details, together with AOCS/GNC interfaces. The use of the DCAP multi-body software has enabled to perform a preliminary design of a robust AOCS design for IXO ; further more, the integration of this dynamics model into a 3-dof AOCS simulator made possible the end-to-end characterization of IXO AOCS performance for the whole science operation mode, proving to be an efficient support for the follow-on of IXO industrial assessment studies, as well as a valuable input for ESA internal Cosmic Vision L-Class review.

Acronyms

AME	Attitude Measurement Error
AOCS	Attitude and Orbit Control Systems
APE	Absolute Pointing Error
ASH	Acquisition and Safe-Hold Mode
ASTOS	Aero-Space Trajectory Optimization Software
CDF	Concurrent Design Facility
CEA	Chemical Equilibrium with Applications
DCAP	Dynamic and Control Analysis Package
ESA	European Space Agency
ESTEC	European Space Research and Technology Centre
FEM	Finite Element Model
FPS	Fine Pointing and Slew Mode
GNC	Guidance, Navigation and Control
GSTP	ESA's General Support Technology Programme
IMU	Inertial Measurement Unit
IOC	Initial Operation Capability
IOV	In-Orbit Validation
IXO	International X-ray Observatory
IXV	Intermediate Expermental Vehicle
JAXA	Japan Aerospace Exploration Agency
LQR	Linear Quadratic Regulator
LV	Launch Vehicle
MDO	Multidisciplinary Design Optimization
NASA	National Aeronautics and Space Administration
NLP	Nonlinear programming solver
OBAX	One-Beam Approximation
PID	Proportional Integrator & Derivative
R&D	Research and Development
S/C	Spacecraft
TCM	Thrusters Control Mode
TEC-ECN	ESA's GNC Section
TEC-MS	ESA's Structures & Mechanisms Division
TVC	Thrust Vector Control

References

1. Baldesi, G., Sciacovelli, D., Portigliotti, S., Dumontel, M.: DCAP (Dynamics and Control Analysis Package) an effective tool for modeling and simulating of coupled controlled rigid flexible structure in space environment. In: 6th Inter. Conf. Dynamics and Control of Systems and Structures in Space 2004, Riomaggiore, Italy (2004)

2. ESA Website, Multibody Dynamics, http://www.esa.int/TEC/mechanisms/SEMZX356JGG_0.html
3. ESA Website, Multibody Analysis Activities, http://www.esa.int/TEC/Mechanisms/SEM4A0GMTGG_0.html
4. Dumontel, M.L., Portigliotti, S., Venugopal, R.: DCAP: A Tool for Analysis and Simulation of Multi-Body Systems. In: 45th IAF Conference, Oslo (October 1995)
5. Franco, R., Dumontel, M.L., Portigliotti, S., Venugopal, R.: The Dynamics and Control Analysis Package (DCAP) - A versatile tool for satellite control. ESA Bulletin Nr. 87 (August 1996), http://esapub.esrin.esa.it/bulletin/bullet87/franco87.htm
6. Sciacovelli, D., Kiryenko, S., Baldesi, G., Thirkettle, A., Redondo, R., Resta, P.D.: Vega Prototype 3D Simulation Software with Time Varying structural characteristics. In: 5th Inter. Conference Space Launchers: Missions, Control and Avionics, Madrid, Spain, November 25-27 (2003)
7. Parin, K., Henkel, E., Majed, A.: Enigma Multibody Loads Analysis: Practical Variational Coupled Loads Analysis for Launch Vehicle/Payloads. In: JPL Spacecraft and Launch Vehicle Dynamic Environments Workshop (June 2003)
8. Baldesi, G., Kiryenko, S., Mendoza, M.: Multi-Payload Separation Analysis Using Multibody Software. In: 11th European Conference on Spacecraft Structures, Materials & Mechanical Testing (ECSSMMTT), Toulouse, France (September 2009)
9. Baldesi, G., Yábar, C., Barbagallo, D.: Complete Lift-Off analysis of a launch vehicle using Multibody Software. In: ECCOMAS Multibody Conference, Warsaw, Poland (August 2009)
10. Möllmann, C., Wiegand, A., Dalheimer, M., Martinez Barrio, A., Kauffmann, J.: Multidisciplinary Design Optimisation of Expendable Launchers in ASTOS. In: 4th ICATT, ESA/ESAC (2010)
11. Wiegand, A., Möllman, C., Martinez Barrio, A.: New concurrent design optimisation models of ASTOS. In: 4th International Workshop on System and Concurrent Engineering for Space Applications. SECESA, October 13-15 (2010)
12. Smet, G., Yorck, M., Baldesi, G., Palladino, M.: Multi-body Dynamics Software Tool: Case study on the International X-ray Observatory. In: 13th European Space Mechanisms and Tribology Symposium (ESMATS 2009), Vienna, Austria (September 2009)
13. CDF ESA Website, Concurrent Design Facility, http://www.esa.int/SPECIALS/CDF
14. Baldesi, G., Scolamerio, L.: Advanced Multibody Simulation Techniques for Large Deplyable Systems: The IXO Telescope Case. In: 61st International Astronautical Congress 2010, Prague, Czech Republic, September 27- October 1 (2010)
15. Stein, G., Doyle, J.C.: Beyond Singular values and Loop Shapes. AIAA Journal of Guidance and Control (January 1991)
16. Chiang, R.: Integrated Robust Control Design Methodology for an Advanced S/C with Large Flexible Structure. In: Proc. of AIAA Guidance, Navigation and Control Conference and Exhibit, Honolulu, Hawaii, August 18-21 (2008), AIAA Paper # 2008-701

Optimal Guidance and Control of Lunar Landers with Non-throttable Main Engine

Thimo Oehlschlägel, Stephan Theil, Hans Krüger, Matthias Knauer,
Jan Tietjen, and Christof Büskens

Abstract. Autonomous soft, safe and precise landing on celestial bodies like the Moon, planets and asteroids is still a challenging task for future exploration missions. To achieve a maximum of payload mass landed on the target body the trajectories of landing vehicles need to be (fuel) optimized. In order to allow an adjustability of the trajectory and a compensation of disturbances for all vehicles so far a thrust modulation is required. The drawback is that currently no main engine is available which allows the needed thrust modulation for an efficient, robust and safe landing on a celestial body like the Moon. The technology of the Apollo missions is not available anymore. Most planned lunar missions rely on the modulation capability of multiple engines where in some cases the thrust of the auxiliary engines for modulation is in the order of main engine thrust. This approach requires a large number of smaller engines to achieve the needed thrust modulation adding complexity and increasing the probability of failure.

This paper shows a different approach to compute and control optimal trajectories for landing vehicles. It provides a new method for computing fuel efficient optimal trajectories which require minimal thrust modulation. A corresponding tracking control scheme is presented which allows the pre-computed optimal trajectory to be followed. The robustness of the method is discussed with results of a simulation.

1 Introduction

In the future more and more exploration missions will include a landing on the surface of a celestial body. Currently missions to the Moon, to Mars and to asteroids

Thimo Oehlschlägel · Stephan Theil · Hans Krüger
DLR Institute of Space Systems, Robert-Hooke-Str. 7, Bremen, Germany
e-mail: `Thimo.Oehlschlaegel@dlr.de`

Matthias Knauer · Jan Tietjen · Christof Büskens
AG Optimierung & Optimale Steuerung, Zentrum für Technomathematik,
Universität Bremen, Germany
e-mail: `knauer@math.uni-bremen.de`

are studied. The most critical phase of these missions is the descent and landing on the surface of the planet, moon or asteroid. In order to achieve an autonomous, safe and precise landing thrust modulation is needed at least for the Moon, Mars and large asteroids or moons. Thrust modulation is necessary for three main reasons:

1. To allow a fuel optimal descent within a specified safety corridor,
2. To enable an approach manoeuvre with low descent rates for landing site inspection, and
3. To allow hovering.

Although not all missions require hovering, a thrust modulation is still needed to fulfil the mission.

For the lunar soft landing missions in the past (Surveyor [10], Apollo [7]) throttable engines were used. Since these engines are not available anymore NASA is developing [1] a new throttable engine for its Altair lander design [4], [3]. Other landing vehicles under study are using two sets of engines. First, a main engine set of a few large thrust engines, whose thrust cannot be modulated, is used for deceleration from orbital velocity. Their thrust can not be modulated. Second, the control engine set comprises smaller pulsable thrusters which allow the modulation of the total thrust. For most of these missions the ratio between total thrust of main engines and total thrust of control engines is about 1 to 1 [8]. This leads to a large number of smaller pulsable engines, thus adding a lot of complexity and risk.

From this situation the motivation arises to investigate the possibility of trajectories and controllers which allow the ratio between main and control engine to shift towards the main engines. Thus the number of smaller control engines can be reduced. In the best case the required modulation is created by the reaction control system (RCS), eliminating the need for the control thruster set.

Keeping this theme, this paper will first present the approach for computing optimized trajectories for landing vehicles which use an RCS and a set of main engines in which single engines can be switched off symmetrically during landing. The result of the optimization will be a reference trajectory in the altitude-downrange plane which minimizes fuel consumption and usage of extra thrust from the RCS for thrust modulation. In the next step a controller is developed to follow this trajectory. Using simulations, the robustness of this approach is analysed and presented.

2 Model of Lunar Lander

In order to calculate optimal trajectories for the powered descent of the lunar lander a mathematical model of its dynamical behaviour has to be established.

2.1 *Coordinate Systems*

For the formulation of the equations of motion of the lunar lander, two different inertial reference frames are defined. First the equations are derived wrt the *xyz*-frame. Since the initial and final position for the descent will be given with respect to locations on the target planet or moon, the equations of motion are converted to downrange d, altitude h and crossrange c coordinates. The definition of the coordinates is shown in Fig. 1. The transformation assumes a reference sphere of radius r. The position can be described in spherical coordinates with the equatorial plane of the spherical coordinates equal to the d-h plane. The angle in the plane is denoted as δ. The angle perpendicular to the plane is γ. Defining crossrange and downrange as the projection on the reference sphere the angles can be expressed as

$$\delta = \frac{d}{r}, \qquad \gamma = \frac{c}{r}. \tag{1}$$

Definition $\mathbf{P}_0$ as the position at zero downrange and zero crossrange, the current position $\mathbf{P}_{xyz} = (x, y, z)^{\mathrm{T}}$ is a function of the angles δ and γ and the altitude h

$$\mathbf{P}_0 = (0, h + r, 0)^{\mathrm{T}}, \qquad \mathbf{P}_{xyz} = \mathrm{Rot}_z(-\delta)\mathrm{Rot}_x(\gamma)\mathbf{P}_0. \tag{2}$$

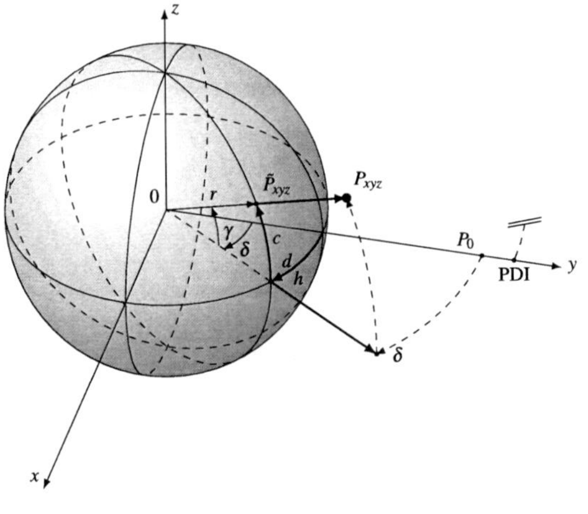

Fig. 1 Coordinate definition of altitude h, downrange d, crossrange c.

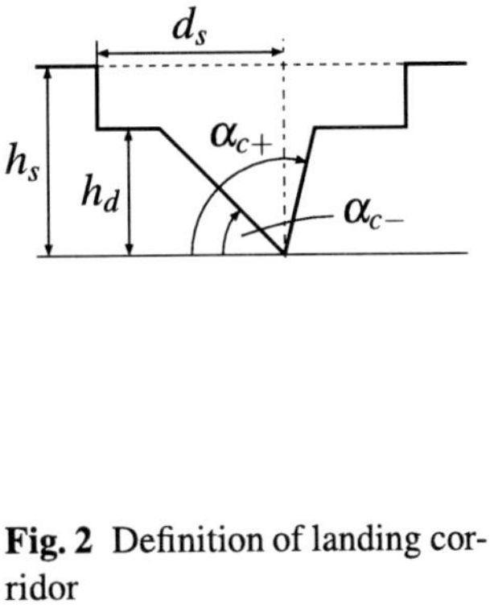

Fig. 2 Definition of landing corridor

2.2 Equations of Motion

The equations of motions of the lunar lander are deduced from equations with respect to the inertial xyz-frame

$$\ddot{\mathbf{P}}_{xyz} = \mathbf{g}_{xyz} + \frac{1}{m}\mathbf{T}_{xyz}, \tag{3}$$

where $\mathbf{g}_{xyz}$ denotes the acceleration due to the gravitational field of the target object and $\mathbf{T}_{xyz}$ equals the effective thrust vector. For simplicity the gravitational field of the target object is assumed to have a spherical potential

$$\mathbf{g}_0 = \left(0,\, -\tfrac{M\cdot G}{(r+h)^2},\, 0\right)^{\mathrm{T}}, \qquad \mathbf{g}_{xyz} = \mathrm{Rot}_z(-\delta)\mathrm{Rot}_x(\gamma)\mathbf{g}_0. \tag{4}$$

The combination of the thrust of the main engines (T_m) and the RCS thrusters (T_u, T_s, T_q) forms the total thrust vector $\mathbf{T} = \left(-T_m - T_u,\, -T_s,\, -T_q\right)^T$. The effective thrust

$$\mathbf{T}_{xyz} = \mathrm{Rot}_z(\beta-\delta)\mathrm{Rot}_y(\chi)\mathbf{T} \tag{5}$$

depends on the orientation of the lander, which is described by the attitude angles β (pitch) and χ (yaw). Under consideration of the total thrust vector $\mathbf{T}$ the latter transformation yields

$$\mathbf{T}_{xyz} = \begin{pmatrix} T_1 \\ T_2 \\ T_3 \end{pmatrix} = \begin{pmatrix} \cos\left(\beta - \frac{d}{r}\right)\left((T_m+T_u)\cos\chi + T_q\sin\chi\right) - \sin\left(\beta-\frac{d}{r}\right)T_s \\ \sin\left(\beta - \frac{d}{r}\right)\left((T_m+T_u)\cos\chi + T_q\sin\chi\right) + \cos\left(\beta-\frac{d}{r}\right)T_s \\ -(T_m+T_u)\sin\chi + T_q\cos\chi \end{pmatrix}. \tag{6}$$

Replacing the left hand side of (3) with the second derivative of (2) wrt time under consideration of (1,4,5) leads to the equations of motion in the *dhc*-frame

$$\begin{pmatrix} \ddot{d} \\ \ddot{h} \\ \ddot{c} \end{pmatrix} = \begin{pmatrix} \frac{r}{m(r+h)\cos\frac{c}{r}}\left(-T_1\cos\frac{d}{r} + T_2\sin\frac{d}{r}\right) + 2\dot{d}\left(\frac{\dot{c}}{r}\cdot\tan\frac{c}{r} - \frac{\dot{h}}{r+h}\right) \\ \frac{1}{m}\left[\left(-T_1\sin\frac{d}{r} - T_2\cos\frac{d}{r}\right)\cos\frac{c}{r} - T_3\sin\frac{c}{r}\right] + \left[\left(\dot{d}\cos\frac{c}{r}\right)^2 + \dot{c}^2\right]\frac{r+h}{r^2} - \frac{M\cdot G}{(r+h)^2} \\ \frac{r}{m(r+h)}\left[\left(T_1\sin\frac{d}{r} + T_2\cos\frac{d}{r}\right)\sin\frac{c}{r} - T_3\cos\frac{c}{r}\right] - \frac{\dot{d}^2}{r}\sin\frac{c}{r}\cos\frac{c}{r} - \frac{2\dot{c}\dot{h}}{r+h} \end{pmatrix}. \tag{7}$$

To ensure that the pitch and yaw angles vary continuously, their first derivatives

$$\dot{\beta} = \omega_\beta, \qquad \dot{\chi} = \omega_\chi \tag{8}$$

are added to the system of differential equations to consider the attitude motion with ω_β and ω_χ as commanded angular rates. Finally,

$$\dot{m} = -|T_m|\cdot\sigma_m - \left(|T_u| + |T_s| + |T_q|\right)\cdot\sigma_{RCS} \tag{9}$$

describes the thrust-dependent fuel consumption depending on σ_m and σ_{RCS}. For the purpose of trajectory optimization and control design, state space representations of the dynamics of the lunar lander has to be derived. All states of the dynamical system defined by (7, 8, 9) are gathered in the state vector

$$\mathbf{x}(t) = \left(\dot{d},\, \dot{h},\, \dot{c},\, d,\, h,\, c,\, \beta,\, \chi,\, m\right)^T. \tag{10}$$

The motion of the lunar lander can be influenced by the translatory and rotatory thrusters, which form the control vector $\mathbf{u}(t) = \left(T_u,\, T_s,\, T_q,\, \omega_\beta,\, \omega_\chi\right)^T$. The thrust of the main engines T_m can take only predefined values. The switching times τ_i, $i = 1,\ldots,k$, at which engines are disabled, can be collected in a vector of free parameters p. With these abbreviations, formulae (7, 8, 9) can be written as

$$\dot{\mathbf{x}}(t) = \mathbf{f}(\mathbf{x}(t), \mathbf{u}(t), \mathbf{p}, t), \qquad \mathbf{x}(0) = \mathbf{x}_0. \tag{11}$$

3 Trajectory Optimization

3.1 Constraints and Landing Corridor

This mission also requires holding terminal conditions at some terminal time point $t_f = \tau_k$

$$\begin{aligned} \mathbf{x}(0) &= (\ \dot{d}_0,\ \ 0,\ \ 0,\ \ 0,\ \ h_0,\ \ 0,\ \ \text{free},\ \ 0,\ \ m_0\ )^T \\ \mathbf{x}(t_f) &= (\ 0,\ \ \dot{h}_f,\ \ 0,\ \ \text{free},\ \ h_f,\ \ 0,\ \ -\tfrac{\pi}{2},\ \ 0,\ \ \text{free}\ )^T. \end{aligned} \tag{12}$$

Further constraints for the states and the control inputs are provided, e.g.

- minimum switch on time for a subset of the main engines,
- maximum auxiliary thrust of the RCS thrusters,
- the boundaries of a landing corridor,
- a minimum time for the landing corridor.

The landing corridor was introduced to ensure that in the final approach phase a defined amount of time is available for the final slow descent to the landing site. The parameters of the corridor are shown in Fig. 2. The safety downrange d_s denotes an area around the landing site, where $h > h_s$ has to be ensured. The corridor height h_c as well as the corridor angles α_{c-} and α_{c+} define the admissible domain for the final landing path. The time from entering the corridor until reaching the final conditions is constrained by a lower bound. For the computation of the optimal reference trajectory, the motion of the lunar is restricted to a two-dimensional motion in the d-h plane. Therefore it is assumed that $T_q = T_s = \omega_\chi = 0$ holds.

3.2 Optimal Control Problem

There exist many ways to bring the lunar lander from its initial state to its desired final position. Optimal control theory provides methods to find exactly that trajectory which minimizes a given objective functional I, given the system of differential equations $\mathbf{f}$, the initial and terminal conditions ω and state and control constraints $\mathbf{g}$

$$\begin{aligned} &\min_{\mathbf{x},\mathbf{u},\mathbf{p}} I(\mathbf{x}, \mathbf{u}, \mathbf{p}) \\ &s.t. \qquad \dot{\mathbf{x}}(t) = \mathbf{f}(\mathbf{x}(t), \mathbf{u}(t), \mathbf{p}, t) \\ &\qquad \omega(\mathbf{x}(0), \mathbf{x}(\tau_1), \ldots, \mathbf{x}(\tau_k)) = 0 \\ &\qquad \mathbf{g}(\mathbf{x}(t), \mathbf{u}(t), \mathbf{p}, t) \leq 0, \quad t \in [0, t_f]. \end{aligned} \tag{13}$$

Here, the free initial value for $\beta(0)$ is also added to the vector of free parameters $\mathbf{p}$. Table 1 gives some possible objective functionals for the optimization of the trajectory of the lunar lander. Empirical evidence shows that a linear combination of these functionals gives the best results.

Table 1 Objective functionals for the trajectory optimization of a lunar lander.

Fuel consumption	Auxiliary thrust (with possible reference thrust $T_{u,\mathrm{ref}}$)	Rotation of the lander	Flight time
$I_1 = \frac{-m(t_f)+m(0)}{m(0)}$	$I_2 = \int_0^{t_f} (T_u - T_{u,\mathrm{ref}})^2\, dt$	$I_3 = \int_0^{t_f} \omega_\beta^2\, dt$	$I_4 = t_f$

By switching from a continuous time axis $t \in [0, t_f]$ to discrete time points $t \in \{0 = t_1 \leq t_2 \leq \cdots \leq t_l = t_f\}$, $l \in \mathbb{N}$, the optimal control problem can be solved numerically. The control function $\mathbf{u}(t)$ is also reduced to a discretized version $\mathbf{u}^i \approx \mathbf{u}(t_i)$, and the discretized state vector $\mathbf{x}^i \approx \mathbf{x}(t_i)$ can be calculated directly from the system of differential equations. After replacing the optimal control problem (13) by the discretized version

$$\begin{aligned}
\min_{\mathbf{x}^1,\ldots,\mathbf{x}^l,\mathbf{u}^1,\ldots\mathbf{u}^l,\mathbf{p}} \quad & I\left(\mathbf{x}^1,\ldots,\mathbf{x}^l,\mathbf{u}^1,\ldots\mathbf{u}^l,\mathbf{p}\right) \\
s.t. \quad & \mathbf{x}^{i+1} = \mathbf{x}^i + (t_{i+1} - t_i)\, f\left(\mathbf{x}^i,\mathbf{u}^i,\mathbf{p},t_i\right),\ i = 1,\ldots,l-1 \\
& \omega\left(\mathbf{x}^1,\ldots,\mathbf{x}^l\right) = 0 \\
& \mathbf{g}\left(\mathbf{x}^i,\mathbf{u}^i,\mathbf{p},t_i\right) \leq 0,\ i = 1,\ldots,l,
\end{aligned} \tag{14}$$

which has the form of a standard NLP problem, common SQP solvers can be applied. The software library NUDOCCCS [2] follows this method to solve optimal control problems. The special structure of this large-scale problem can be exploited by sparse solvers, e.g. WORHP [9].

The consideration of the corridor as constraints in the optimal control problem downgrades the convergence of the iterative solver drastically. To circumvent this effect, the optimization is performed in several steps:

1. Analysis of the corridor: A descent maneuver is calculated within the corridor with free initial velocity to generate reasonable entry values.
2. Reaching the corridor: The trajectory is optimized until the entry of the corridor is reached, where the final state is constrained by the corridor entry values from the previous step.
3. Complete trajectory: By using the previous trajectory as an initial guess and by extending it by the corridor phase, a robust solution for the optimal trajectory can be found.

By treating the main descent and the corridor phase seperately in the discretization, the constraints for the corridor can be implemented easily:

- The minimum flight time within the corridor is not allowed to be less than $60\,s$.
- The flight path angle $\phi = \arctan\left|\frac{\dot{h}}{\dot{d}\cdot\frac{h+r}{r}}\right|$ is constrained by $[\alpha_{c^-}, \alpha_{c^+}]$ during the corridor phase.
- The safety downrange is checked by calculating the possibly overlapping area of the trajectory and the corridor in Fig. 2.

3.3 Results for the Optimal Reference Trajectory

In order to show the quality of the results of the optimization the scenario from [8] was selected for comparison. The initial conditions have been chosen in a similar way. The difference for the thruster system is that the landing vehicle described in [8] has four main thrusters of 500 N thrust, eight pulsable thrusters of 220 N and an RCS with 22 N thrusters. In our example we chose to have a set of eight 500 N thrusters and an RCS with 22 N thrusters for small modulations of the thrust only. It is assumed that the thrusters will be arranged symmetrically such that they can be switched off in pairs, e.g. in a sequence of 8, 6 and 4 thrusters running at the same time. In addition, the landing corridor is defined with the parameters in Table 2. With the given conditions, the optimization of the trajectory provides the solution displayed in Fig. 3. As can be seen in the upper left plot, the downrange is about 550 km and the flight path angle is nearly 90° immediately before landing. The lower left plot shows the attitude angle β of the landing vehicle in the d-h plane which also ends with an upright position of the vehicle of 90° wrt the horizontal plane. The upper right plot shows the corresponding vertical and horizontal velocities. In the lower right plot the main thrust T_m shows the two steps at 553 s and 613 s where a symmetrical pair of main thrusters is switched off. The second graph shows the auxiliary thrust T_u which is needed for the descent through the corridor in order to compensate for the decreasing mass of the landing vehicle. The fuel consumption for the reference trajectory is 45.4 % of the initial mass. In order to check the robustness of the optimization, the initial conditions for the landing trajectory are varied over a range, which is similar to the uncertainty for the PDI state determined in [6]. The variation of downrange by ± 11 km, altitude by ± 700 m, horizontal velocity by ± 0.6 m/s and vertical velocity by ± 0.24 m/s, rendered for all cases an optimal solution fulfilling all constraints. The fuel consumption for all simulated cases varied from 45.3 % to 45.7 %.

Table 2 Parameters of the landing corridor for the reference scenario.

h_s	d_s	h_c	α_{c-}	α_{c+}	t_c	h_f
3000 m	10000 m	1000 m	20°	100°	60 s	50 m

4 Trajectory Control

The trajectory control of the lunar lander (Fig. 4) consist of a combination of feed-forward and feedback controls, commonly referred to as two degree of freedom approach. In order to guide the lunar lander during a landing manoeuvre along a calculated optimal trajectory $\mathbf{x}_{\text{ref}}(t)$ the related control signal $\mathbf{u}_{\text{ref}}(t)$ is used as a feed-forward control for the vehicle. To ensure that the lunar lander follows the optimal trajectory even in the presence of parameter uncertainties and errors related to variations of the initial conditions or external disturbances and despite the fact that the

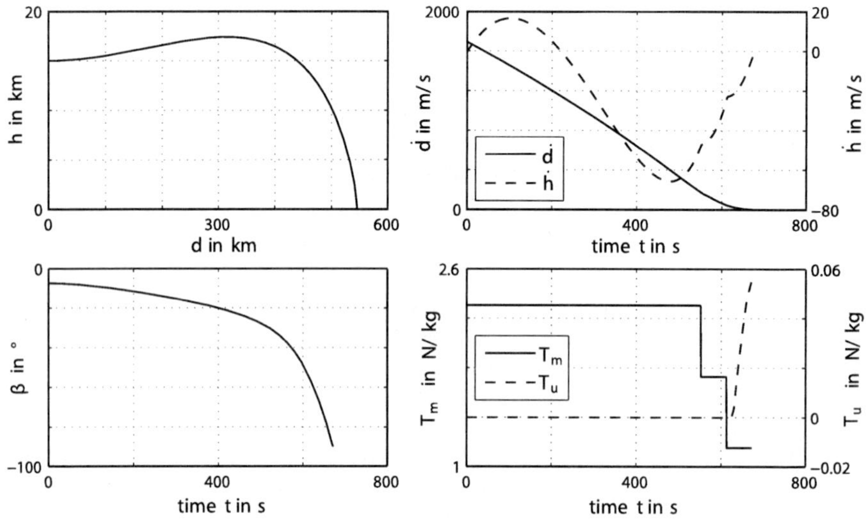

Fig. 3 Result of optimization for reference trajectory

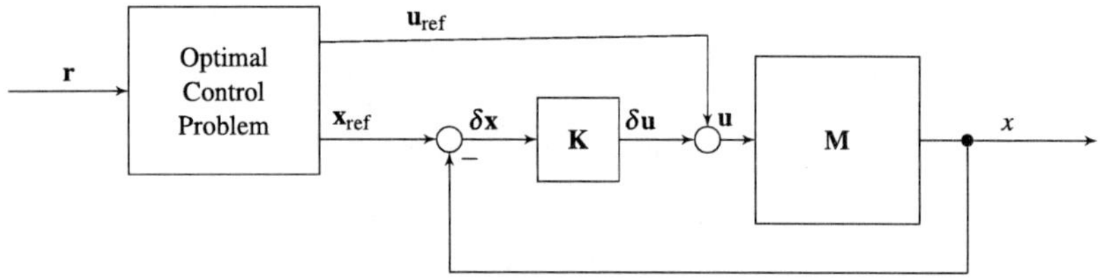

Fig. 4 Structure of the closed loop in case of the provided 2DOF control approach, whereas **M** denotes the plant and **K** the feedback controller.

optimal control $\mathbf{u}_{\text{ref}}(t)$ has been calculated using the assumptions $T_q = T_s = \omega_\chi = 0$, an additional feedback controller is required s.t. the overall control law is described by $\mathbf{u}(t) = \mathbf{u}_{\text{ref}}(t) + \delta\mathbf{u}(t)$.

4.1 Feedback Control Approach

For the purpose of the design of the feedback controller the dynamics of the lunar lander are represented by the nonlinear state space model as shown in Eq. (11). Assuming that the aforementioned state space representation is analytic and that a real landing trajectory is close to the optimal trajectory $\mathbf{x}(t) = \mathbf{x}_{\text{ref}}(t) + \delta\mathbf{x}(t)$, the linearization of (11) along $\mathbf{x}_{\text{ref}}(t)$ and $\mathbf{u}_{\text{ref}}(t)$ yields

$$\dot{\delta\mathbf{x}}(t) = \mathbf{A}(t)\delta\mathbf{x}(t) + \mathbf{B}(t)\delta\mathbf{u}(t) \tag{15}$$

$$\text{with}\quad \mathbf{A}(t) = \left.\frac{\partial \mathbf{f}}{\partial \mathbf{x}}\right|_{\mathbf{x}_{\text{ref}},\mathbf{u}_{\text{ref}}} \in \mathbb{R}^{n\times n} \quad \text{and} \quad \mathbf{B}(t) = \left.\frac{\partial \mathbf{f}}{\partial \mathbf{u}}\right|_{\mathbf{x}_{\text{ref}},\mathbf{u}_{\text{ref}}} \in \mathbb{R}^{n\times m}, t \in [t_0,t_f]$$

and corresponding initial conditions. The resulting linear time variant system motivates the usage of a time variant control law that, considering the above-mentioned overall control law, leads to

$$\delta\mathbf{u}(t) = -\mathbf{K}(t)\delta\mathbf{x}(t) \;\Rightarrow\; \mathbf{u}(t) = \mathbf{u}_{\text{ref}}(t) - \mathbf{K}(t)\left(\mathbf{x}(t) - \mathbf{x}_{\text{ref}}(t)\right),\; t \in [t_0,t_f]\,. \quad (16)$$

The changes in the dynamics of the lunar lander during a landing manoeuvre along a reference trajectory are supposed to be slow compared to the sampling rate of the discretized version of the optimal control problem (14). For this reason the gain matrix of the feedback controller $\mathbf{K}(t)$ is calculated only at each point $t_i, i \in \mathbb{I}$ with $\mathbb{I} = \{1,..,l\}$ of the control discretization by minimization of the cost function

$$J(\delta\mathbf{x},\delta\mathbf{u}) = \int_0^\infty \delta\mathbf{x}(\tau)^T\mathbf{Q}(t_i)\delta\mathbf{x}(\tau) + \delta\mathbf{u}(\tau)^T\mathbf{R}(t_i)\delta\mathbf{u}(\tau)\,d\tau \quad (17)$$

for all $i \in \mathbb{I}$. Assuming the stabilizability of the underlying system, $\mathbf{Q}(t_i) \geq 0$ and $\mathbf{R}(t_i) > 0$ for all $i \in \mathbb{I}$, the feedback matrix that solves the optimal control problem is given as a function

$$\mathbf{K}(t_i) = \mathbf{R}(t_i)^{-1}\mathbf{B}(t_i)^T\mathbf{S}(t_i), \quad i \in \mathbb{I}, \quad (18)$$

of the unique stabilizing solution $\mathbf{S}(t_i)$ of the algebraic Riccati equation

$$\mathbf{A}(t_i)^T\mathbf{S}(t_i) + \mathbf{S}(t_i)\mathbf{A}(t_i) - \mathbf{S}(t_i)\mathbf{B}(t_i)\mathbf{R}(t_i)^{-1}\mathbf{B}(t_i)^T\mathbf{S}(t_i) + \mathbf{Q}(t_i) = 0, \quad i \in \mathbb{I}, \quad (19)$$

s.t. $\mathbf{A}(t_i) - \mathbf{B}(t_i)\mathbf{K}(t_i), i \in \mathbb{I}$ is Hurwitz. To achieve a time-continuous control law, a linear interpolation as in the case of the reference control $\mathbf{u}_{\text{ref}}$ in Section 3 has been applied to the $\mathbf{K}(t_i), i \in \mathbb{I}$.

The main goal for the feedback control is the adherence to the final point regarding the reference trajectory. For this reason it is desirable to have a time-dependent effect on the weighting behavior of the feedback cost function J described by (17). Therefore the weighting matrices $\mathbf{Q}$ and $\mathbf{R}$ are changed during the control of the reference trajectory. At the beginning it is more important to conserve energy than to be precisely on the trajectory. However, at the end of the trajectory it is more important to reach the final state. For this reason the values of the state weighting matrix $\mathbf{Q}$ are chosen to be small at the beginning and increase while approaching the final state, while the values of the control weighting matrix $\mathbf{R}$ are large at the beginning and decrease towards the end.

In case of large perturbations in downrange it is desirable to choose the correct point of the reference trajectory with minimum downrange error. In order to do so, a time delay is calculated to shift the reference trajectory along the time axis. In case of a negative downrange error the lander will fly with zero thrust until the downrange is at the desired value. In case of a positive downrange error the lander skips a part of the reference trajectory and reference control.

4.2 Closed Loop Simulation Results

The closed loop exhibits the structure shown in Fig. 4, where the plant **M** is given by the nonlinear system (11). To proof the performance of the provided trajectory control and to allow a comparison to the results shown in [6], several simulations of the closed loop have been carried out. For this purpose the initial perturbations have been varied along grids within the ranges defined in 3.3. The results of different simulations of the closed loop depending on the choice of the initial conditions are shown in Fig. 5. The contour lines denote the additional mass consumption needed by the feedback algorithm to satisfy the final constraints. A landing is classified as successful if the condition $|\Delta x(t_f)| \leq \Delta x_{\max}$ with

$$\Delta x_{\max} = \left(1\,\mathrm{m/s}\;\; 1\,\mathrm{m/s}\;\; 1\,\mathrm{m/s}\;\; 100\mathrm{m}\;\; 1\,\mathrm{m}\;\; 100\mathrm{m}\;\; 10^\circ\;\; 180^\circ\;\; \text{free}\right)$$

holds, see [3, 5]. Note that the provided trajectory control enables successful landings for each considered initial perturbation. The downrange reference points adaption described in Section 4.1 makes it possible to handle large downrange perturbations and is almost insensitive to negative values.

As an example for the detailed results of the closed loop simulations, a single specific set of initial perturbations, based on the comparatively high additional fuel

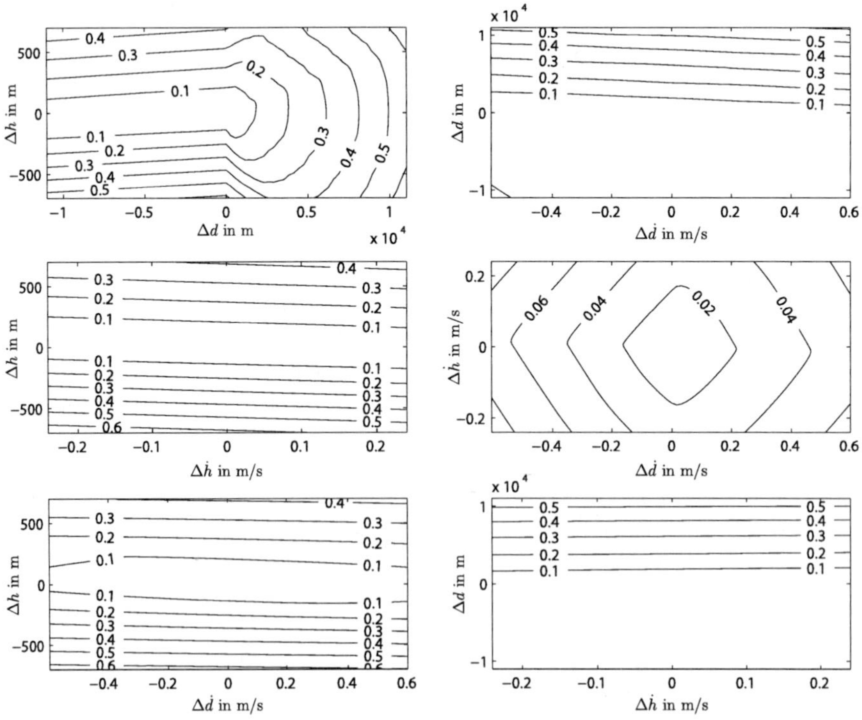

Fig. 5 Additional mass consumption for different perturbations as percentage.

Table 3 Error in the final point.

$\dot{d}_{err}$	$\dot{h}_{err}$	$\dot{c}_{err}$	d_{err}	h_{err}	c_{err}	β_{err}	χ_{err}
0.1421 m/s	0.0349 m/s	−0.0006 m/s	−5.3240 m	−0.0635 m	0.0070 m	−0.0002°	−0.0000°

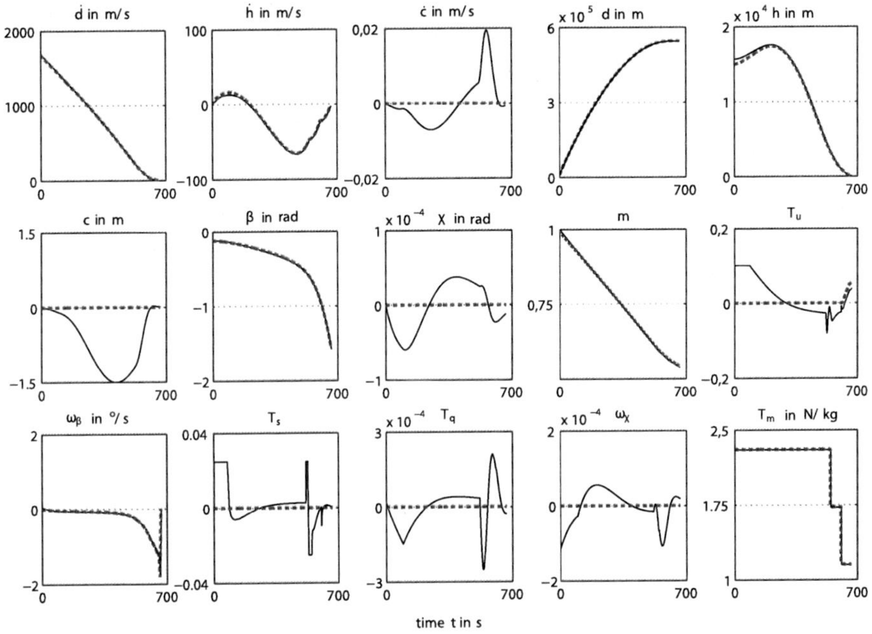

Fig. 6 Result of the feedback control with initial perturbation of 700 m in altitude, 0.24 m/s in vertical velocity, 11000 m in downrange and 0.6 m/s in horizontal velocity (solid lines) and the corresponding reference trajectories (dotted lines).

consumption, has been chosen. Applying the feedback law (16) to the control scenario in Section 3.3, with initial perturbation of 700 m in altitude h, 0.24 m/s in vertical velocity $\dot{h}$, 11 km in downrange d and 0.6 m/s in horizontal velocity $\dot{d}$, shows that the controller is able to reach the final point of the reference trajectory with the final error given in Table 3. To reach this result, 0.7 % of the initial mass was needed additionally as fuel. Fig. 6 shows the controlled trajectory (solid line) and the reference trajectory (dotted line) from Section 4.

5 Conclusions

As demonstrated in the simulations, the proposed methods for optimizing a reference trajectory and applying a feedback control law to follow this trajectory are a solution to the problem of landing a vehicle on the moon with limited thrust modulation capability. It was shown that the technique creating a reference trajectory by solving an optimal control problem is robust against variations of the initial conditions. Furthermore, it was demonstrated that the designed feedback control law is able to correct for errors in the initial conditions. For reasons of comparability the scenario for the demonstration of the performance of the proposed method has been chosen similar to the scenario used in [8]. The worst-case mass consumption for the introduced method is 46.4% of the initial mass of the landing vehicle, resulting from the worst-case mass consumption regarding to the optimal reference trajectories (45.7%) and the worst-case additional mass consumption caused by the feedback controller of (0.7%). The simulation results in [8] show a mean value for the propellant consumption of 51.37% of the initial mass. From this it follows that under the mentioned assumptions lunar landing using the method proposed in this paper requires at least 4.97% less propellant mass. Concerning an initial mass of 1736 kg [8] a landing maneuver requires 86.2 kg less propellant mass than shown in [8]. Although these results are very promising a few more steps have to be taken to make this approach an attractive alternative to today's concepts. One step is to apply the proposed method to one or more planned missions. In order to do so it has to be shown that the method also can work for e.g. initial mass changes and thruster inefficiencies. Furthermore, constraints have to be added for the attitude of the lander, especially during the final approach phase, where the landing site must be visible to the hazard mapping sensors for a given time.

To solve the problems for these next steps, some ideas already exist. First, the feedback controller should be able to change the time when some of the main engines are shut down. Second, more than one reference trajectory can be computed with different initial conditions and parameters (e.g. initial mass, thruster efficiency). Finally the feedback control law can be optimized and designed to guarantee robust stability and performance.

References

1. Altair lander homepage (2010), http://www.nasa.gov/mission_pages/constellation/news/cece.html
2. Büskens, C., Maurer, H.: SQP-methods for solving optimal control problems with control and state constraints: adjoint variables, sensitivity analysis and real-time control. Journal of Computational and Applied Mathematics 120, 85–108 (2000)
3. Davies, J.L., Striepe, S.A.: Advances in post2 end-to-end descent and landing simulation for the alhat project. AIAA-2008-6938. American Institute of Aeronautics and Astronautics (2008)
4. Fisher, J.L., Striepe, S.A.: Post2 end-to-end descent and landing simulation for the autonomous landing and hazard avoidance technology project - aas 07-119. In: Lunar and Planetary Institute Science Conference Abstracts. Advances in Astronautical Sciences, vol. 127 (2007)
5. Houdou, B.: Next Lunar Lander, Phase A Mission Study, Mission Requirements Document. Tech. Rep. NEXT-LL-MRD-ESA(HME)-0001, ESA (2008)

6. Melloni, S., Mammarella, M., Gil-Fernández, J., Colmenarejo, P.: GNC Solutions For Next-Moon Lunar Lander Mission. In: Proceedings of the 21st International Symposium on Space Flight Dynamics, Toulouse, France (October 2009)
7. Multiple. Apollo experience report - mission planning for lunar module descent and ascent. Tech. Rep. NASA TN D-6846, NASA (1972)
8. Neveu, D., Hamel, J.F., Christy, J., de Lafontaine, J., Bilodeau, V.: Next lunar lander: Descent & landing gnc analysis, design and simulations. In: Paper AAS 10-065 Guidance and Control Conference, Breckenridge, CO (2010)
9. Nikolayzik, T., Büskens, C., Gerdts, M.: Nonlinear large-scale Optimization with Worhp. To appear in Proceedings of the 13th AIAA/ISSMO Multidisciplinary Analysis Optimization Conference, AIAA-2010-8108 (2010)
10. Thurman, S.: Surveyor Spacecraft Automatic Landing System. In: Guidance and control 2004: proceedings of the 27th annual AAS Rocky Mountain Guidance and Control Conference, Breckenridge, Colorado, February 4-8, p. 427. American Astronautical Society (2004)

Spiraling Approach for Angles-Only Navigation within On-Orbit Servicing Missions

J. Spurmann

Abstract. On-Orbit Servicing missions possess a navigation problem for the transition from absolute to relative navigation when only using camera based relative navigation. This gap can be covered by the concept of angles-only navigation. To avoid singularities in angles-only navigation specifically trajectory profiles are designed. In this context the concept of the spiraling approach is proposed. The spiraling approach results from an eccentricity/inclination vector separation established during far formation flight superimposed by an along-track drift to initiate far range approach. The benefit for line-of-sight measurements is a change in the formation geometry and thus an alternating measurement profile. As a result specific maneuvers required during two dimensional approaches to overcome singularities within angles-only navigation are no longer necessary. In contrast they are included in the design of far formation flight and approach strategy.

1 Introduction

The capability to rendezvous and dock autonomously to spacecraft in Low Earth Orbit (LEO) would provide many options for future space exploration. On the one hand the increasing problem of Space Debris could benefit from such technology. By autonomously docking to potentially uncooperative satellites at their end of life and by de-orbiting them the collision risk within LEO could be reduced. Another field of interest would be the correction of launch errors. By servicing faulty injected satellites to correct the orbital elements to the desired ones mission success rates could be increased. In a wider context a fleet management of multi satellite constellations could be envisaged. In addition repair of satellites or refueling activities close to their end of life could be targeted. In the most severe case the servicing satellite could also take over the complete attitude and orbit control of a malfunctioning satellite.

J. Spurmann
DLR, German Space Operations Center
DLR Oberpfaffenhofen, D-82230 Wessling, Germany
e-mail: Joern.Spurmann@dlr.de

Due to variety of possible mission scenarios On-Orbit Servicing (OOS) has become part of the space programs of the US, Japan, Canada and Germany. A milestone was set with the successful completion of DARPA's Orbital Express [1] mission in 2007. It demonstrated the ability to autonomously perform Rendezvous & Docking (RvD) operations including maintenance activities.

Proceeding one step further some of the aforementioned on-orbit servicing mission scenarios desire to capture uncooperative client spacecraft in LEO and to deorbit the coupled configuration. In this context the uncooperative client does not provide any attitude or orbit control. Additionally, no docking interfaces or dedicated reflector pattern for vision based navigation will be available.

Trying to incorporate those requirements into a navigation concept ground based absolute navigation can be performed based on GPS data for the servicing spacecraft or radar tracking measurements for the client spacecraft. Those two techniques yield similar accuracies of several tens of meters [2]. With the desire to create a low cost spacecraft for higher reproducibility the selection of the relative navigation sensor is constrained. Low cost leads directly to low mass, low power consumption and reduced complexity. Accordingly, radar or LIDAR systems are not applicable due to their large mass and power penalties. The only option remaining is thus a camera based relative navigation system. Due to the accuracies of the absolute orbit determination the relative navigation has to start at some kilometers relative range. The operative range of camera based sensors however, ends at several hundred meters of relative range [3]. Consequently a gap in the navigation concept results for the handover from absolute to relative navigation.

So far this gap has been covered by a method called Angles-Only navigation or navigation using Line-Of-Sight (LOS) measurements [4]. This concept considers to measure the relative angles between the two spacecraft and to apply those to a Kalman-Filter to estimate the spacecraft state. The state can thereby include different information. The absolute position could be included in different coordinate systems, or the relative position vector in a specific frame. Additionally, attitude and other orbit parameters as for example the drag coefficient could be included. Depending on the relative range however, only a part of the state information can be improved by angles-only navigation [4].

Upon a closer look on earlier missions utilizing angles-only navigation mainly pure along-track separation has been regarded as the driver for the navigation concept. This is especially represented by the so-called one percent rule used for navigation sensor selection due to their achievable accuracy [3]. Due to singularities in the estimation process of angles-only navigation within those navigation concepts specific maneuvers have to be performed to alter the flight profile such that a variation of the LOS measurements is available and accordingly the estimation of relative range with the required accuracy possible [5]. The master thesis of CHARI [6] performed within the mission analysis of Orbital Express considered out-of-plane maneuvers to overcome the singularities for the first time. However, the hybrid trajectories designed by his findings of relative range accuracy within a Kalman filter upon different approach trajectories and applied maneuvers considered basically two dimensional approaches with additional out-of-plane maneuvers to increase relative range accuracy.

The scope of the work at hand will be to apply a three dimensional approach in form of a spiraling trajectory to the concept of angles-only navigation. The spiraling trajectory starts from a safe formation flying ellipse resulting from an eccentricity/inclination vector separation [7][8]. Upon application of an along-track drift maneuver the servicer starts drifting towards the client in form of a spiraling trajectory. As a result the maneuvers to overcome angles-only navigation singularities are incorporated into the approach trajectory. Further, a safe formation flight is guaranteed in the beginning independent of the along-track accuracy since by design radial and cross-track component drive the formation flying ellipse. Upon approach the ellipse can be shrunken according to the improved accuracy resulting from angles-only navigation.

Accordingly, a trajectory profile to overcome angles-only navigation singularities is combined with the formation flying and approach requirements brought up by rendezvous and docking of uncooperative spacecraft.

2 Concept of Along-Track Separation

Prior to going in to detail on the different concepts it should be mentioned that the development of formulas and plots is based on the RTN orbit frame. In this frame the R or radial unit vector is aligned with the radial direction (positive outwards), while the N or normal unit vector is parallel to the servicer angular momentum vector (positive in orbit normal direction). The T or tangential unit vector completes the right-handed coordinate system (positive in chief velocity direction).

To start with the concept description, the gap in navigation accuracy is derived within the two dimensional approach concept mainly driven by the along-track separation. The design of the navigation concept for a rendezvous or formation flying mission is mainly based on the resulting navigation errors. Deriving propagation errors for circular orbits in the relative orbital frame from the Clohessy-Wiltshire equations [3] a radial displacement between the two spacecraft results in the largest error in the along-track component after one orbit propagation [3]. In case of concentric orbits the error is:

$$\Delta r_T = 3\pi \cdot \Delta r_{R,0}$$

In case of orbits with a slightly different eccentricity the two spacecraft have the same initial velocity, which amounts to an even larger error after propagation of one orbit:

$$\Delta r_T = 12\pi \cdot \Delta r_{R,0}$$

Additionally, errors from velocity uncertainties are most severe for along-track inaccuracies. Those from radial difference are only 20% the size.

$$\Delta r_T = 3 \cdot T \cdot \Delta v_{T,0}$$

Thus only regarding a pure along-track separation between the two spacecraft and including the effects of velocity errors [3] a rule was derived stating that the navigation sensors must provide a one percent accuracy of the relative distance between the spacecraft for safe approach [3].

Considering the formation separated by a pure along-track component at the distance of several kilometers the navigation is performed on basis of absolute orbit determination. Ground based orbit determination based on GPS navigation solution for servicer or radar tracking for client yields accuracies of several tens of meters [2]. Hence absolute orbit determination can only be used for phasing and far formation at several kilometers relative distance. The vision based navigation sensors, which are selected to reduce mass, power consumption and complexity are however only applicable up to distances of several hundred meters [3]. Hence a gap in the navigation concept arises during rendezvous with uncooperative spacecraft within the relative distance of several hundred meters to several kilometers. A solution to this problem is the concept of Angles-Only navigation.

3 Angles-Only Navigation

Angles-only navigation can be implemented to acquire the relative range within the mentioned transition zone. The method is well know and widely applied in naval applications, orbit determination, target tracking, lunar and interplanetary optical navigation and homing missile applications [4].

The relative trajectory between client and servicer can be defined by the relative distance r and the line-of-sight (LOS) angles azimuth α and elevation e. The basic principle of the angles-only navigation is then to measure the LOS angles with the according time very accurately (Fig.1), as soon as the client can be detected as a moving star in front of the background star field by the far range camery (FRC). The corresponding range can be derived from the assumptions made above to several kilometers. The obtained measurements are then used to update a Kalman Filter, which propagates the orbit onboard the servicer or on ground. Apart from an initial guess of the relative range the Kalman Filter uses the final spacecraft states from absolute navigation, the orientations of the spacecraft, biases (gyro bias or camera misalignment) and noise terms to determine the state vector of the client spacecraft [4]. By iterative propagation and a continuous update process of the filter, the accuracy of the measurements and most important that of the relative range will improve.

The major problem of angles-only navigation is however, the inherent limitation in determining the relative range with adequate accuracy [5]. If the geometry of the relative motion between servicer and client does not change and equivalently the continuous LOS measurement profile is not altered, a precise determination of the relative range is not possible [6].

When targeting an approach concept based on along-track separation this problem will definitely occur. While resting at hold points or even during hopping approaches in along-track direction an adequate navigation would not be possible [6]. Thus not only a general change of geometry is important but also the direction of applied motion. A solution within this concept is however the application of out-of-plane maneuvers [6]. Those force a change of the relative geometry. As a

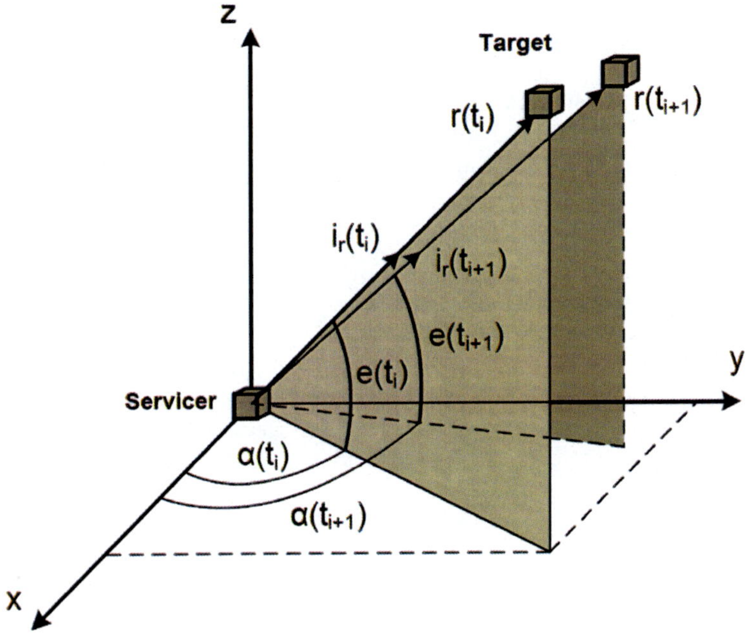

Fig. 1 Geometry of Angles-Only measurements.

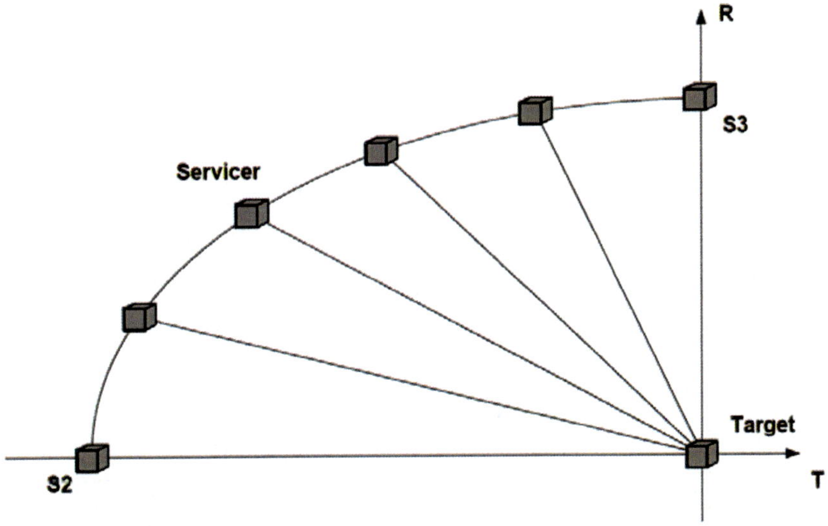

Fig. 2 Angles-Only measurements during fly-around maneuver in RTN frame.

result the relative range can be determined with higher accuracy by planning of specific trajectory profiles at the handover from absolute to relative navigation. As a result from CHARI [6] approach of the client including impulsive maneuvers in cross-track direction is advantageous similar to approaches in radial direction or fly around maneuvers (Fig.2) within this approach strategy. The limitation of increasing navigation errors within the relative range while resting at hold points remains however.

4 Concept of E-/I-Vector Separation and Spiraling Approach

The concept of e-/i-vector separation originally developed for geostationary satellites and adopted to LEO missions [8] considers slightly different eccentricity and inclination of the client and servicer orbit. Accordingly, upon application of parallel eccentricity and inclination vectors a safe separation in radial and cross-track direction is guaranteed to be smaller than

$$\min(a \cdot \delta e, a \cdot \delta i).$$

Thus the servicer is flying in a relative ellipse with the client while being always separated safely in radial and cross-track direction. Accordingly, even if the along-track component vanishes completely due to its imprecise knowledge the configuration would still be safely separated in cross-track and radial direction. As those two components of the relative position vector can on the one hand be determined much more accurately via ground based absolute navigation based on GPS or radar tracking data [2] and the error development is at a much smaller scale on the other hand a large benefit is given to the collision risk.

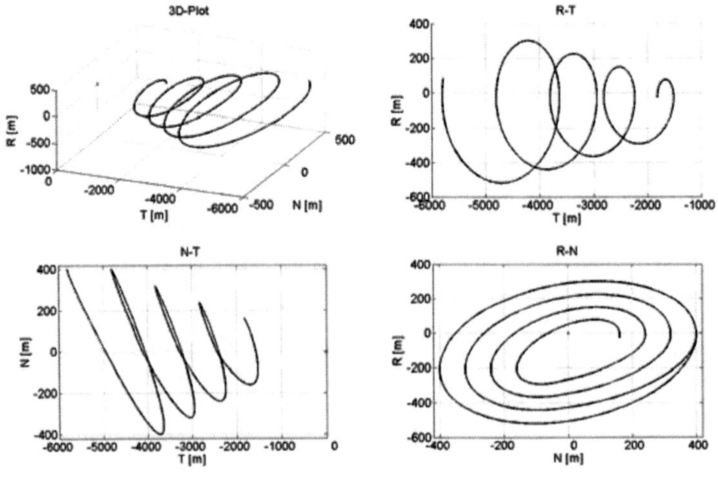

Fig. 3 Spiral approach trajectory in RTN frame.

Upon application of an along-track maneuver the far range approach towards the client spacecraft is initiated. The resulting trajectory has the form of a spiral and is hence called spiraling approach (Fig. 3).

Accordingly, a continuously alternating flight profile in all three dimensions is available by application of one single maneuver. Hence the problem of singularities within angles-only navigation is removed by incorporating one maneuver, which has to be conducted anyhow to start the rendezvous approach. To further improve the relative range determination accuracy upon approaching the client spacecraft in-plane and out-of-plane maneuvers can be applied to shrink the ellipse (Fig. 3).

5 Benefits from Spiraling Approach to Angles-Only Singularities

In contrast to the required trajectory profile during pure along-track separation the e-/i-vector separation provides a huge benefit on angles-only navigation. Considering a configuration with a servicer in a relative ellipse with the client spacecraft at a certain along-track separation a configuration is found for safe far formation flight.

The benefit of this configuration to angles-only navigation is significant. At first the natural motion with in the e-/i-vector separation improves the relative range accuracy [6]. Further, the along-track drift alters the flight profile such that the relative range accuracy should be guaranteed during approach. In case of reduced filter performance the radial and cross-track maneuvers performed to shrink the spiral upon approach improve the performance again. As a result, the singularities can be resolved.

A numerical verification of the benefit has to be investigated in further studies.

6 Conclusion

It is evident that the concept of the spiraling approach, resulting from e-/i-vector separation [7][8] and an applied along-track drift, provides significant benefit to the concept of angles-only navigation [5][6]. With the continuous relative motion between the two spacecraft the line-of-sight measurement profile alternates continuously such that a good set of observations is available for setting up the navigation filter. The inherent limitation of former approach concepts [5] to estimate the relative position with adequate accuracy can thus be overruled and the singularities in angles-only navigation resolved. Especially the required maneuvers to change the trajectory profile [5] have not to be designed specifically anymore, as they are incorporated in the approach strategy. Additionally a passively safe formation is guaranteed prior to availability of angles-only measurements. Concluding the gap in the navigation concept between absolute and relative navigation is elegantly resolved by including the maneuvers required to change the formation geometry in the design of the approach concept in form of the spiraling approach.

References

1. Mulder, T.A.: Orbital Express Autonomous Rendezvous and Capture flight operations, Part 1 of 2 and Part 2 of 2. In: AIAA/AAS Astrodynamics Specialist Conference and Exhibit, Honolulu, Hawaii, August 18 - 21 (2008)
2. Aida, S., Patzelt, T., Leushacke, L., Kirschner, M., Kiehling, R.: Monitoring and Mitigation of Close Proximities in Low Earth Orbit. In: 21st International Symposium on Space Flight Dynamics, Toulouse, France (2009)
3. Fehse, W.: Automated Rendezvous and Docking of Spacecraft. Cambridge Aerospace Series, vol. 16. Cambridge University Press, Cambridge (2003)
4. Woffinden, D.C., Geller, D.K.: Relative Angles-Only Navigation and Pose Estimation For Autonomous Orbital Rendezvous. Journal of Guidance, Control and Dynamics 30(5), 1455 (2007)
5. Woffinden, D.C., Geller, D.K.: Observability Criteria for Angles-Only Navigation. IEEE Transactions on Aerospace and Electronic Systems 45(3) (2009)
6. Chari, R.J.: Autonomous Orbital Rendevous Using Anlges-Only Navigation. Master Thesis, Massachusetts Institute of Technology (2001)
7. D'Amico, S., Montenbruck, O.: Proximity Operations of Formation-Flying Spacecraft Using an Eccentricity/Inclination Vector Separation. Journal of Guidance, Control and Dynamics 29(3), 554–563 (2006)
8. D'Amico, S.: Autonomous formation flying in low earth orbit. PhD thesis, Technical University of Delft (2010), ISBN 978-90-5335-253-3

Author Index